Emerging Capabilities and Applications of Wireless Power Transfer

Alicia Triviño-Cabrera
University of Málaga, Spain

José A. Aguado
University of Málaga, Spain

A volume in the Advances in Wireless Technologies and Telecommunication (AWTT) Book Series

IGI Global
DISSEMINATOR OF KNOWLEDGE

Published in the United States of America by
IGI Global
Information Science Reference (an imprint of IGI Global)
701 E. Chocolate Avenue
Hershey PA, USA 17033
Tel: 717-533-8845
Fax: 717-533-8661
E-mail: cust@igi-global.com
Web site: http://www.igi-global.com

Library of Congress Cataloging-in-Publication Data

Names: Trivino-Cabrera, Alicia, 1978- editor. | Aguado, J. A. (Jose Antonio), editor.
Title: Emerging capabilities and applications of wireless power transfer / Alicia Trivino-Cabrera and Jose A. Aguado, editors.
Description: Hershey, PA : Information Science Reference, [2019] | Includes bibliographical references.
Identifiers: LCCN 2017058222| ISBN 9781522558705 (h/c) | ISBN 9781522558712 (ebook)
Subjects: LCSH: Wireless power transmission.
Classification: LCC TK3088 .E64 2019 | DDC 621.319--dc23 LC record available at https://lccn.loc.gov/2017058222

This book is published in the IGI Global book series Advances in Wireless Technologies and Telecommunication (AWTT) (ISSN: 2327-3305; eISSN: 2327-3313)

British Cataloguing in Publication Data
A Cataloguing in Publication record for this book is available from the British Library.

Advances in Wireless Technologies and Telecommunication (AWTT) Book Series

Xiaoge Xu
Xiamen University Malaysia, Malaysia

ISSN:2327-3305
EISSN:2327-3313

Mission

The wireless computing industry is constantly evolving, redesigning the ways in which individuals share information. Wireless technology and telecommunication remain one of the most important technologies in business organizations. The utilization of these technologies has enhanced business efficiency by enabling dynamic resources in all aspects of society.

The **Advances in Wireless Technologies and Telecommunication Book Series** aims to provide researchers and academic communities with quality research on the concepts and developments in the wireless technology fields. Developers, engineers, students, research strategists, and IT managers will find this series useful to gain insight into next generation wireless technologies and telecommunication.

Coverage

- Wireless Technologies
- Network Management
- Virtual Network Operations
- Broadcasting
- Mobile Technology
- Global Telecommunications
- Grid Communications
- Wireless Sensor Networks
- Cellular Networks
- Digital Communication

The Advances in Wireless Technologies and Telecommunication (AWTT) Book Series (ISSN 2327-3305) is published by IGI Global, 701 E. Chocolate Avenue, Hershey, PA 17033-1240, USA, www.igi-global.com. This series is composed of titles available for purchase individually; each title is edited to be contextually exclusive from any other title within the series. For pricing and ordering information please visit http://www.igi-global.com/book-series/advances-wireless-technologies-telecommunication/73684. Postmaster: Send all address changes to above address.

Titles in this Series

For a list of additional titles in this series, please visit: www.igi-global.com/book-series

Mobile Applications and Solutions for Social Inclusion
Sara Paiva (Instituto Politécnico de Viana do Castelo, Portugal)
Information Science Reference • copyright 2018 • 354pp • H/C (ISBN: 9781522552703) • US $195.00 (our price)

Optimizing E-Participation Initiatives Through Social Media
Laura Alcaide-Muñoz (University of Granada, Spain) and Francisco José Alcaraz-Quiles (University of Granada, Spain)
Information Science Reference • copyright 2018 • 307pp • H/C (ISBN: 9781522553267) • US $195.00 (our price)

Centrality Metrics for Complex Network Analysis Emerging Research and Opportunities
Natarajan Meghanathan (Jackson State University, USA)
Information Science Reference • copyright 2018 • 183pp • H/C (ISBN: 9781522538028) • US $175.00 (our price)

Mobile Technologies and Socio-Economic Development in Emerging Nations
Fredrick Japhet Mtenzi (Dublin Institute of Technology, Ireland & Aga Khan University, Tanzania) George S. Oreku (Open University of Tanzania (OUT), Tanzania) Dennis M. Lupiana (Institute of Finance Management, Tanzania) and Jim James Yonazi (Tanzania Standard (Newspapers) Limited, Tanzania)
Information Science Reference • copyright 2018 • 366pp • H/C (ISBN: 9781522540298) • US $195.00 (our price)

Affordability Issues Surrounding the Use of ICT for Development and Poverty Reduction
Sam Takavarasha Jr. (University of Fort Hare, South Africa & University of Zimbabwe, Zimbabwe) and Carl Adams (University of Portsmouth, UK)
Information Science Reference • copyright 2018 • 319pp • H/C (ISBN: 9781522531791) • US $195.00 (our price)

Handbook of Research on Environmental Policies for Emergency Management and Public Safety
Augustine Nduka Eneanya (University of Lagos, Nigeria)
Engineering Science Reference • copyright 2018 • 393pp • H/C (ISBN: 9781522531944) • US $295.00 (our price)

Positioning and Navigation in Complex Environments
Kegen Yu (Wuhan University, China)
Information Science Reference • copyright 2018 • 577pp • H/C (ISBN: 9781522535287) • US $195.00 (our price)

Examining Cloud Computing Technologies Through the Internet of Things
Pradeep Tomar (Gautam Buddha University, India) and Gurjit Kaur (Gautam Buddha University, India)
Information Science Reference • copyright 2018 • 311pp • H/C (ISBN: 9781522534457) • US $215.00 (our price)

IGI Global
DISSEMINATOR OF KNOWLEDGE

701 East Chocolate Avenue, Hershey, PA 17033, USA
Tel: 717-533-8845 x100 • Fax: 717-533-8661
E-Mail: cust@igi-global.com • www.igi-global.com

To our families

Table of Contents

Detailed Table of Contents

Wireless power transfer (WPT) systems are involved in multiple and heterogeneous applications. This diversity is reflected in several factors such as the amount of power that is transferred or the distance separating the energy source and the receiver. In the current research work, the authors find several groups of technologies that try to adapt the process to the particularities of the application. In this way, wireless power transfer can be achieved with an inductive technology, with a resonant-inductive approach, or with a strongly coupled magnetic resonant configuration. This chapter focuses on strongly coupled magnetic resonant technology, which is appropriate for home applications.

This chapter focuses on mid-range wireless power transfer (WPT) systems applied to electric vehicle (EV) battery chargers. The WPT is recently considered as an efficient electric energy transfer process between two or more points in space, without wiring. The technology associated with each specific process of WPT differs from case to case depending on the distance between those points and the power to be transferred between them. The widely adopted distance categories are named short-range, mid-range, and long-range. The short-range is normally defined as up to a few millimeters range. The mid-range is between a few millimeters and a few meters. The long-range distance is defined as a longer than that of the previous category, stretching up to a few kilometers.

Chapter 3

Alberto Reatti, University of Florence, Italy
Marian K. Kazimierczuk, Wright State University, USA
Agasthya Ayachit, Wright State University, USA
Fabio Corti, University of Florence, Italy

A resonant wireless power system uses capacitors to create resonance between primary and secondary coil at a specific resonant frequency. Thereby, they reduce the reactive power and produce the required transmitted real power over a long range of separation and at higher efficiencies. Depending on the connection between the capacitors and coils, four main topologies can be identified: series-series, series-parallel, parallel-series, and parallel-parallel. In this chapter, these resonance compensations schemes are analyzed, and the advantages and disadvantages of each topology are highlighted.

Chapter 4

Rodrigo Wolff Porto, Universidade Federal do Rio Grande do Sul, Brazil
Lucas Murliky, Universidade Federal do Rio Grande do Sul, Brazil
Valner João Brusamarello, Universidade Federal do Rio Grande do Sul, Brazil
Fernando Rangel de Sousa, Universidade Federal de Santa Catarina, Brazil

The growing number of publications concerning the dynamic compensation of inductive links has demonstrated the importance of the issue upon the WPT system design. There is not still a final and consolidated solution for the case when the coils are in relative movement with each other or misaligned. In this context, this chapter intends to give some contributions to that area, starting with the modeling of the inductive link in Section 2, where the authors show how the magnetic coupling coefficient and the load influence on the PDL. In addition, Section 2 also presents how the excitation frequency and the matching capacitances can improve the PDL. In Section 3, the variable excitation frequency method is explained as well as the hardware needed for the implementation. The voltage-controlled capacitance as a variable element in the matching network is shown in Section 4. Finally, in Section 5, optimization techniques are presented by using both frequency and capacitance to maximize the PDL.

Chapter 5

Stanimir Valtchev, Universidade Nova de Lisboa, Portugal
Svilen S. Valtchev, University of Lisbon, Portugal & Polytechnic Institute of Leiria, Portugal

This chapter presents control methods applied in the operation of the series loaded series resonant (SLSR) power converters in a most efficient operation zone. The choice of the control method is affected by the objective to guarantee suitably the efficiency, being this method in the same time, relatively easy to apply. The first part of the chapter compares three basic principles of regulation: frequency mode (FM), pulse width mode (PWM), and their combination (PWM/FM). Finally, a new method for instantaneous regulation is developed. The proposed technique consists of a simplified observation of a state variable value to limit each portion of supplied energy, depending on the requirement for power in each half period. The result of this regulation is comparable to the current mode (CM) control applied to the hard-switching power converters. The viability of this new regulation method is demonstrated by simulations of its analogue circuit implementations and experimentally proved. The circuit is also prepared for the changes in the magnetic coupling (contactless energy transfer).

The focus of this chapter is the electromagnetic interference (EMI) and the electromagnetic compatibility (EMC) that the wireless power transfer (WPT) systems reveal as problems. The wireless power transfer (WPT) was introduced by Nikola Tesla more than one hundred years ago, and only recently it attracted the attention of specialists, due to the improved technical means. The WPT technology now has many applications, especially for charging of various electronic devices (i.e., mobile phones, laptops, implants, and home appliances), informatics, and electronics equipment. The high-power equipment and installations (e.g., intelligent machining systems, robots, forklift trucks, and electric cars) are also getting wireless. Moreover, much attention has been focused on the electric transportation system for improving the safe and convenient charging of electric vehicle (EV) batteries.

Electromagnetics has an important role in power and energy industry. In this chapter, the concept of rectenna is reviewed. The history of rectenna for wireless energy harvesting and transmission is discussed. Finally, examples are employed to illustrate some rectenna design and measurement issues such as rectenna impedance matching and its conversion efficiency. It is also shown that rectennas can harvest wireless energy efficiently under certain conditions and have the potential to become a power supplier for some special applications.

RFID technology has been widely adapted in industries for uses in logistical tracking, highway tolling, building access, and transportation ticketing. These applications have generally been limited to the original intended use of RFID, that is identification and a replacement for bar codes. Research in this area focuses on increasing the read rate, range, and reliability of their RF tags. The WPT is the enabling technology for realizing a true internet of things. Broad sensor networks capable of monitoring environmental pollutants, health-related biological data, and building utility usage are just a small fraction of the applications which are part of an ever-evolving ubiquitous lifestyle. Realizing these systems requires a means of powering their electronics sans batteries. Removing the batteries from the trillions of these envisioned devices not only reduces their size and lowers their cost, but also avoids an ecological catastrophe. This chapter discusses new theoretical models of RFID, communication standards, radio channel characteristics, RFID readers and tags.

Preface

INTRODUCTION

Wireless Power Transfer (WPT) refers to the technologies that enable powering a device without the need for being connected through a cable to the grid. The powering can be simultaneous to the power transfer or deferred basing on the wireless recharge of a battery in the device. WPT is gaining attention in the recent years due to several advantages it offers. Firstly, WPT adds commodity as users' intervention is eluded. This is particularly convenient in the Internet of Things scenarios where numerous wireless devices are expected to operate and they may not be always be easily accessible. In addition, WPT capability increases the autonomy of the devices so that they can even implement enhanced functionalities or execute them under an augmented set of circumstances. The suppression of the cords may also convert the product in a more robust device.

In spite of this current interest, the first experiment related to WPT dated to 1891, when Nikola Tesla was able to light electric lamps without cables in his laboratories in New York City, but with a reduced efficiency. The recent advances of power electronics and the development of simple controllers have promoted the feasibility of efficient applications of WPT. This is the case of commercially available wireless chargers for devices such as cell phones, laptops, electrical tooth-brushes or identification cards, which are now widespread. There are some companies that already offer such devices. For instance, Starbucks coffee shops are provided with some coffee tables that allow the wireless charge of mobile phones. The popular furniture manufacture IKEA has also developed some furniture with the integrated technology to recharge the mobile phones without cables. They even offer the technology for lighting an entire home without cables. Contactless-cards are also common in hotels, public transport or bank operations.

Some other applications, not as popular for the general consumer as the previous ones, are also a reality nowadays. This is the case of electric vehicles, which can be charged without cables even when they are moving. Some biomedical devices also opt for a wireless power charge, which avoids the hassles and the risks of operating the patient to replace the battery of internal devices. Alternatively, in some industrial applications it is not convenient the presence of operators to change the battery of certain devices. This can be motivated by unhealthy environments or by the tedious task of doing this procedure when there are numerous devices. As can be observed, the application area of WPT is enormous.

Paying attention to the market expectations, we can conclude that this technology will play a relevant role in the near future. According to IHS Markit, the number of wireless chargers sold in 2017 was estimated to be 300 millions of units, which represent an increase of 75% compared to the previous year. Most of the wireless chargers acquired were intended for mobile phones but the use for laptops has experienced an important increment. Navigant Research predicts that the economical benefits of WPT will be 17.9 billion dollars in 2024. This figure is 10 times greater than the benefits obtained in 2015.

The progress of WPT requires a deep understanding of its fundamentals. The basis of WPT is the use of an electromagnetic field to transmit energy from one transmitter to one/multiple receivers. The features of the electromagnetic field and how the energy is used are not homogeneous for all the applications. In fact, it is important to highlight that wireless power transfer refers to heterogeneous techniques, in which the type of the electromagnetic field differs. The convenience of each technology for a particular application depends on several factors such as the distance between the power emitter and receiver, the amount of power to be transmitted and the features of the electromagnetic field involved in the process. The application may also impose some design constraints, which demand the use of a specific technology. For instance, weight restrictions are imposed when WPT is used in drones or high misalignment tolerance is required for wireless sensor networks. One of the main goals of this book is to illustrate the different technological options that exist to transmit power without cables and how they are adapted for some applications.

ORGANIZATION OF THE BOOK

The book is structured in two main parts. The first part addresses the fundamental concepts of the WPT. Alternatively, the second part of the book deals with the issues associated to specific illustrative applications. In particular, the applications considered in this book are: charge of electric vehicles, implementable medical devices, microwave-based transfer, RFID tags and wireless sensor networks. The book is comprised of 13 chapters.

Chapter 1 presents an overview of the technology involved in WPT. Then, it analyses the strongly-coupled magnetic resonant technology, which represent a mechanism to charge electric devices in a home-dimension environment. TVs powered with this technology are commercially available.

The efficiency of these technologies is the focus of Chapter 2. Even without the presence of cables, the wireless power transfer is desirable to be as efficient as the one supported by cords. Several techniques can be implemented to make the system operate efficiently even when the relative positions between the power emitter and the receiver differ from the ones considered in the design process. One of these techniques consists of inserting compensation systems, which are deeply studied in Chapter 3.

Chapter 4 addresses how this efficiency can be maintained even when the power receiver is moving. The movement of the receiver alters its ability to get profit from the electromagnetic field generated by the transmitter. Advanced compensation systems can be incorporated into the WPT system to reduce the decrements on the efficiency when the receiver is moving.

Chapter 5 analyses the control systems that need to be included in the system to ensure that the transfer occurs on specific conditions such as maintaining the power delivered to the load or keeping constant the voltage at the load.

One of the main concerns of the WPT technology is related to the electromagnetic emissions associated to the power transfer. Chapter 6 shows the electromagnetic fields for some processes of transferring power without cables. It also describes some international restrictions that need to be satisfied to develop safe WPT products.

From Chapter 7, it starts the section of the book dedicated to applications of WPT. Chapter 7 focuses on the wireless power transfer based on microwaves signals. This type of signal is specially suitable for the power transmission between two objects that are separated several kilometers.

Chapter 8 describes the well-known RFID technology, which is present in many current applications. The different types of RFID tags are explained and how the information can be transmitted from a passive tag to an active RFID without the use of batteries but with the wireless power transfer from the active tag.

Chapter 9 addresses how the implantable medical equipment can benefit from WPT technology. In particular, it addresses how the antennas should be designed properly for an optimized WPT system.

Chapter 10 reveals how the antenna for wireless sensor networks should be built to cope with misalignment during the power transfer.

The last three chapter of the book focus on the charge of electric vehicles. Chapter 11 describes some models related to wireless chargers for EVs. Chapter 12 explains how the magnetic coupling factor is altered in this kind of application when the transmitter coil and the receiver coil are in a relative position that does not correspond to the supposed one during the design process. This parameter indicates how much of the electromagnetic field generated by the transmitter coil is effectively used to transfer power to the battery in the EV. Finally, Chapter 13 explains the main characteristics of the OLEV system, one of the most relevant dynamic system to transmit power to electric vehicles without cables.

Chapter 1
Analysis of the Strongly Coupled Magnetic Resonant Technology for Wireless Power Transfer

Alicia Triviño-Cabrera
University of Málaga, Spain

José A. Aguado
University of Málaga, Spain

ABSTRACT

Wireless power transfer (WPT) systems are involved in multiple and heterogeneous applications. This diversity is reflected in several factors such as the amount of power that is transferred or the distance separating the energy source and the receiver. In the current research work, the authors find several groups of technologies that try to adapt the process to the particularities of the application. In this way, wireless power transfer can be achieved with an inductive technology, with a resonant-inductive approach, or with a strongly coupled magnetic resonant configuration. This chapter focuses on strongly coupled magnetic resonant technology, which is appropriate for home applications.

1. INTRODUCTION

Wireless power transfer allows the transmission of energy between two objects that are not electrically connected. This idea was pioneered by Nikola Tesla with early experimental attempts over one century ago. However, commercial and efficient solutions have not been available until recent years. This technology is now a reality in several commercial appliances such as tooth brushes, mobile phone chargers, medical devices or tags. As it avoids the hassle of connections, it is being investigated how to apply it to medium and high power scenarios. This is the case of electric vehicles and some home and industrial appliances (Bi et al., 2016). For efficiency purposes, other areas are also getting benefit from the wireless power transfer. This is the case of the wireless communication networks, which try to obtain power from the signals intended for the communication with other receivers.

DOI: 10.4018/978-1-5225-5870-5.ch001

All these applications rely on an electromagnetic wave, which propagates in the air from an emitter to a receiver. The receiver is equipped with the electronics necessary to extract energy from the electromagnetic wave. The wavelength of the electromagnetic wave and its relationship with the distance between the emitter and the receiver is key to decide about which wireless power transfer technology is more convenient for a given scenario. According to this relationship, these technologies can be divided into three groups: near-field, mid-range and far-field technologies. In a near-field scenario, the energy emitter and the receiver are up to one wavelength separated. In contrast, the far-field scenario takes place when the distance between these two elements is greater than two wavelengths. Finally, the mid-range wireless power transfer occurs in the intermediate configuration. It is receiving recent interest as the separation between the transmitter and the receiver is suitable for home appliances.

In each group, there are a set of technologies, which appropriateness mainly depends on the power to be transmitted and on the mobility of the emitter and/or the receiver. The next chart illustrates the classification of the technologies.

Far-field transmission can be achieved by three methods: radio-frequency, microwave and laser. Firstly, radio-frequency is based on a scheme similar to the one employed in telecommunication systems. The signals involved in the process range from 30 MHz to 5 GHz. Theoretically, the power transmission can be omnidirectional or directional. When low powers are sent, then omnidirectional transmission is allowed but for high-power applications a directional scheme should be followed in order to comply with the international restrictions about electromagnetic emissions. The enterprise PowerCast has developed this technology in order to transfer up to 3 W in a directional scheme and up to 1 W for an omnidirectional propagation . Due to the levels of power that are managed, this kind of wireless power transfer can be useful for charging low-power devices in a wide area. This is the case of wireless sensor networks.

Microwave transmission for wireless power transfer relies on a component named rectenna, which is composed of dipoles and diodes that transform the microwave signal into a direct current signal. In the decade of 1980, the SHARP (Stationary High Altitude Relay Platform) project was able to power a small plane. The distance from the sender to the receiver was 21 km. The signal employed for this 500-kW transmission was 5.8 GHz. As can be observed, the power involve in the microwave transmission is

Figure 1. Scheme of the techniques for wireless power transfer

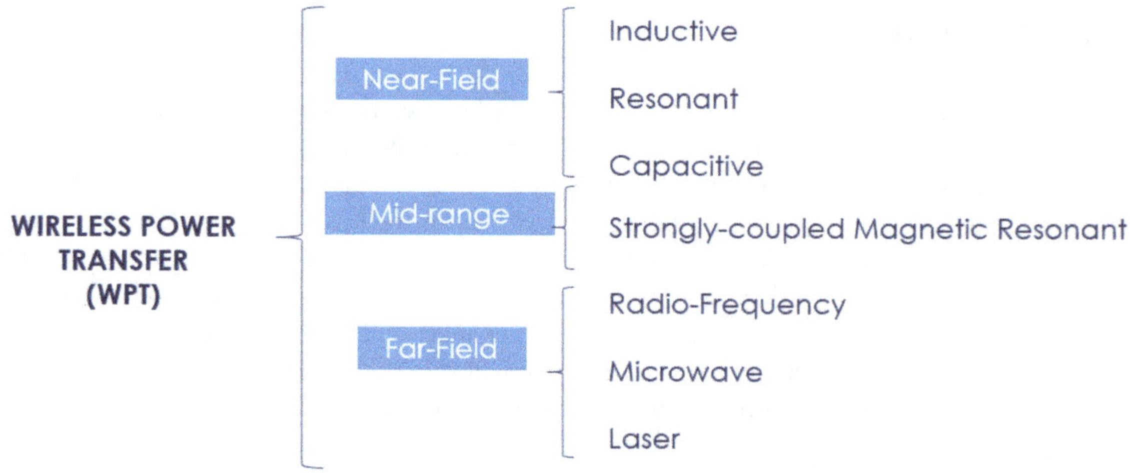

higher than those achieved by the radio-frequency schemes. However, microwave transmission requires a direct line-of-sight without any human-being around due to the potential negative impact on health. A similar problem is present in optical transmissions based on lasers. Nevertheless, they do not cause any interference with other radio systems.

In the near-field group, there are three relevant techniques: inductive, resonant and capacitive. The basis of the inductive systems is a pair of coupled coils. According to Ampère's law, when one of the coil is excited with a time-varying current, it generates a dynamic magnetic field. If this field traverses another coil, a voltage difference is induced between its terminals. This voltage is employed to charge a device. An improved efficiency can be obtained with resonant systems. In these systems, the coils are connected to reactive networks (the simplest solution opts for one capacitor) in order to force the system to work under resonant conditions. These reactive structures are referred to as the compensation topologies. Alternatively, the capacitive systems use to metal plates to transmit the power with an electric field between these two components (Huang & Hu, 2015). The distance and the power that can be transmitted with the capacitive scheme is lower than those achieved by resonant systems (Dai & Ludois, 2015).

Finally, in the mid-range we find the strongly coupled magnetic resonant transmission (Barman, Rez, Kumara, & BakarMunir, 2015). It is also named as magnetic-resonant or based on magnetically-coupled resonance. These schemes are capable of increasing the distance of efficient power transmission that the resonant systems offer. Similarly to the inductive and resonant techniques, the power transfer is not completely avoided by the presence of intermediate objects between the transmitter and the receiver. This is a great advantage that is not provided by the microwave or the laser-based wireless power transfer solutions. These interesting properties are mainly due to the components of the system and the particular configuration that they follow. The original description of this method proposes the use of self-resonant coils, that is, coils that resonates with their parasitic capacitances. The capacitances are mainly due to the effect of the multiple close conductors employed to build the coils. Nevertheless, some recent works propose the inclusion of lumped capacitors and a four-coil topology for strongly-coupled resonant circuits (Liu, Hu, & Georgakopoulos, 2017; Jolani, Yu, & Chen, 2015). This technology allows to transfer energy from a source to a load for distances longer than those permitted by the resonant systems. In particular, the separation of the source and the load can be similar to the coils' dimensions, condition which was not satisfied in resonant circuits. In addition, if properly configured, this type of system can be less prone to variations on the distance between the source and the load. Nearly a constant efficiency can be obtained for different separations when the system is tuned at particular operational frequencies. Control techniques are then a requirement in these systems in order to adjust the frequency according to the instantaneous positions of the load.

Due to its novelty, the chapter focuses on the description of the strongly-coupled magnetic resonant technology. The chapter is structured as follows. In Section II, the electrical configurations of these circuits are described. The properties of the circuit provides some design guidelines that need to be considered as exposed in Section III. Section IV explains the Coupled Mode Theory applied to this WPT technology and how it is related to the Circuit Theory. Finally, Section V draws the main conclusions of the chapter.

2. STRONGLY-COUPLED MAGNETIC RESONANT SYSTEMS

As exposed in the classification of the wireless power transfer techniques, the terminology related to strongly coupled magnetic resonant circuits is not always clear in the literature and it is sometimes referred to as magnetically resonant, based on magnetically-coupled resonance or even as resonant wireless power transfer (Barman et al., 2015). We will opt for the term "strongly-couple magnetic resonant" as it reveals the physical difference of this technology in comparison with some other approaches relying on magnetic fields.

The most relevant advantage of the strongly-coupled magnetic resonant circuits is the possibility to transmit energy between two points when their separation is greater than the one used in the resonant structures. The first experiment of this kind of transmission was conducted by a research group in the Massachusetts Institute of Technology. In this activity, they showed how to transfer energy to a light bulb without cables. Specifically, the energy emitter was 2 meters far from the receiver, which implies that the distance between the objects were three times greater than the coils' diameters. The efficiency was 40%.

As stated in Liu et al. (2017), the increase in the transmission distance is mainly achieved because of the use of a topology named indirect-fed link (Eteng, Rahim, Leow, Jayaprakasam, & Chew, 2017). In this topology, four coils are employed instead of the pair typically used in resonant circuits. Thus, the source and the device to be charged both counts on two coils as shown in Figure 2.

As shown in Figure 2, the coils are usually referred as the drive loop, the transmitter coil, the receiver coil and the load loop (Cheon et al., 2011). The circuits associated to these coils will be referred with the sub-indexes D, T, R and L respectively. Each coil has a resistive component too. The coupling coefficients that are usually considered are K_{DT}, K_{TR}, K_{RD}, being the rest of the coefficients negligible. Z_L corresponds to the impedance offered by the load, that is, the element to be charged. In addition, Z_s models the impedance of the generator.

According to the original ideal of strongly-coupled magnetic resonant circuits, the transmitter and the receiver coils use their parasitic capacitances to resonate. In fact, these two coils impose the operational angular frequency (ω_o), such that:

$$\omega_o^2 = \frac{1}{L_T C_T} \tag{1}$$

$$\omega_o^2 = \frac{1}{L_R C_R} \tag{2}$$

Figure 2. Illustration of an indirect link with generic compensation topologies in strongly-coupled resonant circuits

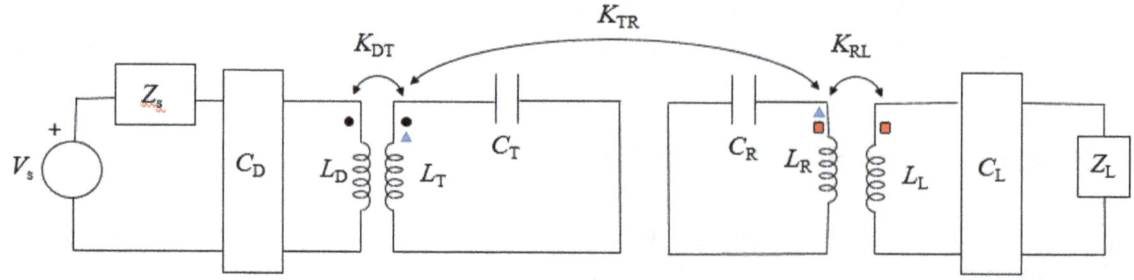

Due to the values of the parasitic capacitances, the operational frequency usually ranges from hundreds of kHz to dozens of MHz (Runhong Huang & Bo Zhang, 2015). Because of this characteristic, the transmitter and the receiver coils are usually referred to as the resonant coils.

Conversely, the drive and the load loop should be complemented with a lumped capacitor to work under resonant conditions. This element could be connected to the circuit in parallel or in series to its corresponding coil. The most popular compensation topology is the Series-Series (Eteng et al., 2017). Nevertheless, there are some solutions which do not include these capacitors explicitly in their scheme but make use of them in the experiments (Liu et al., 2017; Zhang, Zhao, & Chen, 2014).

In order to model the system with an equivalent circuit, it is necessary to analyze the resistances of the coils. The drive and the load coils are built with only one loop, being their internal resistance negligible in comparison with the source and the load impedances. In contrast, the transmitter and the receiver coils counts with more than one turns. This leads to significant internal resistances. Due to the operational frequency, these resistances are computed as the sum of two terms: the ohmic resistance and the radiation resistance (Moorey, Holderbaum, & Potter, 2015).

The analysis of the circuit is based on the concept of the reflected impedance. When applied, two meshes with coupled components are reduced to a single one where an additional component (the reflected impedance) is incorporated. Z_{ij} represents the reflected impedance from j to i where $j,i \in \{D,T,R,L\}$. In this way, we represent the meshes related to the driver, the transmitter, the receiver and the load. In order to derive the input impedance (Z_{in}), the following sequence of reflected impedances is computed:

$$Z_{RL} = \frac{\omega^2 K_{RL}^2 L_R L_L}{j\omega L_L + \frac{1}{j\omega C_L} + Z_L} \tag{3}$$

$$Z_{TR} = \frac{\omega^2 K_{TR}^2 L_T L_R}{j\omega L_R + \frac{1}{j\omega C_R} + Z_{RL}} \tag{4}$$

Figure 3. Illustration of an indirect link with a Series-Series compensation topology in strongly-coupled resonant circuits

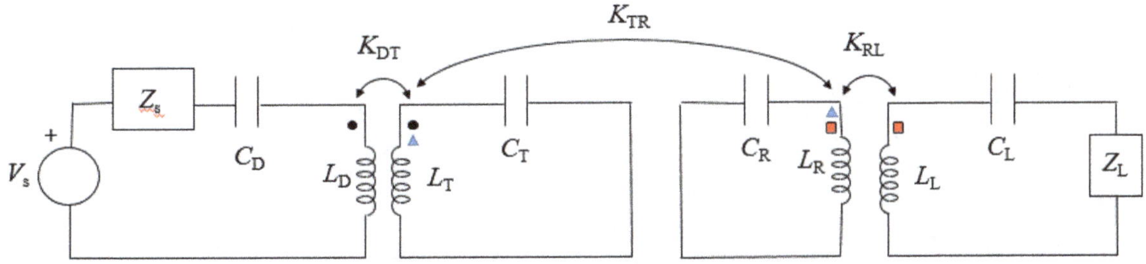

$$Z_{DT} = \frac{\omega^2 K_{DT}^2 L_D L_T}{j\omega L_T + \frac{1}{j\omega C_T} + Z_{TR}} \tag{5}$$

$$Z_{in} = Z_S + j\omega L_D + \frac{1}{j\omega C_D} + Z_{DT} \tag{6}$$

If the transmitter and the receiver coils are identically built, we can assume that their inductance and their parasitic capacitance are equivalent. As the system works at the resonant frequency (ω_o) of these two coils, we can state that:

$$\omega_o^2 L_T C_T = \omega_o^2 L_R C_R \tag{7}$$

Consequently,

$$Z_{TR} = \frac{\omega^2 K_{TR}^2 L_T L_R}{Z_{RL}} \tag{8}$$

$$Z_{DT} = \frac{\omega^2 K_{DT}^2 L_D L_T}{Z_{SR}} \tag{9}$$

$$Z_{in} = Z_S + j\omega L_D + \frac{1}{j\omega C_D} + Z_{DT} \tag{10}$$

The currents in the four coils can also be computed so that:

$$I_D = \frac{V_S}{Z_{in}} \tag{11}$$

$$I_T = \frac{j\omega K_{DT} \sqrt{L_D L_T}}{j\omega L_T + \frac{1}{j\omega C_T} + Z_{TR}} \tag{12}$$

$$I_R = \frac{j\omega K_{TR}\sqrt{L_T L_R}}{j\omega L_R + \frac{1}{j\omega C_R} + Z_{RL}} \tag{13}$$

$$I_L = \frac{j\omega K_{RL}\sqrt{L_R L_L}}{j\omega L_L + \frac{1}{j\omega C_L} + Z_L} \tag{14}$$

In order to verify the need for the compensation topologies, we analyse the efficiency of the system for a similar experiment to that presented in (Zhang et al., 2014). Without including the compensation topologies in the drive and the load loops, the efficiency is $3.29 * 10^{-11}$. The incorporation of the capacitors in these two loops leads to an improved efficiency of 0.5280.

The four-coil topology could be simplified in a very practical equivalent circuit. This simplification be understood by studying the input impedance of the circuit. Assuming that the four coils are adjusted to operate under resonant conditions (the capacitors C_D and C_L are judiciously selected for this purpose), then:

$$Z_{in} = Z_S + Z_{DT} = Z_S + \frac{\omega^2 K_{DT}^2 L_D L_T}{Z_{TR}} \tag{15}$$

If we consider the definition of the reflected impedance, we have that:

$$Z_{in} = Z_S + \frac{K_{DT}^2 L_D L_T}{\frac{K_{TR}^2 L_T L_R}{Z_{RL}}} = Z_S + \frac{K_{DT}^2 L_D L_T}{K_{TR}^2 L_T L_R} \frac{\omega^2 K_{RL}^2 L_R L_L}{Z_L} \tag{16}$$

Working with this last expression, we can obtain that:

$$Z_{in} = Z_S + \omega^2 \frac{K_{DT}^2}{K_{TR}^2} \frac{K_{RL}^2 L_L L_D}{Z_L} \tag{17}$$

Paying attention to this equation, we observe that the input impedance is the association in series of the source impedance and the reflected impedance of the load. This reflected impedance is between the drive and the load coils, being their coupling coefficient $\frac{K_{DT} K_{RL}}{K_{TR}}$. Consequently, the system can be reduced to an equivalent two coupled coil scheme as shown in Figure 4.

Figure 4. Equivalent circuit of the four-coil scheme under resonant conditions as a two-coupled coil scheme

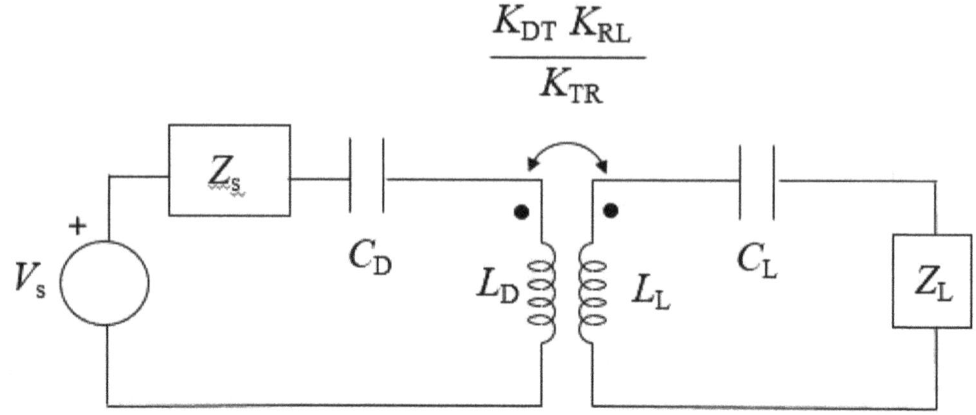

It is important to highlight that this equivalent circuit is only valid under resonant conditions and with no losses in the resonant coils. In order to derive some conclusions about this four-coil based topology, we will further analyse the equivalent scheme based on the two coils. For two coupled coils, the coupling coefficient indicates how much of the magnetic flux the coils concatenate. It is convenient that this parameter is close to 1 to use all the magnetic flux to induce voltage. However, in conventional Inductively Coupled Power Transfer systems, this condition does not usually hold as it requires that the two coils are close. Alternatively, sometimes the two coils are forced to be in contact through a ferromagnetic material in transformers to approximate the coupling coefficient to one. The main advantage of the four-coil scheme is that the ferromagnetic material and the mobility restrictions that this imposes are avoided while the magnetic flux is conveniently concatenated. Under this configuration, we should impose that:

$$\frac{K_{DT}K_{RL}}{K_{TR}} = 1 \tag{18}$$

Even when the individual coupling coefficient are not the unity, the relationship between them expressed in Equation 18 could be 1 and, as a consequence, the power transfer between the drive and the load are efficient. Thus, even if the transmitter and the receiver are far enough so that their coupling coefficient (K_{TR}) is very low, this effected can be compensated by the coupling between the driver and the transmitter (K_{DT}) and the receiver and the load (K_{RL}). By imposing this condition, the distance between the transmitter and the receiver can be enlarged. The next example illustrates this new behavior. We have opted for the data summarized in Table 1.

The driver loop is separated from the transmitter loop 10 cm. This means that the mutual inductance is 1.06 μH. The same distance is set between the receiver and the load coils. Alternatively, the distance between the transmitter and the receiver coils is varied. The coupling coefficient changes accordingly. In particular, we have assumed a circular topology for the coils. As a result of the movement, the efficiency of the system is also modified. In the blue line of the next Figure, we observe how there is an optimum relative position between the resonant coils when the efficiency is the maximum one. The system is quite

Table 1. Data characterizing the scenario under evaluation

	Driver Loop	Transmitter Coil	Receiver Coil	Load Loop
Number of turns	1	5	5	1
Radius [cm]	50	50	50	50
Autoinductance	1.6 μH	28.32 μH	28.32 μH	1.6 μH
Parasitic capacitance	-	55.45 pF	55.45 pF	-

sensible to variations in this position. In contrast, the system exclusively composed of the resonant coils (the efficiency corresponds to the red line in the next Figure) does not exhibit such a strong variation with the distance but it is unable to operate at a long distance as the topology of four coils does.

According to this behavior, the reader could assume that the strongly coupled resonant circuits are very sensible to the position of the load and only a predefined position should be allowed in order to work with an acceptable efficiency. However, one of the features of this technology (named the frequency splitting), will allow the modification of this behavior as it will be explained in the next Section.

Taking into account the equivalent scheme composed of two coils, we can apply the maximum power transfer theorem to decide about the values of the impedances $Z_s = R_s + jX_S$ and $Z_L = R_L + jX_L$. The input impedance of the system (Z_{in}) is:

Figure 5. Efficiency in a wireless power transfer system composed of two and four coils. The operational frequency is kept constant.

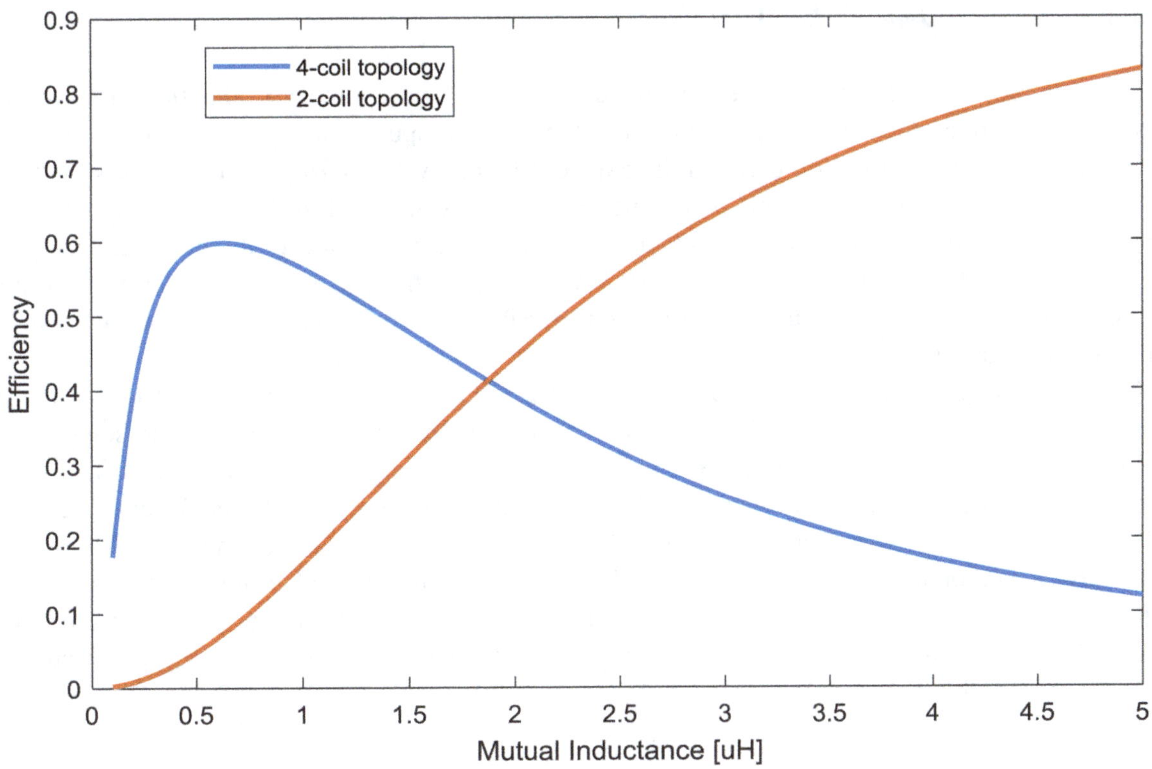

$$Z_{in} = R_S + jX_S + \frac{K_{DT}^2}{K_{TR}^2} \frac{K_{RL}^2 L_L L_D \omega^2}{\left(R_L^2 + X_L^2\right)} \left(R_L - jX_L\right) \tag{19}$$

According to the maximum power transfer theorem, $Z_s = Z_L^*$. So:

$$R_S = \frac{K_{DT}^2}{K_{TR}^2} \frac{K_{RL}^2 L_L L_D \omega^2}{\left(R_L^2 + X_L^2\right)} R_L \tag{20}$$

$$X_S = \frac{K_{DT}^2}{K_{TR}^2} \frac{K_{RL}^2 L_L L_D \omega^2}{\left(R_L^2 + X_L^2\right)} X_L \tag{21}$$

We can observe how the topology composed of four coils lead is able to transfer power at larger distances. However, it requires four coils, which implies higher costs and a less compact system. In order to avoid this last disadvantage, the conformal configuration aims at reducing the space occupied by the indirect-link topology so that the transmitter coil and the receiver coil are both built and place inside the driver and the load loop respectively.

3. DESIGN GUIDELINES IN STRONGLY COUPLED MAGNETIC RESONANT CIRCUITS

In the design process of strongly coupled magnetic resonant systems, some considerations are assumed so that the coil inductance and their parasitic capacitance are computed basing on particular equations. For instance, the relative position between the coils can be one value during the design process but it could be altered in real implementations. Basing on the values obtained in the design, the rest of the circuit is configured to work at the theoretical resonant frequency related to the coil. As previously mentioned, the real implementation could lead to deviations of the coil features, which in turn, may require the adjustment of the circuit configuration if possible. Control techniques are introduced in the system for this purpose.

When designing the control technique, it is important to consider the properties of the circuit. In particular, it is relevant to know how the adjustment of some parameters could impact on its stability. For strongly-coupled resonant circuits, two conditions could change during their operation: modifications of the operational frequency and modifications of the coupling coefficients due to the different positions of the objects. Two effects are consequences of these modifications. They are the frequency bifurcation, commonly studied in inductive systems, and the frequency splitting. Both phenomena are studied next.

In addition, strongly coupled magnetic resonant circuits operate according to the maximum power transfer theorem so an impedance matching is usually performed and included in the design process. This technique is also explained next.

3.1 Impedance Matching

Strongly-coupled magnetic resonant circuits are conventional used in low power applications. Thus, it is of interest that the load (the device to power or charge) receives the maximum power from the source (which could be limited). Losses in the intermediate components are not too relevant as they are not significant in low power applications. This means that the design of strongly-coupled magnetic resonant circuits focus on transferring the maximum power to the load independently of the efficiency of the system. This criterion is completely opposite to the one determining the high power applications. In this kind of system (e.g. a wireless charge for an electric vehicle) the efficiency is the criterion imposed in the design process.

According to the maximum power transfer theorem, the source impedance should be the complex conjugate of the impedance offered by the rest of the circuit. In order to verify this condition, a particular system is added to the source. This system is referred to as the impedance match network. In contrast to the high-power applications, the impedance match network is composed of reactive elements and resistances.

3.2 Frequency Bifurcation

As defined in Runhong Huang and Bo Zhang, (2015), this phenomenon occurs when the impedance seen by the source has no reactive component. According to this definition, bifurcation refers to an input feature of the system. In resonant circuits, the bifurcation phenomenon is usually analysed when designing the coils. In fact, the coils are designed (e.g. determination of their dimensions, the number of turns, etc.) in order to avoid the bifurcation. Towards this goal, the relationship between the quality factors of the coils must satisfy certain conditions, which depend on the compensation topology.

3.3 Frequency Splitting

Frequency splitting is a phenomenon that occurs when the coupling between the transmitter and receiver increases so that it is greater than the splitting coupling. Under these circumstances, the main electrical output magnitudes in the circuit (e.g. power delivered to the load, efficiency) present a different relationship with the changes in the frequency. It is important to highlight that frequency splitting is related to the output performance of the system whereas the bifurcation was associated to the input.

When no frequency splitting occurs, the relationship between an output magnitude and the frequency is a single-peak curve with the maximum efficiency at the resonant frequency. Alternatively, when the coupling exceeds the splitting coupling, the curve is characterized by two peaks. These two types of behavior are illustrated in the next Figure. The configuration of the system is the one presented in Zhang et al. (2014), which is composed of four coils. The resonant frequency is 3.87 MHz. The driver and the load loops are kept at a constant distance from the transmitter and the receiver coils respectively. Only the distance between the two resonant coils (the transmitter and the receivers) is varied to show the frequency splitting phenomenon. In one case, the distance leads to a mutual inductance equal to 0.322 μH (blue line in Figure 6). When the distance is reduced, the mutual inductance decreases to 3.954 μH and the efficiency corresponds to the red line in Figure 6. The maximum efficiency is not associated to the resonant frequency and there are two peaks at different frequencies.

Figure 6. Illustration of the frequency splitting phenomenon

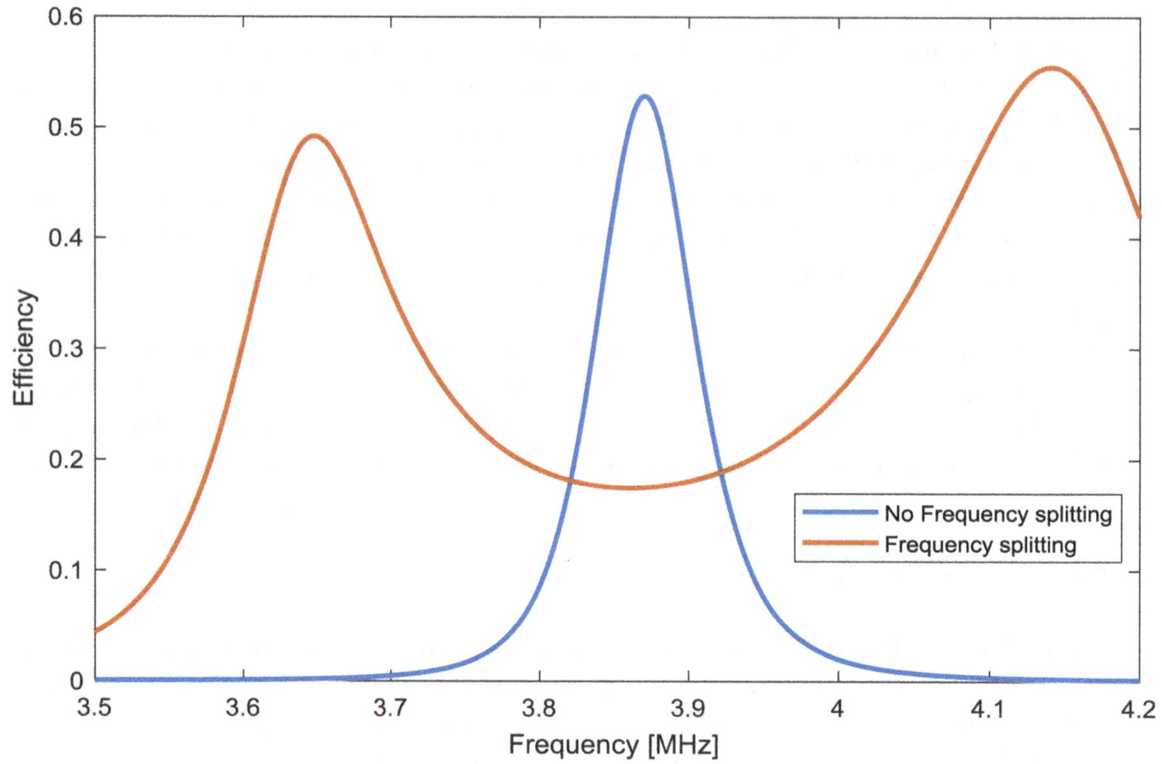

For a static and predefined position of the load, the operation of the system should avoid the frequency splitting as it could lead to several damages in the electrical components if these peaks take place with no control. Thus, it is necessary a deep study about when this phenomenon occurs, that is, what is the maximum coupling coefficient that does not lead to a frequency splitting. This coupling coefficient is referred to as the critical coupling. Once the frequency splitting is forced, it is also desirable to know at which frequency the peaks are.

The frequency splitting is not exclusive of the four-coil topology systems. In this sense, the work in Niu, Chu, Gu, and Shen (2013) presents the analytical study about the basis concerning the frequency splitting for symmetrical and asymmetrical two-coils systems. Firstly, it computes the power delivered to the load and then it maximizes the curve.

The work in Sample, Meyer, and Smith (2011) addresses the definition of the critical coupling for a four-coil topology. Assuming real coils characterized by their internal resistance, the critical coupling depends on their quality factors. In particular, for a topology with the same resonant coils and with the same loops, the critical coupling ($K_{critical}$) is:

$$K_{critical} = \frac{1}{Q_2} + K_{12}^2 Q_1 \tag{22}$$

where Q_1 is the quality factor of the loops and Q_2 corresponds to the quality factor of the resonant coils. These two parameters are defined as:

$$e_s = \left(R_1 + R_s\right) i_1\left(t\right) + \left(L_1 - \frac{M^2}{L_2}\right)\frac{di_1\left(t\right)}{dt} + \frac{M\left(R_2 + R_L\right) i_2\left(t\right)}{L_2} \quad Q_1 = \frac{\omega_1 L_1}{R_1} \tag{23}$$

$$Q_2 = \frac{\omega_2 L_2}{R_2} \tag{24}$$

where R_1 is the internal resistance of the loop, R_2 is the internal resistance of the resonant coil. The angular frequencies employed in these equations correspond to the free resonant frequency of the coils. Thus, $\omega_2 = \omega_o$, that is, the resonant frequency of the system. The resonant frequency is imposed by the topology of the resonant coils in order to exclusively consider the parasitic capacitances. In addition, ω_1 is also the resonant frequency of the loops so that a characterization of their parasitic capacitances (C_{p1}) should be done in order to compute it. Once computed, the free resonant frequency of the loop is computed as:

$$\omega_1 = \frac{1}{\sqrt{L_1 C_{p1}}} \tag{25}$$

Taking into account the critical coupling, a strongly-coupled magnetic resonance system can work in the two following regions:

- The overcoupled region, when $K_{23} > K_{critical}$. Under these circumstances, the transmitter and the receiver coils are very close.
- The undercoupled region, $K_{23} < K_{critical}$, when the two resonant coils are not so close.

If properly configured, the overcoupled region permits to transfer maximum power with small variations for different distances between the transmitter and the receiver. This is achieved by tuning the frequency of operation so that it corresponds to one of the peaks that appear when the frequency splitting takes place. Figure X shows how the efficiency is greatly increased when the frequency tuning is performed. In particular, the system corresponds to the one described previously. As can be observed, the differences on the efficiency are in the overcoupled region, that is, when the transmitter and the receiver coils are closer.

Thus, a frequency adjustment should be performed in order to get nearly a constant efficiency for strongly-coupled magnetic resonant circuits. In particular, the variation of the frequency required for the above distances (related to the mutual inductance between the resonant coils) are illustrated in the next Figure.

Figure 7. Illustration of the efficiency for a perfect frequency tuning and with no frequency adjustment in a 4-coil topology

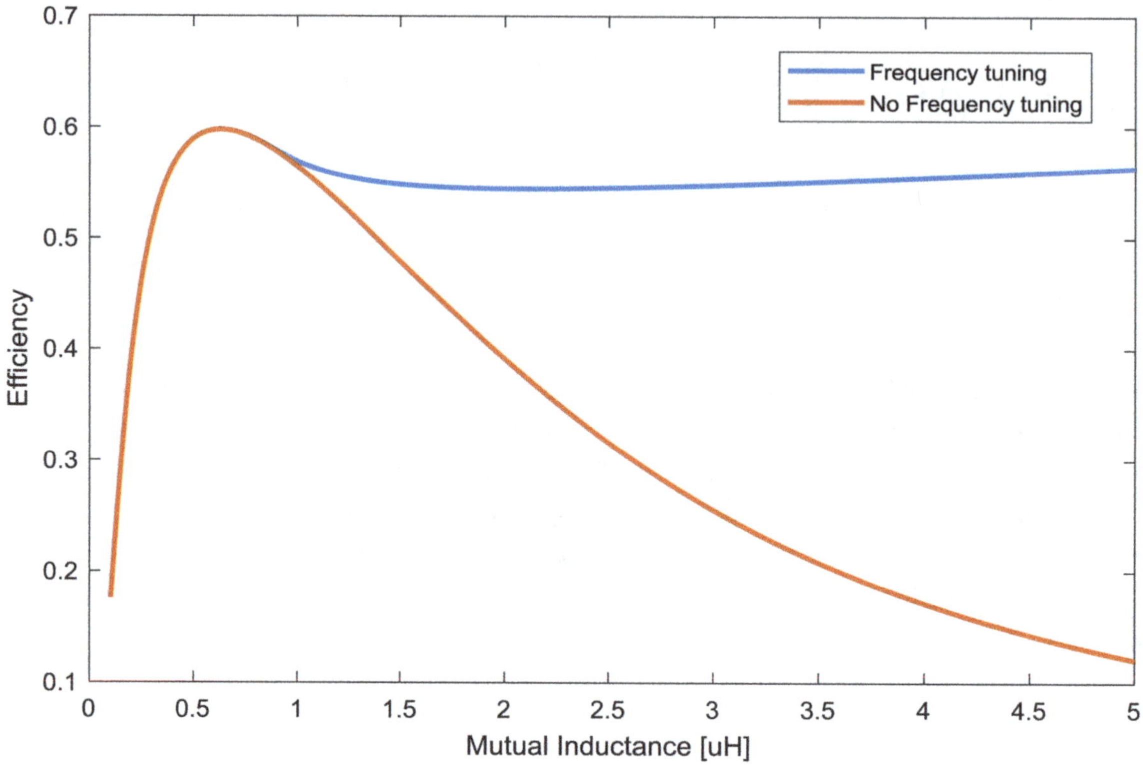

As can be observed, the variations on the frequency are not noticeable so they can be adjusted with conventional power converters in this range.

4. ANALYSIS OF THE STRONGLY-COUPLED RESONANT CIRCUITS

The correct design of a resonant circuit requires its analysis in order to configure the correct parameters that the application requires. In the near-field applications, inductive and resonant circuits are analysed by means of the classical Kirchhoff-based theory. However, the strongly-coupled magnetic resonant circuits are sometimes studied basing on the coupled-mode theory. Actually, this kind of approach was the most frequent initially.

The Coupled Mode Theory (CMT) was initially applied to the microwave domain by 1950s. Its application to wireless power transfer is firstly proposed and justified in PhD Thesis elaborated by A. Kurs at the Massuchusets Institute of Technology. In this work, the author used self-resonating coils for the power transfer conversely to the use of additional capacitors employed in the compensation systems of previous inductive resonant systems. Due to the distributed nature of the capacitors, he defended that it was not appropriate to employ a lumped based analysis as usually done in inductive circuits and the best approach was working with the Coupled-Mode Theory.

Figure 8. Tuned Frequency required for the optimal efficiency in a 4-coil topology

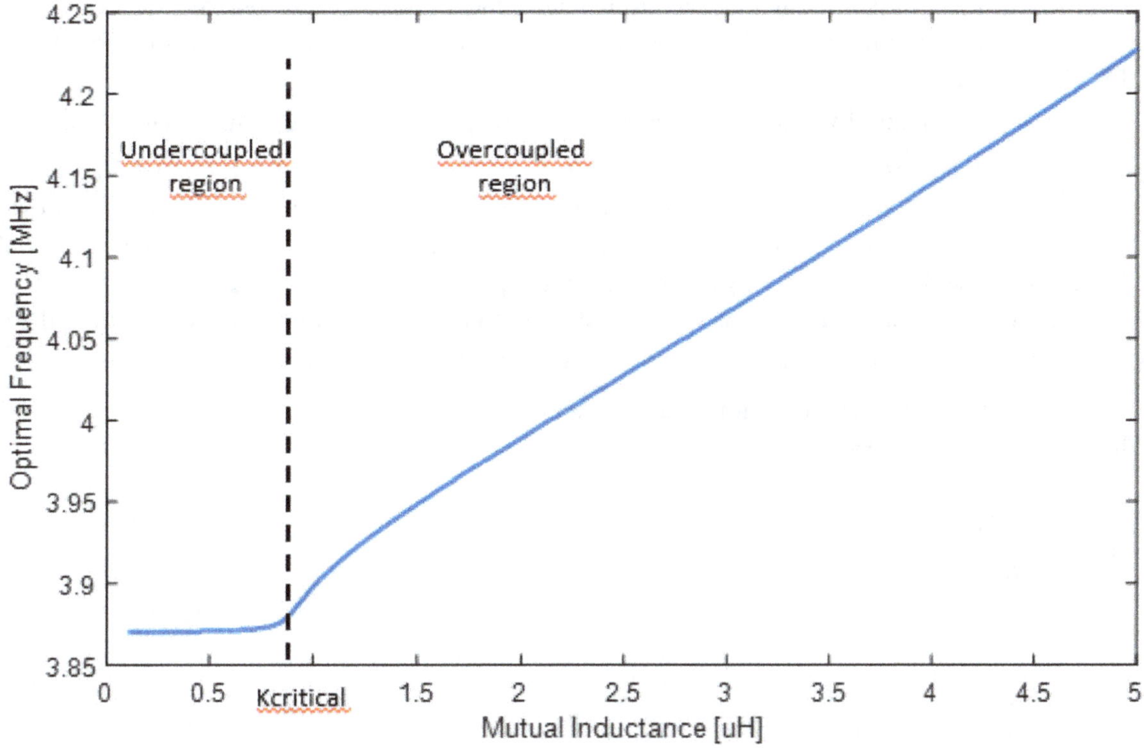

The CMT helps understanding the energy exchange between two resonant objects, that is, between two self-resonant coils (one in the power transmitter and one in the power receiver). When applied to resonant circuits, a mode is related to the electromagnetic energy exchange between two connected reactive components. In particular, a pair composed of one coil and its resonant capacitor (the capacitance could be parasitic as assumed in the strongly coupled magnetic resonant circuits) will be associated to a mode a_i. This mode is defined in such a way that $|a_i|^2$ equals the energy of those components assuming that they are coupled with other elements in the circuit. For a coil with inductance L_1 and the connected capacitance C_1, the associated mode a_1 is:

$$a_1 = \sqrt{\frac{C_1}{2}}V_{C_1} + j\sqrt{\frac{L_1}{2}}I_{L_1} \tag{26}$$

where V_{C_1} and I_{L_1} are the voltage at the capacitance and the current traversing the coil respectively.

From this definition, we observe that:

$$|a_1|^2 = \frac{C_1}{2}V_{C_1}^2 + \frac{L}{2}I_{L_1}^2 = W_{C_1} + W_{L_1} \tag{27}$$

where W_{C_1} and W_{L_1} are the energy due to the electric field and the magnetic field respectively. As we can observe, this theory assumes that the electromagnetic energy described by one mode is independent of the other mode, condition which does not hold in coupled circuits but it is necessary to take it to simplify the computations.

The energy represented by a mode varies along the time as V_{C_1} and I_{L_1} also change with time-varying electrical sources.

The Kirchhoff-based and the CMT theories are both related as we will show basing on the following classical series-series topology of a resonant circuit in Figure 9. Please note that this scheme is a valid simplification of the four-coil topology as exposed in the previous Section.

The goal of the Kirchhoff analysis is to establish four equations with the variables defining the energy in the reactive components (i_1 ,i_2,v_{c_1} and v_{c_2}). In particular, it should relate the derivate of one of the variables to the other variables with no derivations.

The behavior of the capacitors imposes that:

$$i_1\left(t\right) = C_1 \frac{dv_1\left(t\right)}{dt} \tag{28}$$

$$i_2\left(t\right) = C_2 \frac{dv_2\left(t\right)}{dt} \tag{29}$$

From these two equations, it is also to obtain the relationship between the variables $\frac{dv_{c_1}\left(t\right)}{dt}$ and $\frac{dv_{c_2}\left(t\right)}{dt}$ and $i_1\left(t\right)$ and $i_2\left(t\right)$. These two equations constitutes the two first equations of the system to solve. It is still needed to obtain $\frac{di_1\left(t\right)}{dt}$ and $\frac{di_2\left(t\right)}{dt}$.

Figure 9. Electrical scheme used to compare the classic circuit theory and the coupled mode theory

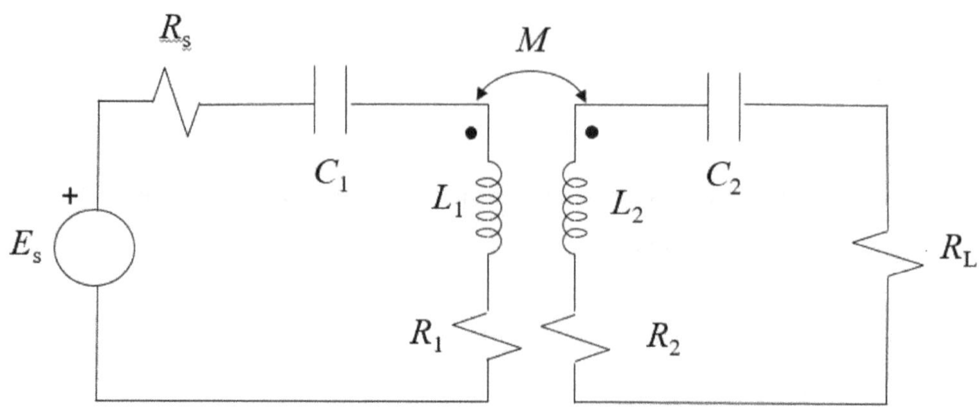

Applying Kirchhoff's voltage law, we observe that:

$$e_s = \left(R_1 + R_s\right)i_1\left(t\right) + v_{L1}\left(t\right) + v_{C1}\left(t\right) \tag{30}$$

Due to the behavior of the coupled coils, we can relate the previous equation to the currents as follows:

$$e_s = \left(R_1 + R_s\right)i_1\left(t\right) + \frac{L_1 di_1\left(t\right)}{dt} + \frac{M di_2\left(t\right)}{dt} + v_{C1}\left(t\right) \tag{31}$$

In the second mesh, the use of the voltage law leads to:

$$0 = \left(R_2 + R_L\right)i_2\left(t\right) + v_{L2}\left(t\right) + v_{C2}\left(t\right) \tag{32}$$

Equation 32 is expressed as:

$$0 = \left(R_2 + R_L\right)i_2\left(t\right) + \frac{L_2 di_2\left(t\right)}{dt} + \frac{M di_1\left(t\right)}{dt} + v_{C2}\left(t\right) \tag{33}$$

From Equation 33, we can conclude that:

$$\frac{di_2\left(t\right)}{dt} = -\frac{R_2 + R_L}{L_2}i_2\left(t\right) - \frac{M}{L_2}\frac{di_1\left(t\right)}{dt} - \frac{v_{C2}\left(t\right)}{L_2} \tag{34}$$

Incorporating this relationship into Equation 31, we obtain:

$$e_s = \left(R_1 + R_s\right)i_1\left(t\right) + \left(L_1 - \frac{M^2}{L_2}\right)\frac{di_1\left(t\right)}{dt} + \frac{M\left(R_2 + R_L\right)i_2\left(t\right)}{L_2} - \frac{M}{L_2}v_{c_2}\left(t\right) + v_{C1}\left(t\right) \tag{35}$$

So we can derive that:

$$\frac{di_1\left(t\right)}{dt} = \frac{L_2}{L_1 L_2 - M^2}\left[e_s - \left(R_1 + R_s\right)i_1\left(t\right) + \frac{M}{L_2}\left(R_2 + R_L\right)i_2\left(t\right) + \frac{M}{L_2}v_{c_2}\left(t\right) - v_{C1}\left(t\right)\right] \tag{36}$$

This equation is inserted in Equation 32 to result in:

$$\frac{di_2\left(t\right)}{dt} = -\frac{R_2 + R_L}{L_2}i_2\left(t\right) - \frac{M}{L_1 L_2 - M^2}\left[e_s - \left(R_1 + R_s\right)i_1\left(t\right) + \frac{M}{L_2}\left(R_2 + R_L\right)i_2\left(t\right) + \frac{M}{L_2}v_{c_2}\left(t\right) - v_{C1}\left(t\right)\right] - \frac{v_{C2}\left(t\right)}{L_2} \tag{37}$$

This last expression can be further simplified as follows:

$$\frac{di_2(t)}{dt} = \frac{M}{L_1L_2 - M^2}v_{C1}(t) - -\frac{L_1}{L_1L_2 - M^2}v_{C2}(t) + \frac{M(R_1 + R_s)}{L_1L_2 - M^2}i_1(t) - \frac{L_1(R_2 + R_L)}{L_1L_2 - M^2}i_1(t) - \frac{M}{L_1L_2 - M^2}e_s \tag{38}$$

Thus, according to the analysis based on the Kirchhoff's laws, the system to solve is composed of 4 equations being the unknowns $v_{c_1}(t)$, $v_{c_2}(t)$, $i_1(t)$ and $i_2(t)$. The equations are Equation 36, 38 and the next two:

$$\frac{dv_{c_1}(t)}{dt} = \frac{1}{C_1}i_1(t) \tag{39}$$

$$\frac{dv_{c_2}(t)}{dt} = \frac{1}{C_s}i_2(t) \tag{40}$$

Therefore, the system counts with 4 equations. The number of equations is reduced with the next analytical approach. We will proceed with an analysis of the system basing on the Coupled Mode Theory. For this, we have to combine the equations in order to obtain the mode definition, which was expressed in Equation 26 as the combination of the voltage in the capacitor and the current in the coil.

Firstly, we multiply Equation 39 and Equation 36 by $\sqrt{\frac{C_1}{2}}$ and j$\sqrt{\frac{L_1}{2}}$, respectively. Then, the two new equations are added resulting in:

$$\frac{d}{dt}\left(\sqrt{\frac{C_1}{2}}v_{c_1} + j\sqrt{\frac{L_1}{2}}i_1\right) = \sqrt{\frac{C_1}{2}}\frac{i_1}{C_1} +$$
$$j\sqrt{\frac{L_1}{2}}\left[-\frac{L_2}{L_1L_2 - M^2}v_{c_1} + \frac{M}{L_1L_2 - M^2}v_{c_2} - \frac{L_2(R_1 + R_s)}{L_1L_2 - M^2}i_1 + \frac{M(R_2 + R_L)}{L_1L_2 - M^2}i_2 + \frac{L_2}{L_1L_2 - M^2}e_s\right] \tag{41}$$

As the first mode is $a_1 = \sqrt{\frac{C_1}{2}}v_{c_1} + j\sqrt{\frac{L_1}{2}}i_1$ and the second one is $a_2 = \sqrt{\frac{C_2}{2}}v_{c_2} + j\sqrt{\frac{L_2}{2}}i_2$, the previous equation can be expressed as:

$$\frac{da_1(t)}{dt} = -j\omega_1 a_1''(t) - \Gamma_1 a_1'(t) + K_{12}a_2''(t) + K_{1e}e_s \tag{42}$$

where

$$a_1^{''} = \frac{L_1 L_2}{L_1 L_2 - M^2}\sqrt{\frac{C_1}{2}}v_{c_1} + j\sqrt{\frac{L_1}{2}}i_1 \tag{43}$$

$$a_1^{'} = j\sqrt{\frac{L_1}{2}}i_1 \tag{44}$$

$$\Gamma_1 = \frac{L_2\left(R_1 + R_s\right)}{L_1 L_2 - M^2} \tag{45}$$

$$K_{12} = j\ \frac{M}{L_1 L_2 - M^2}\sqrt{\frac{L_1}{2}} \tag{46}$$

$$K_{1e} = j\ \frac{L_2}{L_1 L_2 - M^2}\sqrt{\frac{L_1}{2}} \tag{47}$$

$$a_2^{''} = \sqrt{\frac{C_2}{2}}v_{c_2} + \sqrt{\frac{L_1}{2}}i_2 \tag{48}$$

$$\omega_1 = \frac{1}{\sqrt{L_1 C_1}} \tag{49}$$

In order to establish a similarity between Equation 42 and the wave equation, some simplifications are taken in the analysis so that $a_1^{'}$, $a_1^{''}$ and $a_2^{''}$ are reduced to the mode parameters. The simplifications are not obvious. In fact, it is quite arguable if this is a correct methodology because of that.

In the first simplification, $a_1^{''} = a_1$. This is valid if the mutual inductance is negligible when compared with the coils inductances. However, in a four-coil scheme designed to comply with Equation 18, this condition does not hold.

Then, $a_1^{'} = a_1$. This equality implies that the energy stored in the capacitor C_1 is much greater than the energy in the coil L_1. Finally, the last simplification forces that $a_2^{''} = a_2$.

With these simplifications, we obtain the following wave equation:

$$\frac{da_1(t)}{dt} = -j\omega_1 a_1(t) - \Gamma_1 a_1(t) + K_{12} a_2(t) + K_{1e} e_s \tag{50}$$

In a similar way, we can proceed to extract the wave equation of the second mode. Firstly, we multiply Equation 40 and Equation 38 by $\sqrt{\frac{C_2}{2}}$ and j $\sqrt{\frac{L_2}{2}}$ respectively.

$$\frac{d}{dt}\left(\sqrt{\frac{C_2}{2}} v_{C_2} + j\sqrt{\frac{L_2}{2}} i_2\right) = \sqrt{\frac{C_2}{2}} \frac{i_2}{C_2} + j\sqrt{\frac{L_2}{2}}\left[\frac{M}{L_1L_2 - M^2} v_{C_1} - \frac{L_1}{L_1L_2 - M^2} v_{C_2} + \frac{M(R_1 + R_s)}{L_1L_2 - M^2} i_1 - \frac{L_1(R_2 + R_L)}{L_1L_2 - M^2} i_2 - \frac{M}{L_1L_2 - M^2} e_s\right] \tag{51}$$

With similar simplifications to those commented previously for the wave-equation of the first mode, the precedent equation can be approximated to:

$$\frac{da_2(t)}{dt} = -j\omega_1 a_2(t) - \Gamma_2 a_2(t) + K_{21} a_1(t) + K_{2e} e_s \tag{52}$$

Where

$$\Gamma_2 = \frac{L_1(R_2 + R_L)}{L_1L_2 - M^2} \tag{53}$$

$$K_{21} = j\,\frac{M}{L_1L_2 - M^2}\sqrt{\frac{L_2}{C_1}} \tag{54}$$

$$K_{2e} = -j\,\frac{M}{L_1L_2 - M^2}\sqrt{\frac{L_2}{2}} \tag{55}$$

Paying attention to this mathematical system, we state that the number of equations is reduced to half of the ones required by the classical circuit theory. Specifically, the solution of the wave equations is:

$$a_i(t) = A_i e^{-j\omega t} \tag{56}$$

5. CONCLUSION

Wireless Power Transfer makes it possible to transfer energy between to electrical devices through an air gap, without the need for current-carrying wires. This energy transfer can be carried out by using different technologies. Classical near field inductive coupling assume that the magnetic coupling between primary and secondary coils must be high for efficient operation. This forces both coils to be closely aligned to each other. However, whenever the application requires somewhat larger distances or relax the alignment constraint, the strongly-coupled magnetic resonant circuits theory provides a solution.

This chapter has reviewed the technology associated to strongly-coupled magnetic resonant circuits. This is a technology especially suitable for the power transfer without conductors between two electrical circuits that are separated a distance similar to the coils' dimensions. The presence of intermediate objects is also possible as they do not interrupt the power transfer (they can alter the efficiency, though). Thus, it is very well suited for home or domestic appliances.

This technology still needs further contributions from the research and industry communities. For instance, impedance matching is a key problem since efficient power transfer usually comes with complex fine tuning of parameters and a trade-off between efficiency and practical applications must be achieved. System design including control strategies and optimized design of ferromagnetic materials will highly impact the feasibility of these systems.

ACKNOWLEDGMENT

This work has been partially funded by the project PPIT.UMA.B1.2017/27 belonging to the Plan Propio of the University of Málaga under the programme "Proyectos dirigidos por jóvenes investigadores".

REFERENCES

Barman, S., Rez, A., & Kumara, N. (2015). Wireless powering by magnetic resonant coupling: Recent trends in wireless power transfer system and its applications. *Renewable & Sustainable Energy Reviews*, *51*, 1525–1552. doi:10.1016/j.rser.2015.07.031

Bi, Z., Kan, T., Mi, C. C., Zhang, Y., Zhao, Z., & Keoleian, G. A. (2016). A review of wireless power transfer for electric vehicles: Prospects to enhance sustainable mobility. *Applied Energy*, *179*, 413–425. doi:10.1016/j.apenergy.2016.07.003

Cheon, S., Kim, Y.-H., Kang, S.-Y., Lee, M. L., Lee, J.-M., & Zyung, T. (2011). Circuit-Model-Based Analysis of a Wireless Energy-Transfer System via Coupled Magnetic Resonances. *IEEE Transactions on Industrial Electronics*, *58*(7), 2906–2914. doi:10.1109/TIE.2010.2072893

Dai, J., & Ludois, D. C. (2015). A Survey of Wireless Power Transfer and a Critical Comparison of Inductive and Capacitive Coupling for Small Gap Applications. *IEEE Transactions on Power Electronics*, *30*(11), 6017–6029. doi:10.1109/TPEL.2015.2415253

Eteng, A. A., Rahim, S. K. A., Leow, C. Y., Jayaprakasam, S., & Chew, B. W. (2017). Low-power near-field magnetic wireless energy transfer links: A review of architectures and design approaches. *Renewable & Sustainable Energy Reviews*, *77*, 486–505. doi:10.1016/j.rser.2017.04.051

Huang, L., & Hu, A. P. (2015). Defining the mutual coupling of capacitive power transfer for wireless power transfer. *Electronics Letters*, *51*(22), 1806–1807. doi:10.1049/el.2015.2709

Huang, R., & Zhang, B. (2015). Frequency, Impedance Characteristics and HF Converters of Two-Coil and Four-Coil Wireless Power Transfer. *IEEE Journal of Emerging and Selected Topics in Power Electronics*, *3*(1), 177–183. doi:10.1109/JESTPE.2014.2315997

Jolani, F., Yu, Y., & Chen, Z. (2015). Enhanced planar wireless power transfer using strongly coupled magnetic resonance. *Electronics Letters*, *51*(2), 173–175. doi:10.1049/el.2014.4104

Liu, D., Hu, H., & Georgakopoulos, S. V. (2017). Misalignment Sensitivity of Strongly Coupled Wireless Power Transfer Systems. *IEEE Transactions on Power Electronics*, *32*(7), 5509–5519. doi:10.1109/TPEL.2016.2605698

Moorey, C., Holderbaum, W., & Potter, B. (2015). Investigation of High-Efficiency Wireless Power Transfer Criteria of Resonantly-Coupled Loops and Dipoles through Analysis of the Figure of Merit. *Energies*, *8*(10), 11342–11362. doi:10.3390/en81011342

Niu, W.-Q., Chu, J.-X., Gu, W., & Shen, A.-D. (2013). Exact Analysis of Frequency Splitting Phenomena of Contactless Power Transfer Systems. *IEEE Transactions on Circuits and Systems. I, Regular Papers*, *60*(6), 1670–1677. doi:10.1109/TCSI.2012.2221172

Sample, A. P., Meyer, D. A., & Smith, J. R. (2011). Analysis, Experimental Results, and Range Adaptation of Magnetically Coupled Resonators for Wireless Power Transfer. *IEEE Transactions on Industrial Electronics*, *58*(2), 544–554. doi:10.1109/TIE.2010.2046002

Zhang, Y., Zhao, Z., & Chen, K. (2014). Frequency-Splitting Analysis of Four-Coil Resonant Wireless Power Transfer. *IEEE Transactions on Industry Applications*, *50*(4), 2436–2445. doi:10.1109/TIA.2013.2295007

Chapter 2
Efficiency Improvement in Wireless Power System

Luis Romba Jorge
Universidade Nova de Lisboa, Portugal

Elena Baikova
Universidade Nova de Lisboa, Portugal

Stanimir Valtchev
Universidade Nova de Lisboa, Portugal

Rui Melício
Universidade de Lisboa, Portugal

ABSTRACT

This chapter focuses on mid-range wireless power transfer (WPT) systems applied to electric vehicle (EV) battery chargers. The WPT is recently considered as an efficient electric energy transfer process between two or more points in space, without wiring. The technology associated with each specific process of WPT differs from case to case depending on the distance between those points and the power to be transferred between them. The widely adopted distance categories are named short-range, mid-range, and long-range. The short-range is normally defined as up to a few millimeters range. The mid-range is between a few millimeters and a few meters. The long-range distance is defined as a longer than that of the previous category, stretching up to a few kilometers.

1. INTRODUCTION

To summarize the state-of-the-art of the Wireless Power Transfer (WPT), it must be said that one important aim of the WPT is the mobility, i.e. the transport. The evolution of the modern transport sector is determined by two paradigms: the climatic changes driven by pollution and the geopolitical instability in the main geographic areas, where crude oil is currently produced (Turton & Moura, 2008) The oil is still the most important means for the transport propulsion, but now the "green car" is being introduced, driven by the so called "green", non-polluting energy. The name "green cars" is not only tagging the

DOI: 10.4018/978-1-5225-5870-5.ch002

passenger cars, but is also used for other vehicles, like long distance trucks, for example. The meaning of "green" expresses mainly the electrification of any mobility, and the road transport, in particular (Perlo & Meyer, 2012; Eckman, 2011). The final goal of the electrification is the integration of the vehicles into intelligent electrical systems that include the transmission and storage of energy for sustainable applications (Perlo & Meyer, 2012).

This chapter focuses on mid-range WPT systems and especially on their efficiency improvement. The technology, associated with the WPT process varies, depending on the distance and the power to be transferred (Romba et al., 2017). The distance defines three main categories, named: short-range, mid-range and long-range. The short-range is normally defined as up to a few millimetres range. The mid-range is between a few millimetres and a few meters. The long-range distance is defined as a longer than that of the previous category, stretching up to a few kilometres or even more. The technology solutions, used for short-range distances are the inductive power transfer (IPT) and capacitive power transfer. The main part of the solutions for WPT aimed at the mid-range distance is based on magnetic resonance coupling and high switching frequency of the power electronic converter (Romba et al., 2017; Huang et al., 2012; Baikova et al., 2016). The long-range distance WPT requires more narrow concentration of the electromagnetic energy, to avoid the unselective dispersion of the electromagnetic energy in all directions. That beamed power transfer is known since the 1960s, but still it is not enough mature technically for a practical implementation (Glaser, 1968). This technology is also a basis for the idea presented by some researchers and enterprises, known as the solar power satellite system. This concept, introduced by Peter Glaser in 1968, is based on geosynchronous orbit satellites, as transmitters of microwave electromagnetic energy (Glaser, 1968). The energy is obtained from extremely large, kilometric size solar panels, mounted in the space, while the receivers are mounted on the Earth surface. The microwaves beam or the laser beam will be sent to the receivers, supposed to be made as a rectenna system that combines antennas with rectifiers (Glaser, 1968; Bareiss et al., 2013). The different constructions of WPTs depend on the applications, the distances and the power that different classes of vehicles require. The chapter concerns the Wireless Power Transfer Systems efficiency improvement, using the matching of impedances made by a different device, named Magnetic Core Reactor. In addition, a comparative study of the magnetic field behaviour of two type of coils, spiral coil and helicoidal coil, is presented.

2. ELECTRIC VEHICLE CLASSES

The classes of electric vehicles, currently available, as well as their main features, are presented in this section. Every one of those classes is capable to use a WPT:

There are three classes of Electric Vehicles. These classes are:

- Purely Electric Vehicles, Battery Electric Vehicles (BEV) and Fully Electric Vehicles, are three names that have the same meaning;
- Hybrid Electric Vehicles (HEV);
- Plug-in Hybrid Electric Vehicles (PHEV).

The BEV uses only the electric motor to drive the car. The batteries provide the power to the motor for the propulsion. The main advantage of these vehicles is that they do not produce pollution during their locomotion. The main drawback of BEVs, at present, is the limitation of the batteries energy storage capacity. To minimize this drawback, it is necessary to install vehicle charging infrastructure integrated into the emerging smart cities (Fernández et al., 2011).

The HEV includes an internal combustion engine (ICE) and an electric motor. The engine and the electric motor supply the necessary torque to drive the vehicle. In the so called parallel HEV, especially in the most recent ones, the electric motor alone drives the vehicle using power stored in batteries, at low speeds or in traffic. When a high acceleration is necessary, both the IC engine and the electric motor, provide a higher torque to drive the vehicle. In a steady-state, highway cruising, the internal combustion (IC) engine alone drives the vehicle. In the "Series" HEV, the IC engine is charging the batteries, at the best efficient mode of IC engine operation, and the electric motor always is the mover. In some versions, the parallel HEV also can charge its batteries by IC engine, but in the older versions, the charging is only by active braking, when the motor is used as a generator. In these HEV vehicles, the batteries are not charged externally (Hilshey et al., 2013).

The Plug-In (PHEV) is also a HEV with an electric motor and an ICE. The main advantage of PHEV is that it can be recharged from an external electric source. It is still possible to use the IC engine to charge the batteries (Pereirinha & Trovão). The torque creating, in a global point of view, is similar to that of HEV. The HEV and PHEV produce little or no local air pollution. This depends on the type of the construction, battery capacity, motors power, working mode, but those vehicles, as a rule, reduce the IC fuel consumption, and hence, the dependence on crude oil, finally reducing the greenhouse gas emissions from the power source in use (Choi et al., 2015).

3. PREVIOUS STUDY

Near Field Wireless Power Transfer

The near field is generally divided into two areas, i.e. the reactive and the radiative. In the reactive area the magnetic field (H) and the electric field (E) are very strong and predominately reactive fields. This last condition means the H-field and E-field are out of phase by 90° to each other. The condition for propagation or radiation is that the E and H fields, stay orthogonal and in phase with each other. Since these conditions are not maintained in the near vicinity, no radiation occurs. When the distance from the transmitter is short, one field or the other will dominate, depending on the emission, i.e. "antenna" type. A coil, considered as that "antenna", in a most widely used case, is dominated by the magnetic field (H). This coil can be considered to be a primary in a transformer that has not enough good magnetic coupling to the secondary. The primary generates a large intensity of magnetic field, but the flux is not sufficiently influencing the secondary, i.e. the receiver. The fields begin to be radiative in the radiative area, where the far field makes the transmission to resemble a transmission line where E and H are in phase with each other (Johnson et al., 1973; Valtchev & Klaasens, 1990; Valtchev et al., 2009).

The mid-range WPT, where the magnetic coupling is even worse than that in the close vicinity to the transmitter, the High-Frequency (HF) magnetic field is a solution, with a well projected resonance effect. The resonance is also used in all the reactive area, including the close vicinity to the transmitter, as it improves the efficiency of the switches. The resonant converter is a necessary part of the knowledge, as a more efficient operation is concerned. In general, the resonant and especially super-resonant operation is characterised by low switching losses described in (Kurs et al., 2007), while the similar treatment is given in (Valtchev et al., 2009) to the contactless energy transfer.

In this section, three different techniques, used in WPT, are presented.

Techniques Used in WPT

Early before the power grids had appeared, at the beginning of the twentieth century, Nikola Tesla had the dream of transmitting electric power freely through the air. The initial experiments to achieve this goal, e.g. the Tesla coils, produced undesirable electric field. The technology that Tesla used involved dangerous high voltage and electric discharges because Tesla believed that the electric field is the means for transporting energy (Kurs et al., 2007).

With the advent of autonomous electronic devices, i.e. cell phones, PDAs, laptops, has appeared the necessity of the wireless energy transfer, and the electric field was substituted by the safer and much more powerful, magnetic field. The energy level and density, at which the transmission is performed by the magnetic field, is incomparably higher, although the capacitive energy transfer is used too, e. g. for the so called Intelligent Cards. The energy transfer made by the electric field is reported rarely for some higher levels of power, as that published in (Liu et al., 2010). The intensity of the electric field remains dangerously high when the power is higher.

Nowadays, the mid-range wireless energy is transferred by magnetic field using inductive coupling between two coils. One is considered the transmitter another is the receiver. The inductive coupling is considered the most widely used wireless technology. This inductive technique, with resonant loops for compensation was used intensively in the HF inductive heating process since 1960s. As the frequency of inductive heating is quite high, the transformers were made by air core, and rarely, only recently using specific magnetic material of low permeability and low losses. The WPT operation is very similar to an air core transformer. It is used only at very short distances. The electric toothbrushes and electric razors are examples of low power applications, eliminating the danger of electric shock. Today it is considered that the inductive coupling can be divided into three different techniques, as following (Kurs et al., 2007; Kiu et al., 2010; Cannon et al., 2009):

- Magnetic coupling (MC);
- Magnetic resonance coupling (MRC);
- Strongly coupled magnetic resonance (SCMR).

The MC occurs when the primary coil, connected to an energy source, generates a time-varying magnetic field. The power lines of this magnetic field should cover almost entirely the receiver coil to minimize losses. The gap between the two coils must be less than the coil diameter. The main reason is because the module of the magnetic field vector is proportional to $1 / d^3$. The near-field magnetic flux, generated in the primary coil, produces an electromotive force in the secondary coil. When the secondary circuit is closed the transferred electric energy can be used for suppling a device or storing the energy in a system (Kurs et al., 2007). The MC schematic diagram is shown in Figure 1.

In Figure 1 R_{esrp} is the primary equivalent series resistance, L_p is the primary inductance, L_s is the secondary inductance, R_{esrs} is the secondary equivalent series resistance, R_L is the load resistance.

The MRC has a resonant circuit in each branch: transmitter and receiver. The tuning circuitry can be done by connecting capacitors in series or parallel with the coils. The capacitors could be connected in four different configurations: series in the primary (transmitter branch) and secondary (receiver branch)

Figure 1. Magnetic coupling schematic diagram

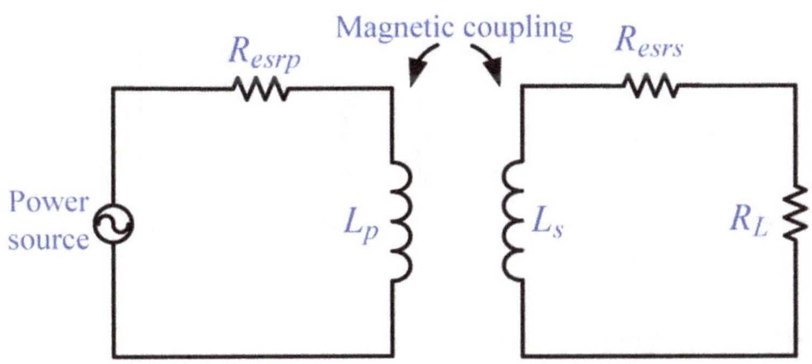

(SS) Figure 2(A), series in the primary and in parallel on the secondary (SP) Figure 2(B), the parallel in the primary and series in secondary (PS) Figure 2(C) and parallel in primary and secondary (PP) Figure 2(D). In Figure 2 the C_p is the primary capacitance, C_s is the secondary capacitance. When the two resonant coils are operating at the same resonant frequency exists a strong coupling and high energy transfer efficiency. The MRC has the additional advantage of immunity to neighbouring environment, due to the property of resonance. The magnetic resonance coupling can be applied between a transmitter (transmitter resonator) and many receivers (receiving resonators), allowing charging of multiple devices (Cannon et al., 2009; Zhu et al., 2008).

The Strongly Coupled Magnetic Resonance (SCMR) technique was proposed in 2007 by an MIT team in the Department of Physics (Zhu et al., 2008). This technique applied to mid-range WPT has a higher efficiency compared to magnetic resonant coupling. The SCMR basic block diagram is shown in Figure 3. This technology is characterized by two pairs of high-quality-factor resonant loops, one for the transmitter, and another for the receiver. The coil directly connected to the transmitting power converter is called a driving coil and it supplies energy to the transmitting high-quality-factor *LC* resonant loop. At the receiver side the situation is similar. The receiver drive coil captures the magnetic field produced in the receiver resonant circuit delivering it to the load circuit. (Zhu et al., 2008; Low et al., 2009).

In Figure 3 L_{td} is the transmitter coil inductance, M_{tdt} is the mutual inductance between the transmitter drive coil and the transmitter resonant circuit, L_t is the transmitter resonant circuit inductance,

Figure 2a. Series-series configuration

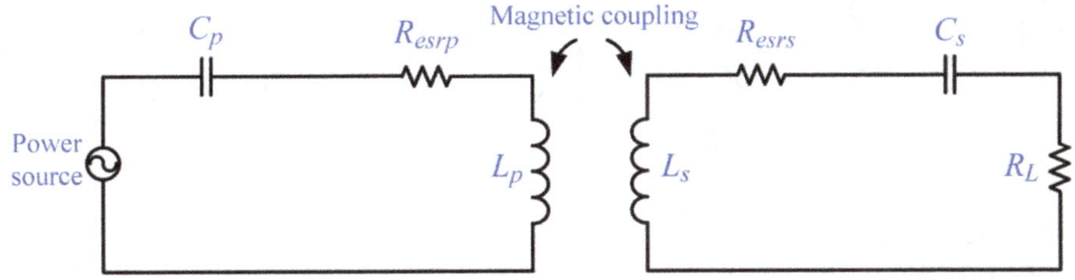

Figure 2b. Series-parallel configuration

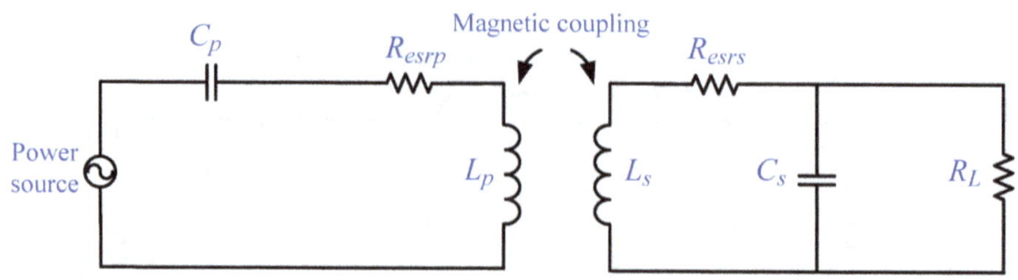

Figure 2c. Parallel series configuration

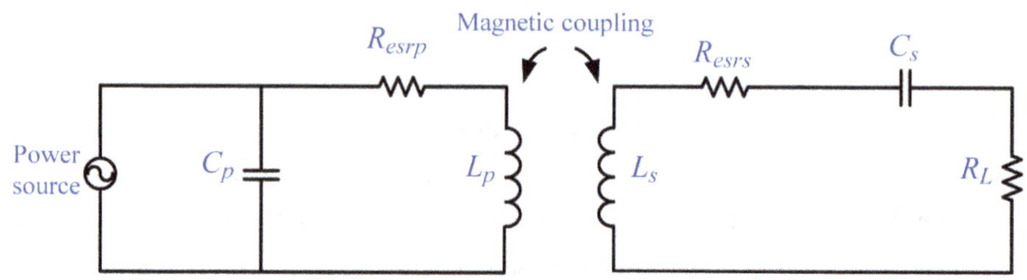

Figure 2d. Parallel-parallel configuration

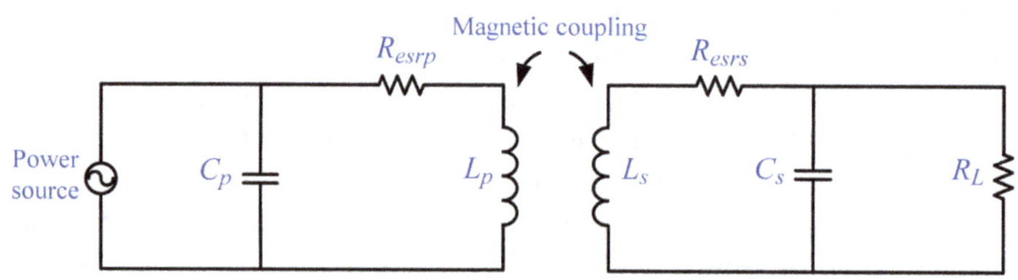

Figure 3. Strongly coupled magnetic resonance basic diagram

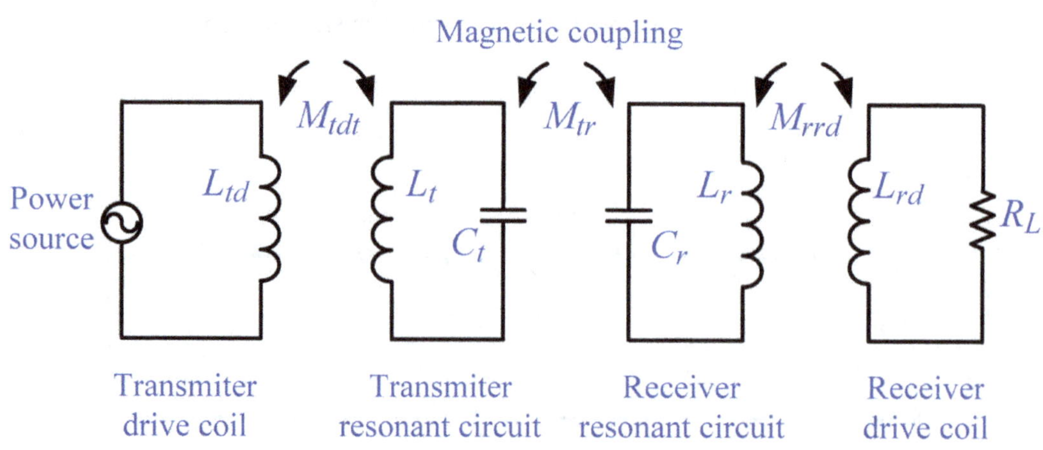

C_t is the transmitter resonant circuit capacitance, M_{tr} is the mutual inductance between the transmitter resonant circuit and the receiver resonant circuit, C_r is the receiver resonant circuit capacitance, L_r is the receiver resonant circuit inductance, M_{rrd} is the mutual inductance between the receiver resonant circuit and the receiver drive coil, L_{rd} is the receiver drive coil inductance, R_L is the load resistance. Both resonant loops are galvanically isolated from the other circuits of the transmitter and the receiver, but the loops are very well magnetically connected by the resonance they maintain mutually. Their connection is exclusively magnetic. The two resonant circuits must be tuned to the same frequency and must have a high Q (Rajagopal & Khan, 2011). However, as the distance between the two resonators may vary, the input and output impedances, seen from the transmitter and from the receiver, may vary too. A similar situation may occur when there is variation in the load. Therefore, to keep an optimum efficiency, the impedances must be adapted to match the transmitter side impedance and the receiver side impedance. The resonant frequency of the loops will vary because of the changes in the load, distance and mutual inductance, so the change is needed to adjust immediately the parameters. To achieve the economically viable wireless charging, the WPT system must satisfy three conditions: high efficiency, large air gaps and capability to transfer enough high power. The WPT through a strong magnetic resonance will satisfy these tree conditions (Rajagopal & Khan, 2011).

4. PROTOTYPE DESIGN

How to Match Impedances

The impedance matching is the process of reducing, or ideally removing mismatch losses. The main goal is to minimize the reflection coefficient, to reduce the power reflected from the load, and maximize the power delivered to the receiver.

As stated in section 3, there must be adapted impedance between the transmitter and the receiver to have the maximum power transfer. When there is impedance mismatching, the tuning frequency in the two resonant circuits (transmitter and receiver) is changed. This change is caused by the load impedance reflection over the receiver resonant circuit, and the receiver resonant circuit over the transmitter resonant circuit.

An innovative process to adjust the tuning frequency is presented in this section (Lu et al., 2016; Austrin & Engdahl, 2004; Trinkhaus, 2000; Xu et al., 2013; Fabbri et al., 2014; Barili et al., 1988; Romba et al., 2017; Romba et al., 2016; Romba et al., 2018). The frequency adjustment is made by a change of the resonant circuit inductance. For this to be made possible, a variable inductance adjusted by direct current (DC) is connected in series. The MCR operation is based on the saturable reactor. The coil inductance is given by:

$$L = \frac{4\pi N^2 \mu S}{\ell} \tag{1}$$

where ℓ is the coil length, N is the turn number, S is the area and μ is the magnetic permeability. Through this equation it can be verified that the inductance L varies if μ is changed. The μ changes are given by:

$$\Delta\mu = \frac{\Delta B}{\Delta H} \tag{2}$$

If the MCR operation point is near the knee of the $B - H$ curve and close to the saturation zone, then significant variations of $\Delta\mu$ will be obtained, as desired. The simplified layout of the MCR with two circuits, i.e., the load circuit and the regulator circuit is shown in Figure 4. In Figure 4 R_C is the adjustment resistance, N_C is the regulator circuit turns number, N_L is the load circuit turns number, R_L is the load resistance.

The flux in the core is regulated by a DC current flowing through the adjustment winding N_C. The load winding is connected in series with the load R_L and the AC generator. When no DC adjustment current is flowing, the load winding presents a maximum of impedance and the load power is at its minimum. On the contrary, when the DC adjustment current flows, the impedance decreases, and a higher output power is delivered to the load from the AC source. When the R_C value changes, the degree of the core magnetization is regulated. As the DC current increases, the flux density increases proportionally, and the saturation of the magnetic core is then partially achieved. The magnetic permeability is regulated through this, described above process. When the desired point of saturation is reached, i.e. the core permeability has the desired value, a certain impedance is achieved, in result. The adjustment current symmetrically influences the lateral ferromagnetic branches of the core, thus minimizing the resonant current distortion.

The following practical case helps to understand how the impedance matching is performed using the magnetic core reactor.

To apply the MCR for a strong magnetic coupling, the WPT system can be similar to the one, shown in Figure 5. It includes two MCRs, one on each system side, i.e., the transmitter and the receiver. The Magnetic Core Reactors are connected in series with each of the resonant circuits, i.e. the transmitter resonant circuit and the receiver resonant circuit.

Figure 4. Magnetic core reactor: load circuit and regulator circuit

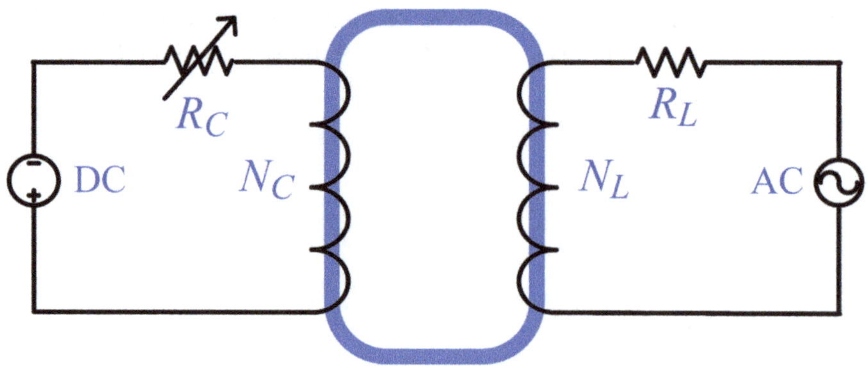

Figure 5. WPT with magnetic core reactor in both sides

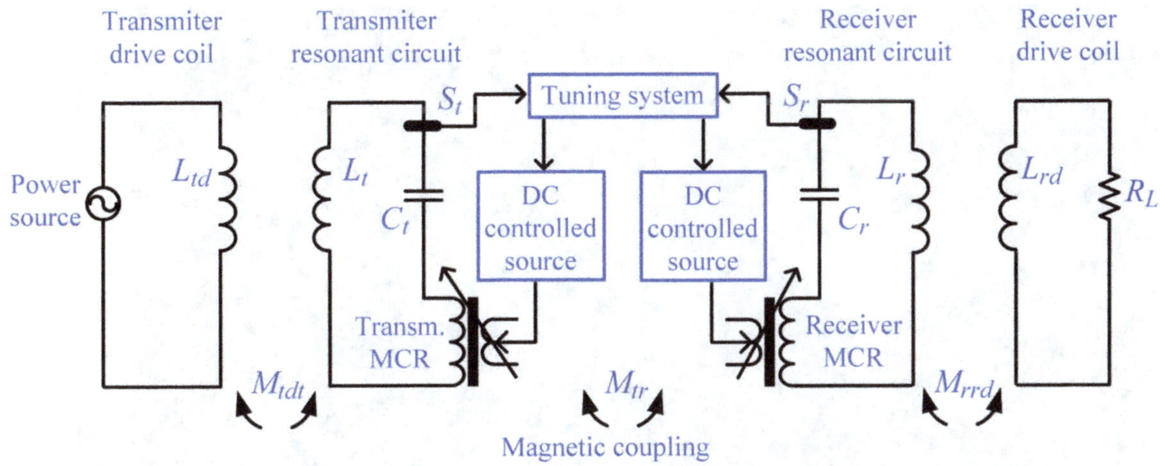

Magnetic Field Created by Spiral Coil and Helicoidal Coil: Comparison

WPT systems can use different coil shapes to perform energy transfer. In this subsection it is shown the comparison of the magnetic field behaviour created by two coils i.e. spiral coil and helicoidal coil. The data for assembly the planar and the helicoidal coils is summarized in Table 1.

The Figures 6 and 7 show the results of the magnetic field dispersion simulation for the spiral coil and the helicoidal coil when are driven by an electrical current value of 15 A. The frequency value used for the simulation is 120 kHz.

5. TUNING SYSTEM

In this section is presented how the tuning system using the MCR connected in series to a spiral coil. This system is based on the MCR connected to a spiral coil to achieve the impedance matching.

Consider the circuit of Figure 3 in which the receiver part is consisted by the receiver resonant circuit (series resonant circuit) and the receiver driver coil connected to the load resistor. Consider further that in this circuit, the load resistor is replaced by an impedance Z_L and the coupling between the coil L_t

Table 1. Data for assembly the planar and the helicoidal coils

	Spiral Coil	Helicoidal Coil
Turns number	35	35
Cable material	copper	copper
Cable section	1.5 mm^2	1.5 mm^2
Outer diameter	240 mm	240 mm
Inner diameter	50 mm	200 mm
Average pitch coil	2 mm	---

Figure 6. Magnetic field helicoidal coil simulation results at the frequency of 120 kHz

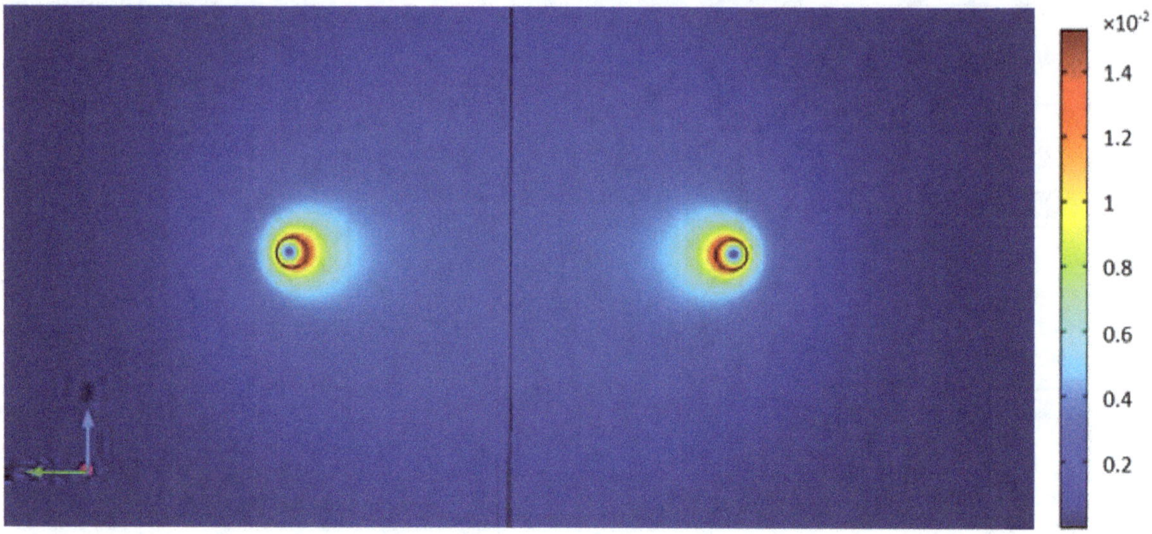

Figure 7. Magnetic field planar coil simulation results at the frequency of 120 kHz

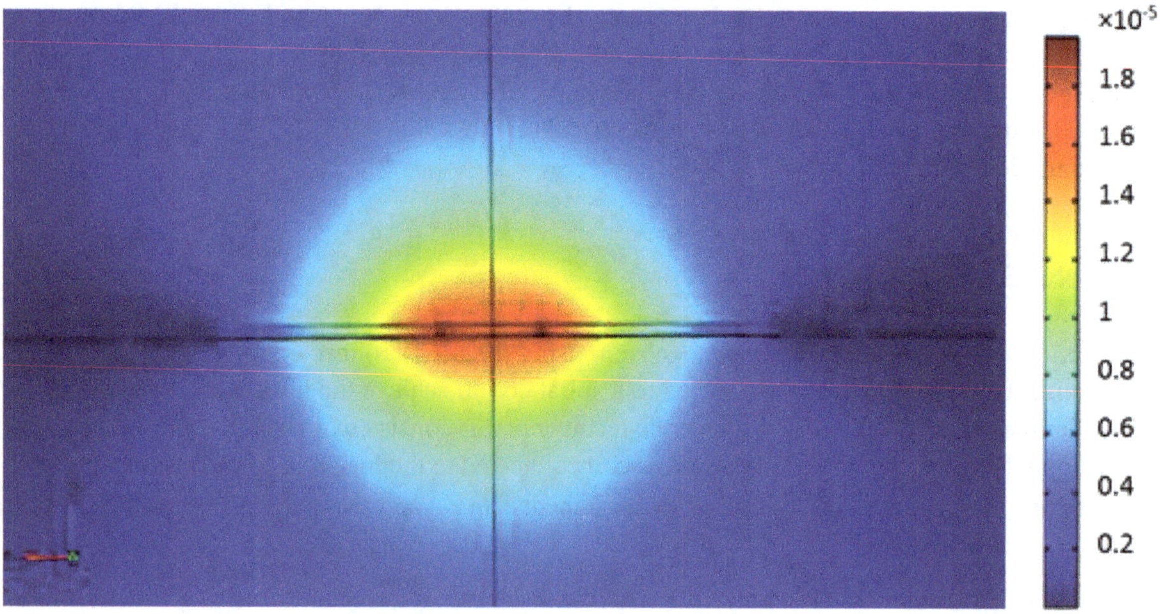

and L_r is replaced by a constant voltage generator with the frequency of $120\,\text{kHz}$. These changes are presented in the circuit of the Figure 8. Assuming that in this circuit, matching impedances exist, then a maximum power transfer is expected.

The T-equivalent circuit is shown in the Figure 9.

The impedance Z_t sense by the generator is given by

Figure 8. Receiver circuit

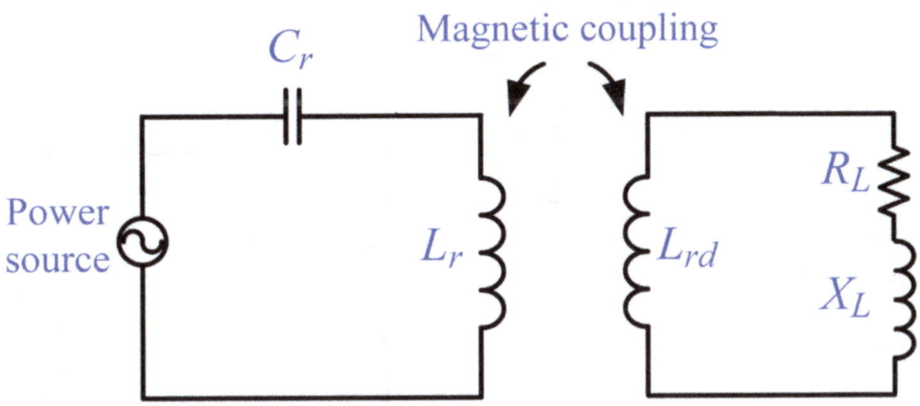

Figure 9. Receiver T-equivalent circuit

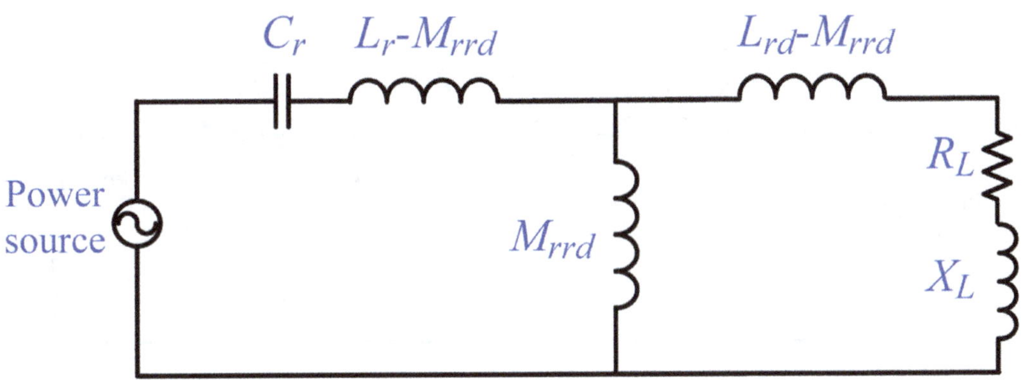

$$Z_t = -j\frac{1}{\omega C_r} + j\omega\left(L_r - M_{rrd}\right) + \left\{\frac{\left[j\omega\left(L_{rd} - M_{rrd}\right) + Z_L\right] \times j\omega M_{rrd}}{\left[j\omega\left(L_{rd} - M_{rrd}\right) + Z_L\right] + j\omega M_{rrd}}\right\} \tag{3}$$

If Z_L increases to a Z_{L_1} value, greater than Z_L, there is an impedance mismatch then the generator to sense an impedance Z_{t_1} greater than Z_t. In these conditions there are no impedance matching and the resonant frequency is out of its value. In the circuit shown in Figure 8, the tuning cannot be performed, since there is no component to make the adjustment.

The Figure 10 shows the same previous circuit, but with inclusion of the MCR. The N_C represents the adjustment coil and the N_l represents the load coil. The MCR inductance is represented by L_{MCR}.

The T-equivalent circuit is shown in the Figure 11.

The initial conditions established at the previous circuit are applied again. The tuning resonant frequency is the same $120\,\text{kHz}$ and there is impedance matching. The impedance Z_t sense by the generator is given by

Figure 10. Receiver circuit with the MCR

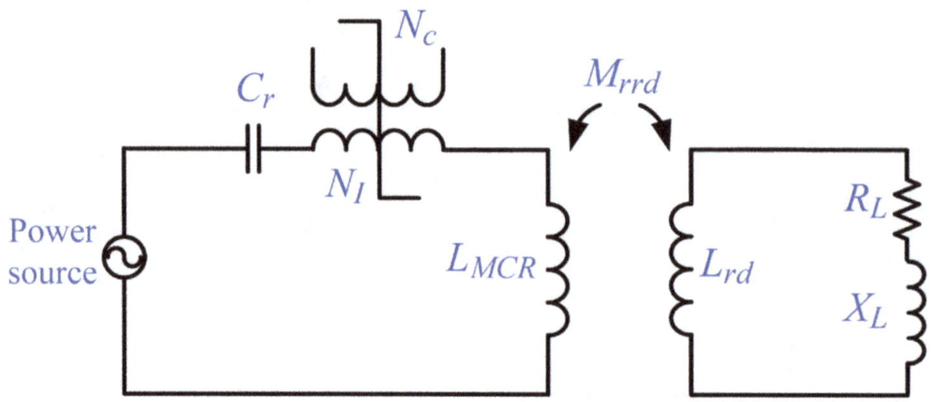

Figure 11. Receiver T-equivalent circuit with MCR

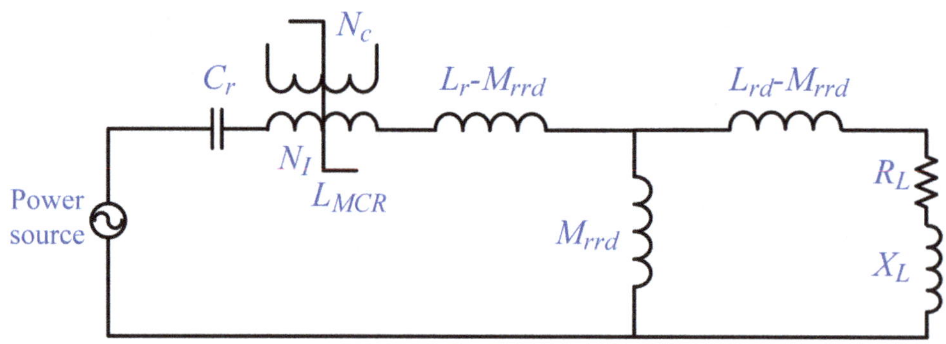

$$Z_t = -j\frac{1}{\omega C_r} + j\omega\left(L_r - M_{rrd}\right) + j\omega L_{MCR} + \left\{\frac{\left[j\omega\left(L_{rd} - M_{rrd}\right) + Z_L\right] \times j\omega M_{rrd}}{\left[j\omega\left(L_{rd} - M_{rrd}\right) + Z_L\right] + j\omega M_{rrd}}\right\} \tag{4}$$

When there is impedance matching, the resonant frequency is adjustable. Since it is a series resonant circuit, the electric current will be at its maximum. Then the maximum magnetic field will be produced and the maximum power transfer will be observed.

Supposing that Z_L increases, the total impedance of the circuit is increased because the numerator of the fourth term of the (4) grows faster than denominator. Under these conditions there is an impedance mismatch, the resonance frequency will be out of its previewed value and the electric current flowing in the circuit will decrease. However, in this circuit, the tuning frequency can be readjusted, and the matching impedance can be achieved through the MCR, reducing the $j\omega L_{MCR}$ value.

6. PROTOTYPE IMPLEMENTATION

The MCR physical data used in the prototype is summarised in the Table 2.

Table 2. MCR prototype data

Load coil	4 turns
Adjustment coil	25 turns
Coils cable material	Copper
Load coil cable section	1.5 mm^2
Adjustment coil cable section	2×0.238 mm^2
Core shape	EE type

The MCR prototype assembled is shown in Figure 12. The red cable forms the load coil and the white-brown cables forms the adjustment coil.

First, it is necessary to determine the MCR inductance as function of DC adjustment current. The experimental setup for measuring the MCR inductance values, as function of the DC adjustment current is shown in Figure 13.

The test bench implemented to measure the MCR inductance values varying the DC adjustment current is shown in Figure 14.

The Figure 15 shows the MCR inductance values obtained by varying the DC adjustment current.

In Figure 15 it is verified that a significant inductance range variation exists. Using the quasi-linear part of the curve, the inductance range is between 0.5 mH to 2.0 mH.

Figure 12. Magnetic core reactor prototype assembled

Figure 13. Schematic diagram of the MCR adjustment inductance value

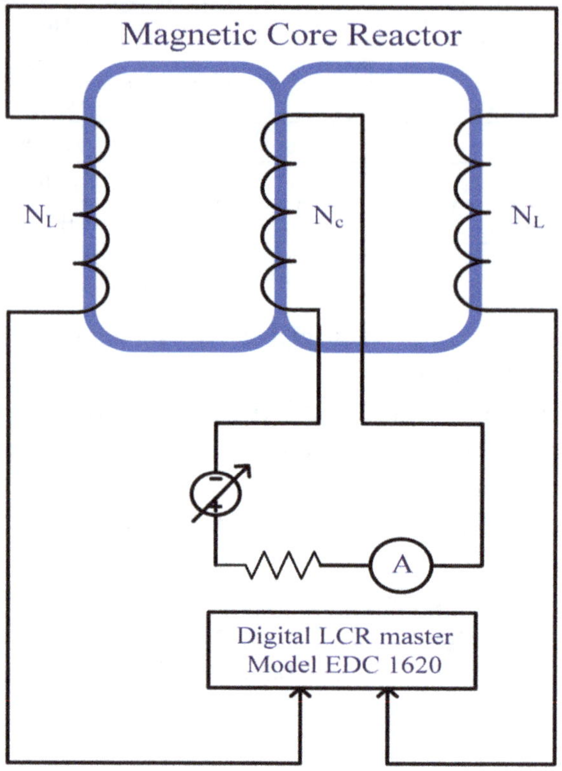

Figure 14. Test bench to measure the MCR inductance values

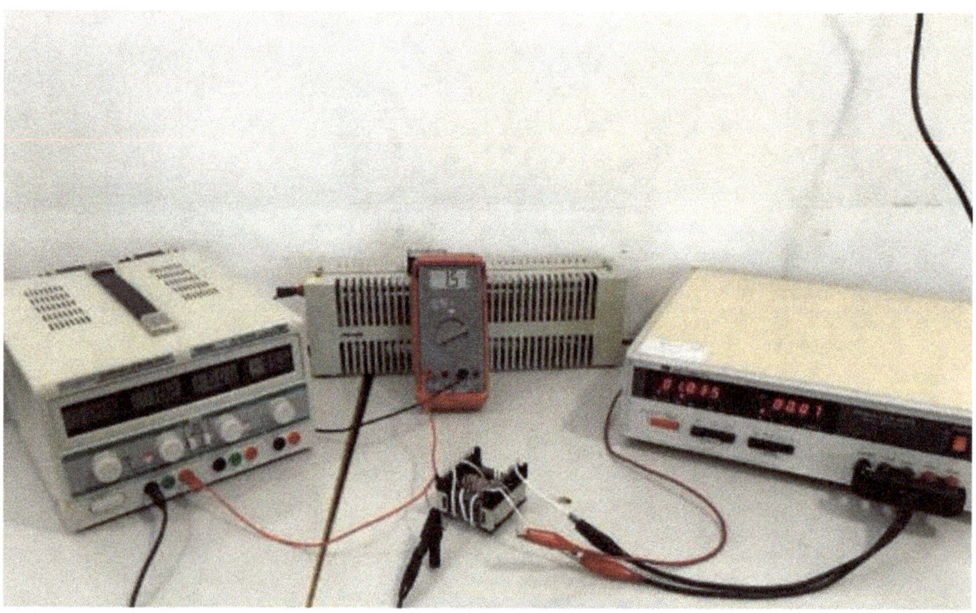

Figure 15. MCR inductance values function of DC adjustment current

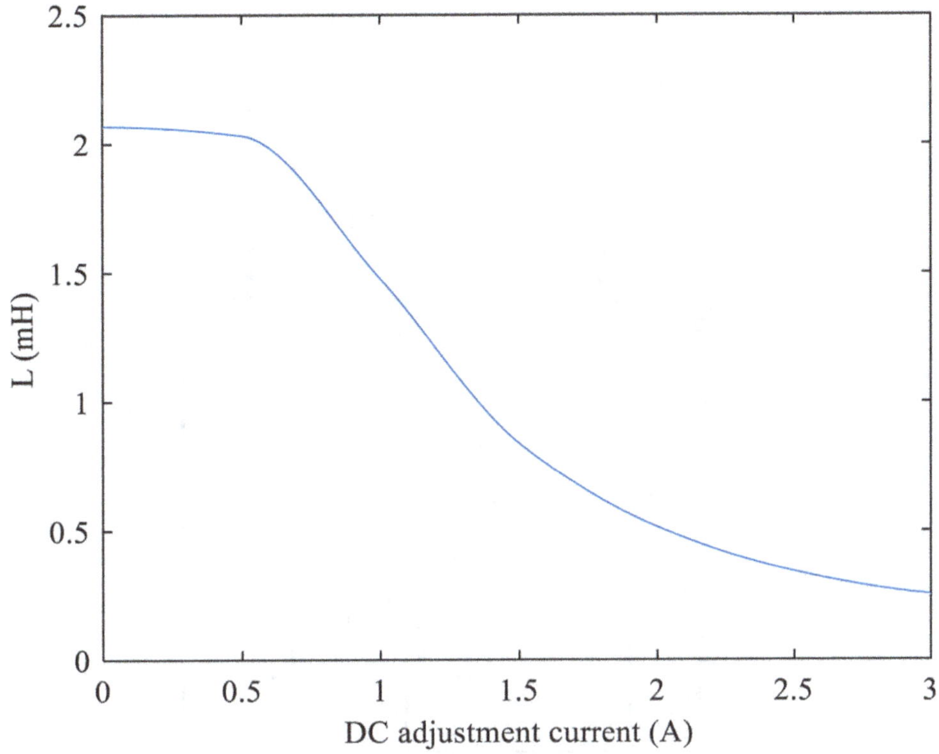

MCR and Helicoidal Transmitter Coil in Series Connected

The MCR and the helicoidal transmitter coil in series connected topology implemented in the tests is shown in Figure 16.

The helicoidal transmitter coil data is summarized in Table 1. The physical characteristics of this coil are the same as the coil characteristics simulated by the finite element program.

The inductance values are connected in series with the transmitter or receiver coil. The total inductance values in function of the DC adjustment current, for the MCR in series connected with the transmitter coil is shown in Figure 17.

For each frequency, there is a pair of values i.e., the inductance and the capacitance, that define the resonant frequency. Assuming the capacitance value is constant, then the inductance value is changed to obtain the desired frequency. The tuning frequency range for this practical case is between $200\,\mathrm{kHz}$ and $600\,\mathrm{kHz}$. The resonant transmitter circuit tuning frequency range, is defined by the L_t and C_t values, as is shown in the Figure 18. The resonant receiver circuit tuning range is defined by the L_r and C_r values, as is shown in Figure 19. To make the adjustment system pre-programmed, values similar to those shown in Figure 18, Figure 19, are previously introduced into the tuning adjustment system memory, together with the initial operating point (pre-set points).

The transmitter and the receiver coils set and 2 MCRs, i.e., the partial view of this practical case is shown in the Figure 20.

Figure 16. MCR and the transmitter helicoidal coil in series connected topology

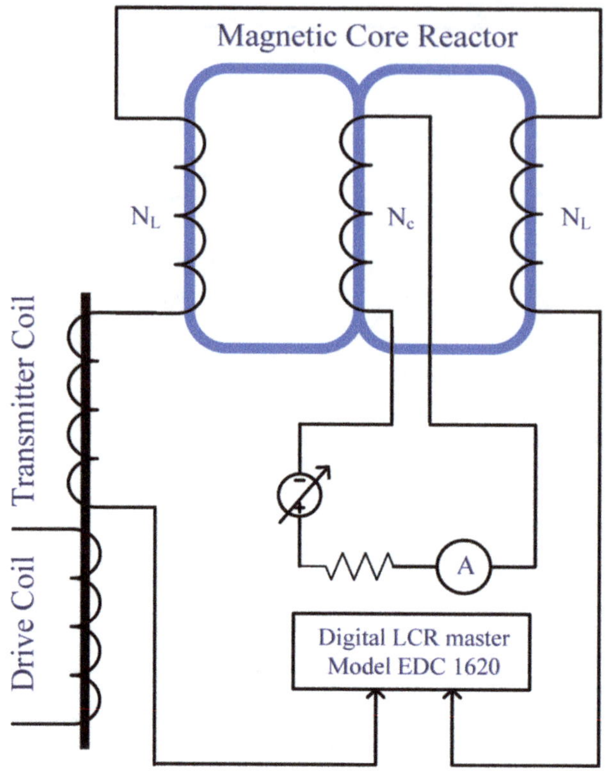

Figure 17. MCR in series connected with transmitter coil as function of the DC adjustment current: inductance values

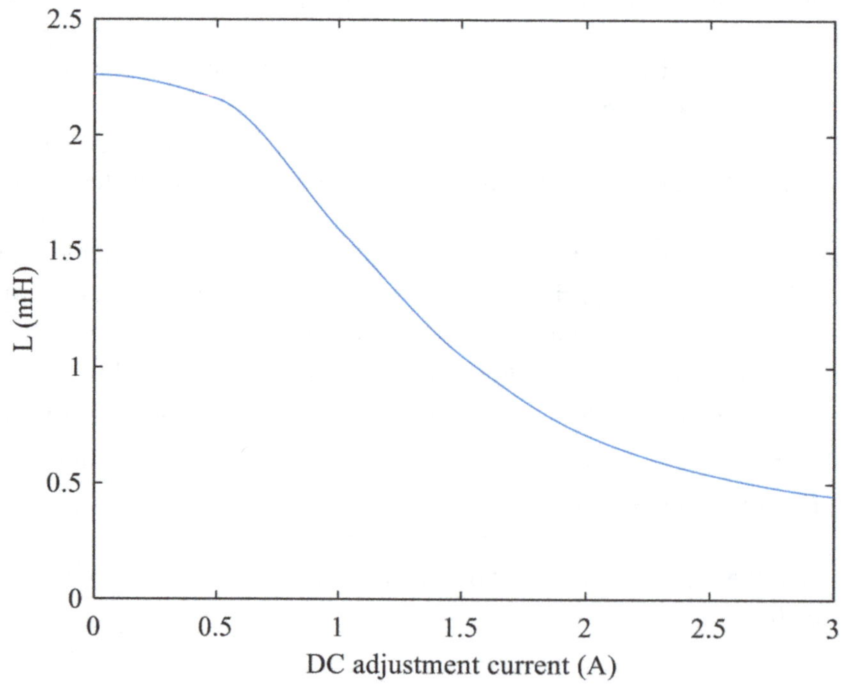

Figure 18. Transmitter resonance values; blue line: capacitance values; orange line: inductance values

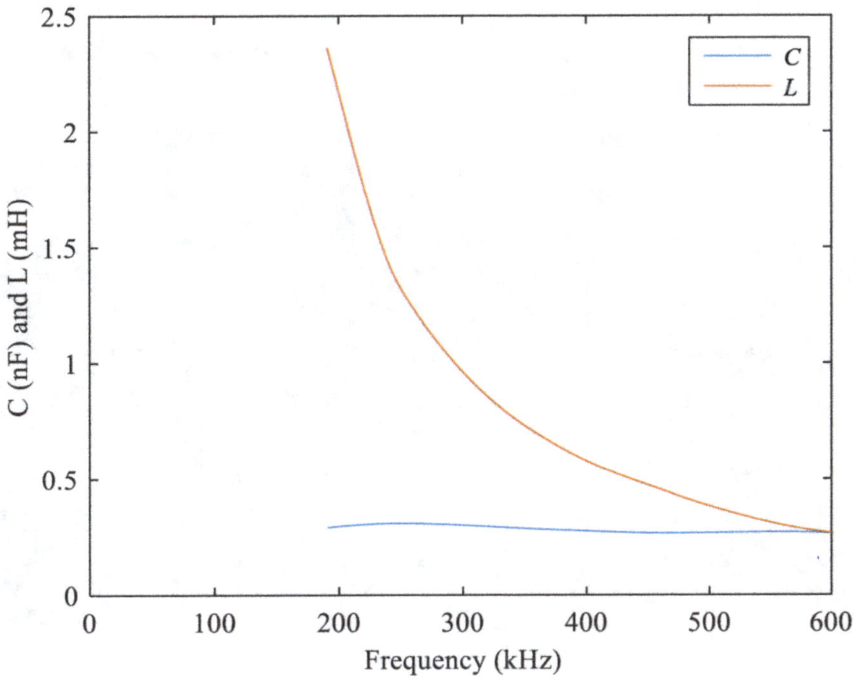

Figure 19. Receiver resonance values; blue line: capacitance values; orange line: inductance values

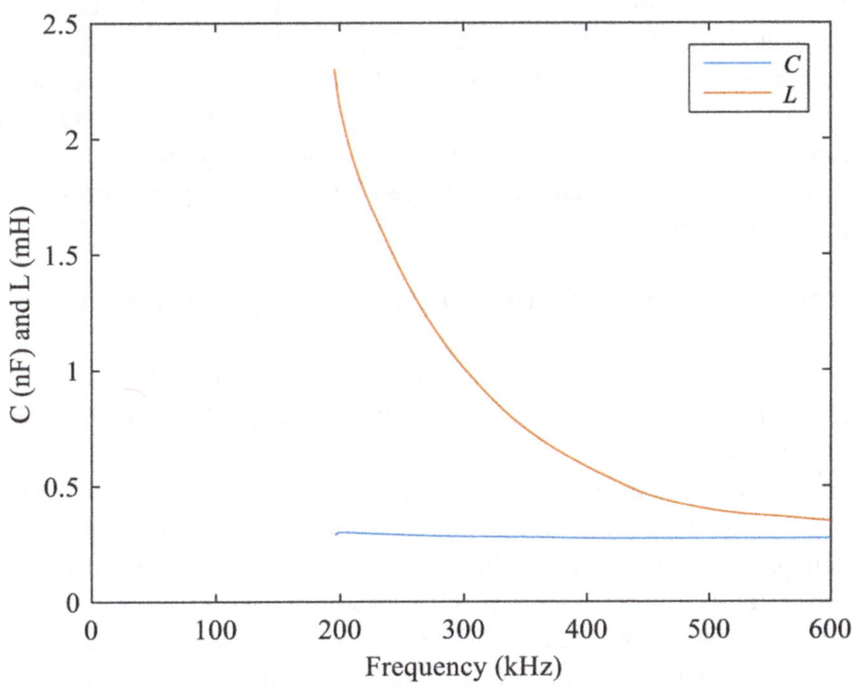

Figure 20. Prototype system partial view

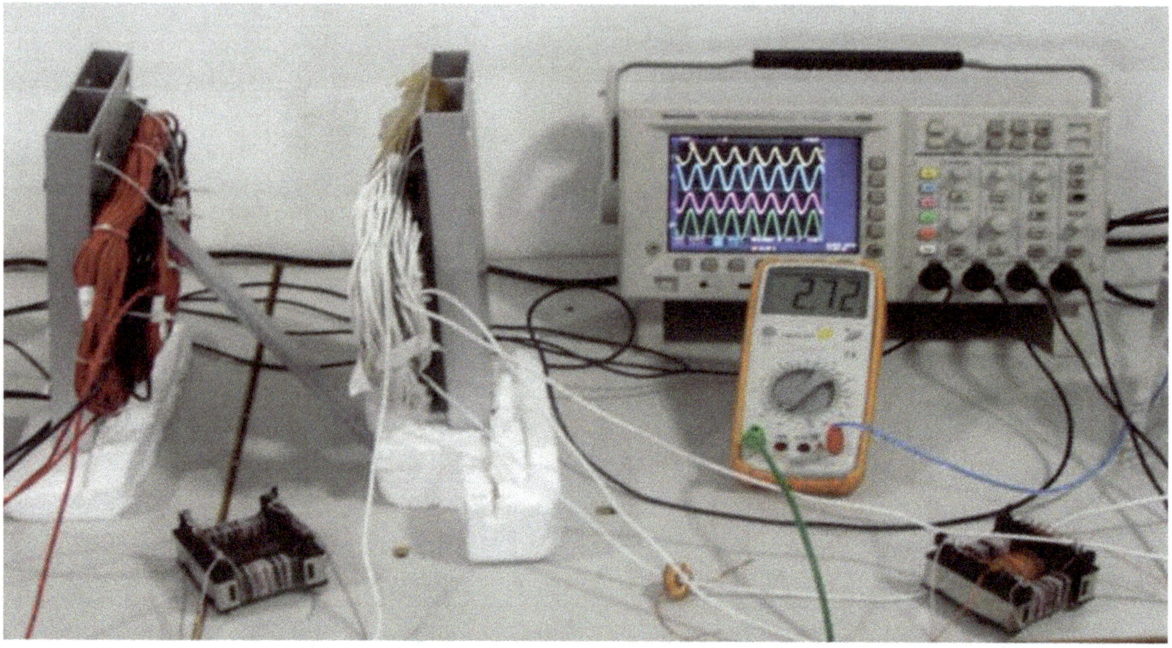

The red coil is the helicoidal transmitter coil and the white coil is the receiver helicoidal coil. The driver coils are embedded in each transmitter and receiver coils set. The transmitter driver coil receives the power from the signal generator. The input power pulses may have the same frequency as the resonant frequency of the circuit formed by the transmitter coil, the MCR and the transmitter resonant capacitor. The receiver circuit formed by receiver coil, the MCR and the receiver resonant capacitor should be tuned to the first harmonic of the primary switching frequency. This condition is essential to obtain the maximum power transmission efficiency. The defined best tuning points of the transmitter and the receiver circuits can be adjusted through the introduced MCR. This explanation summarizes how the adjustment system should work.

MCR and Spiral Coil in Series Connected for Tuning Frequency

Prototype implemented for tuning frequency using the MCR and spiral coil in series connected is shown in Figure 21. The spiral coil data is summarized in Table 1. The spiral coil has the same characterizes of the coil used in previous simulation.

To obtain the inductance and the distributed capacitance values, two different tests are performed. The first with the spiral coil alone, the second with the spiral coil connected in series with the MCR.

The test bench implemented for collected the data for the prototype implemented is shown in Figure 22. The inductances and the distributed capacitances values are measured using a Hioki 3532-50 LCR Hitester meter. In Figure 22 the Hioki 3532-50 LCR Hitester meter is shows on the centre, the spiral coil is on the left side together with the tuning capacitors and the MCR. The DC power supply is next to the Hioki test meter (left side). The MCR DC adjustment current is supplied by the DC power supply. The Laptop is collected the data directly from the test meter. This circuit is also used to perform the tests to

Figure 21. Spiral coil prototype assembled for the tests

Figure 22. Test bench implemented for collected data for the prototype

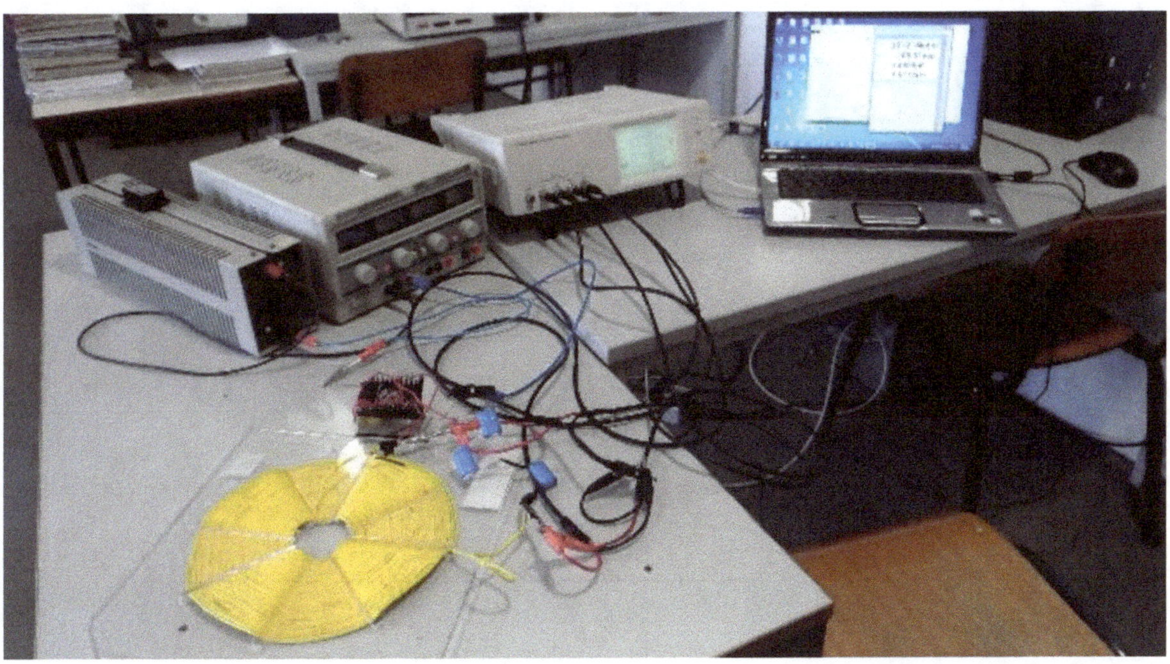

determine the capacitance C_t value (the fixed capacitance). The performed tests result the value of 15.66 nF. This value is achieved with three capacitors of 47 nF connected in series as shown in the Figure 21.

Tuning Frequency System

As referred in the previous sub-section, the test results of the spiral coil alone, for the inductance and the distributed capacitance are shown in the Figure 23 and Figure 24. The working range frequency for this practical case is between $100\,\text{kHz}$ and $140\,\text{kHz}$.

The Figure 25 shows the topology for the transmitter and the receiver resonant circuits. The MCR inductance is represented by L_{MCR} and the distributed capacitance is represented by C_{dc}. The total inductance values, $L_t + L_{MCR}$, are shown in the Figure 27, as function of the DC adjustment current I_c and per frequency The distributed capacitance values C_{dc} are shown in the Figure 28, as function of the DC adjustment current and per frequency too. The capacitance C_t is the fixe capacitance and has the predetermined value of 15.66 nF. This value is function of the working range frequency. For other working frequencies this value would be different.

The tuning circuit is composed by the sensors, the S_t on the transmitter side and S_r on the receiver side, as is shown in Figure 26. These sensors supply the actual frequency value to the tuning system adjustment. Since the capacitor value is constant, the system adjustment detects the corresponding inductance value (Figure 27). The system adjustment compares the actual inductance value with the pre-

Figure 23. Spiral coil: Inductance in function of the frequency

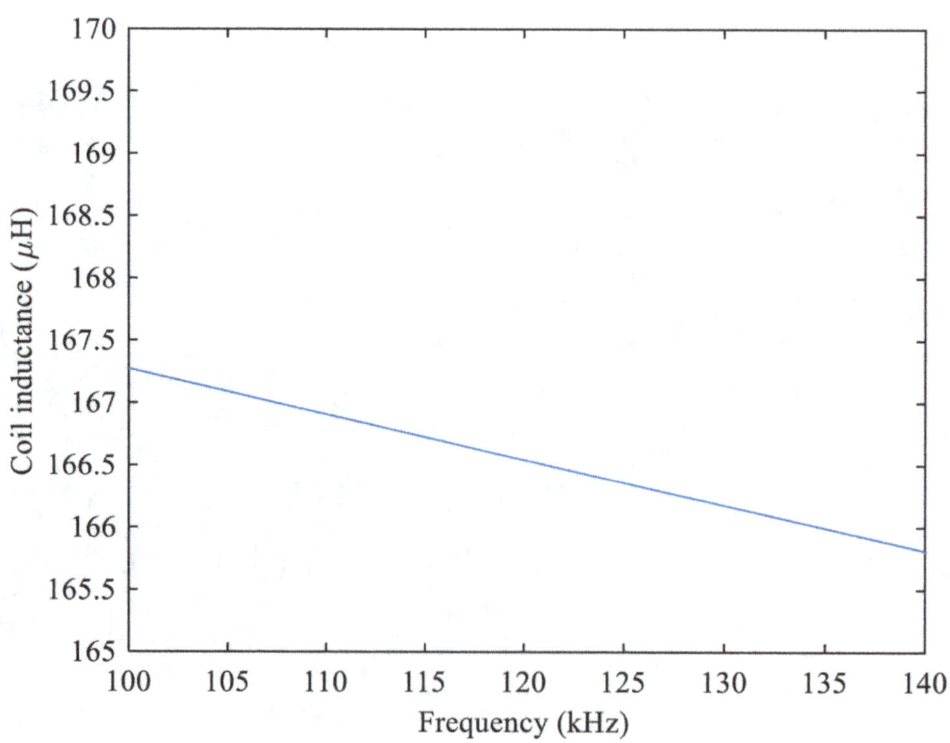

Figure 24. Spiral coil distributed capacitance in function of the frequency

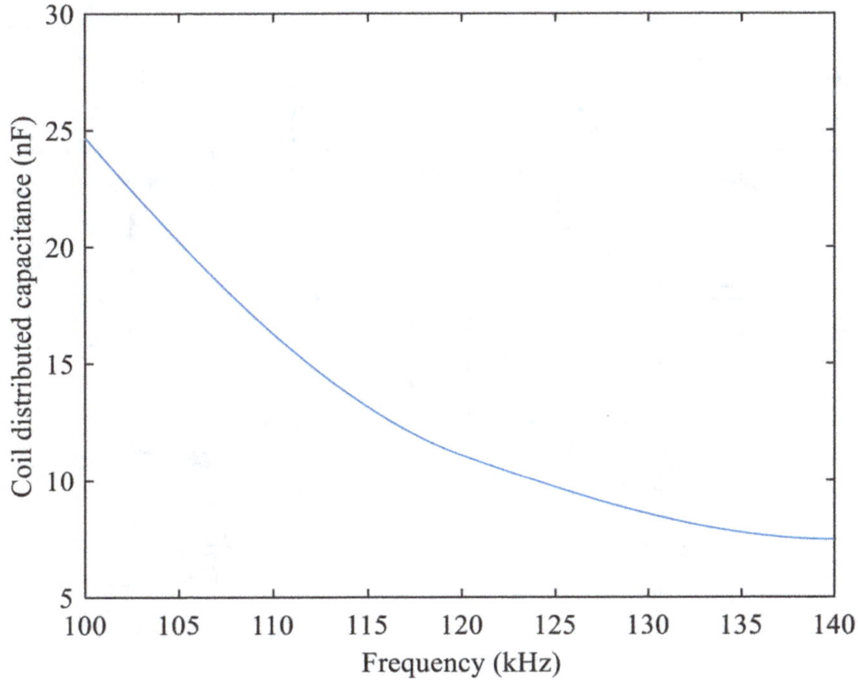

Figure 25. Topology for the transmitter and the receiver resonant circuits

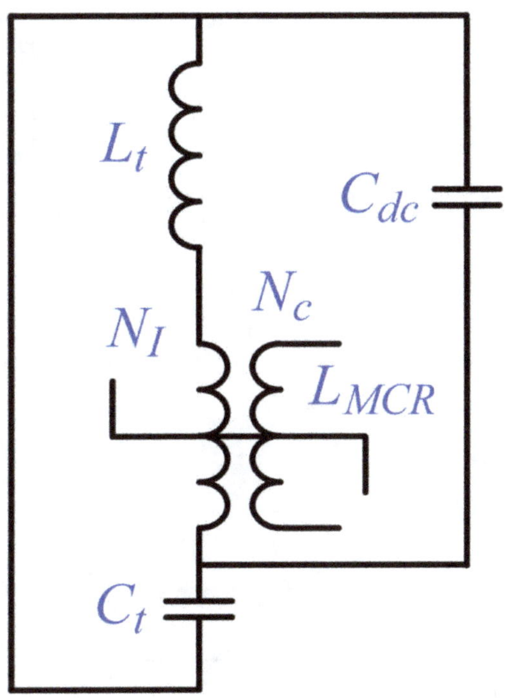

Figure 26. Tuning adjustment system

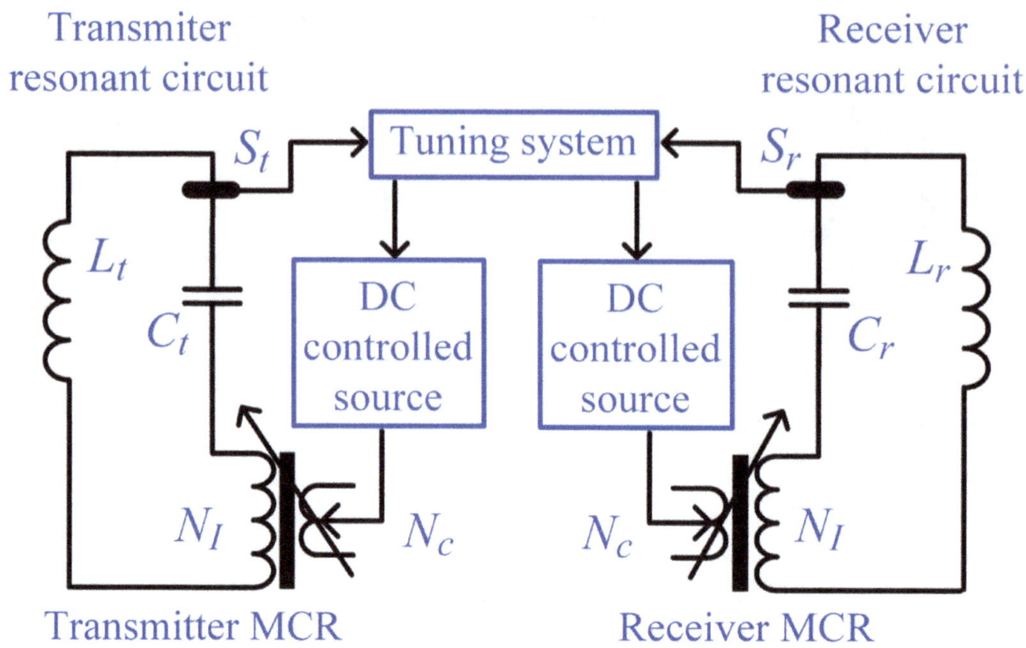

Figure 27. MCR and spiral coil prototype: Inductance versus DC adjustment current

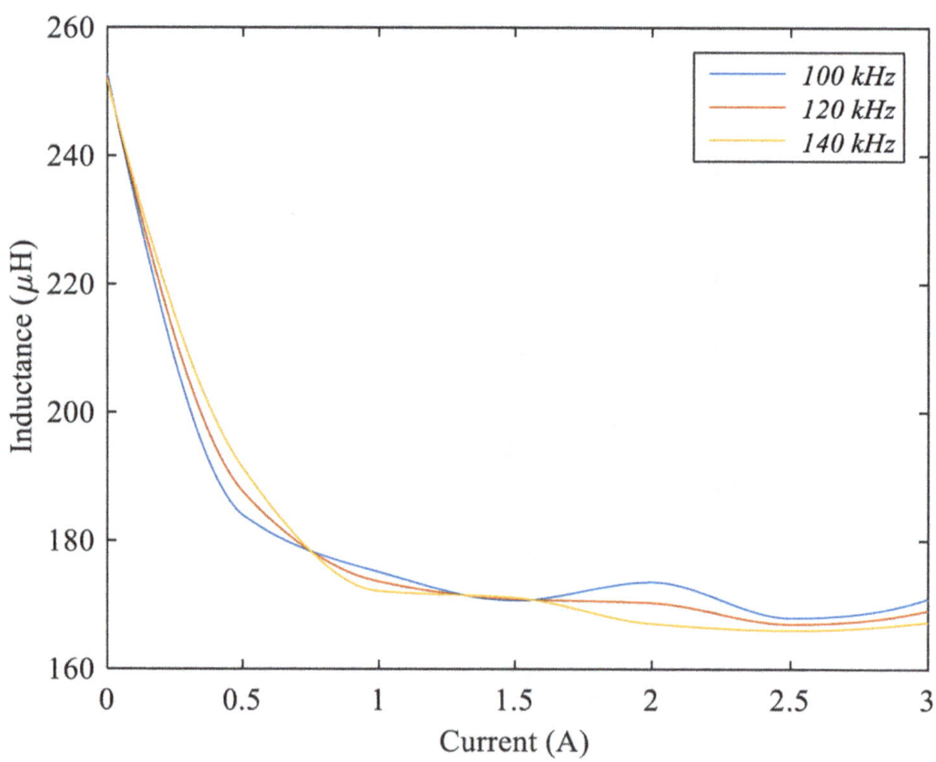

Figure 28. MCR and spiral coil prototype: Distributed capacitance versus DC adjustment current

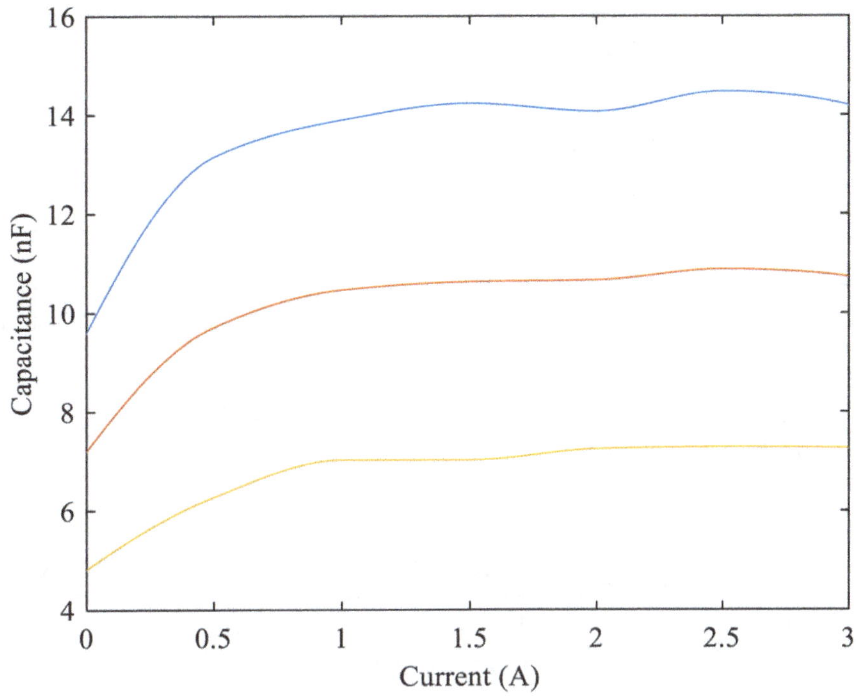

set value for each side. In this way it is determined whether the DC adjustment current should increase or decrease, to place the transmitter and the receiver resonant circuits at the predetermined frequency value.

The Figure 29 shows the test results with this prototype for the frequency range i.e. 100 kHz to 140 kHz. The tuning (resonance) points are obtained for each DC adjustment current value per frequency.

7. CONCLUSION

Comparing the Figure 6 and Figure 7, can be conclude the helicoidal coil magnetic field is more intense close to the coil; the spiral coil magnetic field is more intense in the longitudinal direction i.e. along the coil axis; the helicoidal coil magnetic field dispersion is greater in the transverse direction when compared to the spiral coil.

One the spiral coil advantages is its reduced thickness when compared with others coil shapes. This case study is specifically applied to tuning system regulated by a MCR connected in series to a spiral coil. However, this procedures can be applied to any coil shape.

The tests already made have been shows the functionality of the magnetic core reactor applied to this tuning adjustment system. This process enables the operation frequency adjustments in a WPT system. With this frequency adjustment process, it is possible to match the impedances between the transmitter and the receiver circuits.

Figure 29. Tuning points for each DC adjustment current value per frequency

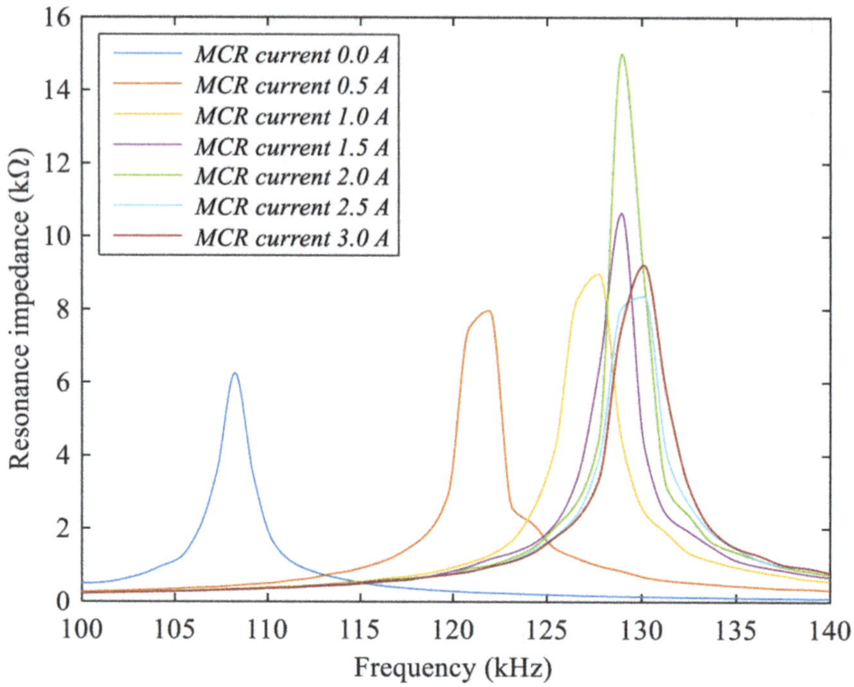

The tuning values are obtained continuously, not by steps, i.e. as it happens if the inductance is fixed and the capacitance is variably in discontinuous way. Other intermediate points can be obtained by continuous variation of the DC adjustment current. The adjustment method proposed here by magnetic core reactor application, allows a significant improvement, compared to other known adjustment methods.

REFERENCES

Austrin, L., & Engdahl, G. (2004). Modeling of a three-phase application of a magnetic amplifier. *Proc. of the 24th International Congress of the Aeronautical Sciences*, 1-8.

Baikova, E. N., Valtchev, S. S., Melicio, R., Krusteva, A., & Pires, V. F. (2016). Study of the electromagnetic interference generated by wireless power transfer systems. *International Review of Electrical Engineering*, *11*(5), 526–534.

Bareiss, M., Krenz, P. M., Szakmany, G. P., Tiwari, B. N., Kalblein, D., Orlov, A. O., ... Lugli, P. (2013). Rectennas revisited. *IEEE Transactions on Nanotechnology*, *12*(6), 1144–1150. doi:10.1109/TNANO.2013.2281373

Barili, A., Brambilla, A., Cottafava, G., & Dallago, E. (1988). A simulation model for the saturable reactor. *IEEE Transactions on Industrial Electronics*, *35*(2), 301–306. doi:10.1109/41.192663

Cannon, B. L., Hoburg, J. F., Stancil, D. D., & Goldstein, S. C. (2009). Magnetic resonant coupling as a potential means for wireless power transfer to multiple small receivers. *IEEE Transactions on Power Electronics*, *24*(7), 1819–1825. doi:10.1109/TPEL.2009.2017195

Choi, S. Y., Gu, B. W., Jeong, S. Y., & Rim, C. T. (2015). Advances in wireless power transfer systems for roadway-powered electric vehicles. *IEEE Journal of Emerging and Selected Topics in Power Electronics*, *3*(1), 18–36. doi:10.1109/JESTPE.2014.2343674

Ekman, C. K. (2011). On the synergy between large electric vehicle fleet and high wind penetration – an analysis of the Danish case. *Renewable Energy*, *36*(2), 546–553. doi:10.1016/j.renene.2010.08.001

Fabbri, G., Tarquini, G., Pasquali, L., Anniballi, L., Odoardi, S., Teodori, S., & Santini, E. (2014). Impact of V2G/G2V technologies on distributed generation systems. *Proc. of the 23th IEEE International Symposium on Industrial Electronics*, 1-4. 10.1109/ISIE.2014.6864867

Fernández, L. P., San Román, T. G., Cossent, R., Domingo, C. M., & Frías, P. (2011). Assessment of the impact of plug-in electric vehicles on distribution networks. *IEEE Transactions on Power Systems*, *26*(1), 206–213. doi:10.1109/TPWRS.2010.2049133

Glaser, P. E. (1968). Power from the sun: Its future. *Science*, *162*(3856), 857–861. doi:10.1126cience.162.3856.857 PMID:17769070

Hilshey, A. D., Hines, P. D. H., Rezaei, P., & Dowds, J. R. (2013). Estimating the impact of electric vehicle smart charging on distribution transformer aging. *IEEE Transactions on Smart Grid*, *4*(2), 905–913. doi:10.1109/TSG.2012.2217385

Huang, X. L., Wang, W., Tan, L. L., Zhao, L. L., & Zhou, Y. L. (2012). Study of transmission performance on strong coupling wireless power transfer system in free position. *Proc. of the 31st Progress in Electromagnetics Research Symposium*, 669-676.

Kurs, A., Karalis, A., Moffatt, R., Joannopoulos, J. D., Fisher, P., & Soljačić, M. (2007). Wireless power transfer via strongly coupled magnetic resonances. *Science*, *317*(5834), 83–86. doi:10.1126cience.1143254 PMID:17556549

Liu, Hu, Nair, & Covic. (2010). 2-D Alignment Analysis of Capacitive Coupled Contactless Power Transfer Systems. *IEEE ECCE Conference*.

Low, Z. N., Chinga, R. A., Tseng, R., & Lin, J. (2009). Design and test of a high power high-efficiency loosely coupled planar wireless power transfer system. *IEEE Transactions on Industrial Electronics*, *56*(5), 1801–1812. doi:10.1109/TIE.2008.2010110

Lu, X., Wang, P., Niyato, D., Kim, D. I., & Han, Z. (2016). D. In Kim, Z. Han, Wireless charging technologies: Fundamentals, standards, and network applications. *IEEE Communications Surveys and Tutorials*, *18*(2), 1413–1452. doi:10.1109/COMST.2015.2499783

Pereirinha & Trovão. (2012). Multiple energy sources hybridization: the future of electric vehicles? In *New Generation of Electric Vehicles*. Intech.

Perlo, P., & Meyer, G. (2012). ICT for the fully electric vehicle, research needs and challenges ahead (2nd ed.). European Union.

Rajagopal, S., & Khan, F. (2011). Multiple receiver support for resonance coupling based wireless charging. *Proc. IEEE International Conference Communications Workshops*, 1-5.

Richard, C. (1973, December). Johnson, H. Allen Ecrer, J. Searcy Hollis, Determination of far-field antenna patterns from near-field-field measurements. *Proceedings of the IEEE, 61*(12), 1668–1694. doi:10.1109/PROC.1973.9358

Romba, Valtchev, & Melicio. (2017). Single-phase wireless power transfer system controlled by magnetic core reactors at transmitter and receiver. *Technological Innovation for Smart Systems*, 419-428.

Romba, L. F., Baikova, E. N., Valtchev, S. S., & Melicio, R. (2016). Electric vehicle battery charger: wireless power transfer system controlled by magnetic core reactor. *Proc. of the Conference on Electronics, Telecommunications and Computers*, 1-2.

Romba, L. F., Valtchev, S. S., Melicio, R., Mudrov, M. V., & Ziuzev, A. M. (2017). Electric vehicle battery charger controlled by magnetic core reactor to wireless power transfer system. *Proc. of the IEEE 17th International Conference on Environment and Electrical Engineering*, 1-6. 10.1109/EEEIC.2017.7977782

Romba, Baikova, Borges, Valtchev, & Melicio. (2018). Wireless battery charger for ev with circular or planar coils: comparison. In *Technological Innovation for Resilient Systems*. Springer.

Trinkaus, G. (2000). *Magnetic amplifiers another lost technology*. High Voltage Press.

Turton, H., & Moura, F. (2008). Vehicle-to-grid systems for sustainable development: An integrated energy analyses. *Technological Forecasting and Social Change, 75*(8), 1091–1108. doi:10.1016/j.techfore.2007.11.013

Valtchev, S., Brandisky, K., Borges, B., & Klaassens, J. B. (2009). Resonant Contactless Energy Transfer with Improved Efficiency. *IEEE Transactions on Power Electronics, 24*(3), 685–699. doi:10.1109/TPEL.2008.2003188

Valtchev, S., & Klaassens, J. B. (1990). Efficient resonant power conversion. *IEEE Transactions on Industrial Electronics, IE–37*(6), 490–495. doi:10.1109/41.103453

Xu, D. Q., Joós, G., Lévesque, M., & Maier, M. (2013). Integrated V2G, G2V, and renewable energy sources coordination over a converged fiber-wireless broadband access network. *IEEE Transactions on Smart Grid, 4*(3), 1381–1390. doi:10.1109/TSG.2013.2253337

Zhu, C., Liu, K., Yu, C., Ma, R., & Cheng, H. (2008). Simulation and experimental analysis on wireless energy transfer based on magnetic resonances. *Proc. IEEE Conference Vehicle Power and Propulsion*, 1-4.

Chapter 3
Resonant Compensations in Inductive Wireless Power System

Alberto Reatti
University of Florence, Italy

Marian K. Kazimierczuk
Wright State University, USA

Agasthya Ayachit
Wright State University, USA

Fabio Corti
University of Florence, Italy

ABSTRACT

A resonant wireless power system uses capacitors to create resonance between primary and secondary coil at a specific resonant frequency. Thereby, they reduce the reactive power and produce the required transmitted real power over a long range of separation and at higher efficiencies. Depending on the connection between the capacitors and coils, four main topologies can be identified: series-series, series-parallel, parallel-series, and parallel-parallel. In this chapter, these resonance compensations schemes are analyzed, and the advantages and disadvantages of each topology are highlighted.

INTRODUCTION

One of the most important feature of an Inductive Wireless Power System (IWPS) is the power transmission efficiency. An IWPS consists of two coupled inductors called primary and secondary coils. If the primary coil is supplied by an alternating current and/or voltage, a time-varying magnetic field B is produced. A part of this magnetic field, according to the coupling coefficient between the primary and secondary coils reaches the secondary coil and, therefore, an electro motive force is induced cross the secondary coil turns. A higher part of the magnetic field produced on the primary reaches the secondary, the higher is the voltage across the secondary coil terminals and, for any given load resistance, the higher is the output power delivered to the load. Unfortunately, non-resonant coupled inductors waste

DOI: 10.4018/978-1-5225-5870-5.ch003

most of energy in the form of conduction losses in the primary coil, also for a very short range and a large coupling (Spanik, FrivaldskyDragona, & Jaros, 2016). The efficiency is actually drastically improved and a wider transmission distance is achieved if the primary and secondary are tuned by utilizing the resonance principle (Chopra, Bauer, 2011; Sirbu, Mandache; & Kazimierczuk, 2014). When an IWPS is considered, the resonance is achieved by utilizing capacitors. This allows Resonant Wireless Power System (RWPS) to be designed and built. These capacitors are designed to create resonance between the primary and secondary coils at a specific resonant frequency; thereby, they reduce the reactive power circulating inside the circuits and can allow the transmission of real power over relatively large distances among the primary and secondary coils with increased efficiencies (Zhang, 2015). More exactly, the main tasks of compensating circuits are as follows:

1. To introduce a secondary side reactance able to resonate with the secondary side reactance of the native circuit. Since the Inductive Wireless Power Systems are based on the magnetic coupling of two coils (which results in equivalent inductances), the resonating reactance is constituted by a capacitor.
2. The reactance introduced on the primary side adds to the native circuit input impedance and must result in a pure resistive equivalent circuit seen by the input voltage source. In Inductive Wireless Power Systems, the secondary side circuit added to the primary side inductance results to be an *R-L* circuit. Therefore, a capacitor is also needed on the primary side to cancel the reactive part of the input impedance.

As a result, the compensating networks studied in this chapter are constituted by one primary side capacitor C_1 utilized on the circuit primary side and a capacitor C_2 on the circuit secondary side (Mohamed & Berzoy, 2017).

Depending on the connection between the capacitors and coils, four main topologies can be identified: Series - Series, Series - Parallel, Parallel - Series, and Parallel - Parallel. In the following section, these resonance compensations schemes are analyzed, and the advantages and disadvantages of each topology are highlighted.

ASSUMPTIONS

The analysis of all the compensations topologies is carried out under the following assumptions:

- The input voltage v_1 supplied by a DC-AC inverter is purely sinusoidal. Its RMS value is V_1 and its frequency f.
- The load is purely resistive.
- The coils used to transmit the power wirelessly are represented in the form of T-model, which includes the mutual inductance. The self-inductance of the primary and secondary coils are L_1 and L_2, respectively. The mutual inductance is defined as $M = k\sqrt{L_1 L_2}$, where k is the coupling coefficient. The parasitic resistances of the windings are denoted by R_1 and R_2.

- Resistances R_1 and R_2 are the parasitic resistances, which are considered in the derivation of the efficiency expressions.
- The compensation C_2 is designed to resonate with the secondary self-inductance L_2 at the resonant angular frequency ω_0. The primary capacitance C_1 is chosen to cancel the imaginary part of input impedance seen from the voltage source V_1 at the resonant angular frequency ω_0.
- The root mean square values of voltages and current are utilized in the expression derivations.

The characteristics of a RWPS are dependent on several parameters such as frequency f, coupling coefficient k, and load resistance R_L. To understand the characteristics of each compensation topology, the expression of the power delivered to the load P_{OUT} and the efficiency η as a function of the above-mentioned parameters are presented. A numerical example is given. The plots of the output power and the efficiency as functions of the operating frequency f and the coupling facto k are shown for the numerical example considered in this chapter. These plots allow for an immediate understanding of the characteristics of the considered compensation circuits, which are series-series, series-parallel, parallel-series, parallel-parallel.

SERIES-SERIES COMPENSATION (SS)

Circuit Analysis

A circuit of a Series-Series compensation topology is shown in Figure 1. The series-series compensation is achieved by using capacitor C_1 connected in series with the ac source and the primary side inductor L_1 on the secondary side, capacitor C_2 is connected in series with the inductance L_2 and load resistance R_L.

Writing the Kirchhoff's voltage law (KVL), and representing them with matrix formulation one obtains

Figure 1. Series-Series compensation topology

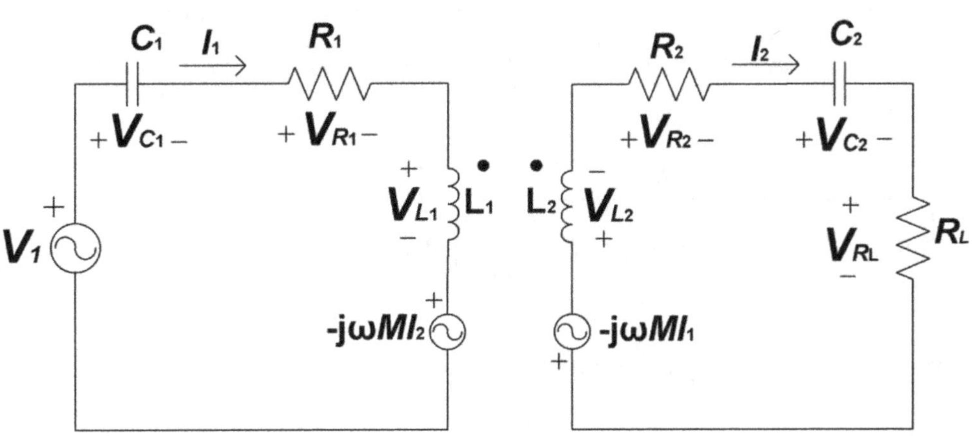

$$\begin{bmatrix} V_1 \\ 0 \end{bmatrix} = \begin{bmatrix} R_1 + j\left(\omega L_1 - \dfrac{1}{\omega C_1}\right) & -j\omega M \\ -j\omega M & R_L + R_2 + j\left(\omega L_2 - \dfrac{1}{\omega C_2}\right) \end{bmatrix} \begin{bmatrix} I_1 \\ I_2 \end{bmatrix}$$

where V_1 is the input voltage phasor RMS module, and I_1 and I2 are the primary and secondary current phasor RMS module, ω is the operating angular frequency of the circuit, R_1 and R_2 are the primary and secondary side parasitic resistances, which in the series-series compensation can be assumes as follows: $R_1 = R_{L1} + R_{C1}$ and $R_2 = R_{L2} + R_{C2}$, where R_{L1} and R_{L2} are the primary and secondary inductor equivalent series resistances, respectively, and R_{C1} and R_{C2} are the primary and secondary capacitor equivalent series resistances, respectively. In practical applications, the Inductive Wireless Power System, according to the nominal power, are operated in frequency ranges starting form some tens of kHz up to some MHz. For automotive applications, the SAE standard SAE TIR J2954 (SAE Standard J2954, 2016). establishes a common frequency band using 85 kHz (81.39 - 90 kHz) for all light duty vehicle systems. In addition, four classes PH/EV of Wireless Power Transfer levels are given today. Future revisions may include even higher power levels: 3.7 kW (WPT 1) specified in TIR J2954:

1. 7.7 kW (WPT 2) specified in TIR J2954,
2. 11 kW (WPT 3) to be specified in revision of J2954,
3. 22 kW (WPT 4) to be specified in revision of J2954,

where WPT 1-4 are four different classes of vehicle to be charged and converter and coils to be utilized.

At these frequencies, the skin and proximity- effects highly increase the inductor equivalent series resistance (Kazimierczuk, 2014; Ayachit, Reatti, Saini, & Kazimierczuk, 2016).Therefore, in this chapter, the capacitor equivalent series resistances are assumed to be negligible with respect to those of inductors.

Inverting Equation 1.1, the analytical expressions for the RMS modules of primary and secondary current phasors are determined as

$$I_1 = \frac{R_2 + R_L + j\left(\omega L_2 - \dfrac{1}{\omega C_2}\right)}{\left[R_1 + j\left(\omega L_1 - \dfrac{1}{\omega C_1}\right)\right]\left[R_2 + R_L + j\left(\omega L_2 - \dfrac{1}{\omega C_2}\right)\right] + \omega^2 M^2} V_1$$

$$I_2 = \frac{-j\omega M}{\left[R_1 + j\left(\omega L_1 - \dfrac{1}{\omega C_1}\right)\right]\left[R_2 + R_L + j\left(\omega L_2 - \dfrac{1}{\omega C_2}\right)\right] + \omega^2 M^2} V_1$$

Thus, the ratio between primary and secondary current is

$$\frac{I_1}{I_2} = \frac{R_2 + R_L + j\left(\omega L_2 - \dfrac{1}{\omega C_2}\right)}{-j\omega M} = \frac{1}{\omega M}\left[-\left(\omega L_2 - \frac{1}{\omega C_2}\right) + j\left(R_2 + R_L\right)\right]$$

The efficiency of the series-series compensation circuit can be calculated as

$$\eta = \frac{R_L \left|I_2\right|^2}{R_L \left|I_2\right|^2 + R_2 \left|I_2\right|^2 + R_1 \left|I_1\right|^2} = \frac{R_L}{R_L + R_2 + R_1 \dfrac{\left|I_1\right|^2}{\left|I_2\right|^2}} = \frac{R_L}{R_L + R_2 + \dfrac{R_1}{\omega^2 M^2}\left[\left(\omega L_2 - \dfrac{1}{\omega C_2}\right)^2 + \left(R_2 + R_L\right)^2\right]}$$

While the output power, under resonance condition is

$$P_o = R_L I_2^2 = \frac{\omega^2 M^2 V_1^2 R_L}{\left[R_1\left(R_2 + R_L\right) + \omega^2 M^2\right]^2}$$

1.3.2 Compensation Design

The secondary capacitance C_2 is chosen to resonate with the secondary coil at the resonant frequency $f_0 = \omega_0 / 2\pi$, thus

$$C_2 = \frac{1}{\omega_0^{\,2} L_2}$$

The impedance seen by the source V_1 is

$$Z_{IN} = R_{IN} + jX_{IN} = R_1 + j\omega L_1 - \frac{j}{\omega C_1} + \frac{M^2\omega^2}{j\omega L_2 - \dfrac{j}{\omega C_2} + R_2 + R_L}$$

Neglecting the parasitic resistance of the primary and secondary coils, the value of C_1 that cancel the imaginary part of the input impedance X_{IN} at the resonance frequency ω_0 is

$$C_1 = \frac{1}{\omega_0^{\,2} L_1}$$

SERIES-PARALLEL COMPENSATION (SP)

Circuit Analysis

A circuit of a Series-Parallel compensation topology is shown in Figure 2. The series-parallel compensation is achieved by using capacitor C_1 connected in series with the ac source and the primary side inductor L_1 as in the series-series compensation. On the secondary side, capacitor C_2 is connected in parallel with load resistance R_L. The parallel combination of C_2 and R_L is connected in series with the inductance L_2 and load resistance R_L.

Applying Kirchhoff's voltage law (KVL) for the circuit in Figure 1, one obtains the voltages in the matrix form as

$$\begin{bmatrix} V_1 \\ 0 \\ 0 \end{bmatrix} = \begin{bmatrix} R_1 + j\left(\omega L_1 - \dfrac{1}{\omega C_1}\right) & -j\omega M & 0 \\ -j\omega M & R_2 + j\left(\omega L_2 - \dfrac{1}{\omega C_2}\right) & \dfrac{j}{\omega C_2} \\ 0 & \dfrac{j}{\omega C_2} & R_L - \dfrac{j}{\omega C_2} \end{bmatrix} \begin{bmatrix} I_1 \\ I_2 \\ I_4 \end{bmatrix}$$

Inverting Equation 1.9, analytical expressions of the primary and secondary currents are obtained as follows

$$I_1 = \frac{\left(R_L - j\dfrac{1}{\omega C_2}\right)\left[R_2 + j\left(\omega L_2 - \dfrac{1}{\omega C_2}\right)\right] + \dfrac{1}{\omega^2 C_2}}{\dfrac{1}{\omega^2 C_2^2}\left[R_1 + j\left(\omega L_1 - \dfrac{1}{\omega C_1}\right)\right] + \left(R_L - j\dfrac{1}{\omega C_2}\right)\left\{\omega^2 M^2 + \left[R_1 + j\left(\omega L_1 - \dfrac{1}{\omega C_1}\right)\right]\left[R_2 + j\left(\omega L_2 - \dfrac{1}{\omega C_2}\right)\right]\right\}} V_1$$

Figure 2. Series-Parallel compensation topology

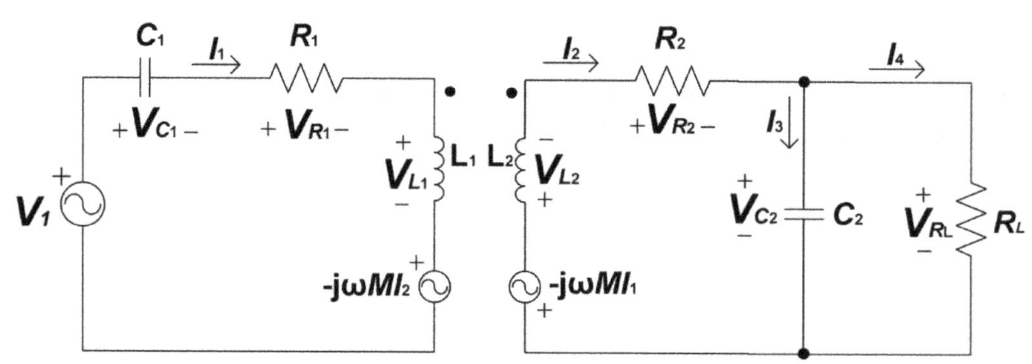

$$I_2 = \frac{-\frac{M}{C_2} - j\omega M R_L}{\frac{1}{\omega^2 C_2^2}\left[R_1 + j\left(\omega L_1 - \frac{1}{\omega C_1}\right)\right] + \left(R_L - j\frac{1}{\omega C_2}\right)\left\{\omega^2 M^2 + \left[R_1 + j\left(\omega L_1 - \frac{1}{\omega C_1}\right)\right]\left[R_2 + j\left(\omega L_2 - \frac{1}{\omega C_2}\right)\right]\right\}} V_1$$

$$I_4 = \frac{\frac{M}{C_2}}{\frac{1}{\omega^2 C_2^2}\left[R_1 + j\left(\omega L_1 - \frac{1}{\omega C_1}\right)\right] + \left(R_L - j\frac{1}{\omega C_2}\right)\left\{\omega^2 M^2 + \left[R_1 + j\left(\omega L_1 - \frac{1}{\omega C_1}\right)\right]\left[R_2 + j\left(\omega L_2 - \frac{1}{\omega C_2}\right)\right]\right\}} V_1$$

Thus, the ratio of the two currents are

$$\frac{I_1}{I_4} = \frac{1}{M}\left\{\left[L_2 + C_2 R_L R_2\right] + j\left[R_L C_2\left(\omega L_2 - \frac{1}{\omega C_2}\right) - \frac{R_2}{\omega}\right]\right\}$$

$$\frac{I_2}{I_4} = 1 + j\omega R_L C_2$$

The overall efficiency can be calculated as

$$\eta = \frac{R_L\left|I_4\right|^2}{R_L\left|I_4\right|^2 + R_1\left|I_1\right|^2 + R_2\left|I_2\right|^2} = \frac{R_L}{R_L + R_2\left[1 + \left(\omega R_L C_2\right)^2\right] + \frac{R_1}{M^2}\left\{\left(L_2 + C_2 R_L R_2\right)^2 + \left[R_L C_2\left(\omega L_2 - \frac{1}{\omega C_2}\right) - \frac{R_2}{\omega}\right]^2\right\}}$$

If the system is working in resonance, the output power can be written as

$$P_o = R_L \cdot \left|I_4\right|^2 = R_L \frac{M^2}{C_2^2}\frac{1}{\left[\frac{R_1}{\omega^2 C_2^2} + R_L\omega^2 M^2 + R_1 R_2 R_L + \frac{M^2 R_2}{C_2 L_2}\right]^2 + \left[\frac{M_2}{\omega L_2 C_2^2} + \frac{\omega M^2 R_2 R_L}{L_2} - \frac{\omega M^2}{C_2} - \frac{R_1 R_2}{\omega C_2}\right]^2} V_1^2$$

Compensation Design

The secondary capacitance C_2 is chosen to resonate with the secondary coil at the resonance frequency $f_0 = \omega_0 / 2\pi$, thus

$$C_2 = \frac{1}{\omega_0^{\,2} L_2}$$

The impedance seen from the source V_1 is

$$Z_{IN} = R_{IN} + jX_{IN} = R_1 - \frac{j}{\omega C_1} + j\omega L_1 + \frac{M^2\omega^2}{R_2 + \dfrac{R_L}{1 + j\omega C_2 R_L} + j\omega L_2}$$

Neglecting the parasitic resistance of the primary and secondary coils, the value of C_1 that cancels the imaginary part of the input impedance X_{IN} at the resonant frequency ω_0 is

$$C_1 = \frac{L_2^2 C_2}{L_1 L_2 - M^2} = \frac{1}{\omega_0^{\,2}\left(L_1 - \dfrac{M^2}{L_2}\right)}$$

In the series-parallel compensation, the primary capacitance depends on the coupling coefficient of the primary and secondary coils. This network design specifications require some information on possible operating coupling coefficient to be known.

PARALLEL-SERIES COMPENSATION (PS)

Circuit Analysis

A circuit of the Parallel-Series compensation scheme is shown in Figure 3. The parallel-series compensation is achieved by using capacitor C_1 connected in parallel with the ac source. Both the power source and the capacitor C_I are connected in series with the primary side inductor L_1. As in the series-series compensation, on the secondary side, capacitor C_2 is connected in series to both to load resistance R_L and secondary inductance L_2.

Applying Kirchhoff's voltage law (KVL) to the circuit in Figure 3 and representing them in matrix form, one obtains

Figure 3. Parallel-Series compensation topology

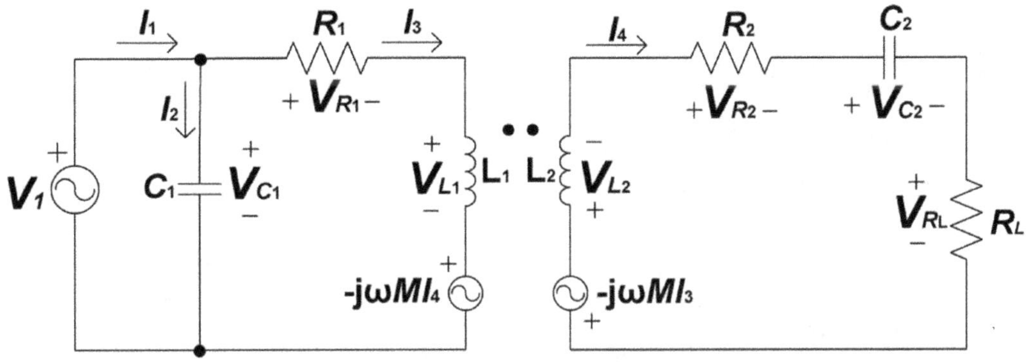

$$\begin{bmatrix} V_i \\ 0 \\ 0 \end{bmatrix} = \begin{bmatrix} -\frac{j}{\omega C_1} & \frac{j}{\omega C_1} & 0 \\ \frac{j}{\omega C_1} & R_1 + j\left(\omega L_1 - \frac{1}{\omega C_1}\right) & -j\omega M \\ 0 & -j\omega M & R_2 + R_L + j\left(\omega L_2 - \frac{1}{\omega C_2}\right) \end{bmatrix} \begin{bmatrix} I_1 \\ I_3 \\ I_4 \end{bmatrix}$$

Inverting this relationship, analytical expressions of the primary and secondary current are derived

$$I_1 = j\omega C_1 \frac{\omega^2 M^2 + \left[R_1 + j\left(\omega L_1 - \frac{1}{\omega C_1}\right)\right]\left[R_L + R_2 + j\left(\omega L_2 - \frac{1}{\omega C_2}\right)\right]}{\omega^2 M^2 + \left(R_1 + j\omega L_1\right)\left[R_L + R_2 + j\left(\omega L_2 - \frac{1}{\omega C_2}\right)\right]} V_1$$

$$I_3 = \frac{R_2 + R_L + j\left(\omega L_2 - \frac{1}{\omega C_2}\right)}{\omega^2 M^2 + \left(R_1 + j\omega L_1\right)\left[R_L + R_2 + j\left(\omega L_2 - \frac{1}{\omega C_2}\right)\right]} V_1$$

$$I_4 = \frac{-j\omega M}{\omega^2 M^2 + \left(R_1 + j\omega L_1\right)\left[R_L + R_2 + j\left(\omega L_2 - \frac{1}{\omega C_2}\right)\right]} V_1$$

Thus, the ratio of the two currents is

$$\frac{I_3}{I_4} = \frac{1}{\omega M}\left\{\left(\omega L_2 - \frac{1}{\omega C_2}\right) - j\left(R_2 + R_L\right)\right\}$$

The efficiency can be determined as

$$\eta = \frac{R_L\left|I_4\right|^2}{R_L\left|I_4\right|^2 + R_1\left|I_1\right|^2 + R_2\left|I_3\right|^2} = \frac{R_L}{R_L + R_2 + \frac{R_1}{\omega^2 M^2}\left\{\left[\left(\omega L_2 - \frac{1}{\omega C_2}\right)^2 + \left(R_L + R_2\right)^2\right]\right\}}$$

Under resonance condition, the power delivered to the load is

$$P_o = R_L \frac{\omega^2 M^2 V_1^2}{\left[\omega^2 M^2 + R_1\left(R_2 + R_L\right)\right]^2 + \left[\omega L_1\left(R_2 + R_L\right)\right]^2}$$

Compensation Design

The secondary capacitance C_2 is chosen to resonate with the secondary coil at the resonance frequency $f_0 = \omega_0 / 2\pi$, thus

$$C_2 = \frac{1}{\omega_0^{\ 2} L_2}$$

The impedance seen by the source V_1 is

$$Z_{IN} = \frac{1}{j\omega C_1 + \frac{1}{R_1 + j\omega L_1 + \frac{\omega^2 M^2}{R_2 + R_L + j\omega L_2 - \frac{j}{\omega C_2}}}}$$

Neglecting the parasitic resistance of the primary and secondary coils, the value of C_1 that cancels the imaginary part of the input impedance X_{IN} at the resonant frequency ω_0 is

$$C_1 = \frac{L_1 L_2^2 R_L^2 C_2^2}{M^4 + L_1^{\,2} L_2 R_L^2 C_2} = \frac{1}{\omega_0^{\,2}\left(L_1 + \frac{\omega_0^2 M^4}{L_1 R_L^2}\right)}$$

In the series-parallel compensation, the primary capacitance depends on the coupling coefficient of the primary and secondary coils. Moreover, in this case, the value of primary side capacitance also depends on the resonance frequency and load resistance. Even more information than in the previous series-parallel compensating circuit is needed to fix this network design specifications.

PARALLEL-PARALLEL COMPENSATION

A circuit of a Parallel-Series compensation scheme is shown in Figure 4. The parallel-parallel compensation is achieved by using capacitor C_1 connected in parallel with the ac source, as in the parallel-series compensations. The parallel combinations of these two components is connected in series with the primary side inductor L_1. On the secondary side, capacitor C_2 is connected in parallel to load resistance R_L. The parallel combination of C_2 and R_L is connected in series with the inductance L_2 and load resistance R_L as in the series-parallel compensation circuit.

Using Kirchhoff's voltage law (KVL) of the circuit in Figure 4, results in the voltage in the matrix form as

$$\begin{bmatrix} V_1 \\ 0 \\ 0 \\ 0 \end{bmatrix} = \begin{bmatrix} -\frac{j}{\omega C_1} & \frac{j}{\omega C_1} & 0 & 0 \\ \frac{j}{\omega C_1} & R_1 + j\left(\omega L_1 - \frac{1}{\omega C_1}\right) & -j\omega M & 0 \\ 0 & -j\omega M & R_2 + j\left(\omega L_2 - \frac{1}{\omega C_2}\right) & \frac{j}{\omega C_2} \\ 0 & 0 & \frac{j}{\omega C_2} & R_L - \frac{j}{\omega C_2} \end{bmatrix} \begin{bmatrix} I_1 \\ I_3 \\ I_4 \\ I_6 \end{bmatrix}$$

Figure 4. Parallel-Parallel compensation topology

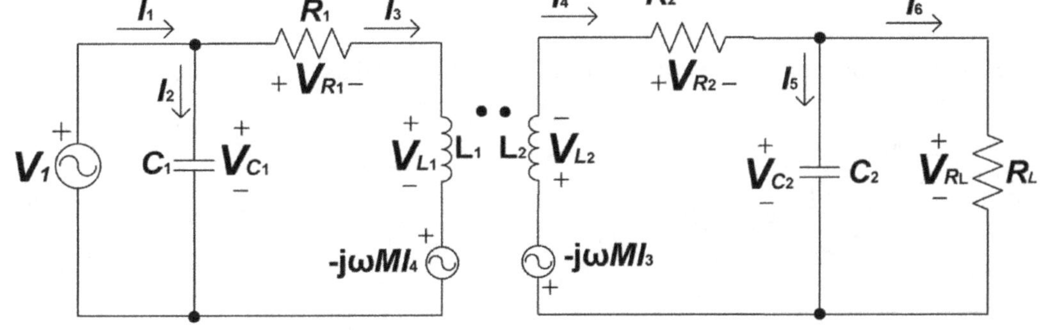

Inverting the above relationship, analytical expressions for the primary and secondary currents are

$$I_1 = j\omega C_1 \frac{\omega^2 M^2\left(\dfrac{j}{\omega C_2} - R_L\right) + \left[R_1 + j\left(\omega L_1 - \dfrac{1}{\omega C_1}\right)\right]\left\{\dfrac{1}{\omega^2 C_2^2} + \left[R_2 + j\left(\omega L_2 - \dfrac{1}{\omega C_2}\right)\right]\left(R_L - \dfrac{j}{\omega C_2}\right)\right\}}{\left(R_L - \dfrac{j}{\omega C_2}\right)\left\{\omega^2 M^2 - \left[R_2 + j\left(\omega L_2 - \dfrac{1}{\omega C_2}\right)\right]\left(R_1 + j\omega L_1\right)\right\} - \dfrac{1}{\omega^2 C_2^2}\left(R_1 + j\omega L_1\right)} V_1$$

$$I_3 = \frac{\dfrac{1}{\omega^2 C_2^2} + \left[R_2 + j\left(\omega L_2 - \dfrac{1}{\omega C_2}\right)\right]\left(R_L - \dfrac{j}{\omega C_2}\right)}{\left(\dfrac{j}{\omega C_2} - R_L\right)\left\{\omega^2 M^2 - \left[R_2 + j\left(\omega L_2 - \dfrac{1}{\omega C_2}\right)\right]\left(R_1 + j\omega L_1\right)\right\} + \dfrac{1}{\omega^2 C_2^2}\left(R_1 + j\omega L_1\right)} V_1$$

$$I_4 = \frac{\dfrac{M}{C_2} + j\omega M R_L}{\left(\dfrac{j}{\omega C_2} - R_L\right)\left\{\omega^2 M^2 - \left[R_2 + j\left(\omega L_2 - \dfrac{1}{\omega C_2}\right)\right]\left(R_1 + j\omega L_1\right)\right\} + \dfrac{1}{\omega^2 C_2^2}\left(R_1 + j\omega L_1\right)} V_1$$

$$I_6 = \frac{\dfrac{M}{C_2}}{\left(\dfrac{j}{\omega C_2} - R_L\right)\left\{\omega^2 M^2 - \left[R_2 + j\left(\omega L_2 - \dfrac{1}{\omega C_2}\right)\right]\left(R_1 + j\omega L_1\right)\right\} + \dfrac{1}{\omega^2 C_2^2}\left(R_1 + j\omega L_1\right)} V_1$$

Therefore, the efficiency can be calculated as

$$\eta = \frac{R_L\left|I_6\right|^2}{R_L\left|I_6\right|^2 + R_2\left|I_4\right|^2 + R_1\left|I_3\right|^2} = \frac{R_L}{R_L + R_2 + R_2 R^2{}_L \omega^2 C_2^2 + \dfrac{R_1}{M^2}\left\{\left(R_2 C_2 R_L + L_2\right)^2 + \left[\dfrac{R_L}{\omega} + R_2 - R_L L_2 \omega C_2\right]^2\right\}}$$

Under resonance condition, the output power can be calculated as

$$P_0 = R_L\left|I_6\right|^2 = R_L \frac{M^2}{C_2^2} \frac{1}{\left[\dfrac{R_2 L_1}{C_2} + R_1 R_2 R_L + \dfrac{R_1}{\omega^2 C_2^2} - \omega^2 M^2 R_L\right]^2 + \left[\dfrac{\omega M^2}{C_2} + \omega R_2 R_L L_1 + \dfrac{L_1}{\omega C_2^2} - \dfrac{R_1 R_2}{\omega C_2}\right]^2} V_1^2$$

Compensation Design

The secondary capacitance C_2 is chosen to resonate with the secondary coil at the resonant frequency $f_0 = \omega_0 / 2\pi$ to give

$$C_2 = \frac{1}{\omega_0^{\ 2} L_2}$$

The impedance seen by the source V_1 is

$$Z_{IN} = \frac{1}{j\omega C_1 + \cfrac{1}{R_1 + j\omega\left(L_1 - M\right) + \cfrac{1}{\cfrac{1}{j\omega M} + \cfrac{1}{R_2 + j\omega\left(L_2 - M\right) + \cfrac{1}{j\omega C_2 + \cfrac{1}{R_L}}}}}}$$

Neglecting the parasitic resistance of the primary and secondary coils, the value of C_1 that cancels the imaginary part of the input impedance X_{IN} at the resonant frequency ω_0 is

$$C_1 = \frac{1}{\omega_0^2 \left[\left(L_1 - \frac{M^2}{L_2}\right)^2 + \cfrac{\cfrac{M^4 R_L^2}{L_2^4}}{\omega_0^2 \left(L_1 - \cfrac{M^2}{L_2}\right)}\right]}$$

Also the parallel-parallel compensation results in a design expression for the primary capacitance, which depends on the coupling coefficient of the primary and secondary coils, the resonance frequency, and load resistance. Even more information than in the previous considered series-parallel compensating circuit is need to fix this network design specifications.

DESIGN EXAMPLE

In this section, a numerical example is shown. All the compensations are calibrated to resonate at the angular frequency $\omega_0 = 85$ kHz, at the coupling coefficient $k_0 = 0.2$ and at a load resistance $R_{L0} = 10\ \Omega$. The primary and secondary self-inductance are assumed to be identical $L_1 = L_2 = 100\ \mu\text{H}$. The parasitic resistance of the coil is $R_1 = R_2 = 0.1\ \Omega$. The systems are supplied by a sinusoidal voltage source with a RMS value $V_1 = 50\ \text{V}$. Since in real application some parameters can vary, plots of the output power P_{OUT} and the efficiency η for different value of load resistance R_L, frequency f, and coupling coefficient k are shown. Figure 5 shows the behavior of each topologies for a coupling coefficient $k = 0.2$ and three different value of load $R_L = 1\ \Omega$, $R_L = 10\ \Omega$ and $R_L = 100\ \Omega$.

Figure 5 shows that at low load resistances, the series-series and parallel-series compensations results in high efficiency peaks, but efficiency varies a lot over the considered frequency range. The series-parallel and parallel-parallel compensations result in nearly constant efficiencies which, in contrast, are low over the entire considered frequency range. When the load resistance is increased, the parallel-parallel compensation results in a very low efficiency. The series-series and series-parallel efficiency variations with frequency reduce and the series-series compensation efficiency is higher than 75% over the entire considered frequency range. The plots of the output power versus frequency show that parallel-parallel compensation results in a very low output power at any load resistance. This explain the low efficiencies achieved. At higher load resistances, the double resonance frequencies, which are typical for series-series compensation at high load currents disappear, while the maximum output power can be delivered in a narrow frequency range. Figure 5(b) shows that, for certain load resistances, the output power results from a series-series compensation remains nearly constant for frequencies from 8 to 9.5 kHz. The efficiency and the output power varying the coupling coefficient at the resonant frequency $\omega_0 = 85\ \text{kHz}$ are shown in Figure 6.

The plots of the efficiency as function of the coupling factor k show that the series-series and parallel-series compensations allows for a nearly constant and relatively high efficiency operation over a wide range of coupling coefficient for all the three considered load resistance values. In practical application, these two compensations circuits are suitable to be utilized in cases, where a perfect alignment of primary and secondary coil cannot be guaranteed. Because this happens for small load resistance values and large load resistance values as well, high efficiency operation can be achieved during the entire battery charging cycle. While the parallel-parallel compensation results in efficiency strongly dependent on coupling coefficient at any load resistance, the series-parallel compensation efficiency increases, when load resistance increases. For high values of load resistances, the series-parallel compensations results in a nearly constant efficiency over the entire coupling coefficient range and the achieved efficiency is even higher than that of the series-series and parallel-series compensation circuits.

Also, when the coupling coefficient is considered, the power delivered to the load with a PP compensation circuit remains low over the entire range of the coupling coefficient. The series-series compensation network results in high output power levels at very low coupling factor values ($k < 0.1$ for load resistances lower than 10 Ω and $k = 0.1$ for a load resistance as high as 100 Ω.). The series-parallel compensation circuit delivers an output power slightly higher than the series-series compensations at low load resistance. The delivered maximum power decreases as the load resistance increases and this maximum occurs at coupling coeffcient close to 0.2 for low and medium values of load resistance and decreases toward 0.17 at high values of load resistance.

Figure 5. Efficiency as function of operating frequency. (a) Load resistance R_L=1 Ω. (b) Load resistance R_L=10 Ω. (c) Load resistance R_L=100 Ω.

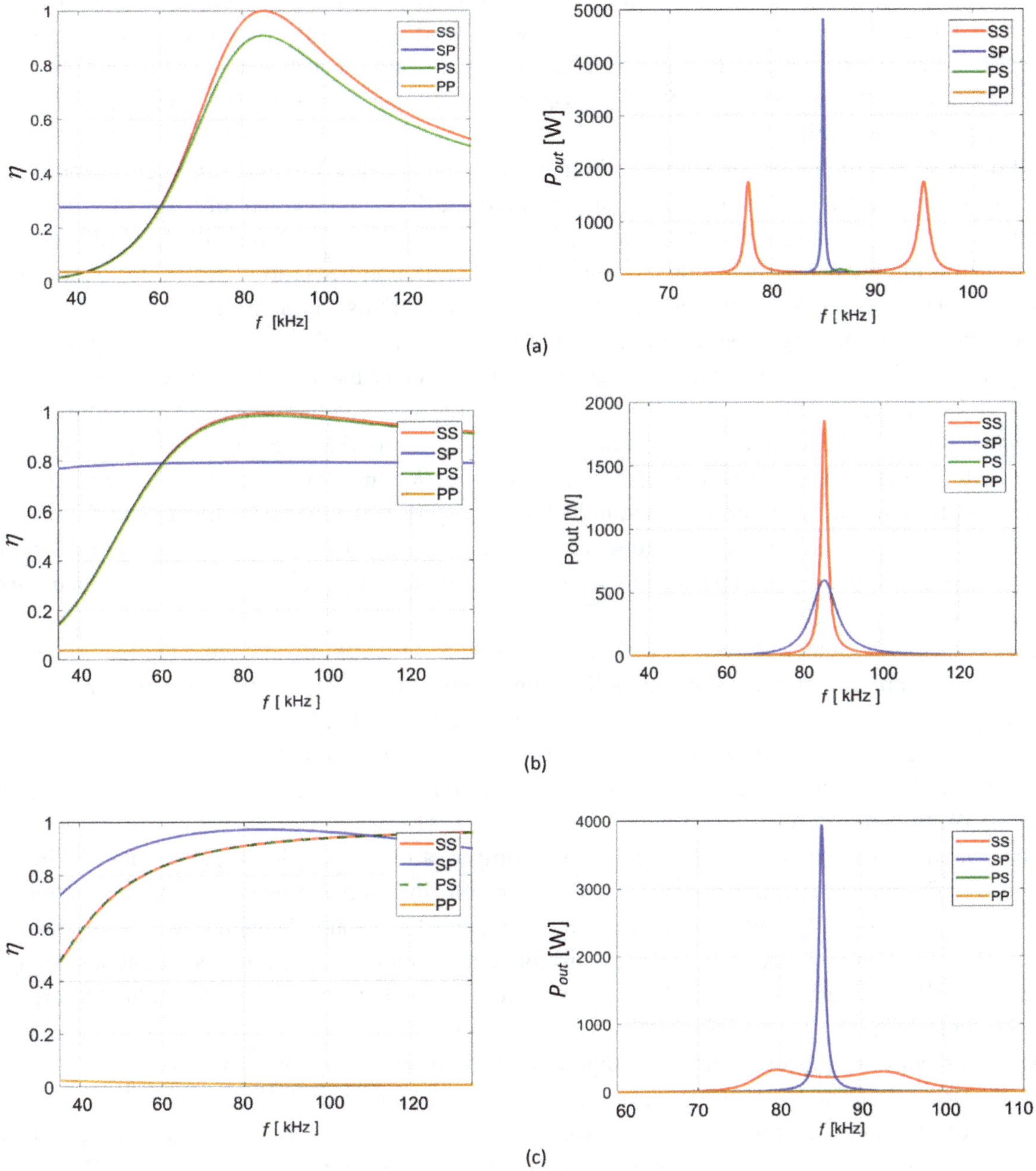

CONCLUSIONS AND TOPOLOGIES FIELD OF APPLICATION

The analysis, design equations and their applications for a practical design example have been given in this chapter. The studies considered in this text allows us to arrive at some important conclusions about the performance of the four possible capacitor connection inside Inductive Wireless Power Transfer circuits that constitute the series-series (SS), series-parallel (SP), parallel-series (PS), and parallel-parallel (PP) compensation circuits.

All the compensation circuits operate at higher efficiencies, when the coupling factor k increases.

In the SS compensation circuit, the primary capacitance C_1 is independent of the load and coupling coefficient. Therefore, the resonant frequency of the circuit does not depend on the load variations and neither on misalignment between coils. This property of the SS compensation is useful in all the applications affected from coupling coefficient variations. The ability of this compensation to work in resonance independently from misalignment and load variation and the possibility of achieving high efficiency at long distance between coils, makes this compensation widely used for static and dynamic electric vehicle wireless charging (D. Patil, V. Agarwal, April 2015). On the other hand, the main drawbacks of this compensation is the behavior when an open circuit at the secondary side occurs. In fact, in SS compensation the output power increase with the load resistance R_L. In this condition the primary current is limited only from the primary capacitor and inductance parasitic resistances, leading to unsafe condition.

In general, a series compensated secondary capacitance is used to supply a stable secondary load voltage, where as a parallel compensated secondary is used to supply a stable secondary load current (Aditya & Williamson, 2014).

The primary capacitance of Series-Parallel compensation depends on the mutual inductance between the coils. This implies that the circuit can deliver high power only at the coupling coefficient for which it was designed. The SP compensation is able to transfer power at high efficiency also for long distance, but it is more sensitive to coil misalignment than the SS compensation. This topology is still used for high power and low coupling coefficient applications such as electric vehicle wireless charging (Chinthavali & Wang, Sept. 2016).

The primary capacitance of the Parallel-Series compensation topologies are dependent on both the coupling coefficient and the load resistance. And this results in cumbersome design of these circuits and satisfactory operation only under certain load, coupling coefficient and frequency value combinations. Moreover, one major drawback is it requires current source input to avoid any instantaneous change in voltage. Usually, to overcome this problem, an additional inductor at the primary is added to create an LCL resonant tank.

Another important aspect regarding compensation topology is the component voltage and current stresses. In Ravikiran and Keshri (2017) a comparison between stresses on coils, capacitors and switches in SS and PS is studied. Results shown that voltage stress on capacitor and coil in SS topology is high compare to PS topology. Also observed that current stress on capacitors and coil is more in PS topology compared to SS topology. Therefore, the PS compensation is more suitable for low and medium power applications.

Finally, the PP compensation is characterized from high load voltage of the parallel secondary and large current source requirements of the parallel primary and for this reason it is not widely used.

Figure 6. Efficiency and output power as function of coupling coefficient. (a) Load resistance R_L=1 Ω. (b) Load resistance R_L=10 Ω. (c) Load resistance R_L=100 Ω.

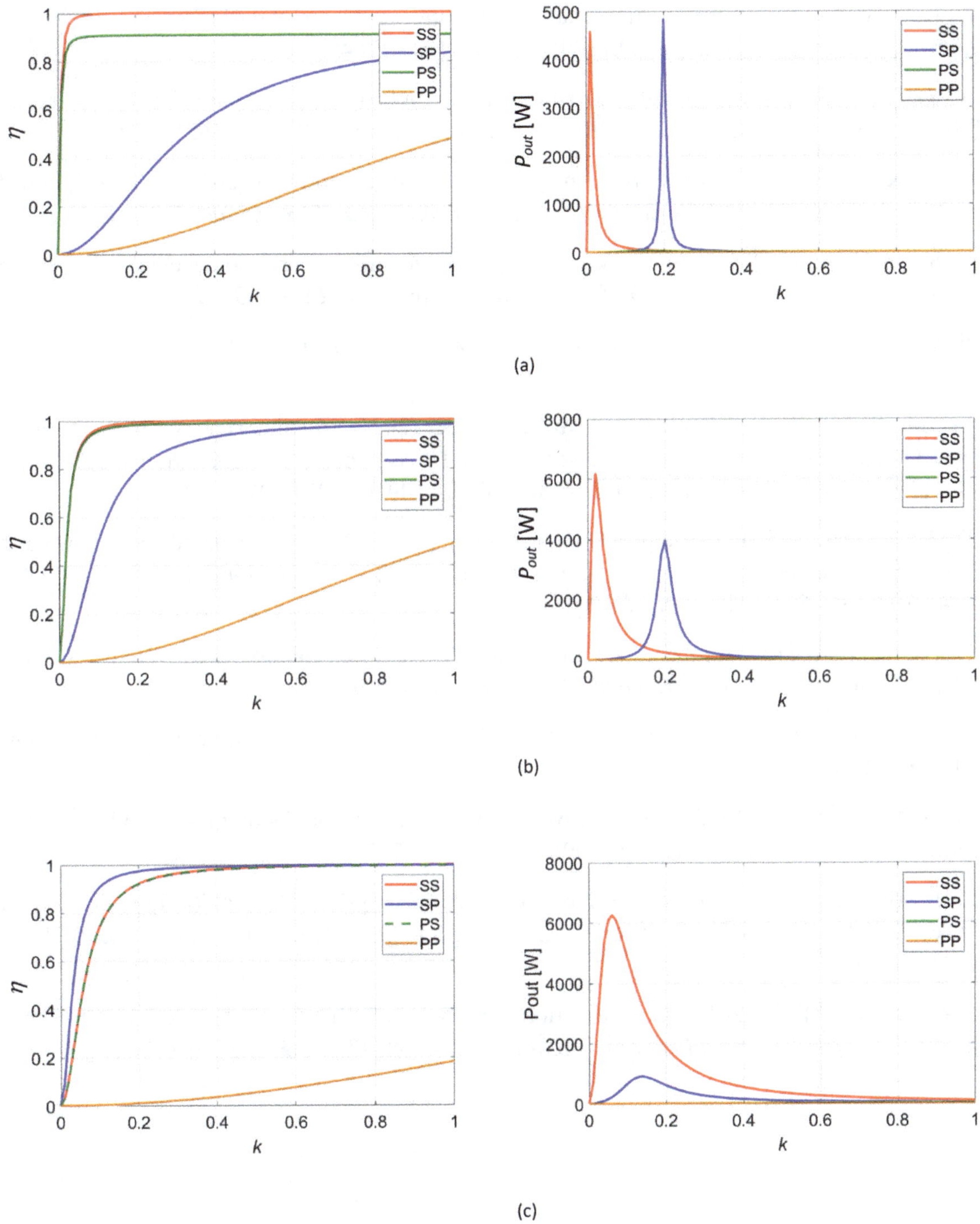

REFERENCES

Aditya, K., & Williamson, S. S. (2014). Design Considerations for Loosely Coupled Inductive Power Transfer (IPT) System for Electric Vehicle Battery Charging – A Comprehensive Review. *IEEE Transportation Electrification Conference*. 10.1109/ITEC.2014.6861764

Ayachit, Reatti, Saini, & Kazimierczuk. (2016). Design of choke inductor in class-E ZVS power amplifier. *Proc. of 42nd Annual Conference of IEEE Industrial Electronics Society*, 5621-5626.

Chinthavali & Wang. (2016). Sensitivity Analysis of a Wireless Power Transfer (WPT) System for Electric Vehicle Application. *IEEE Energy Conversion Congress and Exposition (ECCE).*

Chopra, S., & Bauer, P. (2011). Analysis and design considerations for a contactless power transfer system. *Telecommunications Energy Conference*. 10.1109/INTLEC.2011.6099774

Kazimierczuk, M. K. (2014). *High-Frequency Magnetic Components* (2nd ed.). Chichester, UK: John Wiley.

Mohamed, A., Berzoy, A., De Almeida, F., & Mohammed, O. (2017). Modeling and Assessment Analysis of Various Compensation Topologies in Bidirectional IWPT System for EV application. *IEEE Transactions on Industry Applications*, *53*(5), 4973–4984. doi:10.1109/TIA.2017.2700281

Patil, D., & Agarwal, V. (2015). Compact Onboard Single-Phase EV Battery Charger With Novel Low-Frequency Ripple Compensator and Optimum Filter Design. *IEEE Transactions on Vehicular Technology*, *65*(4), 1948–1956. doi:10.1109/TVT.2015.2424927

Ravikiran & Keshri. (2017). Comparative Evaluation of S-S and P-S Topologies for Wireless Charging of Electrical Vehicles. *IEEE Industrial and Electronics society Conference.*

SAE Standard J2954 "Wireless Power Transfer for Light-Duty Plug-In/ Electric Vehicles and Alignment Methodology" 2016.

Sirbu, I., & Mandache, L. (2017). Comparative Analysis of Different Topologies for Wireless Power Transfer System. *EEEIC Conference*. 10.1109/EEEIC.2017.7977863

Spanik, P., Frivaldsky, M., Dragona, P., & Jaros, V. (2016). Analysis of Proper Configuration of Wireless Power Transfer System for Electric Vehicle Charging. *ELEKTRO Conference*. 10.1109/ELEKTRO.2016.7512071

Zhang, W., & Chris Mi, C. (2015). Compensation Topologies of High Power Wireless Power Transfer System. *IEEE Transactions on Vehicular Technology*, *65*(6), 4768–4778. doi:10.1109/TVT.2015.2454292

APPENDIX

Matrix Inversion

The relationship between two vectors is

$$\begin{bmatrix} V_1 \\ 0 \\ 0 \end{bmatrix} = \begin{bmatrix} A & B & 0 \\ B & C & D \\ 0 & D & E \end{bmatrix} \begin{bmatrix} I_1 \\ I_2 \\ I_4 \end{bmatrix}$$

Inverting this relationship, one obtains

$$\begin{bmatrix} I_1 \\ I_2 \\ I_4 \end{bmatrix} = \begin{bmatrix} \frac{EC - D^2}{-AD^2 - E(B^2 - AC)} & \frac{-EB}{-AD^2 - E(B^2 - AC)} & \frac{BD}{-AD^2 - E(B^2 - AC)} \\ \frac{-EB}{-AD^2 - E(B^2 - AC)} & \frac{EA}{-AD^2 - E(B^2 - AC)} & \frac{-AD}{-AD^2 - E(B^2 - AC)} \\ \frac{BD}{-AD^2 - E(B^2 - AC)} & \frac{-AD}{-AD^2 - E(B^2 - AC)} & \frac{B^2 - AC}{-AD^2 - E(B^2 - AC)} \end{bmatrix} \begin{bmatrix} V_1 \\ 0 \\ 0 \end{bmatrix}$$

Thus,

$$I_1 = \frac{EC - D^2}{-AD^2 - E(B^2 - AC)} V_1$$

$$I_2 = \frac{EB}{-AD^2 - E(B^2 - AC)} V_1$$

$$I_4 = \frac{BD}{-AD^2 - E(B^2 - AC)} V_1$$

The relationship between the two vectors is

$$\begin{bmatrix} V_i \\ 0 \\ 0 \\ 0 \end{bmatrix} = \begin{bmatrix} A & -A & 0 & 0 \\ -A & B & C & 0 \\ 0 & C & D & E \\ 0 & 0 & E & F \end{bmatrix} \begin{bmatrix} I_1 \\ I_3 \\ I_4 \\ I_6 \end{bmatrix}$$

Inverting this relationship, one obtains

$$\begin{bmatrix} I_1 \\ I_3 \\ I_4 \\ I_6 \end{bmatrix} = \begin{bmatrix} \frac{C^2F + B\left(E^2 - DF\right)}{A\left\{F\left[C^2 - D\left(B - A\right)\right] + E^2\left(B - A\right)\right\}} & \frac{E^2 - DF}{F\left[C^2 - D\left(B - A\right)\right] + E^2\left(B - A\right)} & \frac{CF}{F\left[C^2 - D\left(B - A\right)\right] + E^2\left(B - A\right)} & \frac{-CE}{F\left[C^2 - D\left(B - A\right)\right] + E^2\left(B - A\right)} \\ \frac{E^2 - DF}{F\left[C^2 - D\left(B - A\right)\right] + E^2\left(B - A\right)} & \frac{E^2 - DF}{F\left[C^2 - D\left(B - A\right)\right] + E^2\left(B - A\right)} & \frac{CF}{F\left[C^2 - D\left(B - A\right)\right] + E^2\left(B - A\right)} & \frac{-CE}{F\left[C^2 - D\left(B - A\right)\right] + E^2\left(B - A\right)} \\ \frac{CF}{F\left[C^2 - D\left(B - A\right)\right] + E^2\left(B - A\right)} & \frac{CF}{F\left[C^2 - D\left(B - A\right)\right] + E^2\left(B - A\right)} & \frac{F\left(A - B\right)}{F\left[C^2 - D\left(B - A\right)\right] + E^2\left(B - A\right)} & \frac{E\left(B - A\right)}{F\left[C^2 - D\left(B - A\right)\right] + E^2\left(B - A\right)} \\ \frac{-CE}{F\left[C^2 - D\left(B - A\right)\right] + E^2\left(B - A\right)} & \frac{-CE}{F\left[C^2 - D\left(B - A\right)\right] + E^2\left(B - A\right)} & \frac{E\left(B - A\right)}{F\left[C^2 - D\left(B - A\right)\right] + E^2\left(B - A\right)} & \frac{C^2 - D\left(B - A\right)}{F\left[C^2 - D\left(B - A\right)\right] + E^2\left(B - A\right)} \end{bmatrix} \begin{bmatrix} V_1 \\ 0 \\ 0 \\ 0 \end{bmatrix}$$

Thus,

$$I_1 = \frac{C^2F + B\left(E^2 - DF\right)}{A\left\{F\left[C^2 - D\left(B - A\right)\right] + E^2\left(B - A\right)\right\}} V_1$$

$$I_3 = \frac{E^2 - DF}{F\left[C^2 - D\left(B - A\right)\right] + E^2\left(B - A\right)} V_1$$

$$I_4 = \frac{CF}{F\left[C^2 - D\left(B - A\right)\right] + E^2\left(B - A\right)} V_1$$

$$I_6 = \frac{-CE}{F\left[C^2 - D\left(B - A\right)\right] + E^2\left(B - A\right)} V_1$$

Chapter 4
Multivariable Compensation for Wireless Power Transfer Systems in Dynamic Environments

Rodrigo Wolff Porto
Universidade Federal do Rio Grande do Sul, Brazil

Lucas Murliky
Universidade Federal do Rio Grande do Sul, Brazil

Valner João Brusamarello
Universidade Federal do Rio Grande do Sul, Brazil

Fernando Rangel de Sousa
Universidade Federal de Santa Catarina, Brazil

ABSTRACT

The growing number of publications concerning the dynamic compensation of inductive links has demonstrated the importance of the issue upon the WPT system design. There is not still a final and consolidated solution for the case when the coils are in relative movement with each other or misaligned. In this context, this chapter intends to give some contributions to that area, starting with the modeling of the inductive link in Section 2, where the authors show how the magnetic coupling coefficient and the load influence on the PDL. In addition, Section 2 also presents how the excitation frequency and the matching capacitances can improve the PDL. In Section 3, the variable excitation frequency method is explained as well as the hardware needed for the implementation. The voltage-controlled capacitance as a variable element in the matching network is shown in Section 4. Finally, in Section 5, optimization techniques are presented by using both frequency and capacitance to maximize the PDL.

DOI: 10.4018/978-1-5225-5870-5.ch004

INTRODUCTION

The Wireless Power Transfer (WPT) has been studied since the early experiments of Heinrich Hertz in 19th century (Brown, 1984). However, it was from the Tesla's work that WPT has been brought to the light and the theoretical basis has been developed, which have allowed constructing practical WPT systems. From the first Tesla's work until the decade of 1930, many attempts have been made to transmit energy without wires in an efficiently way. A WPT system working at 100 MHz has been developed at Westinghouse's labs, where hundreds of watts have been transmitted over a distance of eight meters (Tesla, 1914).

The technology available at the first half of 20th century did not allow fast and significant development of WPT systems. In fact, Brown (1984) states that a lack of interest in WPT during the first half of 20th century was related to the way that engineers and scientists believed to transmit energy without wires. They believed that for transmitting energy from one point to another, the energy in the magnetic field should be concentrated in a narrow beam. Thus, the implementation would require small wavelengths (radio frequencies) and suitable antennas. However, electronic devices at that time could operate with just a few miliwatts at those wavelengths. A greater step up has been made after the second Great War with the development of high power vacuum tubes (Brown, 1957). Thus, it was possible to transmit energy wirelessly at greater distances by using microwaves (Brown, 1974; Goubau & Schwering, 1961; Degenford, Sirkis, & Steier, 1964).

The WPT systems operating at low frequencies, that is, not involving electromagnetic wave propagation became popular from 1970 decade in the biomedical engineering area (Ko, Liang, & Fung, 1977; Hochmair, 1984). Such systems are considered safe in applications inside or in contact with the human body. Instead of using wave propagation theory, the transferred energy was based on the magnetic coupling between two or more coils and mathematical modeling can be done through the electric circuits theory (Hui, Zhong, & Lee, 2014). Other application fields have benefited from the WPT, such as the induction heating (Hurley & Kassakian, 1979), recharging of mobile devices (Choi et al., 2004; Jang & Jovanovic, 2003; Kim et al., 2001; Hui & Ho, 2005; Liu & Hui, 2007a), and other high power applications for electric vehicles (Green & Boys, 1994; Boys, Covic, & Green, 2000; Boys, Hu, & Covic, 2000; Elliot et al., 2006; Kissin, Boys, & Covic, 2009).

Since the 2000 decade, the consumer electronics and mobile devices have become an important field of application for WPT. After the remarkable results published by researches from Massachusetts Institute of Technology (MIT), WPT has gained much more interests from the scientific community (Kurs et al., 2007). In 2008, eight companies founded an open industrial group in order to standardize solutions for mobile phone recharging. In August 2010, the Qi interface was published, containing the specifications and test procedures for certifications of products with WPT embedded into them. In 2015, more than 200 companies have been already participating in the WPT consortium and more than 700 different products have been already certified with the Qi interface (Treffers, 2015).

The Qi interface specifications are mainly concerned with applications dealing with battery recharging of mobile electronic devices. They describe the bounds of the power transferred, transfer efficiency, standby power consumption, safety issues, maximum working distance and ease of use (Wireless Power Consortium [WPC], 2016a, 2016b; Wageningen & Staring, 2010). Among these specifications, ease of use is described in terms of the positioning degree of freedom of the mobile device. There is a tradeoff between the ease of use and costs. Wireless battery chargers with large charging areas tend to be more complicated and expensive to manufacture. Therefore, it is an important issue shared by manufacturers and consumers (Treffers, 2015). Figure 1 summarizes the history of the WPT with some remarkable developments.

One of the first solutions to improve the 2D positioning freedom of the mobile devices under charge was the use of multiples transmitting coils side-by-side or overlapped. This technique ensures greater magnetic flux distribution along the coils' plane (Hui & Ho, 2005; Liu & Hui, 2007b). The multiple coils form a matrix structure, where the coils can be connected in series or parallel, according to the power load requirements. The transferred power can be controlled generally through variable excitation frequency, voltage amplitude on the transmitting coil or by variable matching network (Kissin, Boys, & Covic, 2007; Choi et al., 2010; Dieckerhoff, Ryan, & Doncker, 1999).

In the last years, the concept of ease of use has been extended to control the Power Delivered to the Load (PDL) in dynamical environments, that is, situations where the transmitting and receiving coils are in relative movement each other or the load is not constant. Variable excitation frequency was one of the first techniques to improve the PDL and the Power Transfer Efficiency (PTE) of the inductive link when the load and the magnetic coupling coefficient are variables (Sample, Meyer, & Smith, 2011; Brusamarello et al., 2013; Aldhaher, Luk, & Whidborne, 2014). Furthermore, Pantic, Lee & Lukic (2014) proposed a WPT system with multiple simultaneous frequencies. Even though such system has presented low losses, and thus good PTE results, the design of the matching networks are more complicated.

In addition to the variable frequency methods, the compensation networks or matching networks can be also used as a variable element in order to maintain the PDL or PTE when the magnetic coupling and the load are not constant. Such technique can be crucial to keep the power transferred when the WPT must operate in a narrow bandwidth. Lim et al. (2014) proposed an adaptive compensation network based

Figure 1. Time line of WPT development

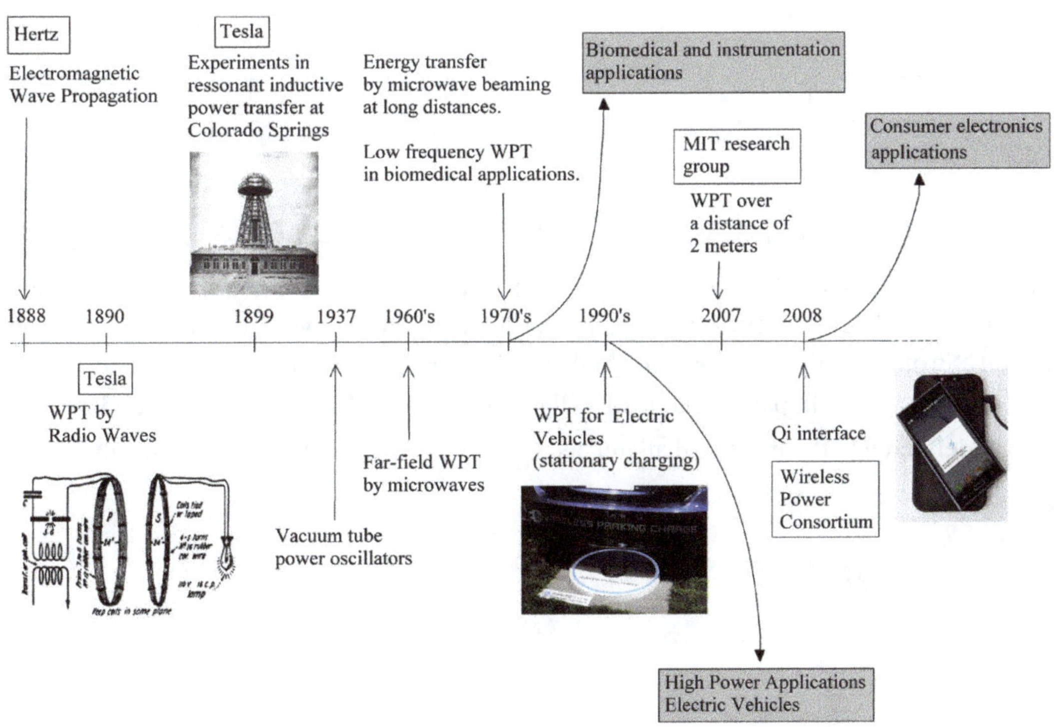

on a capacitor array, where a switching system selects the capacitance needed to match the impedances. Ahn and Hong (2015) proposed a methodology to match impedances by modulating the load value on the secondary side of the inductive link. Their system can deliver 20 W to a load located 15 cm far from the transmitter source with some advantages over other methods that use variable matching networks.

The growing number of publications concerning the dynamical compensation of inductive links demonstrate the importance of such issue upon the WPT system design. Also, there is not a final and consolidated solution for the case when the coils are in relative movement each other yet. In this context, this chapter intends to give some contributions to the dynamic compensation for low power WPT systems, starting by the modeling of the inductive link in section II, where it is shown how the magnetic coupling coefficient and the load influence on the PDL. Section II also presents how the excitation frequency and the matching capacitances can improve the PDL. In section III the variable excitation frequency method is explained as well as the hardware needed for the implementation. The voltage-controlled capacitance as a variable element in the matching network is shown in section IV. Finally, in section V optimization techniques are presented by using both frequency and capacitance to maximize the PDL.

MODELING AND CHARACTERIZATION OF AN INDUCTIVE LINK

A typical wireless power transfer system can be reduced to the model illustrated in Figure 2. The model contains an inductive link composed by two magnetically coupled lossy coils (L_1, L_2, R_1, R_2) connected to an Alternating Current (AC) voltage source v_g and to a load $Z_L = R_L + jX_L$. The PDL can be calculated after the observation of the power consumed by R_L, which is given by $P_L = \eta\, P_{in}$, where P_L is the PDL, P_{in} is the inductive link input power and η is the PTE, obtained from Porto et al. (2017a) (considering that $\Im\left\{Z_L\right\}$ cancels L_2):

$$\eta = \frac{1}{1 + p + \frac{1}{A}\left(2 + p + \frac{1}{p}\right)}, \tag{1}$$

where $p = R_2 / R_L$, and $A = k^2 Q_1 Q_2$, being $k = M\big/\sqrt{L_1 L_2}$ the magnetic coupling factor, M is the mutual inductance and Q_1 and Q_2 the quality factor of the link coils. From Equation 1 it is easily figured out how the power delivered to the load depends on the coupling between the inductors as well as on the load itself. Strong coupled links lead to higher values of A, which increases η, and consequently P_L. On the other hand, the impact of the load will depend on its relationship with A and R_2.

The power delivered to the link is calculated as $P_{in} = V_g^2 \big/ \Re\left\{Z_{in}\right\}$, where Z_{in} is the input impedance given by:

$$Z_{in} = R_1 + j\omega L_1 + \frac{\left(\omega M\right)^2}{Z_{2s}} \tag{2}$$

Figure 2. Equivalent Circuit Model of an Inductive Link

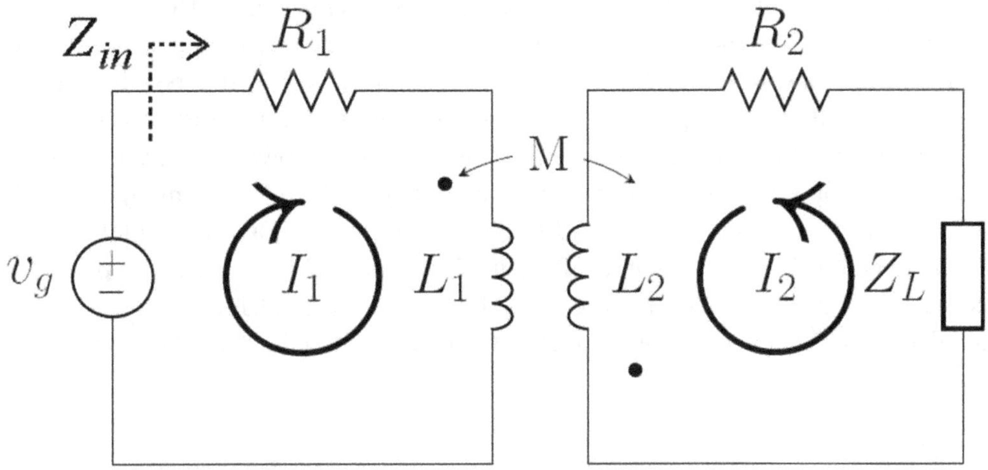

where Z_{2s} is the series equivalent impedance of the link's secondary side. From Equation 2 one can infer how the input power is related to the value of the load, as well as to the coupling level of the link. This fact is going to be used in the next sections for providing a mean of controlling the PDL by observing only the input impedance of the link.

Impedance Matching and Compensation Network

The resonant inductive coupling is a special case of the induction phenomenon, where the capacitive reactances counterbalance the effects of inductive reactances of the coils. The capacitive matching network is of utmost importance for the development of a WPT system, since the PDL is strongly dependent on the resonance conditions of the inductive coupling. The capacitive matching network seeks to tune the circuit, which is an important feature of the WPT, because the chosen capacitors should maximize the PDL, the PTE or a combination of both. Figure 3 shows the placement for the matching network at each side of the inductive link.

A capacitive compensation network can be implemented in many different ways. There are several topologies reported in the literature to perform the compensation. The possibility of applying the concepts simultaneously to the transmitter and receiver has led to the development of circuits with two or more compensation capacitors. By using two compensation capacitors, one can connect each capacitor

Figure 3. Basic WPT compensated system

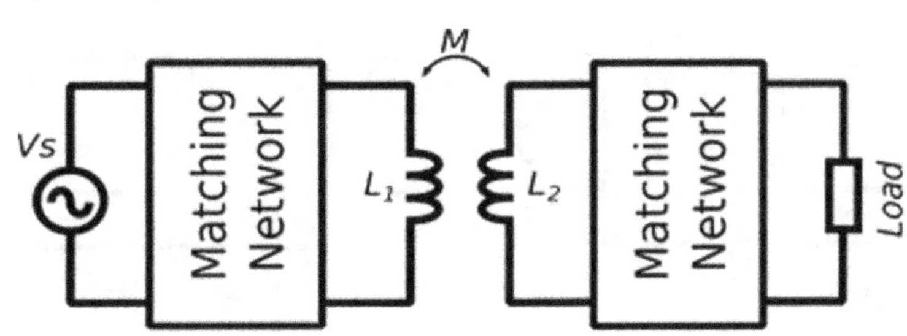

in series or in parallel with the respective transmitter and receiver coil (Wang, Stielau, & Covic, 2005). That leads four possibilities of connections: series-series (SS); series-parallel (SP); parallel-series (PS); and parallel-parallel (PP). The more compensation capacitors in the circuit, the more complex become the PDL and PTE equations. However, the use of four capacitors has proved to be advantageous at fixed load. In comparison to other topologies of one or two compensation capacitors, it is possible to obtain a gain of up to 5% in transmission efficiency (PTE) and 20% in the power delivered to the load (PDL), using four capacitors, for a given load and magnetic coupling coefficient (Azambuja et al., 2014).

A SP capacitive network has two degrees of freedom and can transform a complex impedance Z_L as demonstrated here below. Let us consider a SP capacitive network as a series and a parallel capacitors cascaded as illustrated in Figure 4, each represented by their *ABCD* matrix.

The equivalent *ABCD* network is the product of the two individual ones, given by:

$$ABCD_{SP} = \begin{pmatrix} 1 + \frac{C_{P1}}{C_{S1}} & \frac{1}{C_{S1}} \\ sC_{P1} & 1 \end{pmatrix} \tag{3}$$

The *ABCD* matrix can be easily transformed to the *Z* matrix (Ludwig & Bretchko, 2000), resulting in:

$$Z_{SP} = \begin{pmatrix} 1 + \frac{C_{P1}}{C_{S1}} & 1 \\ 1 & 1 \end{pmatrix} \tag{4}$$

Figure 4. Capacitive matching network and its two-port equivalent matrix for: (a) Serial-Parallel configuration, and (b) Parallel-Series configuration

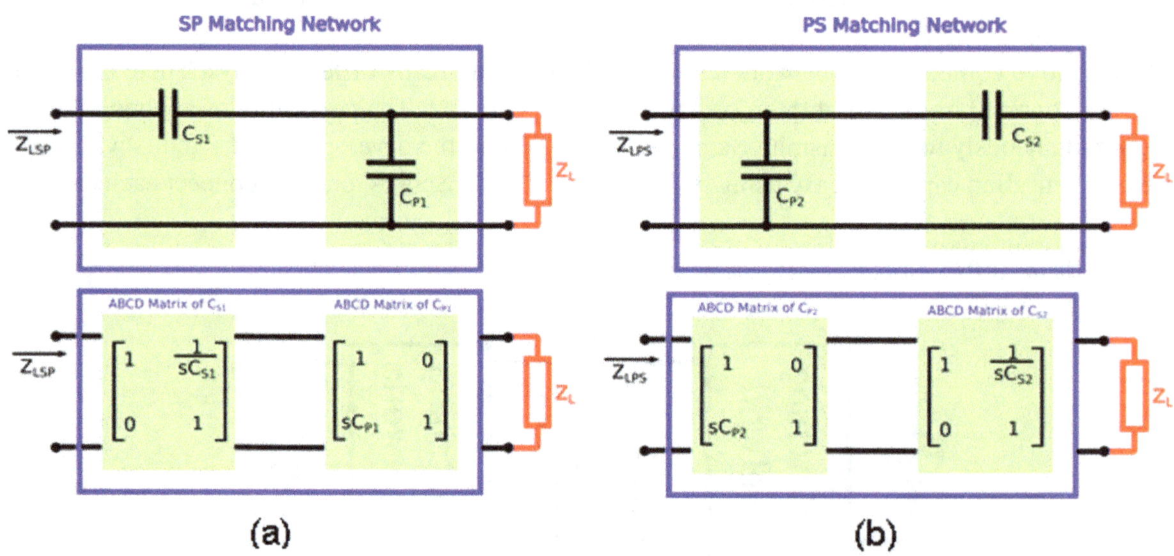

It is well known that the input impedance of a two-port terminated by a load Z_L is readily obtained from its Z-matrix (Ludwig & Bretchko, 2000), so that the SP matching network can be written as:

$$Z_{inLSP} = \frac{\Re\left\{Z_L\right\}}{\left|1 + j\omega C_{P1} Z_L\right|^2} + j\left(\frac{\Im\left\{Z_L\right\} - \omega C_{P1}\left|Z_L\right|^2}{\left|1 + j\omega C_{P1} Z_L\right|^2} - \frac{1}{\omega C_{S1}}\right) \tag{5}$$

For the particular case where $\Im\left\{Z_L\right\} = 0$,

$$Z_{inLSP} = \frac{\Re\left\{Z_L\right\}}{1 + Q^2_{C_{P1}Z_L}} - j\left[\frac{1}{\omega C_{S1}} + \frac{1}{\omega C_{P1}\left(\frac{1 + Q^2_{C_{P1}Z_L}}{Q^2_{C_{P1}Z_L}}\right)}\right] \tag{6}$$

where $Q_{C_{P1}Z_L} = \omega C_{P1}\Re\left\{Z_L\right\}$. Thus, for a given frequency, the real part of Z_L is transformed to a value of choice with a careful choice of C_{P1}. On the other hand, the imaginary part is compensated by choosing an adequate series capacitor. As the equations shown, the SP configuration is not useful to increase $\Re\left\{Z_L\right\}$. Its counterpart, the PS configuration, fixes it as the equations below indicate:

$$Z_{inLPS} = \left[\frac{\omega^2 C_{S2}^2 \Re\left\{Z_L\right\}}{\left|1 + j\omega C_{S2} Z_L\right|^2} + j\left(\frac{\omega C_{S2} - \omega^2 C_{S2}^2 \Im\left\{Z_L\right\}}{\left|1 + j\omega C_{S2} Z_L\right|^2} + \omega C_{P2}\right)\right]^{-1} \tag{7}$$

Similarly to the SP network, the particular case where $\Im\left\{Z_L\right\} = 0$ and $Q_{C_{S2}Z_L} = \frac{1}{\omega C_{S2}\Re\left\{Z_L\right\}}$ the input impedance with PS network is:

$$Z_{inLPS} = \left[\frac{1}{\Re\left\{Z_L\right\}\left(1 + Q^2_{C_{S2}Z_L}\right)} + j\omega\left(C_{P2} + C_{S2}\frac{Q^2_{C_{S2}Z_L}}{1 + Q^2_{C_{S2}Z_L}}\right)\right]^{-1} \tag{8}$$

Figure 5 shows an extension of the SP and PS configurations, which uses four compensation capacitors.

Through the mesh equations of the circuit of Figure 5 it is possible to obtain the input and output current equations, which are respectively given by:

$$I_S = \frac{Z_{C4}D + Z_L G}{Z_{C4}\left(H + R_S D\right) + Z_L\left(N + R_S G\right)} V_S \tag{9}$$

Figure 5. Equivalent T-circuit of the inductive link with compensation capacitors C_1, C_2, C_3 and C_4. R_{C1}, R_{C2}, R_{C3} and R_{C4} are the respectively Equivalent Series Resistance (ESR) of the compensation capacitors. R_1 and R_2 are the intrinsic resistances of the coil inductances L_1 and L_2, respectively. V_S represents the excitation voltage source with internal resistance R_S, and R_L is the load.

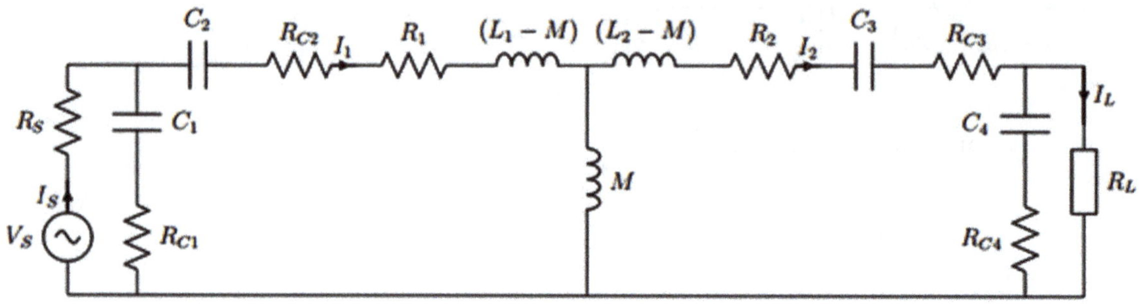

$$I_L = \frac{Z_{C1} Z_{C2} X_M}{AB\left(R_S + Z_{C1}\right)\left(Z_{C4} + R_L - Z_{C4}^2\right)} V_S \tag{10}$$

where $Z_{C1} = R_{C1} - j\omega C_1$; $Z_{C2} = R_{C2} - j\omega C_2$; $Z_{C3} = R_{C3} - j\omega C_3$; $Z_{C4} = R_{C4}$ - $j\omega C4$; $Z_{L1} = R1 + j\omega L1$; $Z_{L2} = R2 + j\omega L2$; $X_M = j\omega M$, $Z_L = R_L + j\omega X_L$ and

$$A = Z_{C1} + Z_{C2} + Z_{L1} - \frac{Z_{C2}^2}{R_S + Z_{C1}} \tag{11}$$

$$B = Z_{L2} + Z_{C3} + Z_{C4} - \frac{X_M^2}{A} \tag{12}$$

$$D = X_M^2 - (Z_{C1} + Z_{C2} + Z_{L1})(Z_{C3} + Z_{L2}) \tag{13}$$

$$G = X_M^2 - \left(Z_{C1} + Z_{C2} + Z_{L1}\right)\left(Z_{C3} + Z_{C4} + Z_{L2}\right) \tag{14}$$

$$H = Z_{C1}\left[X_M^2 - \left(Z_{C2} + Z_{L1}\right)\left(Z_{C3} + Z_{L2}\right)\right] \tag{15}$$

$$N = Z_{C1}\left[X_M^2 - \left(Z_{C2} + Z_{L1}\right)\left(Z_{C3} + Z_{C4} + Z_{L2}\right)\right] \tag{16}$$

Once obtained I_L and I_S, the PDL (P_L) and the PTE (η) can be calculated by:

$$P_L = \frac{1}{2}\left|I_L\right| \cdot \Re\left\{Z_L\right\} \tag{17}$$

$$\eta = \frac{P_L}{\frac{1}{2} \cdot \Re\left\{V_S \cdot I_S^*\right\}} \tag{18}$$

where V_S is the amplitude of the input voltage source and I_S* is the complex conjugate of the input current.

DYNAMIC COMPENSATION BY VARIABLE FREQUENCY

The WPT system should operate in resonance for transferring energy from primary to secondary circuit efficiently. A frequency control methodology is a technique that can be used to tune the inductive link to the maximum output power capability for a given magnetic coupling coefficient *k,* whose value depends on, for instance, the misalignment between the coils. The control strategy discussed in this section monitors continuously the amplitude and phase of the input current at the primary side of the inductive link, and estimates the load voltage at the secondary side.

Misalignment Effects

Drifts in the capacitance values or in the inductance of the coils might detune the resonance condition of the circuit. In addition, the magnetic coupling coefficient *k* also modifies the input equivalent impedance of the circuit. Each time the coils are moved, a different magnetic coupling coefficient is presented, resulting in a variable input impedance. Moreover, environmental factors such as humidity and temperature variations might change the capacitors values. Both situations require fine-tuning the resonance point of the circuit.

Figure 6 shows the behavior of the output voltage (seen by the load) as a function of working frequency for some values of t *k*. One should notice that for each value of *k* there is a working frequency capable to tune the inductive link. Thus, the frequency compensation strategy can be applied as long as there are no constraints on the working frequency and misalignments are supposed to happen during the system operation.

Dynamic Frequency Tuning Strategy

In order to reach the resonance operation point, the frequency adjustment has been explored using a microprocessed system with wireless communication as presented in Brusamarello et al. (2013). Thus,

Figure 6. Dependence of the resonance frequency on the coupling factor k

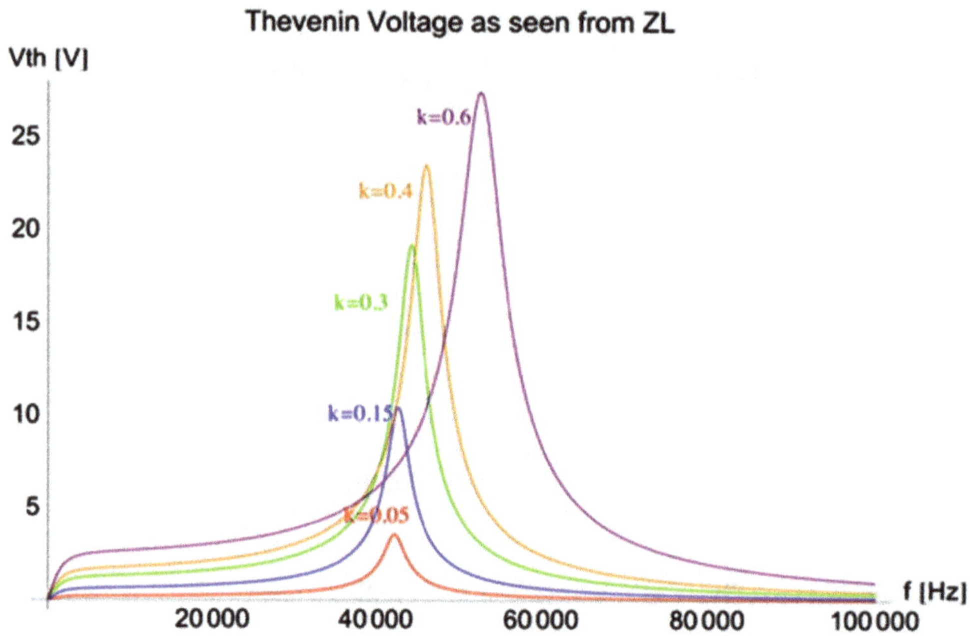

it is possible for a microcontroller connected at the receiver coil of an inductive link to send information about the circuit parameters to the transmitter coil, where another microprocessor system can receive such information and change the working frequency. Figure 7 shows the block diagram of the proposed variable frequency wireless charger.

The microcontroller (MCU) on the measuring device monitors the voltage on the load and communicates with the MCU, which is responsible for controlling and adjusting the frequency at the transmitter coil. The proposed algorithm is shown in the diagram of Figure 8. The MCU of the transmitter coil initializes the communication channel with the MCU at the receiver. Thus, the control system on the transmitter base starts a frequency scan synchronized with the measurements on the load. At each frequency point, a measurement of the PDL is performed. The algorithm scans for the resonance frequency and starts charging the battery.

Search for New Frequency

Several techniques can be used to determine the resonant frequency that maximizes the PDL. In Figure 9 is shown the search method for finding the function $f(k)$, which allow maximizing the PDL for each magnetic coupling coefficient k, by Newton's method. Initially the output power $P_L(\omega, k)$ of the inductive link is calculated.

In order to use the Newton's method, the first and the second derivatives of P_L with respect to the frequency ω are calculated. Finally, the initial values are assigned to the constants: k_0, k_{M+1}; f; M; and ε. The coupling factors k_0 and k_{M+1} are, respectively, the lower bound and the upper bound of k. The frequency f is the initially designed tuning frequency of the WPT system, which is 500 kHz in this example, being the initial value to search the resonance that maximizes the PDL for a given k. The constant M is the number of divisible parts of k and the constant ε is the stop error of the algorithm.

Figure 7. Block diagram of the variable frequency wireless charger
Source: Brusamarello et al., 2013

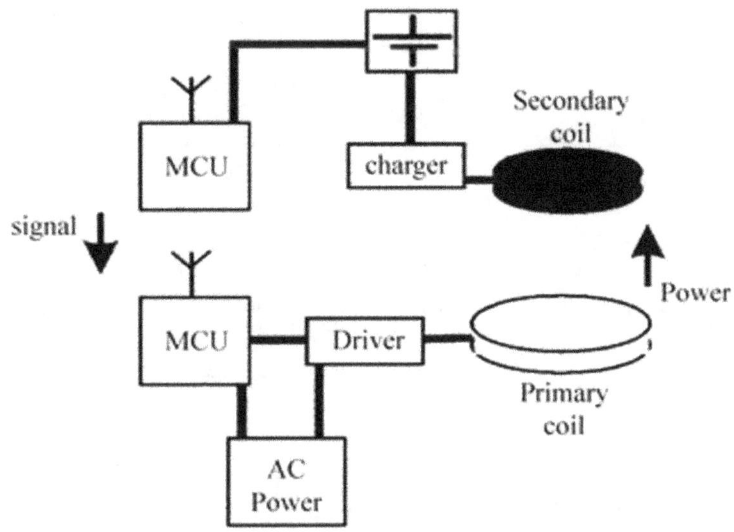

Figure 8. Block diagram of the charger with wireless communication
Source: Brusamarello et al., 2013

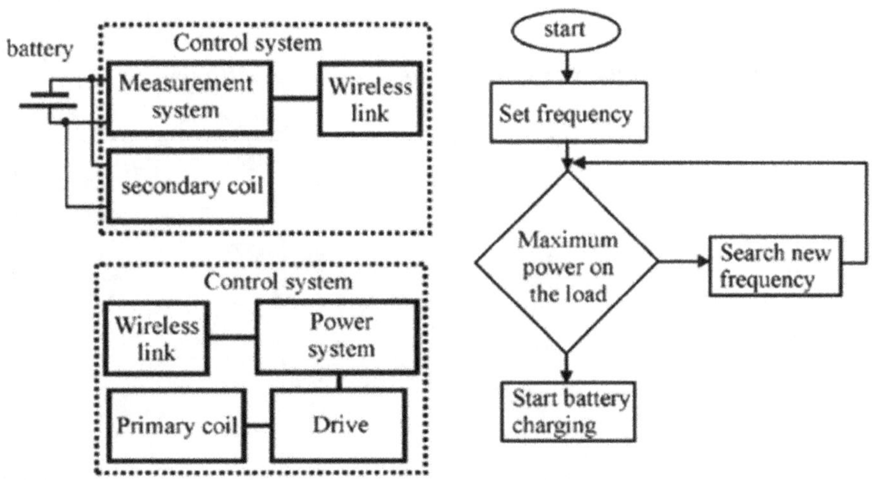

Frequency Splitting

One of the greatest difficulties of frequency tuning is the occurrence of the splitting frequency phenomenon (Hui, Zhong, & Lee, 2014). This phenomenon causes the division of the resonance peak into a pair of peaks, considerably reducing the $P_L(\omega)$ power in the charge, as illustrated on Figure 10. The output power is expressed as a function of the frequency for some magnetic coupling coefficients k.

Figure 10 shows that by increasing the magnetic coupling coefficient k (the transmitter and receiver coils are increasingly close together), it modifies the resonance point. For small values of k, the maxi-

Figure 9. Flowchart of f(k) which maximizes the power delivered to the load
Source: Brusamarello et al., 2013

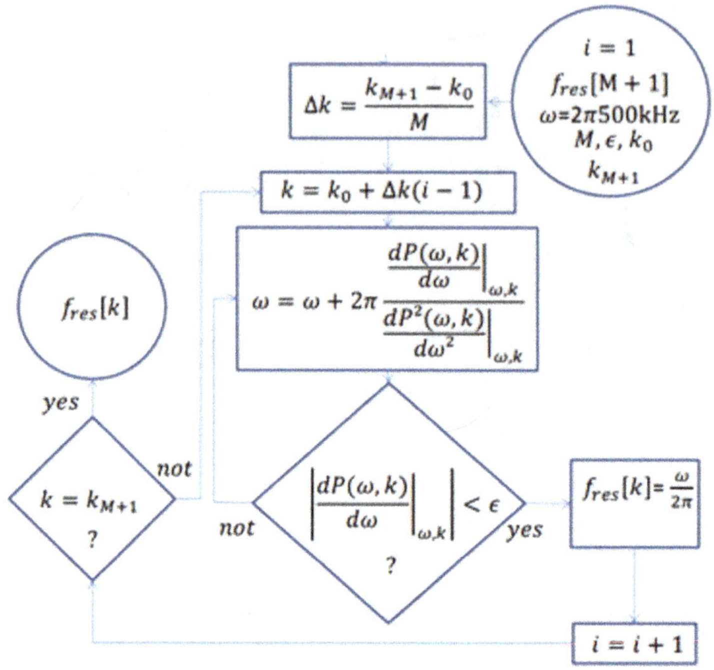

Figure 10. Frequency splitting with the variation of the coupling coefficient

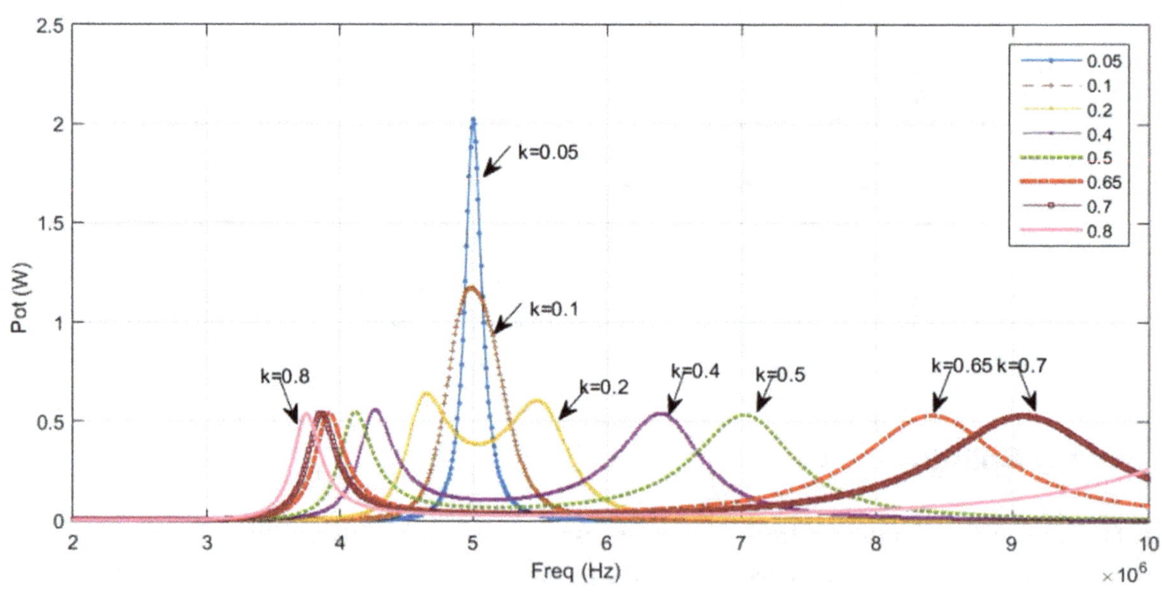

mum power occurs at the exact point of the designed resonance frequency (500 kHz). However, as the coupling is increased beyond $k > 0.1$, two points of local magnitudes for output power arise, which points gradually away from the designed resonance frequency. Furthermore, the maximum values of the PDL decay with the increase of k. This decay, which initially seems to be contradictory since increasing the magnetic coupling of the coils should increase the value of the maximum power, is in fact a direct consequence of the capacitive compensation. In values of k where the system is not design to work on, one can observe a decay of the PDL caused by the combined reactance of capacitors and inductors.

DYNAMIC COMPENSATION BY VARIABLE CAPACITANCE

Many WPT optimized designs have been presented in the literature, considering static coils positioning and constant load. In the last years, the dynamic optimization of inductive links has been a concern. When the transmitting and receiving coils are misaligned or even in relative movement, the transferred energy is highly affected and the output power decays sharply. In such situations, the inductive link can dynamically adapt itself through a variable matching network. In this section, a voltage-controlled capacitance is presented as a key role to a variable matching network implementation.

Dynamic Tuning Strategy

It is well known that the maximum output power occurs when the circuit reaches a resonance point, which can be detected by measuring the input impedance of the inductive link. Figure 11 shows a strategy for measuring the input impedance and to control the capacitance C_2, which is the variable element in the matching network at the primary side of the link (Porto et al., 2017b).

Variable Capacitance Circuit Design

A convenient way to implement a variable capacitance is through the capacitance multiplier or the Negative Impedance Converter (NIC) concepts (Franco, 2001). The total input equivalent capacitance can be altered by adjusting the voltage gain of the amplifier A in Figure 12.

A first-order approximation of the open-loop transfer function of the amplifier $A = f(V_{ctrl})$ is given by:

$$A = \frac{A_0}{1 + j\omega\tau} \tag{19}$$

where A_0 is the DC voltage gain and $1/\tau$ is the dominant pole of the amplifier. If one considers infinite input impedance of the amplifier A, then the closed-loop input impedance of the circuit of Figure 12 is

$$Z_{in} = \frac{1}{j\omega C_0} \cdot \frac{1 + j\omega\tau}{\left(1 - A_0\right) + j\omega\tau} \tag{20}$$

where C_0 is the reference capacitor. The pole influence can be neglected if the working frequency is much lower than the cut off frequency. Therefore, Equation 20 can be rewritten as:

Figure 11. Dynamic tuning strategy by means of a variable capacitance at the primary side of the inductive link

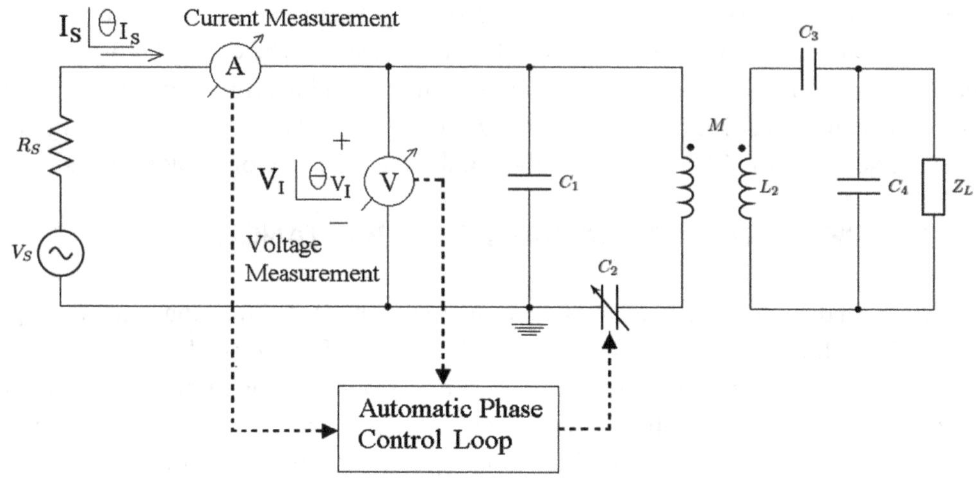

Figure 12. Voltage-Controlled capacitor topology

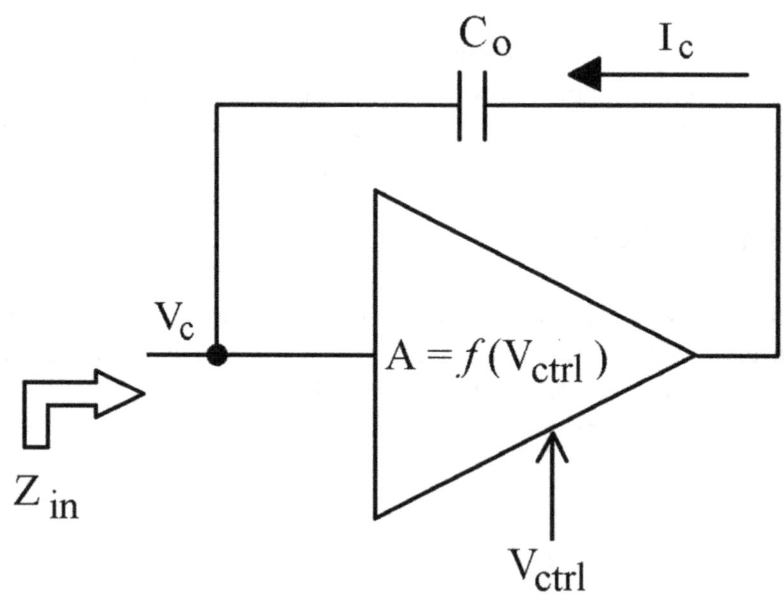

$$Z_{in} = \frac{1}{j\omega C_0 \left(1 - A_0\right)} \tag{21}$$

Thus, the model of Figure 12 can be seen as a capacitive load from its input terminals, whose value is calculated by $C_{adj} = C_0 \cdot (A - A_0)$. The voltage-controlled capacitor can be implemented accordingly to the block diagram of Figure 13. There are many ways to implement a Variable Gain Amplifier (VGA). One of them uses a 4-quadrant analog multiplier (Franco, 2001). A voltage divider (R_A and R_B) can be used to keep the input voltage amplitude inside the limits for safe operation of the analog multiplier. In addition, a phase correction block might be necessary in order to keep the voltage gain $A_0 = V_o/V_c$ a real number. Therefore, no phase shift should be accepted between the voltages V_o and V_c. The ideal transfer function of the voltage-controlled capacitor implementation is expressed by

$$A_0 = \frac{R_B}{R_A + R_B} G_0 V_{ctrl} \tag{22}$$

where G_0 is the combined gain of the multiplier, phase correction and output stage blocks.

The proposed circuit to implement the block diagram of Figure 13 is shown in Figure 14. The analog multiplier MPY634 has been used due to its dual supply voltage and bandwidth enough to operate with working frequencies in excess of 500 kHz. The operational amplifier AD817 implements an all-pass filter with unity gain and central frequency of 500 kHz. This stage adjusts the phase shift between the signals V_c and V_o. The output stage uses a high-speed power operational amplifier TI3584.

Phase Control Circuit Design

The dynamic tuning strategy of the inductive link is completed with the automatic phase control block (Porto et al., 2017b). The phase shift comparison between the signals V_I and I_S is important to detect the maximum power transfer capability. An output control voltage (V_{ctrl}) is produced in order to change the value of C_{adj} (and consequently the capacitor C_2) and to keep the phase shift as close to zero as possible. Figure 15 represents a block diagram for the automatic phase control. A simple Proportional-Integral (PI) controller should be included in the signal chain in order to minimize the steady-state phase error.

The phase detection circuit is show in Figure 16, which is based on the Integrated Circuit (IC) AD8302. The phase difference φ between the signals V_A (sample of voltage V_I) and V_B (sample of current I_S) is processed by the IC AD8302 and an output DC voltage V_{phs} is produced at the rate +10 mV/°. The input current of the inductive link (I_S) is measured by a shunt resistor R_{sh} and converted to a voltage by means

Figure 13. Block diagram of the voltage-controlled capacitor

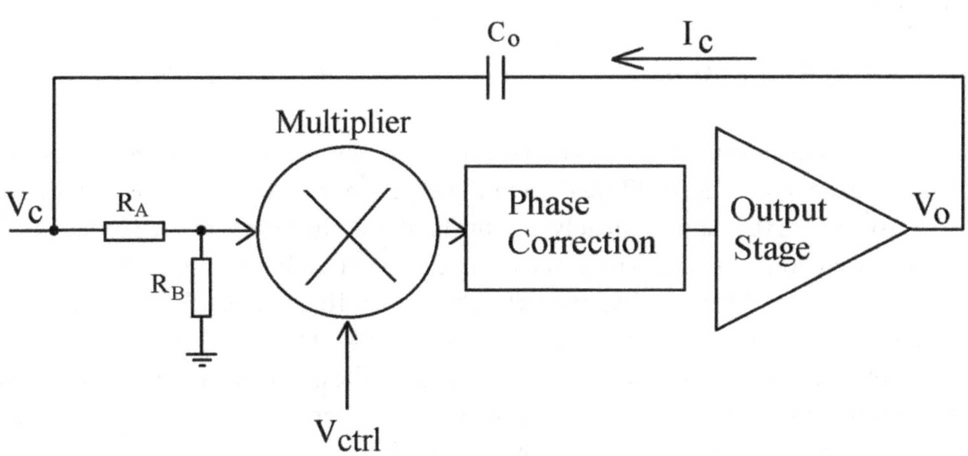

Figure 14. Voltage-controlled capacitor circuit implementation
Source: Porto et al., 2017b

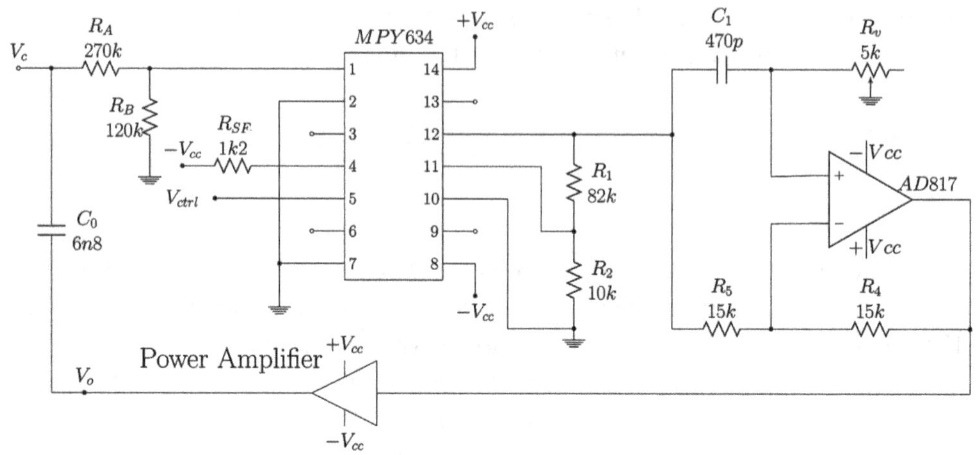

Figure 15. Automatic phase control loop

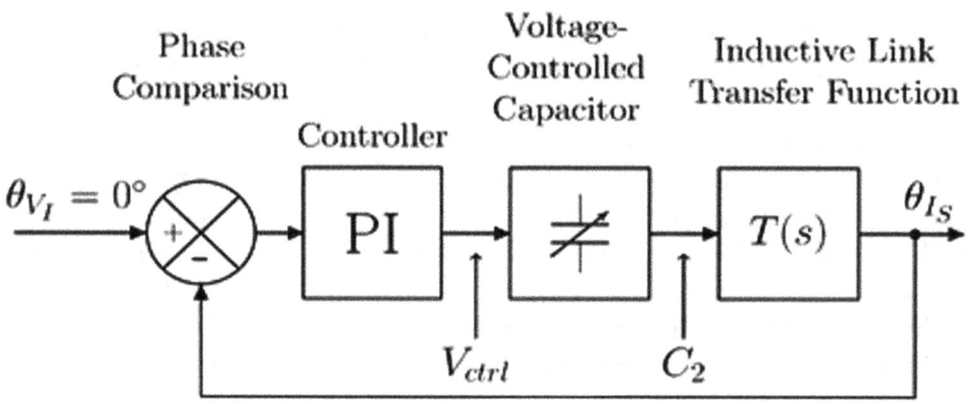

of a differential amplifier. The voltage-controlled capacitor presented in Figure 14 is connected to the node V_c in Figure 16.

Automatic Tuning Evaluation

The automatic tuning capability of the inductive link has been evaluated by measuring the voltage amplitude (V_L) at a fixed load of $R_L = 47\ \Omega$. During the experiments, the magnetic coupling coefficient has been varied from 0.2 to 0.8, approximately. Figure 17 shows the comparative results of V_L when the inductive link has been compensated with a fixed matching network and using the proposed voltage-controlled capacitor as a variable matching network. There is a difference of 2 V, approximately, between the simulation and the experiments. One of the causes is the ESR of the compensation capacitors, which were neglected in the simulation model. Furthermore, there are the nonidealities of the voltage-controlled capacitor. Depending on the value of V_{ctrl}, the input equivalent series resistance, R_{in}, can produce very different values. Such behavior has been also neglected in the simulations.

Figure 16. Circuit implementation for phase detection and the input impedance of the inductive link
Source: Porto et al., 2017b

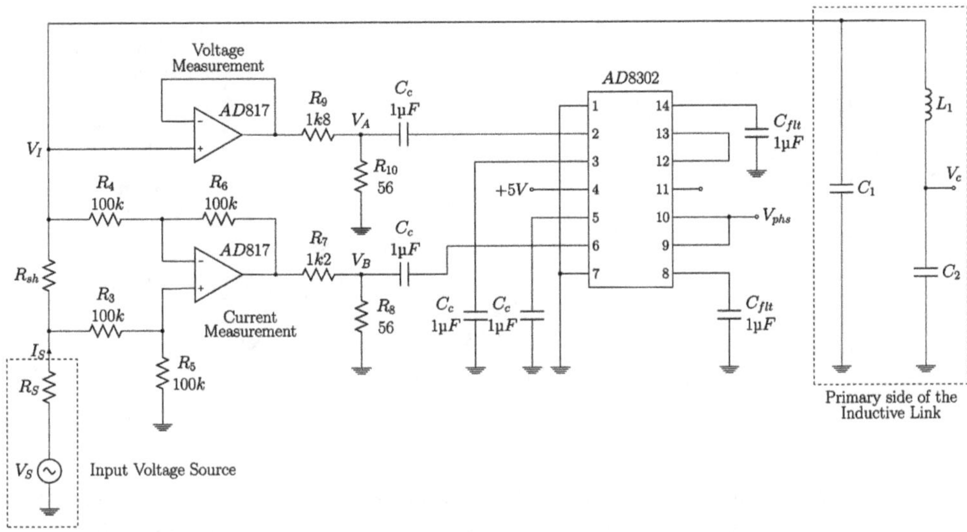

Figure 17. Comparative results of output voltage V_L as a function of the magnetic coupling coefficient k
Source: Porto et al., 2017b

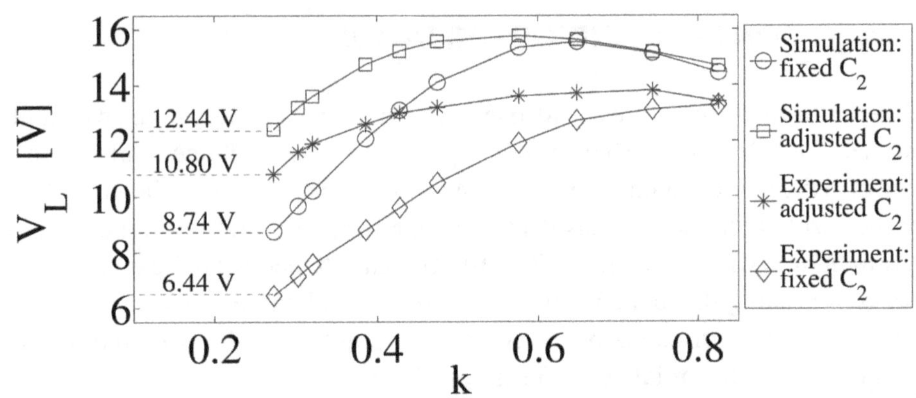

Although there are differences between the simulation and experimental results, the voltage-controlled capacitance connected at the primary side of the inductive link can be advantageous in terms of output voltage regulation. By comparing a fixed matching network with the presented design, the output voltage can be 67% higher when the voltage-controlled capacitor is applied at the worst case of magnetic coupling.

Figure 18 shows the comparative results of the phase shift φ in the simulation and experiments. The inductive link has been initially designed for $k = 0.65$. Therefore, the simulation and experimental results tend to be close at that operation point. However, φ should be zero at that point. Even in the simulation results for fixed C_2, the phase shift is 5° approximately. Such error is mainly because the set of optimal solutions of the matching capacitors is not the same of the commercial standardized values. When k varies the input impedance can be capacitive or inductive. That means there is no phase control actuating in the circuit. In contrast, for adjustable C_2 case, there is an input impedance's phase control.

Figure 18. Simulation and experimental results of phase shift for fixed and adjustable C_2 cases
Source: Porto et al., 2017b

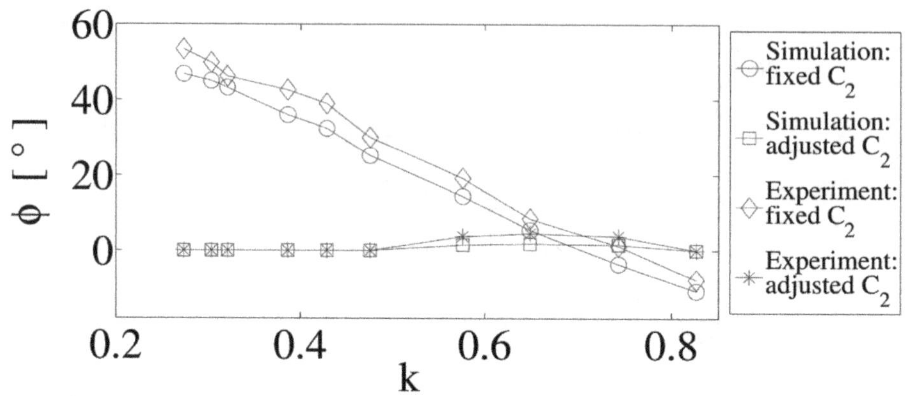

The automatic phase control was also evaluated in a dynamic way through the experimental set up shown in Figure 19. The distance between the coils was varied between 3 mm and 9 mm periodically. When the coils are being separated, V_{ctrl} increases to compensate the phase shift that also is increasing. Figure 20 shows the dynamic of the signals V_{ctrl} and V_{phs} when the coils are in relative movement.

OUTPUT POWER CONTROL AND OPTIMIZATION

It has been well reported in the literature and also in this chapter that the inductive link should work tuned in order to optimize the PDL, or the PTE (or a composition of both). Some works have used variable frequency to perform the inductive link tuning as long as k varies (Brusamarello et al., 2013; Aldhaher, Luk, & Whidborne, 2014). Other works consider fixed excitation frequency but variable capacitive matching networks (Lim et al., 2014; Porto et al., 2017b). The simultaneous application of both techniques can be investigated for tuning the inductive link under dynamical environments.

The PDL analysis can be carried out on the compensated inductive link with four capacitors, as previously shown in Figure 5. The initial design uses $C_1 = 30$ pF, $C_2 = 2.4$ nF, $C_3 = 4.7$ nF, $C_4 = 30$ pF, $R_L = 47\ \Omega$, $k = 0.3$ and $R_S = 1\ \Omega$. The excitation frequency ω is used to complement the performance of the variable capacitor method. Figure 21 shows the comparative simulation results of the PDL when the frequency ω and capacitance C_2 are used to tune the circuit. By using only C_2 as a variable element, the PDL cannot remain constant for $k > 0.3$. That similar behavior also happens by using only ω for tuning the inductive link. However, by tuning the circuit with both C_2 and ω the PDL remains constant for $k > 0.3$.

Multivariable Compensation for Inductive Links

The inductive link can be designed by following a design procedure usually based on constraints from the application (Brusamarello et al., 2013; Porto et al., 2015; Ramrakhyani, Mirabbasi, & Chiao, 2011; Leung et al., 2013). The dimensions of the coils are highly dependent on the application constraints (Jow & Ghovanloo, 2007). In addition, the frequency range and the PDL also affect the dimensions of the coils (Bosshard et al., 2013). Table I summarizes the design of an inductive link, which uses a pair

Figure 19. Experimental set up used to characterize the voltage-controlled capacitor connected to an inductive link
Source: Porto et al., 2017b

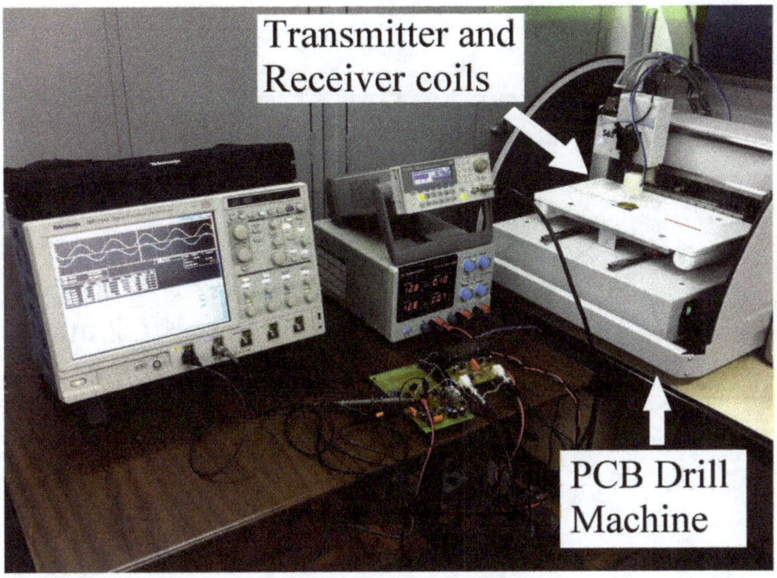

Figure 20. Adapted figure from the Tektronix TDS1001C oscilloscope's screen. V_{phs} and V_{ctrl} waveforms when k is varying
Source: Porto et al., 2017b

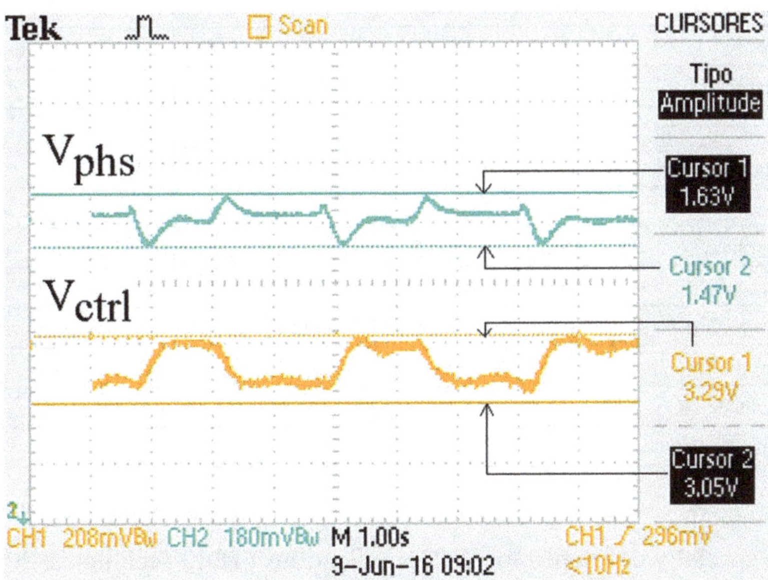

Figure 21. Comparative simulation results of PDL (P_o) as function of magnetic coupling coefficient k for three tuning methods
Source: Murliky, Porto, & Brusamarello, 2017

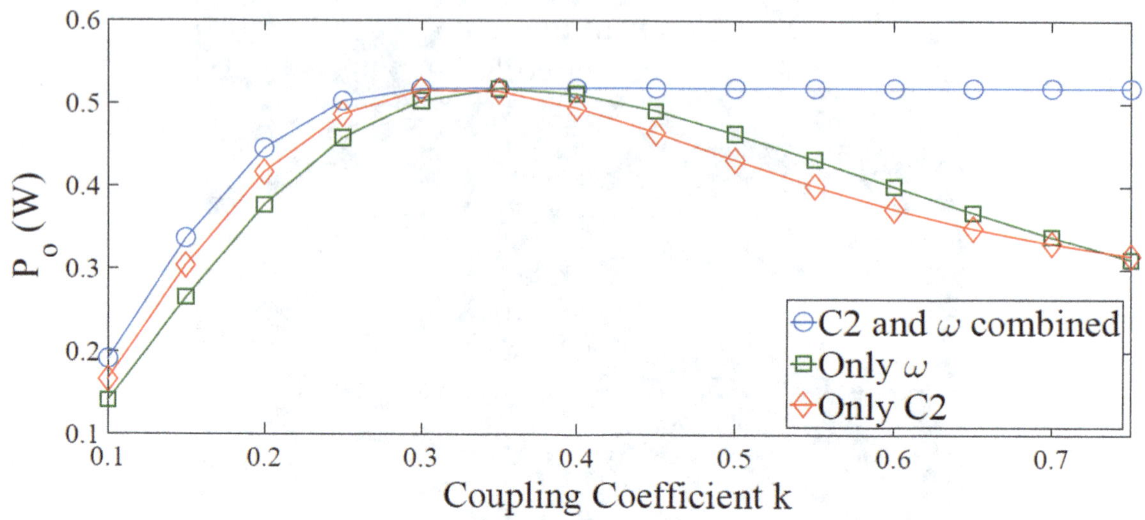

Table 1. Design parameters of an inductive link

Parameter Description	Symbol	Value
Transmitting Coil Inductance	L_1	24 µH
Transmitting Coil DC Resistance	R_1	0.1 Ω
Receiving Coil Inductance	L_2	13 µH
Receiving Coil DC Resistance	R_2	0.2 Ω
Load Resistance	R_L	47 Ω
Magnetic coupling coefficient	k	0.1
Working Frequency	f	500 kHz
Primary side matching capacitor	C_1	1.24 nF
Primary side matching capacitor	C_2	4.26 nF
Secondary side matching capacitor	C_3	18.4 nF
Secondary side matching capacitor	C_4	11.5 nF
Input Voltage Source Amplitude	V_S	12 V
Input Voltage Source Output Resistance	R_S	1 Ω

of commercial coils specially developed for WPT applications (TDK Tech notes, 2017). The matching capacitors are chosen in order to maximize the PDL for $k = 0.1$ according to the procedure described in Azambuja et al. (2014).

Figure 22 shows the simulated frequency response of the inductive link for some discrete values of k using the respectively optimized values of C_2. The optimum values of C_2, f and the respectively maximum PDL for each given k are shown in Table 2.

Figure 22. Parametric curves for PDL (P_O), phase shift φ and amplitude $|I_S|$ of the input current as a function of f for some discrete values of k
Source: Porto et al., 2017b

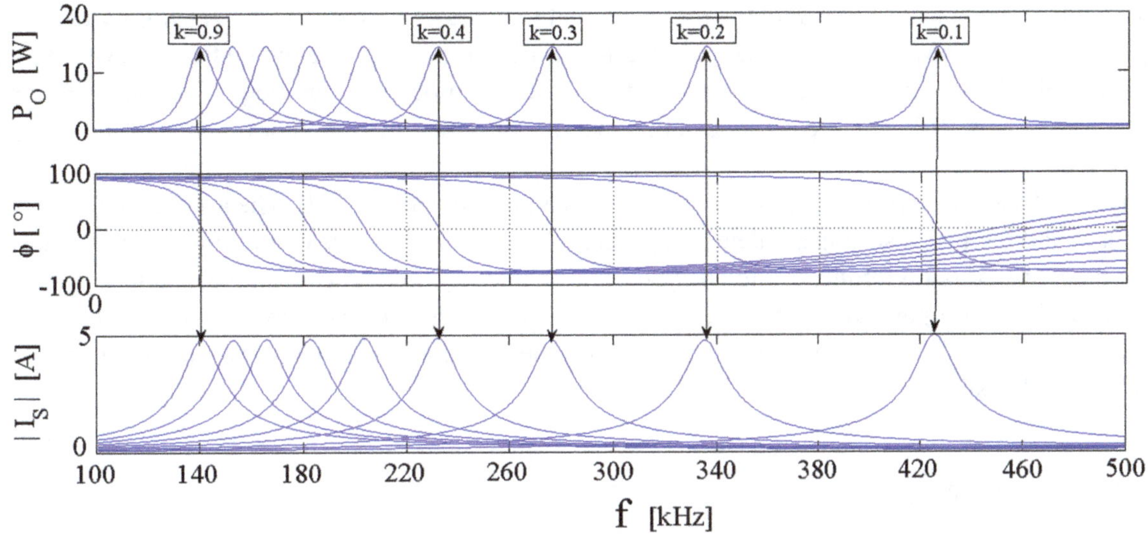

Table 2. Optimal Values of C_2, f and the Respective P_O for some Discrete Values of k. [A] Initial Design Parameters

k	C_2 [nF]	f [kHz]	Po [W]
0.1[A]	4.26[A]	500[A]	12.61[A]
0.1	5.76	428.5	13.97
0.2	9.13	333.5	14.09
0.3	13.30	273.0	14.16
0.4	18.66	233.5	14.20
0.5	24.01	203.5	14.20
0.6	29.57	181.7	14.23
0.7	35.52	165.0	14.24
0.8	41.27	151.7	14.25
0.9	48.22	141.0	14.26

One can notice that for each curve of P_O the maximum PDL point can be detected when $|I_S| \approx 5$ A and $\varphi \approx 0$ °. In general, the maximum PDL point is reached (or at least very close to) when $\varphi(\omega) = 0°$, $\nabla |I_S| = 0$ and $|I_S|_{\max} = K$, where K is a constant that depends on the PDL desired level.

The values shown in Table 2 can be used on a curve fitting procedure in order to predict the relationship between C_2 and f for optimizing the PDL at any k within the interval of 0.1 – 0.9. The fitted curve to a power law shown in Figure 23 is given by:

$$C_2 = 250.9f^{-1.887} - 2.267 \cdot 10^{-10} \tag{23}$$

A practical air-core WPT system with k of 0.9 or higher is hard to reach. Thus, the part of the fitted curve for $k > 0.9$ is just to illustrate the mathematical tendency. For $k < 0.1$, the sensibility of C_2 with respect to f is very low. In practice, the optimization of PDL is solely made by the frequency for loosely coupled coils. Thus, for any k between 0.1 and 0.9 a search procedure can be performed on Equation 23 for maximizing PDL (the procedure can be repeated for optimizing another objective function, such as efficiency).

Figure 24 shows the proposed hardware to control and optimize the PDL. The amplitude of the input current I_S and the phase shift φ are detected by the IC AD8302, which is commonly used for gain and phase detection of radio frequency signals (Yee & Chakrabarty, 2007). The analog output V_{mag} is a DC voltage, which is proportional to the rate of the voltage and current measurement channels. Furthermore, the analog DC output voltage V_{phs} depends on the phase angle φ. These two analog outputs can be respectively estimated by:

$$V_{mag} = 0.9 + 20\log\left(\frac{\alpha V_I}{\beta I_S}\right) \tag{24}$$

$$V_{phs} = 1.8 - 10 \cdot 10^{-3}\left|\varphi\right| \tag{25}$$

where α is a constant representing a voltage divider, which is necessary to keep the voltage at the input terminals of the IC AD8302 within the allowed range (<0.6 V_{pp}), and β is the transimpedance gain of the current sensing measurement.

Figure 23. Fitted curve from data of Table 2

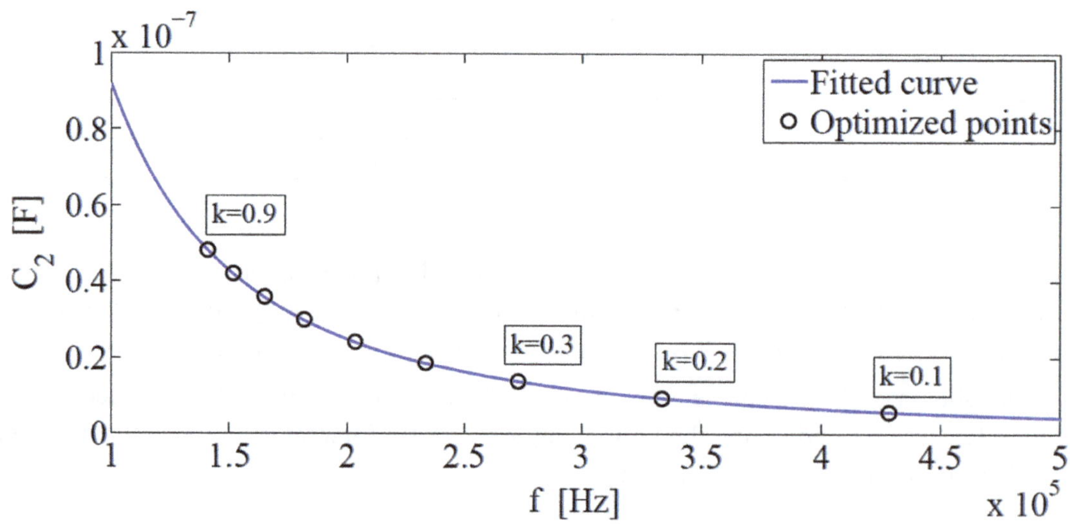

Figure 24. Proposed hardware to control and optimize the PDL

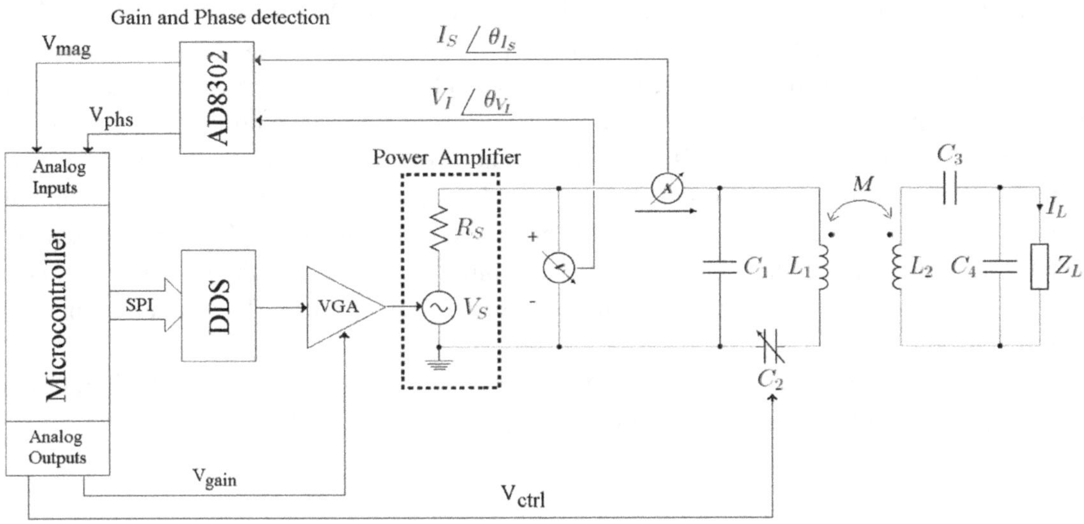

The microcontroller reads and process the input analog signals V_{mag} and V_{phs}. The control outputs are generated according to a proposed algorithm, which searches a pair of values (C_2 ; *f*) given by Equation 23 saved in a vector. The step search is given by fuzzy logic, which is implemented for reaching the maximum PDL conditions indicated in Figure 23. Thus, the maximum PDL is ideally achieved when V_{mag} = 0.9 V and V_{phs} = 1.8 V. The analog output V_{ctrl} controls the equivalent capacitance through a similar scheme described in Porto et al., 2017b. The operating frequency is properly adjusted by the Direct Digital Synthesis (DDS) IC AD9834, which is controlled by a serial communication port (SPI). An additional analog output V_{gain} is provided to adjust the output power level, but it has no influence on the maximum PDL search procedure.

Fuzzy Logic Controlling

Fuzzy logic controllers can be an advantageous option when the system to be controlled is nonlinear or some of its parameters are not well defined (Mendel, 1995). The PDL maximization problem in WPT systems is clearly nonlinear. In addition, the magnetic coupling coefficient *k* is not precisely known and some parameters of the equivalent circuit present uncertainties in their values, such as the capacitance and the equivalent series resistance of such matching capacitors. A fuzzy logic controller has been proposed in order to select the step size variation of a parallel tuning capacitor in the secondary side of an inductive link. The sampled load voltage is used to compare with a reference voltage (Hsu, Hu, & Swain, 2012). Despite the advantages reported by the authors, such control scheme might not be suitable in situations where the receiver circuit have space constraints, as in biomedical implants for instance. Similar control scheme has been presented in Wang et al. (2015), where the fuzzy logic controller optimization is performed by using genetic algorithm. The proposed WPT system samples the load voltage and feeds it to the fuzzy logic controller, which generates the control signals for the inverter at the primary side of the link. The power transfer is regulated by using constant frequency phase-shift pulse width modulation.

In order to implement a fuzzy logic controller fully at the primary side, as shown in Figure 24, a response of the inductive link when excited with the condition stated in Equation 23 is shown in Figure 25. However, just the excitation frequency is shown in the horizontal axis. Although the response has been obtained for $k = 0.3$, the pattern depicted remains valid for $k > 0.1$. The effect of changing k in that response is just to shift the pattern in the frequency axis. Thus, one can roughly divide such response in regions according to the proximity of the optimum point (region A). From Figure 26, for frequencies less than the optimum point (maximum PDL) the signal V_{phs} can indicate how much close the inductive link is to that point (E and F regions). On the other hand, the V_{mag} remains stable at that frequency range (region B). For frequencies higher than the optimum point, the signal V_{phs} cannot inform where is the maximum PDL (region G). However, the signal V_{mag} can be used to localize the optimum point and how far is that point (regions C and D).

Instead of directly processing these signals, the error of V_{mag} and V_{phs} are more suitable for a closed loop control system implementation. Thus, the error signals are defined by:

$$\varepsilon_{mag} = 0.9 - V_{mag} \tag{26}$$

$$\varepsilon_{phs} = 1.8 - V_{phs} \tag{27}$$

Figure 25. Response of the inductive link for k = 0.3 when tuned with both C_2 and f
Source: Porto et al., 2017b

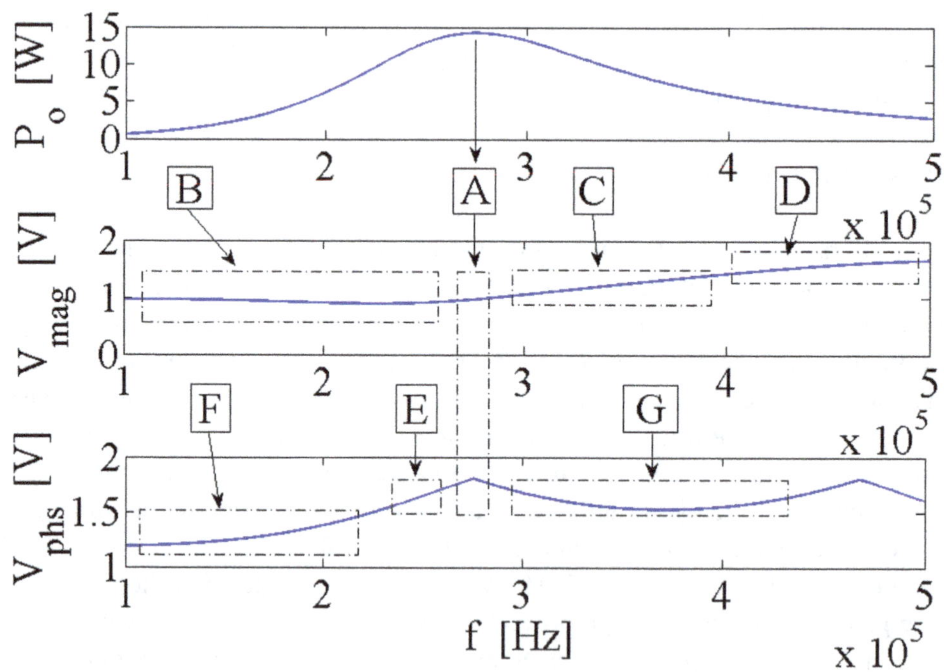

The membership functions is a way to segment the error signal in the presence of uncertainties in the equivalent circuit model. By using such functions, one can evaluate the error functions (26) and (27), calculate the degree of membership of each membership function and detect whether the maximum PDL has been reached or not. The regions roughly depicted in Figure 25 are represented in the form of membership functions in Figure 26.

The membership functions of the inputs ε_{mag} and ε_{phs} are combined in the logic function AND. The output signal of the fuzzy logic block controls a step search Δ, which can be either positive or negative. This step is used to search the optimum point (maximum PDL) in the excitation vector (C_2 ; *f*), whose values are given by Equation 23. The more values are included in that vector, the more resolution it gets and closer to the maximum PDL point it reaches. In the presented implementation it has been used a excitation vector of 800 elements of *f* ranging from 100 kHz to 500 kHz with 500 Hz of resolution (Porto et al., 2017c).

Table 3 shows the rules summary for the inference process in the fuzzy logic block. The step search size (Δ) has divided into the following fuzzy output sets: *Z* - Zero; *S* - Small; *N* - Normal; and *B* - Big. The maximum Δ is bounded in 10% of the size of the excitation vector and the signal of each output fuzzy set is properly chosen in order to minimize the error functions and to stabilize the PDL into the region A.

Figure 26. Membership functions proposed for ε_{mag} and ε_{phs}
Source: Porto et al., 2017c

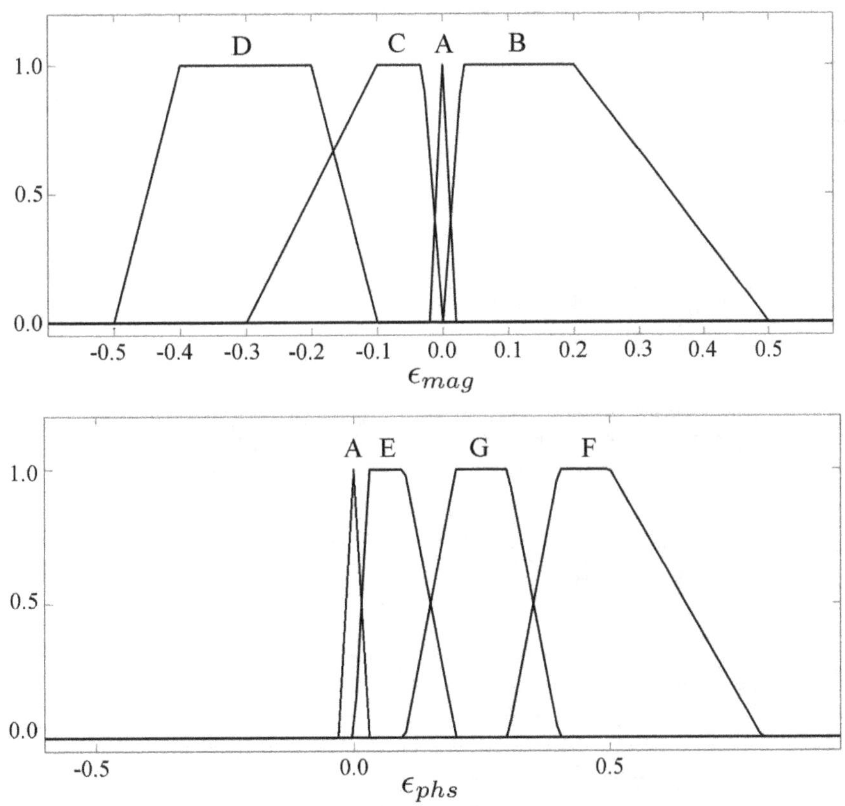

Table 3. Fuzzy matrix and rules for inference process

		ε_{phs}			
		A	**E**	**G**	**F**
ε_{mag}	**B**	*Z*	*S(+)*	*N(+)*	*B(+)*
	A	*Z*	*S(+)*	*N(+)*	*B(+)*
	C	*S(-)*	*N(-)*	*N(-)*	*N(-)*
	D	*B(-)*	*B(-)*	*B(-)*	*B(-)*

The flowchart in Figure 27 summarizes the proposed methodology. The inductive link is powered with the initial excitation parameters $f(0) = 100$ kHz and $C_2(0) = 91.92$ nF (first position in the excitation vector). As the magnetic coupling coefficient k is not precisely known, the acquired variables V_{mag} and V_{phs} are probably far from the setpoint (0.9 V and 1.8 V, respectively). Then, the error functions are calculated according to Equations 26 and 27. The next step is to process the errors ε_{mag} and ε_{phs} into the fuzzy logic block, and then the output is provided in a form of a step search Δ. Finally, an updated value of frequency and capacitance are calculated according to the step provided by the fuzzy logic controller. The feedback loop tends to minimize the error leading the PDL at his maximum operating point.

The effect of the magnetic coupling coefficient k on the PDL and the response of the fuzzy logic controller have been evaluated by means of the experimental setup previously shown in Figure 19, where the coils have been positioned parallel each other and attached on a holder. By varying the distance

Figure 27. Flowchart of the PDL control strategy
Source: Porto et al., 2017c

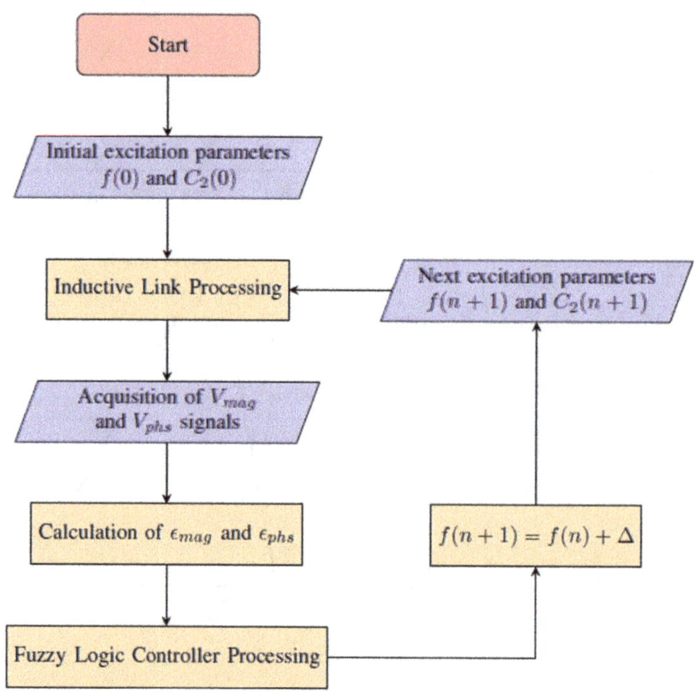

between the coils, and consequently k, the output power (PDL) was measured and the performance was compared with single variable compensation methods and fixed matching network. Figure 28 shows the comparative results of the PDL.

When using only C_2 as a variable element with a fixed frequency of 500 kHz, the PDL cannot remains constant for $k > 0.1$. That similar behavior also happens when using only f for tuning the inductive link with a fixed C_2 of 4.26 nF. However, when both C_2 and f are used for tuning the inductive link, the PDL remains approximately constant for $k > 0.1$. The response of the fuzzy logic control for $k = 0.3$ is shown in Figure 29. One should notice the saturated amplitude steps Δ at the first 6 samples. After that, the maximum PDL is fastly reached and Δ tends to zero, as expected. In case of higher values of k the PDL response reaches the maximum value faster.

In case of $k = 0.1$ the proposed fuzzy logic controller needs more samples to stabilize in the maximum PDL, as shown in Figure 30. Between the samples 10 and 30 approximately, the combination of the errors ε_{mag} and ε_{phs} falls in a region similar to the maximum PDL point and the step Δ remains zero. For $k < 0.1$ the fuzzy logic controlling cannot maintain constant the PDL as predicted in the simulation results shown in Figure 28.

CONCLUSION

This chapter has presented some methodologies and techniques for tracking the maximum output power transfer. The performance of the variable matching network is clearly superior to the other methods that use fixed circuit parameters. As previously shown in Figure 22, the use of only one variable leads a unique value of k that optimizes the PDL. By using both C_2 and f, it is possible to maximize the PDL value over a large range of k. The analysis of fuzzy logic controlling in a WPT system has proved to

Figure 28. Comparative results of PDL. The points highlighted are experimental data for some evaluated values of k
Source: Porto et al., 2017c

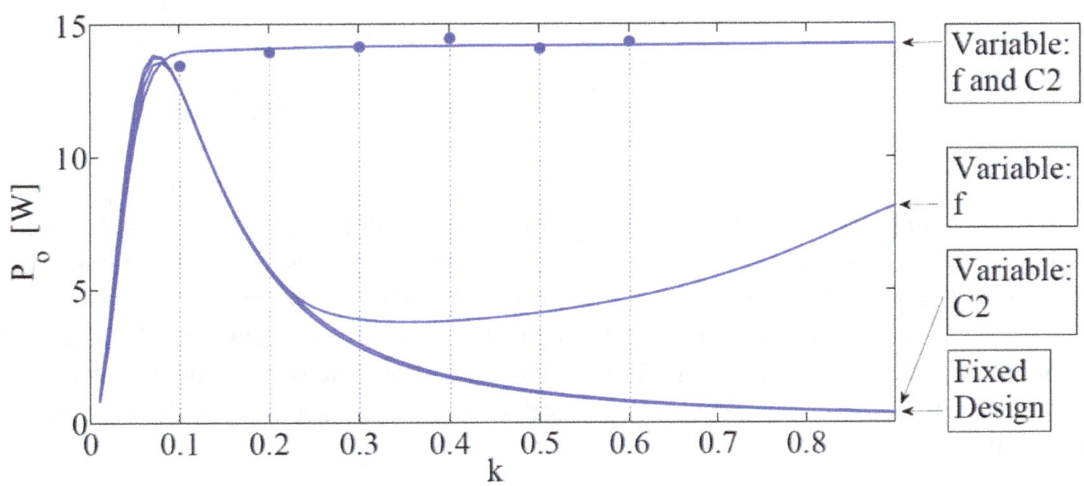

Figure 29. PDL Sample results of the fuzzy logic control for k = 0.3
Source: Porto et al., 2017c

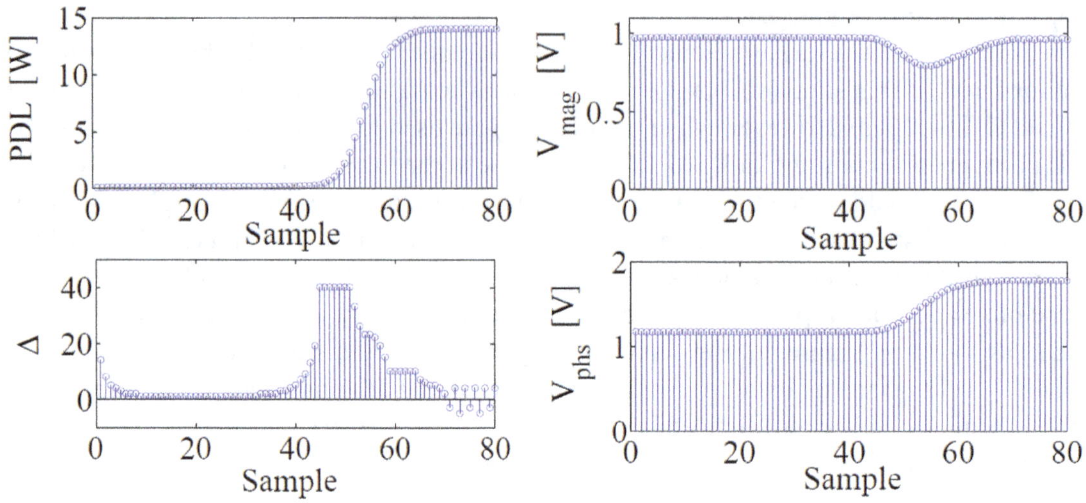

Figure 30. PDL Sample results of the fuzzy logic control for k = 0.1
Source: Porto et al., 2017c

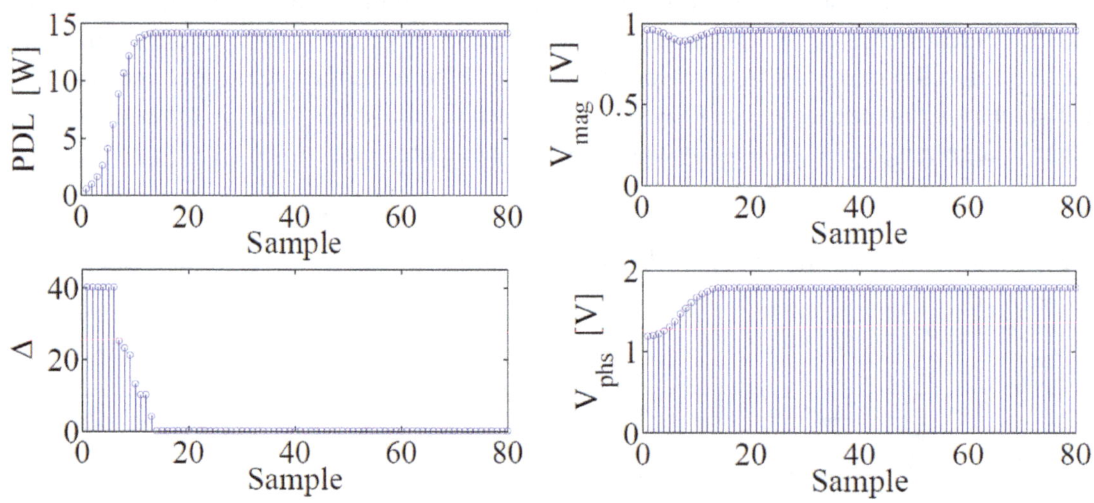

be efficient in keeping constant the PDL when k varies. The membership functions play an important role in that process, because the measured values of amplitude and phase of the input current cannot precisely indicate how much power is actually delivered to the load. That uncertainty occurs because the value of k is unknown. Many relative positions between the coils can outcome the same value of k. The role of membership functions is to discriminate operating regions based on a knowledge priori taken from the equivalent circuit. That operating regions can spread and overlap themselves when C_2 or f are varied with low resolution.

The proposed techniques presented in this chapter is not fully suitable for high power WPT applications. The voltage-controlled capacitor shown in Figure 14 is not capable to produce voltage and current amplitude at high power resonance condition, such as in electric vehicles charging. Furthermore, variable frequency methods may violate the allowed bandwidth descripted in the standards. However, the presented multivariable optimization method can be useful for biomedical implants, powering wireless sensor nodes, and consumer electronics. Therefore, it offers more degrees of freedom in positioning the remote devices for powering and or charging purposes.

REFERENCES

Ahn, D., & Hong, S. (2015). Wireless Power Transfer Resonance Coupling Amplification by Load Modulation Switching Controller. *IEEE Transactions on Industrial Electronics*, *62*(2), 898–908. doi:10.1109/TIE.2014.2336627

Aldhaher, S., Luk, P. C. K., & Whidborne, J. F. (2014). Electronic Tuning of Misaligned Coils in Wireless Power Transfer Systems. *IEEE Transactions on Power Electronics*, *29*(11), 5975–5982. doi:10.1109/TPEL.2014.2297993

Azambuja, R., Brusamarello, V. J., Haffner, S., & Porto, R. W. (2014). Analysis and Optimization of an Inductive Power Transfer with a Randomized Method. *IEEE Transactions on Instrumentation and Measurement*, *63*(5), 1145–1152. doi:10.1109/TIM.2013.2296397

Bosshard, R., Muhlethaler, J., Kolar, J. W., & Stevanovic, I. (2013). Optimized magnetic design for inductive power transfer coils. In *Proceedings of the 28th Applied Power Electronics Conference and Exposition (APEC 2013)*. Long Beach, CA: IEEE. 10.1109/APEC.2013.6520541

Boys, J. T., Covic, G. A., & Green, A. W. (2000). Stability and Control of Inductively Coupled Power Transfer Systems. *IEE Proceedings. Electric Power Applications*, *147*(1), 37–43. doi:10.1049/ip-epa:20000017

Boys, J. T., Hu, A. P., & Covic, G. A. (2000). Critical Q Analysis of a current-fed resonant converter for ICPT applications. *Electronics Letters*, *36*(17), 1440–1442. doi:10.1049/el:20001030

Brown, W. C. (1957). Description and Operating Characteristics of the Platinotron – A New Microwave Tube Device. *Proceedings of the IEEE*, *45*(9), 1209–1222.

Brown, W. C. (1974). The Technology and Application of Free-Space Power Transmission by Microwave Beam. *Proceedings of the IEEE*, *62*(1), 11–25. doi:10.1109/PROC.1974.9380

Brown, W. C. (1984). The History of Power Transmission by Radio Waves. *IEEE Transactions on Microwave Theory and Techniques*, *MTT-32*(9), 1230–1242. doi:10.1109/TMTT.1984.1132833

Brusamarello, V. J., Blauth, Y. B., Azambuja, R., Müller, I., & Sousa, F. R. (2013). Power Transfer with an Inductive Link and Wireless Tuning. *IEEE Transactions on Instrumentation and Measurement*, *62*(5), 924–931. doi:10.1109/TIM.2013.2245041

Choi, B., Nho, J., Cha, H., Ahn, T., & Choi, S. (2004). Design and Implementation of Low-Profile Contacless Battery Charger Using Planar Printed Circuit Board Windings as Energy Transfer Device. *IEEE Transactions on Industrial Electronics*, *51*(1), 140–147. doi:10.1109/TIE.2003.822039

Choi, W. P., Ho, W. C., Liu, X., & Hui, S. Y. R. (2010). Comparative Study on Power Conversion Methods for Wireless Battery Charging Platform. In *Proceedings of 14th International Power Electronics and Motion Control Conference (EPE-PEMC 2010)*. IEEE. 10.1109/EPEPEMC.2010.5606675

Degenford, J. E., Sirkis, M. D., & Steier, W. H. (1964). The Reflecting Beam Waveguide. *IEEE Transactions on Microwave Theory and Techniques*, *12*(4), 445–453. doi:10.1109/TMTT.1964.1125845

Dieckerhoff, S., Ryan, M. J., & Doncker, R. W. D. (1999). Design of an IGBT-Based LCL-Resonant Inverter for High-Frequency Induction Heating. In *Conference Record of the 1999 IEEE Industry Applications Conference. Thirty-Forth IAS Annual Meeting (Cat. No.99CH36370)*. IEEE. 10.1109/IAS.1999.806017

Elliot, G. A. J., Covic, G. A., Kacprzak, D., & Boys, J. T. (2006). A New Concept: Asymmetrical pick-ups for inductively coupled power transfer monorail systems. *IEEE Transactions on Magnetics*, *42*(10), 3389–3391. doi:10.1109/TMAG.2006.879619

Franco, S. (2001). *Design with Operational Amplifiers and Analog Integrated Circuits*. New York: McGraw-Hill.

Goubau, G., & Schwering, F. (1961). On the Guided Propagation of Electromagnetic Wave Beams. *IRE Transactions on Antennas and Propagation*, *9*(3), 248–256. doi:10.1109/TAP.1961.1144999

Green, A. W., & Boys, J. T. (1994). *10 kHz Inductively Coupled Power Transfer Concept and Control*. London: Institution of Electrical Engineers.

Hochmair, E. S. (1984). System Optimization for Improved Accuracy in Transcutaneous Signal and Power Transmission. *IEEE Transactions on Biomedical Engineering*, *BME-31*(2), 177–186. doi:10.1109/TBME.1984.325404 PMID:6546736

Hsu, J.-U. W., Hu, A. P., & Swain, A. (2012). Fuzzy logic-based directional full-range tuning control of wireless power pickups. *IET Power Electronics*, *5*(6), 773–781. doi:10.1049/iet-pel.2011.0364

Hui, S. Y. R., & Ho, W. W. C. (2005). A New Generation of Universal Contacless Battery Charging Platform for Portable Consumer Electronic Equipment. *IEEE Transactions on Power Electronics*, *20*(3), 620–627. doi:10.1109/TPEL.2005.846550

Hui, S. Y. R., Zhong, W., & Lee, C. K. (2014). A Critical Review of Recent Progress in Mid-Range Wireless Power Transfer. *IEEE Transactions on Power Electronics*, *29*(9), 4500–4511. doi:10.1109/TPEL.2013.2249670

Hurley, W. G., & Kassakian, J. G. (1979). Induction Heating of Circular Ferromagnetic Plates. *IEEE Transactions on Magnetics*, *MAG-15*(3), 1174–1181. doi:10.1109/TMAG.1979.1145874

Jang, Y., & Jovanovic, M. M. (2003). A Contactless Electrical Energy Transmission System for Portable Telephone Battery Chargers. *IEEE Transactions on Industrial Electronics*, *50*(3), 520–527. doi:10.1109/TIE.2003.812472

Jow, U. M., & Ghovanloo, M. (2007). Design and Optimization of Printed Spiral Coils for Efficient Transcutaneous Inductive Power Transmission. *IEEE Transactions on Biomedical Circuits and Systems*, *1*(3), 193–202. doi:10.1109/TBCAS.2007.913130 PMID:23852413

Kim, C. G., Seo, D. H., You, J. S., Park, J. H., & Cho, B. H. (2001). Design of a Contacless Battery Charger for Cellular Phone. *IEEE Transactions on Industrial Electronics*, *48*(6), 1238–1247. doi:10.1109/41.969404

Kissin, M. L. G., Boys, J. T., & Covic, G. A. (2009). Interphase Mutual Inductance in Polyphase Inductive Power Transfer Systems. *IEEE Transactions on Industrial Electronics*, *56*(7), 2393–2400. doi:10.1109/TIE.2009.2020076

Ko, W. H., Liang, S. P., & Fung, C. D. F. (1977). Design of Radio-Frequency Powered Coils for Implant Instruments. *Medical & Biological Engineering & Computing*, *15*(6), 634–640. doi:10.1007/BF02457921 PMID:203785

Kurs, A., Karalis, A., Moffatt, R., Joannopoulos, J. D., Fisher, P., & Soljacic, M. (2007). Wireless Power Transfer via Strongly Coupled Magnetic Resonances. *Science*, *317*(5834), 83–86. doi:10.1126cience.1143254 PMID:17556549

Leung, H. Y., McCornick, D., Budgett, D., & Hu, A. P. (2013). Design methodology for inductive power transfer systems targeting high power implantable devices. In *Proceedings of the IEEE International Symposium on Circuits and Systems (ISCAS 2013)*. Beijing: IEEE. 10.1109/ISCAS.2013.6572457

Lim, Y., Tang, H., Lim, S., & Park, J. (2014). An Adaptive Impedance Matching Network Based on a Novel Capacitor Matrix for Wireless Power Transfer. *IEEE Transactions on Power Electronics*, *29*(8), 4403–4413. doi:10.1109/TPEL.2013.2292596

Liu, X., & Hui, S. Y. R. (2007a). Equivalent Circuit Modeling of a Multilayer Planar Winding Array Structure for use in a Universal Contacless Battery Charging Platform. *IEEE Transactions on Power Electronics*, *22*(1), 21–29. doi:10.1109/TPEL.2006.886655

Liu, X., & Hui, S. Y. R. (2007b). Simulation Study and Experimental Verification of a Universal Contactless Battery Charging Platform with Localized Charging Features. *IEEE Transactions on Power Electronics*, *22*(6), 2202–2210. doi:10.1109/TPEL.2007.909301

Ludwig, R., & Bretchko, P. (2000). *RF Circuit Design – Theory and Applications*. Prentice Hall.

Mendel, J. M. (1995). Fuzzy logic systems for engineering: A tutorial. *Proceedings of the IEEE*, *83*(3), 345–377. doi:10.1109/5.364485

Murliky, L., Porto, R. W., & Brusamarello, V. J. (2017). Multivariable Optimization Method for Inductive Power Transfer in Wireless Sensors Nodes. In *Proceedings of the IEEE International Instrumentation and Measurement Technology Conference (I2MTC 2017)*. Torino: IEEE. 10.1109/I2MTC.2017.7969690

Pantic, Z., Lee, K., & Lukic, S. M. (2014). Multifrequency Inductive Power Transfer. *IEEE Transactions on Power Electronics*, *29*(11), 5995–6005. doi:10.1109/TPEL.2014.2298213

Porto, R. W., Brusamarello, V. J., Müller, I., Riano, F. L. C., & Sousa, F. R. (2017a). Wireless power transfer for contactless instrumentation and measurement. *IEEE Instrumentation & Measurement Magazine*, *20*(4), 49–54. doi:10.1109/MIM.2017.8006394

Porto, R. W., Brusamarello, V. J., Müller, I., & Sousa, F. R. (2015). Design and characterization of a power transfer inductive link for wireless sensor network nodes. In *Proceedings of the IEEE International Instrumentation and Measurement Technology Conference (I2MTC 2015)*. Pisa: IEEE. 10.1109/I2MTC.2015.7151454

Porto, R. W., Brusamarello, V. J., Pereira, L. A., & Sousa, F. R. (2017b). Fine tuning of an inductive link through a voltage-controlled capacitance. *IEEE Transactions on Power Electronics*, *32*(5), 4115–4124. doi:10.1109/TPEL.2016.2598284

Porto, R. W., Murliky, L., Oliveira, G., & Brusamarello, V. J. (2017c). Multivariable Automatic Compensation Method for Inductive Power Transfer System. In *Proceedings of the 2017 IEEE International Conference on Environment and Electrical Engineering and 2017 IEEE Industrial and Commercial Power Systems Europe (EEEIC / I&CPSEurope)*. Milan: IEEE. 10.1109/EEEIC.2017.7977659

Ramrakhyani, A. K., Mirabbasi, S., & Chiao, M. (2011). Design and optimization of resonance-based efficient wireless power delivery systems for biomedical implants. *IEEE Transactions on Biomedical Circuits and Systems*, *5*(1), 48–63. doi:10.1109/TBCAS.2010.2072782 PMID:23850978

Sample, A. P., Meyer, D. T., & Smith, J. R. (2011). Analysis, Experimental Results, and Range Adaptation of Magnetically Coupled Resonators for Wireless Power Transfer. *IEEE Transactions on Industrial Electronics*, *58*(2), 544–554. doi:10.1109/TIE.2010.2046002

TDK Tech Notes. (2017). *Wireless Power Transfer Products*. Retrieved from: https://product.tdk.com/info/en/products/wirelesscharge/ technote/tpo/index.html

Tesla, N. (1914). *Apparatus for Transmitting Electrical Energy*. US patent n. 1119732.

Treffers, M. (2015). History, Current Status and Future of the Wireless Power Consortium and the Qi Interface Specification. *IEEE Circuits and Systems Magazine*, *15*(2), 28–31. doi:10.1109/MCAS.2015.2418973

Wageningen, D., & Staring, T. (2010). The Qi Wireless Power Standard. In *Proceedings of the International Power Electronics and Motion Control Conference (EPE-PEMC 2010)*. IEEE. 10.1109/EPEPEMC.2010.5606673

Wang, C., Stielau, O. H., & Covic, G. A. (2005). Design Considerations for a Contacless Electric Vehicle Battery Charger. *IEEE Transactions on Industrial Electronics*, *52*(5), 1308–1314. doi:10.1109/TIE.2005.855672

Wang, W. C., Tai, C. C., Wu, S. J., & Liu, Z. Y. (2015). A hybrid genetic algorithm with fuzzy logic controller for wireless power transmission system of electric vehicles. In *Proceedings of 2015 International Conference on Industrial Technology*. IEEE. 10.2991/itms-15.2015.255

Wireless Power Consortium (WPC). (2016a). *The Qi Wireless Power Transfer System Power Class 0 Specification Parts 1 and 2: interface definitions*. Version 1.2.2. Wireless Power Consortium. Retrieved from: https://www.wirelesspowerconsortium.com/

Wireless Power Consortium (WPC). (2016b). *The Qi Wireless Power Transfer System Power Class 0 Specification Part 4: reference designs*. Version 1.2.2. Wireless Power Consortium. Retrieved from: https://www.wirelesspowerconsortium.com/

Yee, Y. F., & Chakrabarty, C. K. (2007). Phase detection using ad8302 evaluation board in the superheterodyne microwave interferometer for line average plasma electron density measurements. *Measurement*, *40*(9), 849–853. doi:10.1016/j.measurement.2007.02.005

Chapter 5
Control for the Contactless Series Resonant Energy Converter

Stanimir Valtchev
Universidade Nova de Lisboa, Portugal

Svilen S. Valtchev
University of Lisbon, Portugal & Polytechnic Institute of Leiria, Portugal

ABSTRACT

This chapter presents control methods applied in the operation of the series loaded series resonant (SLSR) power converters in a most efficient operation zone. The choice of the control method is affected by the objective to guarantee suitably the efficiency, being this method in the same time, relatively easy to apply. The first part of the chapter compares three basic principles of regulation: frequency mode (FM), pulse width mode (PWM), and their combination (PWM/FM). Finally, a new method for instantaneous regulation is developed. The proposed technique consists of a simplified observation of a state variable value to limit each portion of supplied energy, depending on the requirement for power in each half period. The result of this regulation is comparable to the current mode (CM) control applied to the hard-switching power converters. The viability of this new regulation method is demonstrated by simulations of its analogue circuit implementations and experimentally proved. The circuit is also prepared for the changes in the magnetic coupling (contactless energy transfer).

1. INTRODUCTION

This chapter presents control methods, applied in the effort to maintain the operation of the Series Loaded Series Resonant (SLSR) power converters in their most efficient operation mode, including wireless energy transmission. This is not an exhaustive study of all possible techniques of SLSR converter control but rather a demonstration of the possibility to achieve high efficiency operation, following the principles and ideas suggested in previous publications, e.g. Valtchev and Klaassens (1990). The choice of the control method is affected by the objective to guarantee a suitably efficient contactless conversion of energy, being this method in the same time, relatively easy to apply.

DOI: 10.4018/978-1-5225-5870-5.ch005

The present chapter starts with a comparison of three basic ways of regulation for SLSR converters: frequency mode (FM), pulse width mode (PWM), and their combination (PWM/FM). Finally, a separate, new method for instantaneous regulation of the energy delivered to the resonant circuit, is presented. The proposed technique is based on a direct measurement of the instantaneous value of a state variable. Based on simple calculations, the switching is commanded, to regulate each one portion of supplied energy, sent through the resonant circuit. In this way, the necessary energy is decided and sent in each individual half period. The result of this regulation may be compared to the effect produced by the current mode (CM) control applied to the hard-switching power converters. The viability of this new regulation method is demonstrated by simulations of scaled in time analogue circuit implementations, and by measured results. The circuit is provided with another feedback that permits the changes in the magnetic coupling to be compensated for. This allows the operation of the contactless power converter to be commanded correctly and the power switches to be protected accordingly.

2. THE CONTROL OF THE SLSR AND CONTACTLESS CONVERTERS

Several control methods are known to have been developed during the last 30-35 years and many of them are already considered classical, implemented in factory-made resonant converters, e.g. Hewlett-Packard (1982). The problem of the resonant converter physics is that the energy is flowing from magnetic to electric and vice versa, so the instantaneous control is extremely difficult. It is not possible to cut off the process in the right moment. The difficulty is even greater if contactless energy is to be transferred through a loosely coupled magnetic transmitter/receiver, in which case the reaction should be fastest, to avoid damaging the power switching devices. In the effort to maintain the operation of the SLSR converter stable and efficient and especially, to keep the contactless converter in the best efficiency operation region, the modes of operation were considered, and some graphical methods are already in use, e.g. (Jovanovic, Liu, Oruganti, & Lee, 1987). The safe behaviour of the SLSR converter, especially the transmitting part, is very important for the mass inclusion of the wireless energy in the electric vehicle charging technology (Valtchev, Jorge, Craciunescu, & Brandisky, 2012) and some initial ideas are discussed in (Medeiros, Valtchev, & Valtchev, 2012).

2.1. The Regulation Methods and the Difficulties Up to Now

The above mentioned graphical, state-plane method looks reasonable, but its implementation rises some concern. There are methods that include calculation of normalized phase-plane trajectory as in Nguen and Dhyanchand (1987), Nguyen, Dyanchand, and Thollot (1986), and Rossetto (1992). Those calculations are complicated and need multiple points of real-time measurements. In the works of Jovanovic et al. (1987), Souesme et al. (1989), Kim et al. (1991), etc. many measurements are also required as an input for the calculation of the instantaneous conditions (phase-plane graphics), to recalculate the moments of necessary commutations. Since the beginning of their work on the SLSR power converter, the intention of the authors was to find a simple and reliable solution. The beginning is maybe in the simplification of the TLO problem (Klesser, & Klaassens, 1986), a heavy problem from the thyristor times and still existing in the super-resonant converter when the resonant could provoke a back-discharge (Valtchev, 1990). In more recent publications, e.g. (Amghar, Darcherif, Barbot, & Boukhetala, 2014), the modelling of the converter is complex. The calculations of Lyshevski (2000) are even more complex, involving large matrixes, and the results are not obvious. The experimental data looks odd and not convincing even at the claimed frequency of 41 kHz. The complexity in the newer article by Jang et al. (2015), is not lower and the simplification began with the master thesis work Wahjudi (1992) in TU Delft, when S. Valtchev worked there as a university lecturer. The modelling was simplified, and the formulas became applicable in fast decisions. The work presented at a local conference ICEST showed how to make a fast and simplified almost instant control to the resonant converters (Valtchev, Miletiev, Arnaoudov, & Valtchev, 2009).

Gratitude must be expressed to all other authors that worked on the solution of this resonant converter control problem, but this chapter will be mostly dedicated to the proper solution, considered here as a new one. It is worth mentioning that some publications already embrace this method as a useful tool for other resonant power converters (Dias, Kim, & Jang, 2011).

The method is described here in the way it was developed initially: from the frequency mode (FM), to pulse width mode (PWM), and their combination (FM/PWM). Finally, the instantaneous method is described which applies other combined FM/PWM control, based on the calculated individual energy portions delivered to the resonant circuit.

3. THE SWITCHING FREQUENCY (FM) REGULATION

The switching frequency is the first method of control, already in use for a long time and implemented in factory made resonant converters, e.g. (HP, 1982). The problem here is in the change of the frequency in a very rapid manner. Every rapidly applied frequency change leads the operation to a non-symmetrical waveform, thus creating more problems.

3.1. FM Regulation Applied to SLSR Converters and to Contactless Converters

The application of a frequency mode (FM) regulation is the simplest way to guarantee the zero-voltage switching (ZVS) of the power switches. An idealized schematics diagram of the end stage is shown in Figure 1. The operation is generically executed as switching on of the devices in diagonal: M_1 with M_4 and M_2 with M_3, described in Valtchev and Klaassens (1990).

As it is stated in Valtchev and Klaassens (1990), the ZVS is guaranteed for switching frequency higher than the resonance (super-resonant operation). It was pointed out in the graphical results of the same work that FM regulation is applicable only when the relation between the required maximum and minimum output power is not large (maximum variation of the output power is inversely proportional to the switching frequency variation, i.e. 8-10 times). Usually the required relation maximum/minimum is much larger and to avoid the necessity of very high switching frequency for achieving a very low output power in FM regulation, one possible solution is to provide some constantly existing minimum of output consumption. This solution is not possible for high power converters as it reduces the originally excellent efficiency of the FM operation. To avoid this drawback, the simple frequency controller is later described as a combination with another power regulation (i.e. limited PWM).

As it is stated in Valtchev and Klaassens (1990), the best efficiency operation of the SLSR converter requires a relatively low switching frequency (only 10-20% higher than the resonant frequency). The closeness to the resonance is a hard to solve problem of stability of FM regulation because of the inherently instable resonant frequency. The changes in the load, temperature, magnetic coupling, etc.,

Figure 1. Series loaded resonant converter power stage with loosely coupled transformer

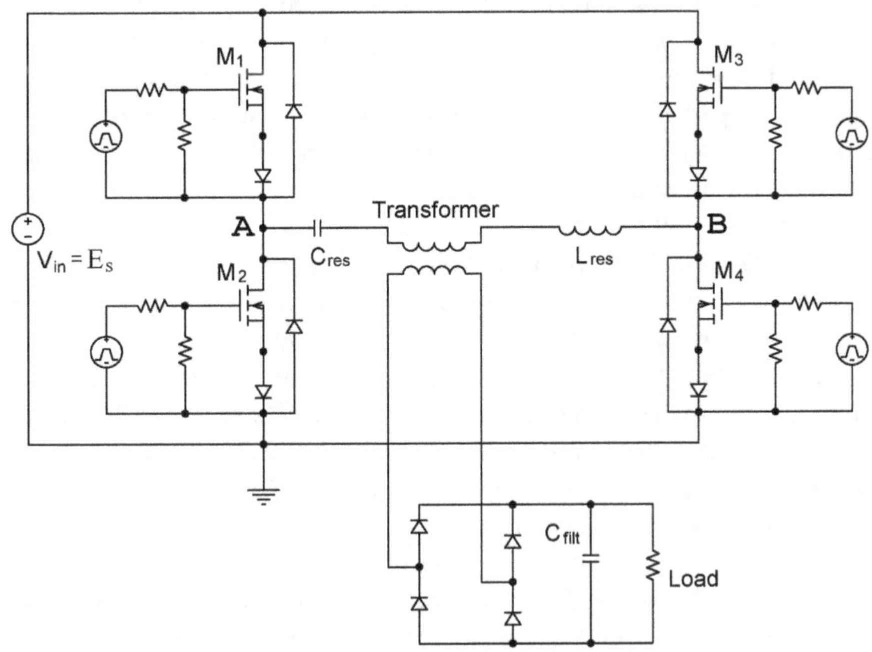

influence strongly and rapidly the resonant frequency. It is impossible for the switching frequency oscillator to react immediately and correctly to those changes. As a result, the unstable resonant frequency puts at risk the ZVS operation mode, making possible the switching to become sub-resonant, destroying the power switches by short circuit and high di/dt value. The fast semiconductor switches operating in the super-resonant mode are not prepared for limiting the di/dt value and should not be obliged to operate in the sub-resonant operation mode, thus the super-resonant mode of operation should be guaranteed always.

In the process of contactless power transfer, the variation of the magnetic coupling causes an even larger variation of the resonant frequency, compared to the classical SLSR converter case. For this reason, the turn-on pulse of the power transistor, must be additionally guaranteed to be applied before its own (antiparallel) diode finishes its conduction (Valtchev & Klaassens, 1990). The frequency-regulated (FM) operation guarantees the zero-voltage switching, but a safe change of the switching frequency is difficult. A type of compulsory super-resonant switching operation is implemented by a PLL circuitry (shown later) that guarantees the ZVS even in the close neighbourhood of the resonance. However, the reaction of the PLL circuit is slow because of the necessity of a low pass filter in the loop.

3.2. Modified FM Regulation

The FM regulation of the output power was discussed in Valtchev and Klaassens (1990). Its current limiting characteristics guarantee the safe operation in case of the short circuit at the output. As it is shown in the same article, the "current source" characteristics are better defined at a higher switching frequency, i.e. far from the resonant frequency. The necessity to achieve a higher output power however, obliges to operate at switching frequency closer to the resonance, e.g. only 1.1 or even 1.05 times higher than f_{res}. In this case the precision of the switching frequency generator and its fast adjustment becomes very important. The situation is more difficult when the contactless converter is considered as its resonant frequency changes. Even if the change is slight, they require a highly precise readjustment of the switching frequency oscillator, to avoid entering in sub-resonant mode of operation (Valtchev & Klaassens, 1990). To avoid the sub-resonant mode of operation it is compulsory to have the conduction of the anti-parallel diode still not finished, when the transistor is to be turned on, i.e. to guarantee Zero-Voltage Switching (ZVS). The implementation of this kind of control would require a sensor of the rear slope of the resonant current or a detector of its zero crossing. A straightforward way to control the converter in super-resonant mode can be the PLL loop presented in the block diagram of Figure 2.

The circuit in Figure 2 is based on a closed PLL which maintains the required phase shift between the zero-crossing of the rear slope of the resonant current and the end of the control pulse for the corresponding diagonal of switches. The end of the control pulse shown in Figure 3 (marked as "Transistor") must be kept before that zero-crossing. The block of Figure 2, named Zero-Crossing Detector (ZCD) is detecting the instant when the resonant current waveform is crossing the zero. This zero crossing marks the end of the half-period of the resonant current pulse. This point is designated by x_o, shown also in Figure 3.

The output of the block VCO (Figure 2) produces a train of pulses, each ending at the instant x_k (Figure 3). The switching frequency f_{sw} produced by the VCO will ensure the required phase shift between x_k and x_o. The corresponding transistor receives the VCO signal as a command pulse applied to its gate (marked in Figure 3). This pulse finishes at x_k. Identical pulses (finishing at x_k) are applied to both transistors in the first diagonal, T_1 and T_4 (Figure 2). The opposite diagonal (T_2, T_3) receives the next pulse, starting from x_k and maintained during the other half period of switching. The pulse length will be kept the same, $\psi_1+\psi_2$, if no necessity for change occurs, i.e. a "steady-state" operation.

The minimum phase shift between the rear slope zero crossing of the current (x_o) and the end of the transistor conduction interval (x_k) corresponds to a minimum length of the diode conduction interval ψ_r (identified in Figure 3 as ψ_2). This minimum phase shift corresponds to a switching frequency close to the resonant one. If a zero-phase difference (ψ_2=0) would be approached, it will be when the switching frequency is equal to the resonant frequency. This lowest switching frequency should be avoided as being dangerously close to the short-circuit (sub-resonant) operation (Valtchev & Klaassens, 1990).

This danger is prevented by applying the voltage V_{add} produced in the Comparator block presented in Figure 2. The voltage V_{add} must have a certain guaranteed minimum value to produce the necessary minimum phase-shift. This phase shift will guarantee the ZVS: the transistor is switched on, always

Figure 2. Simplified PLL loop for the straightforward control circuit

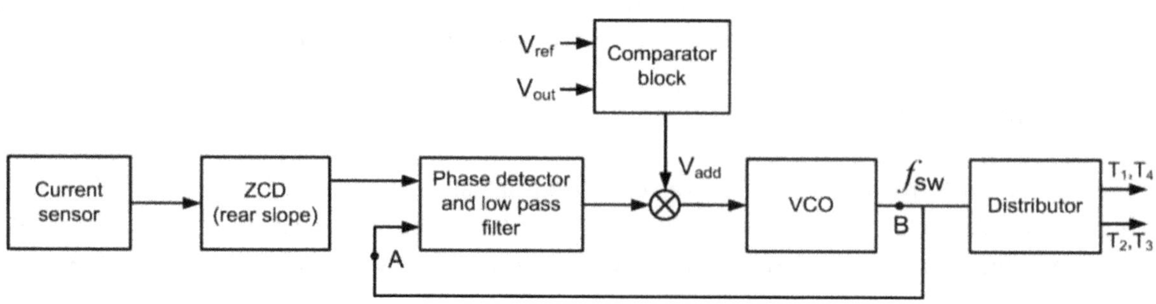

Figure 3. Resonant current i_r intervals: $-\psi_2$, ψ_1, ψ_2 and the power switch command

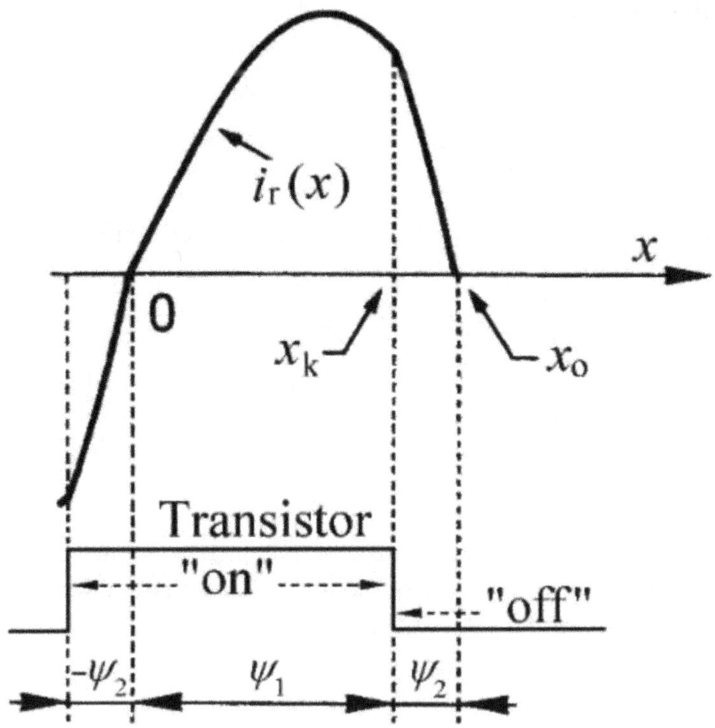

before its own antiparallel diode cuts off its current. The normalized time (angular value in radians) between the turn-off of the conducting transistor, and the zero-crossing of the resonant current is the value marked as ψ_2. It is the value ψ_2 that is the difference between the resonant half-period and the switching frequency half-period. The regulation of the output is accomplished by the Comparator block through further increase of the voltage V_{add}, thus increasing further the phase-shift ψ_2 (over the minimum ψ_2 necessary for a ZVS operation). The increased phase-shift ψ_2 will be inversely proportional to the output power, which means producing a current source or voltage source behaviour, depending on the operational point. Here the variables are considered normalized (dimensionless) for a simpler calculation, in a steady-state operation. The output voltage V_o is given as a relation to the supply voltage, i.e. $q = V_o/V_{in}$, and the regulation function for the Frequency Mode (FM) operation $q = f(I_o,\psi_2)$ is calculated from the equation (36) of Valtchev and Klaassens (1990) that connects the maximum value of the resonant capacitor voltage, the switching period or frequency and average of the output current (1), namely:

$$I_o = \frac{2v_{c\max}}{x_0} = 2v_{c\max} \cdot \omega_{SW} = 2v_{c\max} \cdot 2\pi f_{SW}$$

Here the values are normalized as in Valtchev and Klaassens (1990), i.e. the period is in radians, the frequency is angular, and the amplitude of the resonant capacitor voltage v_{cmax} is also normalized, although not always it is explicitly mentioned. The normalized v_{cmax} is calculated in Valtchev and Klaassens (1990) as a function of the conduction angles from Figure 3, where the angle ψ_2 is the regulating variable and further only ψ_2 is used:

$$v_{cmax} = \frac{(1-q)(1-\cos\psi_1)}{\cos\psi_1 + q} = \frac{(1+q)(1-\cos\psi_2)}{\cos\psi_2 - q} \tag{2}$$

To obtain the half period of switching $x_o = {}^{1}\!/_{\omega_{sw}} = \psi_{1+}\psi_2$, ψ_1 is recalculated by ψ_2, and (3) is obtained:

$$x_o = \Psi_1 + \Psi_2 = \arccos\left(\frac{1-q-q\dfrac{(1+q)(1-\cos\Psi_2)}{\cos\Psi_2 - q}}{1-q+\dfrac{(1+q)(1-\cos\Psi_2)}{\cos\Psi_2 - q}}\right) + \Psi_2 \tag{3}$$

Substituting this expression in (1), the normalized output current is finally obtained as a function of the normalized output voltage q at fixed values of the normalized phase shift ψ_2 between the command for turn-off and the zero crossing of the resonant current:

$$I_o = \frac{2\dfrac{(1+q)(1-\cos\Psi_2)}{\cos\Psi_2 - q}}{\Psi_2 + \arccos\left(\dfrac{1-q-q\dfrac{(1+q)(1-\cos\Psi_2)}{\cos\Psi_2 - q}}{1-q+\dfrac{(1+q)(1-\cos\Psi_2)}{\cos\Psi_2 - q}}\right)} \tag{4}$$

The regulation function $q=f(I_0,\psi_2)$ is obtained from (4) and is illustrated graphically in Figure 4.

The phase shift ψ_2 is an important control parameter. This phase shift, as Figure 3 shows, is the delay that must be kept between the turn-off command to the transistors and the zero-voltage crossing of the resonant current. The characteristics of the resonant power converter, controlled by the phase shift ψ_2, are shown in Figure 4. They exhibit a behaviour of the fold-back protection of a power supply. When the output voltage is very low, and the operation is close to a short circuit, the output current is drastically diminishing, if the phase shift is kept constant. It is seen also, that if the load is demanding a fixed output power (qI_0=Const.) or the load is approximately resistive, as shows the line in Fig.4 (q/I_0=Const.), then the necessary operational point (corresponding to a combination of output voltage and current) is achieved by approximately linear change of the interval ψ_2. At a fixed value of the output voltage q the interval ψ_2 changes in a non-linear manner: at the lowest demand for output current the change of ψ_2 is stronger. Having in mind that the phase shift interval and the switching half-period x_o are connected by (3), the frequency variation in this "lowest-current" will be too large. This confirms the solution adopted in Valtchev (1990) to achieve the lowest output power by shortening the transistor command pulses, e.g. by Pulse-Width Mode (PWM) application.

Figure 4. Output voltage as a function of the output current at fixed values of ψ_2 (in radians), with a load line

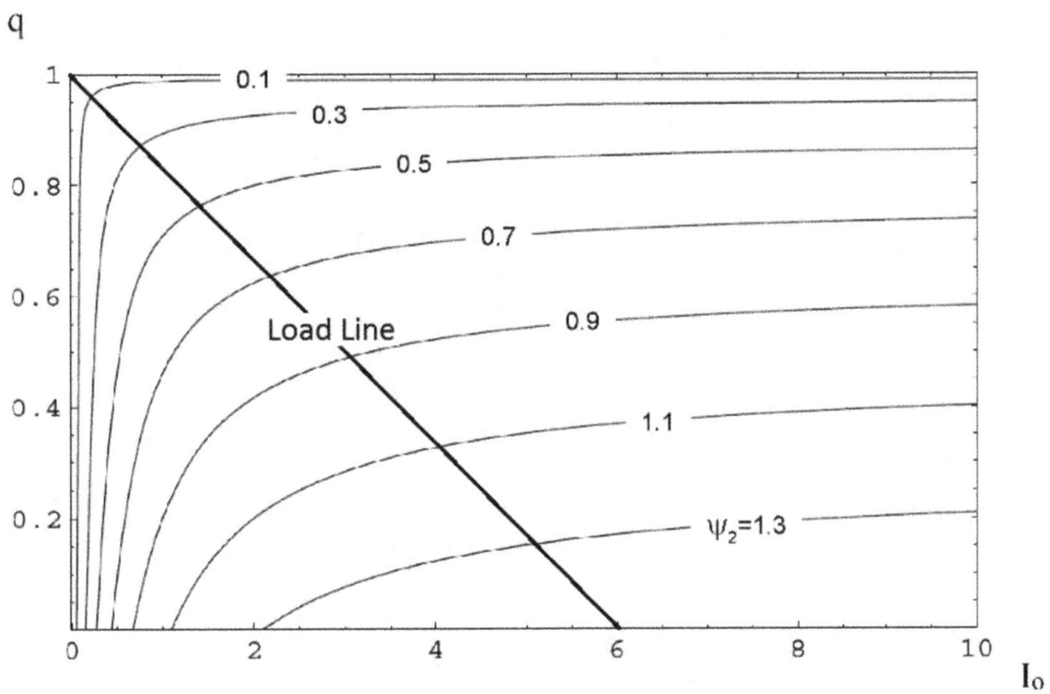

To ensure that the command pulse turns immediately ON or turns OFF (turning OFF the power switch is always the most difficult), some measures must be taken. If a Voltage-Controlled Oscillator (VCO) is the basis of the control circuit, a correction of the driver pulse length must be applied. The correction must introduce pauses (dead-time intervals) corresponding to the turn-off and turn-on times of the power switches to guarantee a stable operation with Zero-Voltage Switching (ZVS).

The implementation of this idea is shown in Figure 5 as an addition to the PLL circuit of Figure 2. The circuit from Figure 5 substitutes the feedback (wire) connection between the points marked as "B" and "A" in Figure 2 and introduces the necessary pauses between the turn off command to the first diagonal and the turn on command applied to the next (to be switched on) diagonal.

The pulses are taken from the VCO, and an equal delay in the rising and falling times (τ_1 and τ_2) is applied by two RC groups. The delayed pulses are combined in logical "AND" gates. In the output of each logical "AND" the displaced pulses applied at the input result in shortened pulses. This is shown in Figure 6, where the active level of the outputs E and F is positive.

The circuit shown in Figure 5 permits two versions of connection to the circuit (Figure 2). The connection to the phase detector (point "A") can be taken from the "Version 1" or from "Version 2" terminal. The pulses, produced in the points indicated by: "A", "B", "C", "D", "E" and "F" are shown in the diagram of Figure 6, illustrating the "Version 1".

The diagrams in Figure 6 show the shortened driver pulses obtained at the outputs "F" and "E", connected then to the transistors "T_1, T_4" and "T_2, T_3", correspondingly. The required fixed pause can be pre-calculated and applied to guarantee the ZVS. To make sure that the super-resonant operation is guaranteed, it is defined a minimum phase-shift by supplying a fixed minimum value of V_{add} (Figure

Figure 5. Delay circuit for additional command pulse displacement

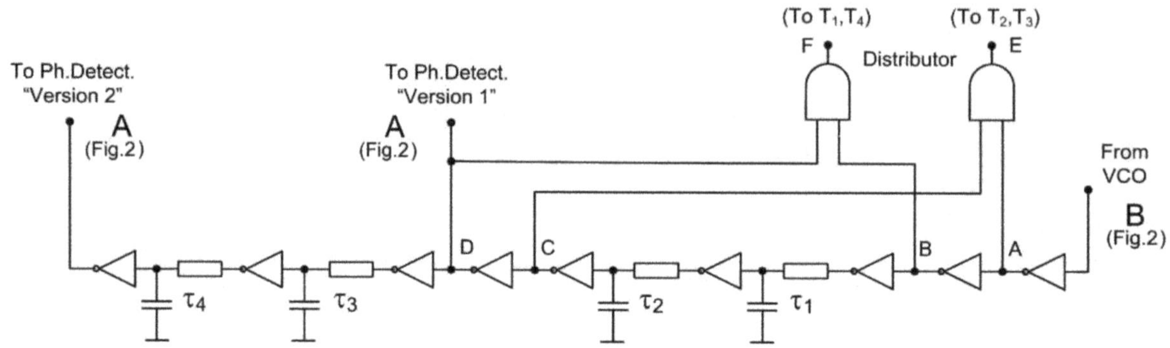

Figure 6. Time diagram of signals in characteristic points of Figure 5

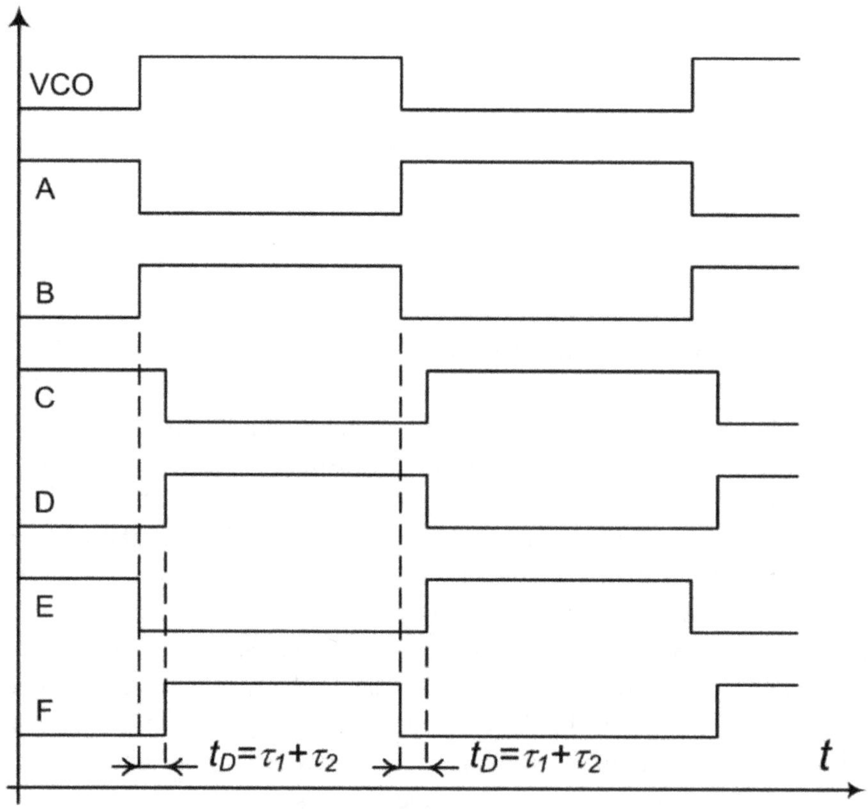

2). The voltage V_{add} is always higher than the minimum calculated voltage, corresponding to the minimum angular value ψ_2. This "normalised" angle corresponds to different "real" phase shifts in the time domain, depending on the specific operation point, and mainly, depending on the switching frequency.

In this case, the use of the output "Version 2", as shown in Figure 5, represents a better solution. In this case, the minimum ψ_2 will not vary in proportion to the voltage V_{add} and the frequency. The phase shift in the time domain will be defined directly by the shift time ($\tau_3+\tau_4$). It will not be necessary to guarantee any minimum for the control voltage V_{add}, because the transistors will switch off always slightly before (i.e. $\tau_3+\tau_4$) the zero-crossing of the current.

This type of regulation circuit was used in the induction heating converter (Valtchev, 1980) because it avoids the overlapping of the command pulses and allows to keep the widely changing switching frequency only in the super-resonant zone. Other method, not based on the circuit from Figure 2, that regulates the switching frequency to obtain the required output power is also known since 1980s, e.g. published by Hewlett-Packard (1982). The main problem of this factory-made resonant power converter remains the same: slow reactions, overshoot and instability at high speed disturbances.

When the pure FM regulation is not deep enough as a regulation method, the pulses should be shortened or displaced by other methods. The integrated circuits (IC), available in the market, that produce Pulse-Width regulation and Pulse-Shift regulation (here, PWM and PSM), are a universal solution. For a simple FM regulation those ICs possess their internal VCOs. This solution was already applied before, for example, in Valtchev and Klaassens (1990), and Valtchev (1988, 1995), etc. Usually the circuit combined FM with PWM and this combined circuit is presented later. In the ICs for PWM regulation, e.g. SG1524, the internal oscillator frequency is fixed (PWM regulation) of forced to be regulated by external command.

There are other integrated circuits, that can regulate the switching frequency, but this is accomplished by varying the width of the switching half-period, maintaining the pulse width fixed (useful for the quasi-resonant converters). Those ICs can be (and are) used, but with the necessary additional circuitry to shape the pulses, using the original circuit only as a VCO. The examples can be UC1860 produced by UNITRODE (TI), MC33066 by MOTOROLA, CA1523 by HARRIS, etc.

The limitation of the FM regulation was already discussed in Valtchev and Valtchev (2012). For example, the speed of FM regulation is limited by the IC used as a VCO and additionally by the PLL heavy low-pass filtering. Because of the possible change of the resonant frequency (temperature, magnetic coupling, etc.), the contactless converter risks either not to achieve the full output power capacity or to enter in the dangerous sub-resonant mode of operation (Valtchev & Klaassens, 1990).

To guarantee that the minimum switching frequency will be always higher than the resonant frequency, and there is no expectation for resonant frequency fluctuations, the switching frequency may be fixed. Then a PWM regulation exclusively may be applied, or another fixed frequency regulation is used: the phase-shift mode (PSM) regulation. Unfortunately, the PSM has a narrower range of regulation than the PWM, but some converters operate in PSM successfully, e.g. (Valtchev, Neves-Medeiros, Krusteva, Gigov, & Avramov, 2014; Gigov, Krusteva, & Valtchev, 2016).

4. PWM REGULATION

The pulse width modulation (PWM) is an appropriate control method over a large range of output power. It is analysed in depth and considered stable and classic, studied in university level textbooks (Mohan, Undeland, & Robbins, 2003). Unfortunately, it presents lower efficiency in case of resonant operation, and in some situations, can have the same problem presented by the sub-resonant operation: a forced recovery of the anti-parallel diodes through short-circuiting of the power bridge (Valtchev & Klaassens, 1990). The way to avoid this problem, and the possibility of TLO (Klesser, & Klaassens, 1986), was already clarified in Valtchev and Valtchev (2012) and Valtchev (1990). It was discovered that the maximum resonant capacitor voltage v_{cmax} needs to be limited to the calculated value, usually to the power supply voltage to the power bridge (Valtchev, 1990). If exclusively PWM regulation will be used in any operation mode and in any moment, the limitation of the resonant capacitor voltage must be respected always. In this case, by limiting the v_{cmax} value, the output power will be limited as well, which does not allow to obtain the maximum possible power of the converter. From the efficiency point of view, applying solely the PWM regulation produces very high current form factor ρ_i with its relatively low efficiency. This low efficiency is characteristic for all the versions of the discontinuous current mode of operation.

The problems of the PWM regulation do not forbid this method from being applied in real circuits. Though, the resonant converter is more efficiently and in a wider power scale controlled by a combined method, that should be preferred (Valtchev, 1990). An example can be shown that applies the widely spread integrated circuit SG 1524/3524, used in Valtchev (1988) both as a PWM and a FM regulator.

5. PWM-FM MIXED CONTROL

The most important mode of operation of that combined regulation (FM and PWM) is considered the FM, as it can obtain the maximum output power (Valtchev & Klaassens, 1990; Valtchev, 1990; Valtchev, Klaassens, & Wesenbeeck, 1995; Valtchev & Valtchev, 2012). The FM guarantees always the safe operation, maintaining ZVS, low switching losses and high efficiency in the best operation zone (Valtchev & Klaassens, 1990).

As it was pointed out in Valtchev (1990), the switching frequency in FM must be chosen roughly between the $f_{sw} = f_{res}$ and $f_{sw} = 1.5f_{res}$, or higher. That allows the converter to deliver its maximum power in a safe mode, in FM. The maximum voltage charged at the resonant capacitor is not enough to trigger a short-circuiting recovery of the anti-parallel diodes of the switches. In this case, at $f_{sw} > 1.5f_{res}$, the switching frequency may be fixed and the PWM operation starts safely (Valtchev, 1990). In the designing of the control, the maximum switching frequency, at which the FM is stopped and the PWM starts, must be relatively low, between $1.5f_{sw}$.and $2f_{sw}$. If not, the power transistors will be forced, in PWM to function at very short gate pulses. The range of variation of the pulses produced by the PWM regulator circuit is mainly limited by the power transistors, which switching times limit the operation zones. So, the frequency variations must be chosen for the specific power switching devices.

As an example of mixed FM/PWM operation, the integrated circuit UC1825, used in Valtchev (2001) and Valtchev and Valtchev (2012) is presented as a basis of that simplified regulator circuit. This is not the only possible solution. In other circumstances, other similar IC's were used as well, e.g. TDA1060 and SG3524 (applied in high frequency induction heating converter (Valtchev, 1980)).

The circuit UC1825 is a PWM regulator which internal oscillator frequency may be controlled externally, as it is presented in Figure 7. The comparator block "comp" inside the IC produces PWM regulated pulses at its output. A distributor, not shown in Figure 7, is like that of Figure 5. The pulses, in full or shortened, produced by the circuit UC1825, are applied alternatively to each corresponding diagonal of the power bridge in Figure 1, to guarantee the symmetrical operation of the resonant circuit.

The original error amplifier "E.A." of the integrated circuit UC1825 is blocked by external connection turning it into a voltage follower. The necessary external error amplifier is not shown in Figure 7. The output voltage from that external error amplifier is applied to the "error voltage" input of the circuit, i.e. to the internal modified error amplifier and to the external amplifier, identified as "FM/PWM".

Figure 7. Example of mixed operation (FM/PWM) regulation circuit

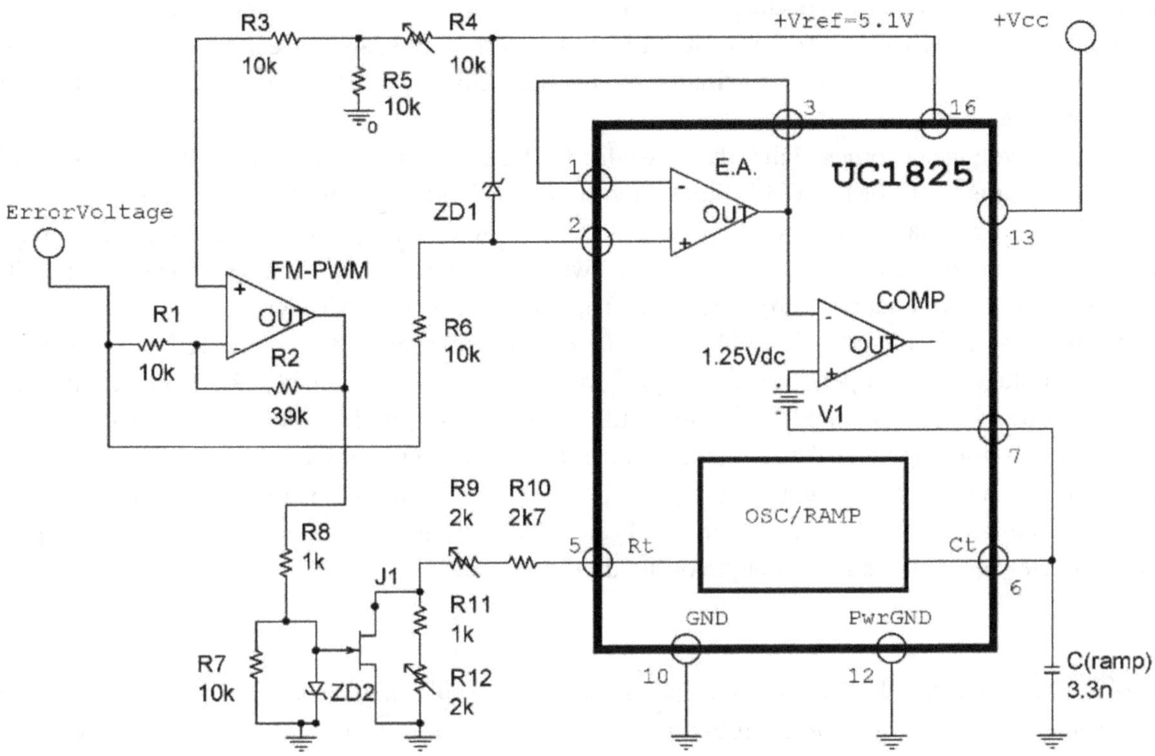

The integrated circuit UC1825 can operate as a normal PWM regulator for the lowest error voltages (this is the proper mode of operation when the demand of power is low). The error signal (amplified by the external error amplifier) is applied to the internal amplifier "E.A." (i.e. the voltage follower) and goes directly to the comparator.

To avoid the application of very low or negative value of error voltage, the internal amplifier is protected by a Zener diode ZD_1 (in this case V_Z=4.7 V). The excluding of the negative voltage values and even the positive voltage values lower than 1 V is not critical and produces no disturbance, because the ramp voltage from the internal oscillator applied to the internal comparator will never be lower than 1.25 V.

The lower (but positive) error voltage corresponds to a lower demand of output power, i.e. the operation mode must be kept exclusively PWM. In this case, the external amplifier indicated as "FM-PWM" in Figure 7 will amplify the difference between this low error voltage and the voltage from the divider R_5/R_4. This divider defines the voltage level that is the threshold between the PWM and FM. In case the error voltage is low, this results in a very good conduction of the JFET transistor "J_1". The transistor will have a gate-source voltage close to zero or positive (limited by the Zener diode ZD_2 in its positive direction). In consequence, the switching frequency of the integrated circuit UC1825 is kept at a pre-set maximum, e.g. 160 kHz, which finally means f_{sw}=80 kHz, having in mind the distribution of the pulses afterwards. The maximum frequency is defined by the resistor R_{10} and adjusted by the trimmer R_9. At this frequency, the only regulated pulse parameter is its width (PWM).

If a higher value of error voltage appears, the amplifier “FM-PWM” will compare it to the threshold stated by the trimmer R_4 of the divider R_5/R_4. If the error voltage is still not higher than the threshold, the output of the “FM-PWM” amplifier remains low and the transistor J_1 keeps its good conduction, maintaining the frequency f_{sw} at its maximum and the operation is still PWM, changing the pulse width to a larger value.

If the error voltage becomes higher than the adjusted threshold mentioned above, the negative output voltage of the amplifier “FM-PWM” (limited by the Zener diode ZD_2) will command the transistor J_1 to start rising its resistance. This will provoke a lower switching frequency, i.e. the regulation passes to FM. With the increase of the error voltage the switching frequency f_{sw} continues further to fall. The lowest limit for that frequency is adjusted by the trimmer R_{12} together with R_{11}.

The operation in the mixed FM and PWM regulation can be demonstrated by the measured waveforms at the power stage of the super-resonant SLSR converter of 3 kW (Valtchev, 1995). The oscilloscope images shown in Figures 8, 9 and 10, illustrate the turn-on pulse (the lower beam) addressed to one of the power transistor diagonals. The command pulses are shown together with the voltage drop at one of the switches in the opposite diagonal. It must be pointed out that the real gate pulse is not shown here, being applied by an isolated driver. That internal gate pulse is slightly different from the command pulse, i.e. it is delayed and enlarged by a couple of hundred ns.

Figure 8. Operation in FM. Upper beam: voltage across a transistor/diode switch (200 V/div), lower beam: control pulses at the opposite IGBT gate (10 V/div); fsw=56.8 kHz, Vo=15.33 V, Io=64.8

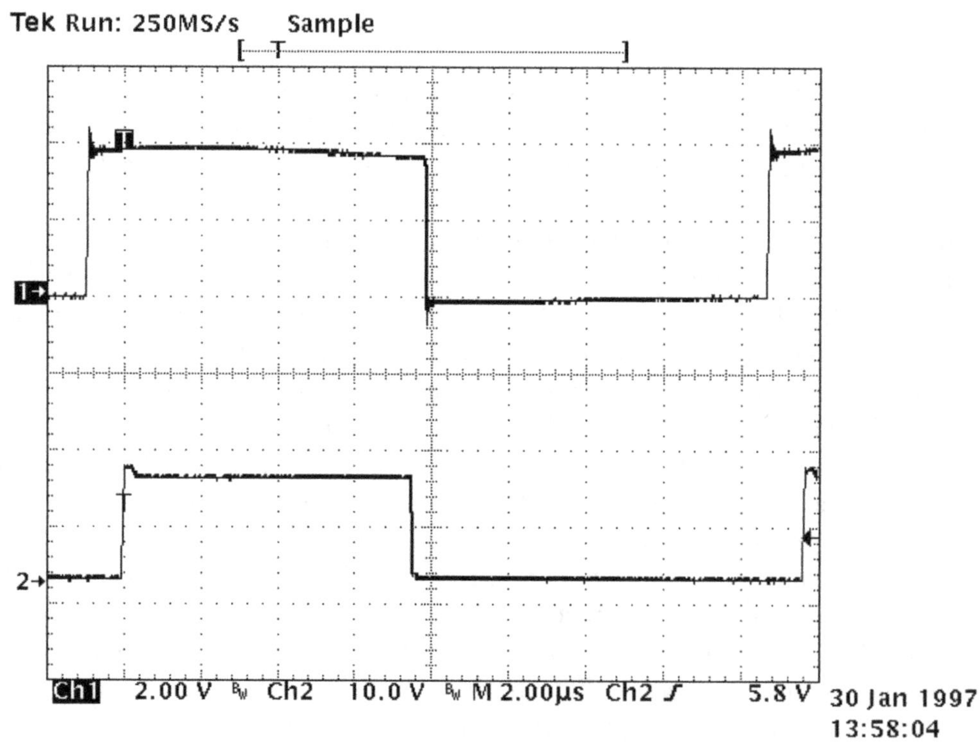

Figure 9. Operation in PWM close to the boundary FM/PWM. Upper beam: voltage across a transistor/ diode switch (200 V/div), lower beam: control pulses at the opposite IGBT gate (10 V/div); fsw=69.4 kHz, Vo=8.13 V, Io=35.71 A

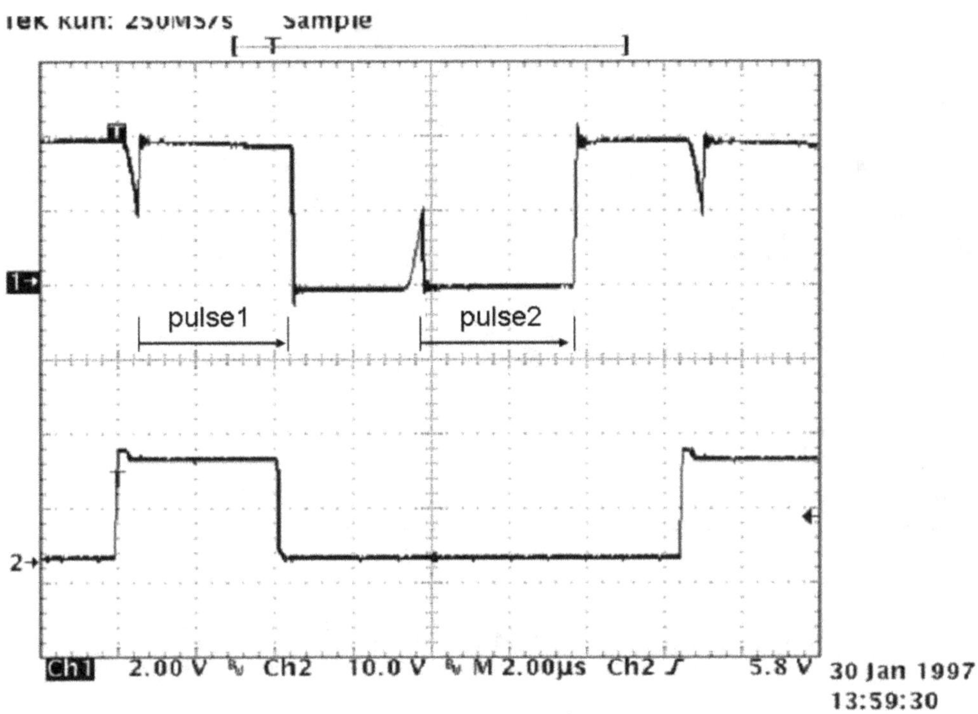

The operation mode in Figures 8, 9 and 10 is gradually changing from FM to PWM. As the circuit does not clamp the resonant loop during the time when the power switches (including their anti-parallel diodes) are all switched off, some oscillations are observed (Figure 10). Those oscillations are harmless to the converter and its efficiency. They are originated by recharging parasitic elements with relatively low reactance and low stored energy. The parasitic oscillations during the pauses illustrated by Figure 10 are much higher than the illustrated by simulation graphics presented in Valtchev (1990), because the parasitic capacitance in the simulation was considered lower.

The image in Figure 8 is taken at roughly 67% of the converter full power (993.4 W) with control pulses at their full length and switching frequency f_{sw}=56.8 kHz. The operation remains in FM. The lower beam shows the control pulses to one of the transistors of the corresponding diagonal. The upper beam is the voltage over the opposite transistor (receiving a control pulse during the pause of the first transistor pulse). The cursor on the left side indicates the zero voltage. The power switch has slightly negative voltage drop in the beginning which is the voltage drop on its anti-parallel diode conduction ψ_2 of Figure 3. As it is seen from the upper beam the voltage across the switch is very smooth with almost no spikes at the transitions.

In Figure 9 the FM regulation passes to PWM at a maximum switching frequency fixed by the control at a value f_{sw} = 69.4 kHz. The control starts to shorten the pulses. Figure 9 represents a PWM operation slightly over the boundary FM/PWM. The time intervals indicated as "pulse1" and "pulse2" denote the real reaction of the power transistors to the command pulses depicted in the lower beam image. The

Figure 10. Operation in deeper PWM. Upper beam: voltage across a transistor/diode switch (200 V/div), lower beam: control pulses at the opposite IGBT gate (10 V/div); fsw=69.4 kHz, Vo=1.19 V, Io=5.51 A

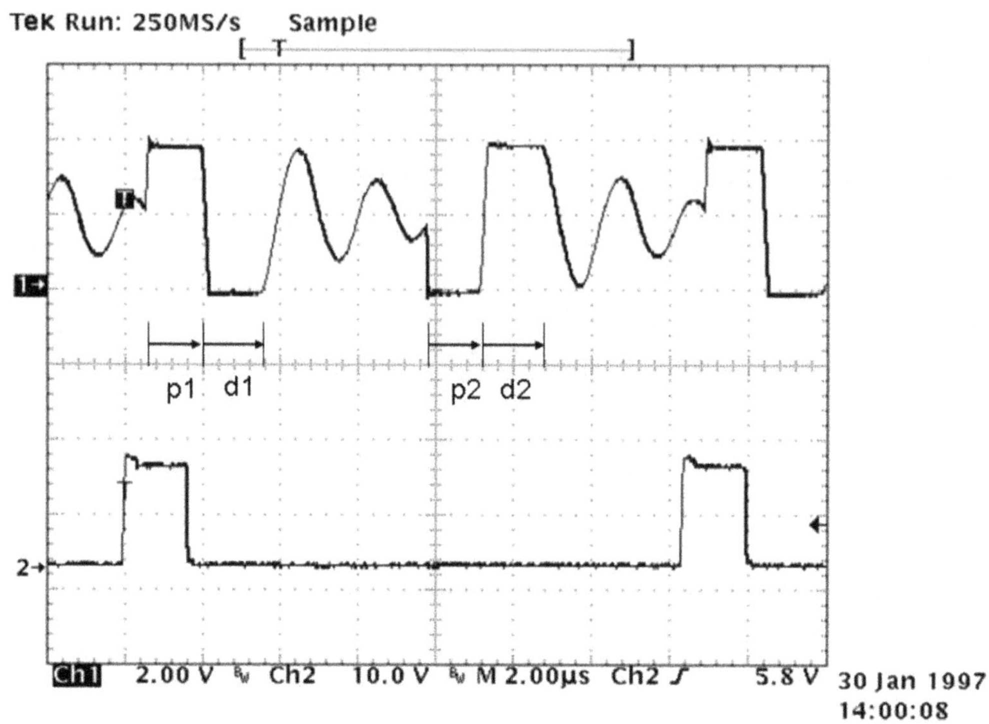

notch in the voltage across the switch is observed after the anti-parallel diode current falls to zero and the resonant circuit is left disconnected. As it was mentioned above this oscillatory behaviour of the circuit in PWM is not dangerous and does not present considerable additional losses.

The output power of the circuit in Figure 9 is 20% of the full power, i.e. 290.3 W. The pulses in Figure 9 are much shorter than the ones shown in Figure 8. This corresponds to the described in Valtchev (1990) behaviour of the pulses at the boundary FM/PWM: only when the pulse becomes shorter than the calculated transistor conduction interval ψ_1, the operation is really entering PWM. When the pulses are not short enough, the diode conduction interval ψ_2 permits for the converter to remain in FM operation. This lagging of the resonant circuit reaction to the gate command is a specific problem of the mixed PWM-FM control. The problem is minimized if the converter has heavy output filters and a slow reaction. Another, better solution is the instantaneous control presented further on.

The time intervals marked as "p_1" and "p_2" in Figure 10 correspond to the intervals "pulse1" and "pulse2" from the previous figure, i.e. they are the real time correspondence to the normalized conduction intervals ψ_1 of the switch. In Figure 10, the pulse width ("p_1" or "p_2") is 1.6 µs, i.e. 4.7 times shorter, compared to the pulse width presented in Figure 8. Following the transistor conduction interval ψ_1, named here "p_1", the diodes take over the conduction during ψ_2, corresponding to the rear slope of the resonant current, named "d_1". The conduction intervals produce the horizontal lines of smooth voltage drops (Figure 10), followed by the above-mentioned oscillation of the disconnected resonant circuit.

As it is expected, the shorter control pulses shown in Figure 10 produce a lower level of output power: 6.56 W, i.e. less than 0.44% from the full power. The output power in the same time is reduced 229 times, comparing again Figure 10 to the operation situation shown in Figure 8. The PWM regulation shows its capacity to regulate the output power in a very wide range. The wide range of regulation that the real mixed FM/PWM operation demonstrates is also predicted in Valtchev (1990).

The main draw-back of this regulation method remains the lagging effect of the boundary PWM/FM. The heavier output filtering cannot be a good solution for the contactless converter, because the scale of changes and the speed of the change is expected to be high. It is necessary to apply a method of regulation that reacts immediately to the error signal, producing resonant energy portions, defined by power pulses which length is defined by the transistor conduction time, having of course, at least a slight additional time to assure the ZVS. The immediate pulse control will introduce an internal sub-loop of regulation, like the speed control of the robotic arm positioning or the current-mode control (CMC) in the hard-switching power converters.

6. INSTANTANEOUS CONTROL OF THE SLSR CONVERTER

The SLSR power converter is a system characterized by a heavy inertial element, i.e. the resonant LC tank. The main difficulty of its control is in applying instantaneous changes in its internal processes. The controllable power switches do not permit neither to stop immediately nor to change instantaneously the direction of the energy flux through the LC tank. The resonant current continues (almost unchanged) in the resonant inductor and the voltage continues (almost unchanged) across the resonant capacitor at the instant when the power switches are commutated.

The resonant inductor current and the resonant capacitor voltage variables form the state vector. Those two variables (with their corresponding devices) contain the information about the total (magnetic and electric) energy of the LC loop. The energy contained in the tank is gradually interchanged with the outside circuits and its level cannot be rapidly changed. During each next half switching period the stored energy in the resonant circuit is changing its sign and repeating its absolute value in case of a steady-state operation. The energetic aspect of the SLSR converter operation is also discussed in Valtchev and Valtchev (2012).

The search for an instantaneous method of regulating the SLSR converter was started with the MSc thesis of Wahjudi (1992). Those initially used calculations were heavy and not practically applicable and here the energy balance is simplified, i.e made faster.

The general idea of the instantaneous control is to produce portions of energy dispatched during each half period from the input to the output. This must be done by defining the exact moment of commutation of the power switch to control (and limit) the transferred energy portion. The control loop defines the size of the energy portion that is necessary for each half period and commands the transfer of exactly the calculated portion through the resonant loop. It is expected that this method will make the transient reaction of the resonant converter faster and smoother, qualities that the SLSR converters are still missing and for which reason (mainly) this type of converters is often avoided by the industry.

6.1. Energy Balance of the SLSR Converter and Control by Pre-Calculated Energy Portions

To control the energy portions transferred through the resonant loop of the converter, it will be necessary to find a faster method to assess (and regulate) the energy stored in the reactive elements. The energy balance in the steady-state, i.e. the equality between the incoming and outgoing energy of the SLSR converter circuit was already discussed in Valtchev and Klaassens (1990). The distortion of the energy balance will be used as the instrument that regulates the above-mentioned energy portions transferred by the converter.

The SLSR converter, which energy balance will be further evaluated, continues to be considered lossless. This approximation is consistent with the calculations and the presumption that the circuit is purely reactive is proved by implementations, e.g. (Valtchev, 1988, 2001;Valtchev & Valtchev, 2012).

The equivalent circuit (-s) of the contactless SLSR converter, permit for a recalculation of the transformed parameters. As it was described in (Valtchev Medeiros, Valtchev, and Klaassens (2011), the primary circuit is the most important to be considered, as it contains the power switches. For the presence of the resonant loop and the switches, it is more convenient that the primary will be chosen as the base for the energy measuring and control. It is more likely to have the regulation circuitry placed in the primary side of the real converter. It will be easier for the converter control to use a relatively slow feedback signals from the secondary, but the fast driver reaction to be produced in the primary. It is not impossible to modify the circuit for execution of all calculations and producing commands from the isolated secondary (receiver), but it will be a relatively slow process, or it will increase the complexity of the circuit.

The excitation voltage (i.e. the voltage mirror), indicated in Valtchev and Klaassens (1990) has the value $V_{LC1}>0$ during the interval ψ_1 (transistor conduction time t_1, also normalized as ψ_f) and the value $V_{LC2}<0$ in the interval ψ_2 (diode conduction time t_1, normalized as ψ_r). Thus, the energy of the resonant circuit repeats its value after each half period of switching and, in a steady-state operation, no increase or decrease is observed. Equation (5) corresponds to time domain values:

$$\Delta E_{LC} = \int_0^{t_1} V_{LC1} i_r dt + \int_{t_1}^{\frac{T_{sw}}{2}} V_{LC2} i_r dt = C_r \frac{1}{C_r} [\int_0^{t_1} V_{LC1} i_r dt + \int_0^{\frac{T_{sw}}{2}} V_{LC2} i_r dt] \tag{5}$$

The symbol $\Delta\xi_{LC}$ in (5) represents the total energy added to the resonant tank portion during the half switching period. When $\Delta\xi_{LC}$ remains zero, the energy portion extracted from the input voltage source E_s corresponds exactly to the energy portion delivered to the output voltage source V_o, i.e. the operation is a steady-state one. If it is necessary to increase the energy portion in the tank, then the circuit will be commanded by a longer conduction time t_1 (longer than the necessary for its steady state), and hence $\Delta\xi_{LC}$ will be positive, adding more energy to the next portion. When a diminished energy will be required, then the time t_1 is decreased and $\Delta\xi_{LC}$ will be negative, reducing the next energy portion.

In equation (5) i_r is the resonant current, T_{sw} is the switching period and t_1 is the instant of time when the power switch (transistor) stops conducting. This point of commutation in (Valtchev & Klaassens, 1990) corresponds to the normalized point x_k. The second-time interval corresponds to the diode conduction, which finishes at the end of the switching half period $T_{sw}/2$. The capacitance value of the resonant capacitor is included in (5) to prepare its integration.

It is easier to observe the value of the total energy portion in the resonant current zero crossing point. Considering that at this time instant the total resonant tank energy consists only of the stored in the resonant capacitor $\frac{C_r v_{c\max}^2}{2}$, the capacitor voltage maximums v_{cmax} can be used as a measure for that energy. After integrating expression (5), it can be rewritten in capacitor voltage terms:

$$\Delta E_{LC} = V_{LC1}\left[v_c(t_1) - v_c(0)\right]C_r + V_{LC1}\left[v_c\left(\frac{T_{sw}}{2}\right) - v_c(t_1)\right]C_r \tag{6}$$

If the (absolute value) of the voltage $v_c(0)$ is marked as v_{cmax1} and the voltage $v_c\left(\frac{T_{sw}}{2}\right)$ is marked as v_{cmax2} then it can be written:

$$\Delta E_{LC} = C_r\left[V_{LC1}\left(v_{cmax1} + V_c(t_1)\right) + V_{LC2}\left(v_{cmax2} - v_c(t_1)\right)\right] \tag{7}$$

The expressions of V_{LC1} and V_{LC2} can be substituted in (7) from their normalized forms $V_{LC1} = 1-q > 0$ and $V_{LC2} = -1-q < 0$. The normalized output voltage q is supposed to be a constant in steady-state and then the capacitor energy change $\Delta\xi_{LC}$ (in normalized form) is:

$$\Delta\xi_C^N(0, x_0) = \Delta\xi_{LC}^N = [2v_c(t) - q(v_{c\max 1} + v_{c\max 2})] = 2\left[v_c(t_1) - q\frac{v_{c\max 1} + v_{c\max 2}}{2}\right] \tag{8}$$

Expression (8) can be used to predict the capacitor voltage $v_c(t_1)$ when it will be necessary to switch off the transistor (t_1 time point corresponds to the angular point x_k). The decision about this command will need a calculation of the predicted value $v_c(t_1)$ and then the comparison of the predicted to the measured capacitor voltage, in steady-state. In case of a necessity to change the output power, the prediction must consider another, corrected value of $v_c(t_1)$.

In the case of contactless power transfer, the converter operation is described by the same expression (8) where v_{cmax1} and v_{cmax2} correspond to the resonant capacitor voltage amplitudes in the primary. In this case, the normalized output voltage q in (8) will be substituted by a value q, recalculated to the primary. The expression for the recalculated q considers the magnetic coupling K of the contactless transformer: $q^T = Kq$. This expression is correcting the output voltage as it is seen from the primary side of the converter when the magnetic coupling coefficient is not unitary.

Predicting the exact value of necessary energy portion and the addition to it $\Delta\xi_{LC}$, that must be transferred during each half period of operation, will be the aim of this fast regulation strategy.

6.2. Simplified Strategy for the Regulated Transfer of Energy in SLSR Converters

From the beginning, it must be clear that the simplification cannot go beyond certain limits, i.e. it cannot be simpler than it is necessary to stay safe. The SLSR converter may keep its steady-state operation during long enough time (i.e. many switching periods) if there is no variation of the load and no change of the input power source parameters. When a transition from one steady-state level to another is required, the internal variables must be prevented from reaching dangerous levels or surpassing the targeted next steady-state level. This danger must be avoided by the regulation strategy. The converter that has a good magnetic coupling between the primary and the secondary side, has the possibility to make a filtering. The heavy filtering can bring up another problem, that of overshooting, but it can be solved by classical PID solutions. In the case of unstable and loosely coupled primary and secondary, no filtering can save the situation. The filtering can get the problem to a dangerous level, as the feed-back will respond too late to an instant change inside the switches and the resonant loop.

A similar danger exists in the hard-switching converters: when a rapid transition is needed from one steady-state to another, the reaction can be exaggerated, and the variables can assume dangerous values. The solution found for the hard-switching converter is the Current-Mode operation: the output voltage error defines the height of the inductor current at which the power switch must stop or start its conduction. This strategy limits the maximum inductor energy transferred per cycle at a level corresponding to the required output power (expressed as a proportion to the measured error voltage of the output). In the hard-switching converter, the variable serving to indicate (and regulate) the height of the energy portions each cycle, is the inductor current (usually the filtering inductor current).

It is important to choose a convenient variable which level can serve as an indicator for the energy transferred at each half period in the SLSR converter. This variable in steady-state must keep its value unchanged, reflecting proportionally the size of the energy portions transferred in each half period. During the transient process by the value of this variable, it must be possible to predict and change the size of the energy portions circulating in the resonant tank.

In the hard-switching Current Mode Control (CMC), the inductor current transferring the periodically repeated portions of energy is the instantaneous power indicator. On the contrary, the resonant current in the SLSR converter is not suitable to be the main indicator. The first reason is that the peak value of the resonant current does not correspond to the maximum stored energy in the resonant loop. Another reason is that after the switching instant x_k the resonant current abruptly changes its derivative (the slope).

The current is chosen as the internal measure of energy portions of the hard-switching CMC power converter. In fact, the hard-switching converter (in its continuous current mode) has output characteristics that are like a voltage source, as shown for example, in Mohan, Undeland, and Robbins (2003), and the convenient internal variable is convenient to be the current that supplies the output.

In the case of the resonant converter, in contrary, the SLSR converter shows output characteristics like a current source, with curves that fall to limited current values as in Valtchev and Klaassens (1990) and Valtchev (1990). The internal, resonant inductor current cannot be stopped. The resonant current is a measure of the energy but also the resonant capacitor voltage is. The inductor current has its maximums, but they appear in most cases before the switching off of the switches, so they are impossible control and influence. This can be seen in Figure 3. In this case, the measured state-variable is better to be the resonant capacitor voltage, proportional to the energy portion level. The resonant capacitor voltage has a shape of a smooth cosine waveform, having maximums exactly at the resonant current zero crossings. At the zero crossing the LC tank, the energy is concentrated only in the resonant capacitor and that maximum serves as a good indicator. The smooth character of the curve permits also for an easier comparison to the calculated reference.

From the practical point of view, the only variable that the power switches can commutate, is the resonant current passing through the transistors. The current cannot be stopped, but it can be commutated, from transistors to diodes. The commutation controls the resonant capacitor voltage maximums.

At the moment of switching, the resonant capacitor has also its electrical energy, proportional to the square of its voltage, and both its voltage and energy will continue to rise. The end of the transistor conduction comes earlier than the stored electric energy in the resonant capacitor achieves its maximum. So, the moment of the transistors switching off must be predicted by calculation. In the same time the prediction must be achieved by simplified calculations because the operation frequency is expected to be high, at the order of tens and hundreds of kilohertz.

The interchange of inductive and capacitive energy does not allow for an easy measuring of the total energy in every operation instant, but the suggestion here is to measure the total energy, represented by the resonant capacitor voltage in the points of resonant current zero crossings. The condition for the SLSR converter steady-state operation is to repeat symmetrically the variable values, so the two amplitude values of the resonant capacitor voltage stated in eq. (8), v_{cmax1} and v_{cmax2}, are equal seen in their absolute value v_{cmax}. The value v_{cmax}, is proportional to the output current, in compliance with (1). Consequently, for the steady-state operation, the expression (8) becomes:

$$\Delta\xi_{LC}^{N} = 2\left[v_c(t_1) - qv_{c\max}\right] = 0 \tag{9}$$

This equation can be rewritten as (10) and illustrated by Figure 11.

$$V_c(t_1) = qv_{cmax} \tag{10}$$

In Figure 11 the time instant at which the transistor diagonal is switched off (t_1) in steady-state is defined by the previous amplitude v_{cmax1} and defines the next amplitude of the resonant capacitor voltage v_{cmax2}.[1]

The waveform in Figure 11 proves the steady-state equation (10). To assure the identical amplitudes of the resonant capacitor voltage v_{cmax} it is necessary to keep the switching off at the defined value $v_c(t) = qv_{c\max}(previous)$. . By defining the pulse width in this way, the energy portions are kept exactly at the necessary level. Thus, in steady-state a comparator needs to assure the switching off exactly in the instant when the equation (10) holds. This will require a calculation of the multiplication between the normalized output voltage and the last measured amplitude voltage of the resonant capacitor.

The transition process from a lower output power consumed in an initial steady-state, to a higher level of output power, will require several portions of additional energy to be supplied by the resonant tank. The amplified error voltage (proportional to the difference between the really obtained and the required output voltage) will define the necessary portions of energy for the transition (similarly to the Current Mode Control).

The requirement for a positive increment $\Delta\xi$, in the energy portions corresponds to a higher output power. When this requirement is to maintain a stabilized output voltage (i.e. the usual case of DC-DC converter), the control reaction will be to produce the turn-off of the transistor diagonal at a higher level of resonant capacitor voltage expressed in (11):

Figure 11. SLSR converter resonant current commutation at the time point t_1

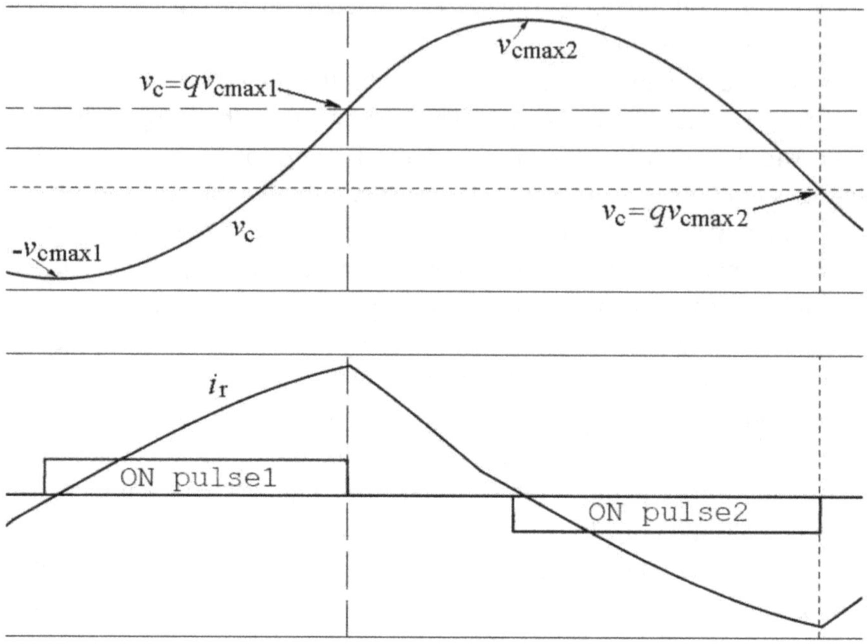

$$v_c(t_1)_{new} = qv_{c\max} + \Delta\xi_{LC}^N = qv_{c\max 2} \tag{11}$$

The higher value of the calculated voltage $v_c(t_1)$ will increase the next amplitude v_{cmax}. In (11) the voltage $v_c(t_1)_{new}$ is that new (recalculated) value applied for the transient process. This (new) reference level will permit the resonant capacitor voltage to surpass the steady-state amplitude value v_{cmax1}, calculated before in (10) and to reach the new level v_{cmax2}. If this change of amplitude is expressed as $\Delta v_{cmax}=v_{cmax2}-v_{cmax1}$ then from eq. (8) it can be derived that:

$$v_c(t_1)_{new} = qv_{cmax1} + q\Delta v_{cmax1} = q(v_{cmax1} + \Delta v_{cmax1}) \tag{12}$$

The expression (12) is an approximation but it is more practical for the resonant regulation process. It expresses the energy increment $\Delta\xi_{LC}$ as a linear proportion to the capacitor voltage amplitude change Δv_{cmax}. To apply a more accurate method to determine the necessary energy portion will be more difficult as an implementation.

The calculation of (12) is easier to implement by storing the initial value of the resonant capacitor voltage v_{cmax1} in a memory. After applying the calculations corresponding to the steady-state (10), the voltage Δv_{cmax} (proportional to the amplified output voltage error signal) is added to the calculated by (10) initial v_{cmax1}, then the sum is multiplied by q in order to correspond to the expression (12). The resulted new level $v_c(t_1)_{new}$ is applied in the comparator, substituting the initially calculated reference voltage $v_c(t_1)$. When the resonant capacitor voltage reaches the reference $v_c(t_1)_{new}$ the transistor diagonal is switched off as it is illustrated in Figure 12.

Figure 12. Switching off at the resonant capacitor voltage level q(vcmax1+Δvcmax1)

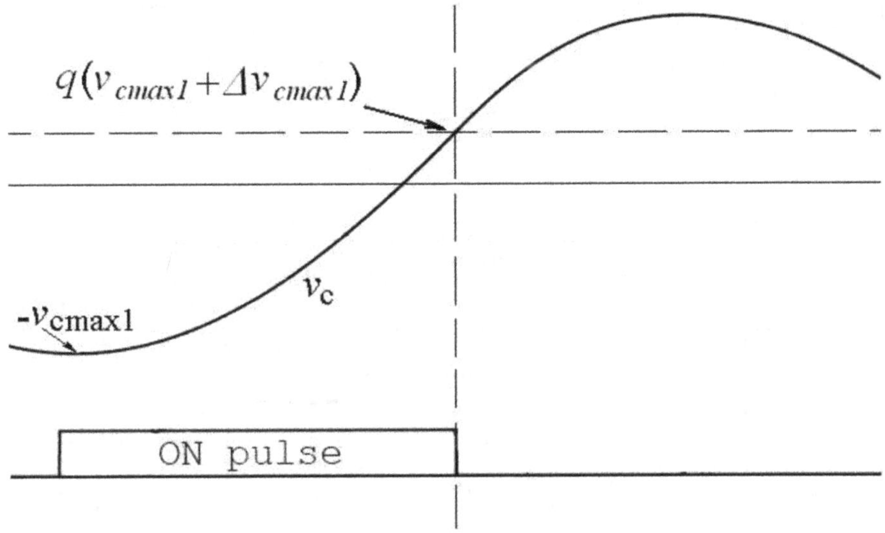

After the transistors are switched off, the capacitor voltage is expected to reach the new amplitude value v_{cmax2}. As the process is limited by the already running co-sinusoidal waveform, the maximum reference level $q(v_{\mathrm{cmax1}}+\Delta v_{\mathrm{cmax1}})$ cannot be required to be higher than the initial capacitor voltage value v_{cmax1}.

The full transition process will need several half periods because the next resonant capacitor voltage amplitude v_{cmax} which can be reached in only one switching half period is limited. The newly calculated $v_c(t_1)_{\mathrm{new}}$ cannot be higher than the already known steady-state amplitude v_{cmax1} of the capacitor voltage (as the switching off of the transistors cannot be required after the half-sinusoid of the resonant current is completed). It cannot be even required to reach values close to the steady-state amplitude v_{cmax} because of the necessary protection against excessive resonant capacitor voltages and resonant currents (or because the error signal is already lower).

In a process of producing this step-by-step increase of the resonant capacitor voltage amplitude, the necessary value of output power is reached, and the resonant capacitor voltage reaches the new steady-state amplitude value v_{cmax}.

The limitation for the calculated value $v_c(t_1)_{\mathrm{new}}$ to a specified percentage of v_{cmax1}, when the transition is climbing to a higher power, is substituted by another limitation, when the transient process requires lower amplitude v_{cmax2}. In that case, the symmetrical operation requires that the $v_c(t_1)_{\mathrm{new}}$ is at least slightly higher than zero, otherwise the oscillation of the resonant current will stop.

6.3. Implementation of the Method for Instantaneous SLSR Control

The implementation of the calculation idea described above will require several electronic blocks to memorize, calculate, compare and limit the values of the variables that the internal sensors will measure. The simplified block diagram of the whole power converter is presented in Figure 13.

The simplified blocks correspond to the most important operations that will be executed. Those operations do not depend on the practical schematic implementation. The diagram in Figure 13 shows the general idea of the operation, based on the above-mentioned multiplication of the capacitor voltage

Figure 13. Block diagram illustrating the method for instantaneous SLSR control

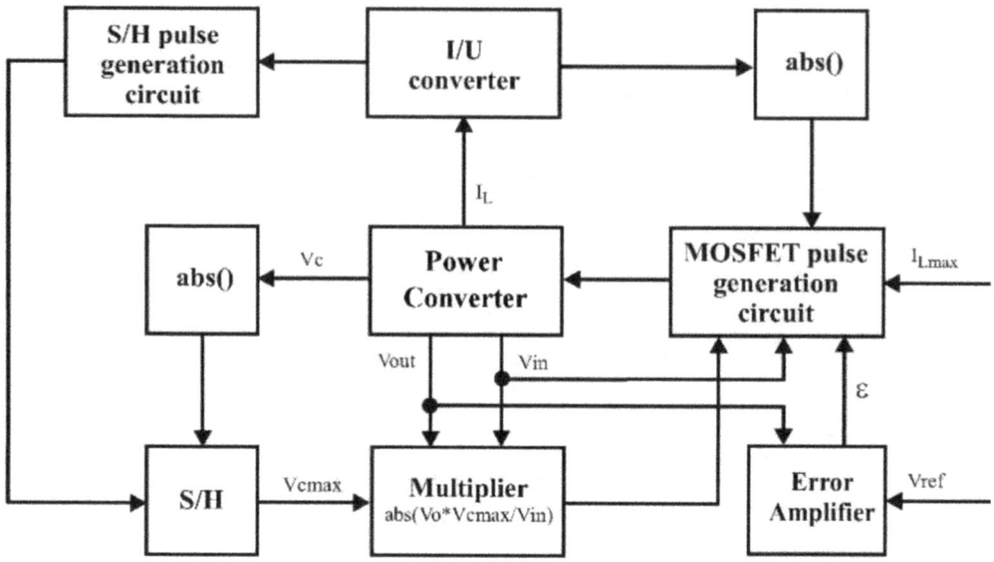

(measured by sensors) by the calculated ratio q. The control circuit is applicable to any SLSR converter in super-resonant mode of operation, including the contactless operation.

The variables that characterise the block "Power Converter" in Figure 13 are the input voltage E_s, the output voltage V_o, the resonant current i_r and the resonant capacitor voltage v_c. The zero crossing of the resonant current activates the sample/hold circuit "S/H" which memorizes the resonant capacitor amplitude value v_{cmax1} (indicated in Figure 11). The output voltage is divided by the input voltage to produce the ratio q which is then multiplied by the memorized v_{cmax1} in the block "Multiplier". The value qv_{cmax1} is the level at which the transistors are turned off, as it is illustrated in Figure 12.

The block "Error Amplifier" is detecting the error signal ε comparing the output voltage to the reference voltage V_{ref} (as the usual voltage control does). The amplified error signal is defined to be the voltage increment $q\Delta v_{cmax1}$ in (12).

There is a protection that uses an imposed current limit reference "I_{Lmax}" which is compared to the absolute value of the resonant current. If by some reason the resonant current tends to rise higher than the permitted value, the transistors are switched off immediately. The final command pulse is produced by the "MOSFET pulse generation circuit" block.

The block diagram presented in Figure 14 exhibits in more details the regulation idea and the necessary circuits for its implementation. This more detailed diagram indicates the functions to be implemented.

The block "DC/DC Converter" is representing the power stage. The large arrow "Inductive Power (I.P.) Transfer" denotes the contactless power transfer to the "Converter Secondary". The "Converter Secondary" block is the secondary (receiver) side of the converter including the rectifier and the output filter.

The block "Sensor of the resonant capacitor voltage" can have different schematic implementations[2]. The measured and scaled capacitor voltage is rectified with the precision rectifier block "Abs()" and its maximum value v_{cmax} is memorized by the "Sample/Hold" block. The memorized v_{cmax} value is kept unchanged until the next amplitude is detected.

Figure 14. Detailed functional diagram illustrating the method for instantaneous SLSR control

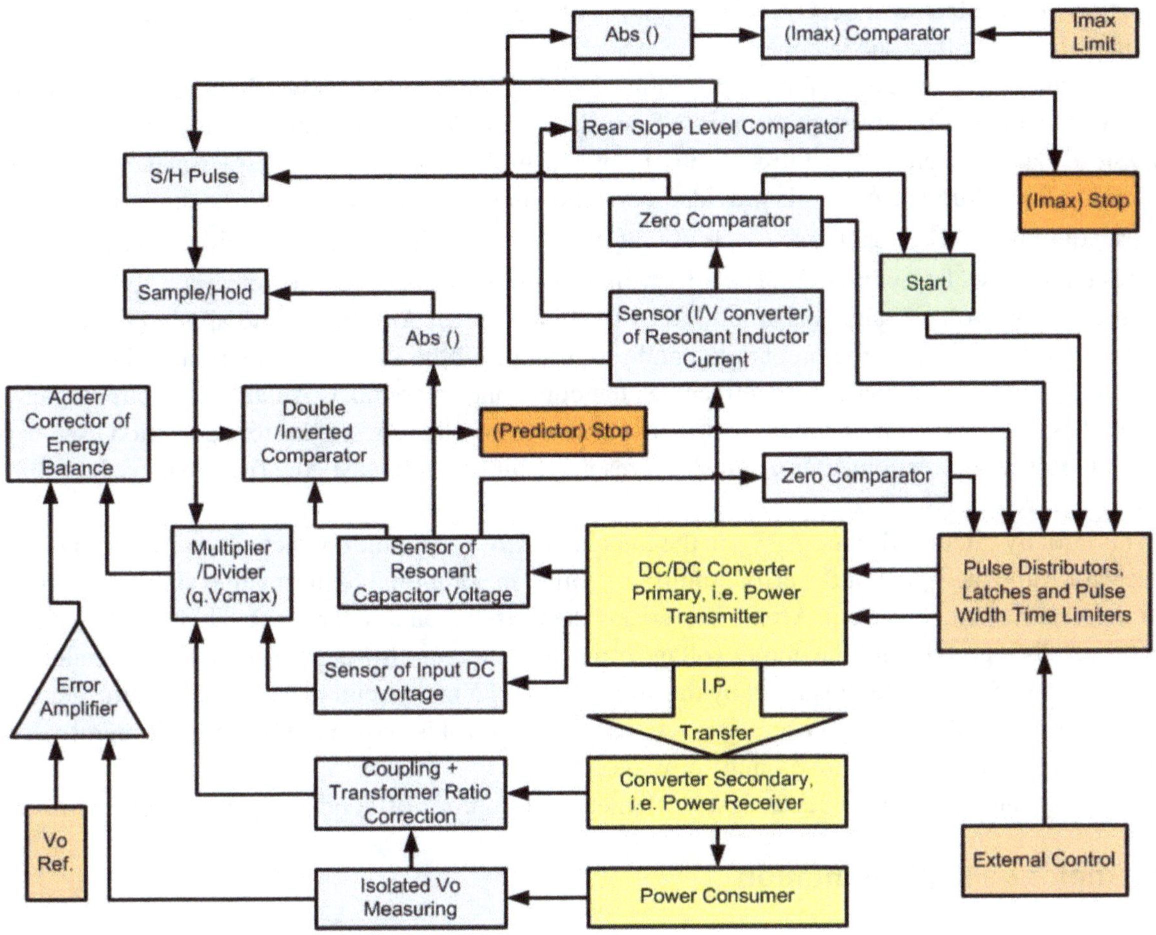

The inductor current is measured by the "Sensor of Resonant Inductor Current (I/V converter)" which can be in practice based on a sensor of Hall. The "Zero comparator" produces a pulse at the zero-crossing of the resonant current (applied as a command to the previously mentioned "Sample/Hold" circuit). Another, slightly advanced pulse is produced before the rear slope of the resonant current reaches the zero. This pulse is obtained from the "Rear Slope Level Comparator" and is applied to start the next transistor gate pulse before the anti-parallel diode current reaches zero (the ZVS condition). The block "Pulse Distributors, Latches and PW limiters" can ignore the start signal and delay the start of the gate pulse if PWM regulation is necessary (at lowest power). The current sensor signal (rectified by the "Abs()" block) is also compared to the "I_{max} Limit" in order to stop immediately the pulse in case of a dangerous height.

The block named "Isolated Vo Measuring" is measuring the output voltage and is sending its value as a signal through the isolation between the secondary and primary sides of the converter. In the rotary construction, an optocoupler can be used but usually the signal is sent by high frequency (capacitive) link. When larger differences in the magnetic coupling coefficient K are expected, a block "Coupling + Transformation Ratio Correction" is necessary to correct automatically the received output voltage signal in proportion to the changing coupling coefficient.

Another necessary measurement is provided by the block “Sensor of Input DC Voltage”. The implementation of this block will depend on the supply voltage value and the scale of its changes. Generally, a voltage divider will be sufficient.

The value qv_{cmax1} necessary for the equation of energy balance (12) is obtained in the “Multiplier/Divider” block. In a steady-state the result from this block defines the end of the conduction of the transistors that were active at that time. From the block “Adder/Corrector of Energy Balance” the amplified and scaled addition $q\Delta v_{\mathrm{cmax1}}$ is provided correspondingly to the increase or decrease of the output power requirement. The signal $q\Delta v_{\mathrm{cmax1}}$ is not reflecting exactly the necessary additional energy $\Delta\xi_{LC}$, but it is quite close, following (12). The error voltage (difference between the reference voltage and the measured output voltage) signal is amplified by the block “Error Amplifier” and applied as $\Delta v_{\mathrm{cmax1}}$.

The operation starts by the preparation and resetting of all the supply and auxiliary circuits. Then, the first diagonal of the power bridge receives the command to conduct. An analogue circuit can use the slight asymmetry of the channels, to start the first auto-oscillation, similar to the method used in the classical “Royer” power converter. At this first resonant current half wave, the first memorized value of the v_{cmax} is zero or close to zero.

This initially calculated value qv_{cmax1}, or the calculated Kqv_{cmax1}, for the contactless converter, is zero but in the same time the signal of the error amplifier applies an additional highest value and the result is an almost full half-wave of current. After a short delay the next diagonal will receive the turn-on command. The process is repeated until the output voltage error becomes very low and the converter is maintained in steady-state. This can be illustrated by the simulations of a real circuit implementation further on.

Another way to illustrate the necessary actions of the control is shown in Figure 15. It represents the actions that the blocks of Figure 14 must provide.

The instantaneous control idea is first implemented as a proof-of-concept by an analogue circuit.

6.4. Analogue Implementation

Analogue integrated circuits (combined with some digital logic ICs) are chosen to demonstrate the described idea in more realistic operation. The simulation of the circuit operation is sufficiently illustrative. In this version, there is no digital comparator or counter which makes the regulation difficult in discontinuous current mode (PWM).

The real power converter will require switching frequency not lower than 100 kHz and hence the regulation will be also very fast (e.g. the reaction of the power switches must not be slower than 100 ns). Not all the variables need to be considered as rapidly changing. For example, the output voltage will vary slowly (due to the filter capacitor). Even though, the change of the error voltage signal will be noticeable during a few periods. The operations that need fast measuring, decision and reaction, concern mainly the resonant capacitor voltage.

The block-diagram of Figure 14 is developed into a circuit of discrete devices, discrete components and low scale integrated circuits as presented in Figure 16. To simulate this circuit built of practically available components, it was necessary to scale down the switching frequency (and the resonant frequency defined by the respective components) exactly 100 times. By reducing the operation frequency, the influence of the parasitic effects and delays in the numerous integrated circuits is ignored and as a result the simulation plots are cleaner.

Figure 15. Functional diagram illustrating the entities involved and the actions of the control in SLSR converter

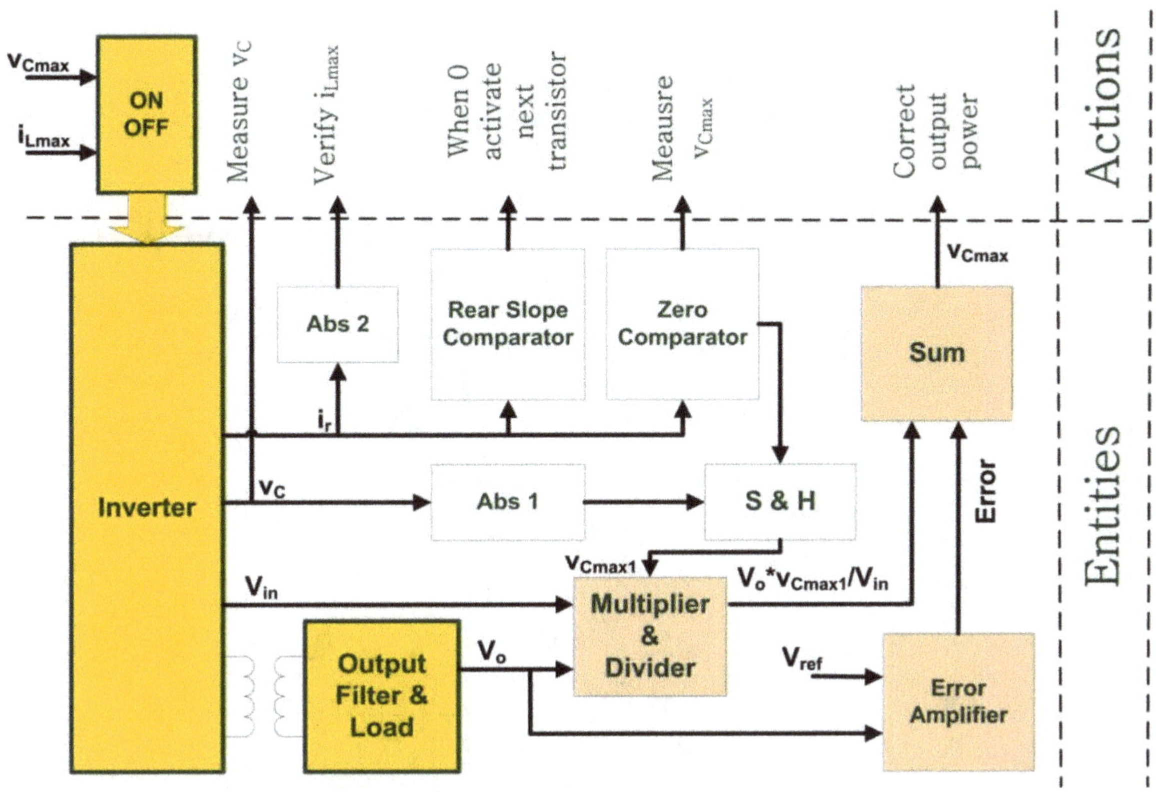

The converter control circuit shown in Figure 16 includes the power bridge (MOSFETs) and the loosely coupled transformer (K=0.8). All the connections between the elements are fully developed with one exception: the feedback is opened at the point of the command pulses and the transistors receive command pulses from pulse generators.

During this fixed switching frequency operation, the control circuit sensors measure the necessary variable values and produce the pulses which should be applied to the gates. The result from the simulation is illustrated in Figure 17.

Figure 17 shows the produced (through calculations) control pulses ("ON pulse1" and "ON pulse2"). Those pulses have their rear slope positioned exactly where the power transistors switch off, although the pulses are not the command given to the transistors. In fact, the transistor gates receive other command pulses, with a full length of 49.99% duty cycle. This is done by external, fixed frequency oscillators. The current and voltage waveforms observed in the converter, controlled by those external pulses, are exactly the same as they would be if the internally calculated command pulses were applied.

By Fig.17 it is proven that the operation of the circuit will be predictable, and the produced pulses will have exactly the necessary parameters. For example, the calculated level of resonant capacitor voltage at which the transistors should be switched off is $v_c(t_1) = Kqv_{cmax} = (0.8)(0.4)(105) = 33.6$ V and this agrees very well with the actual values shown by the cursors A1 (33.443 V) and A2 (33.849 V). This means that the really applied pulse finishes at the same time as the calculated one. The only difference

Figure 16. SLSR converter instantaneous control circuit with connected latches and disconnected feedback

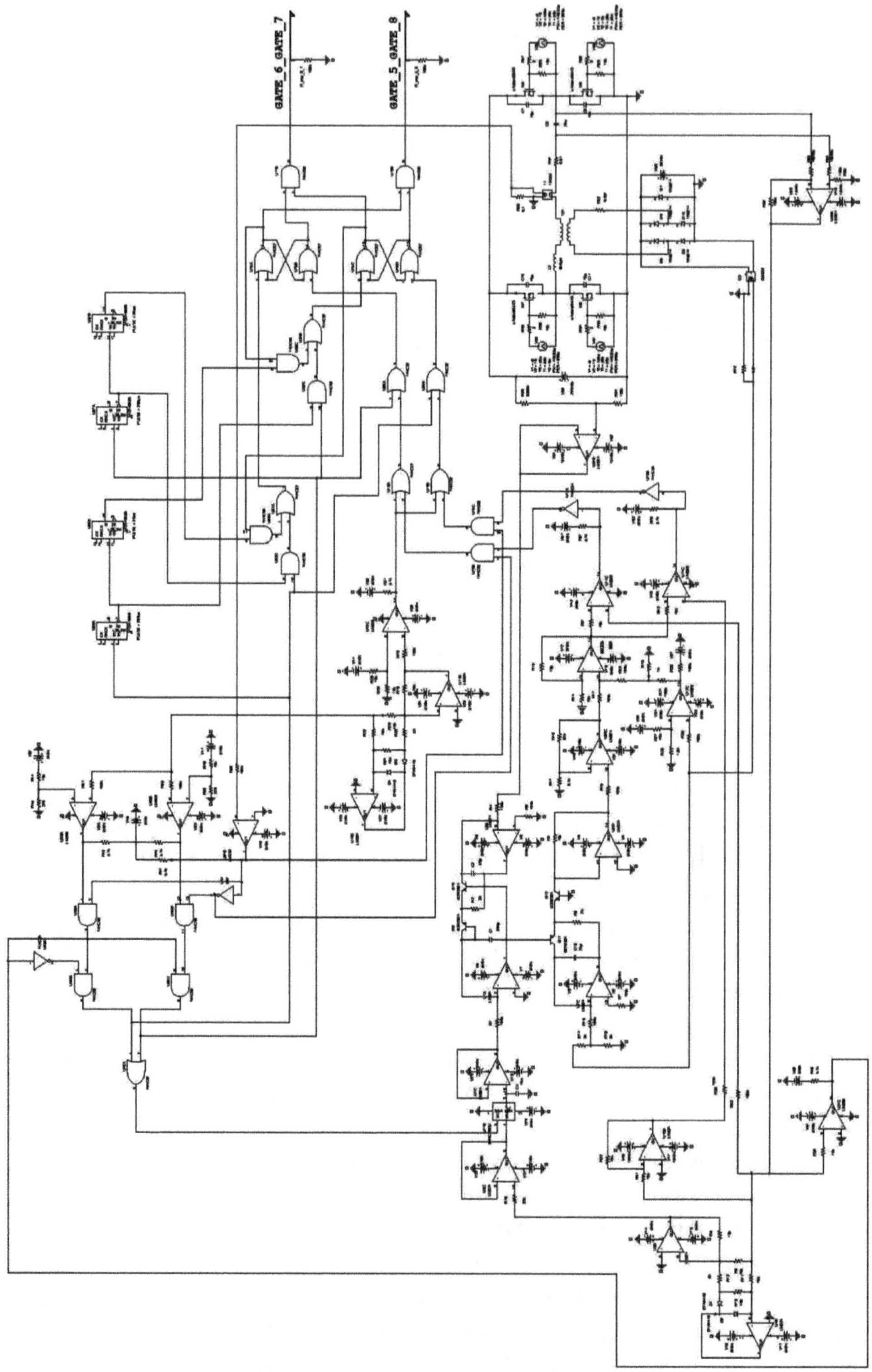

Figure 17. Simulation results from circuit of Fig.16. Upper plot: v_c; Middle plot: i_{res} and control pulses; Lower plot: amplified voltage v_c(sensor) and control pulses

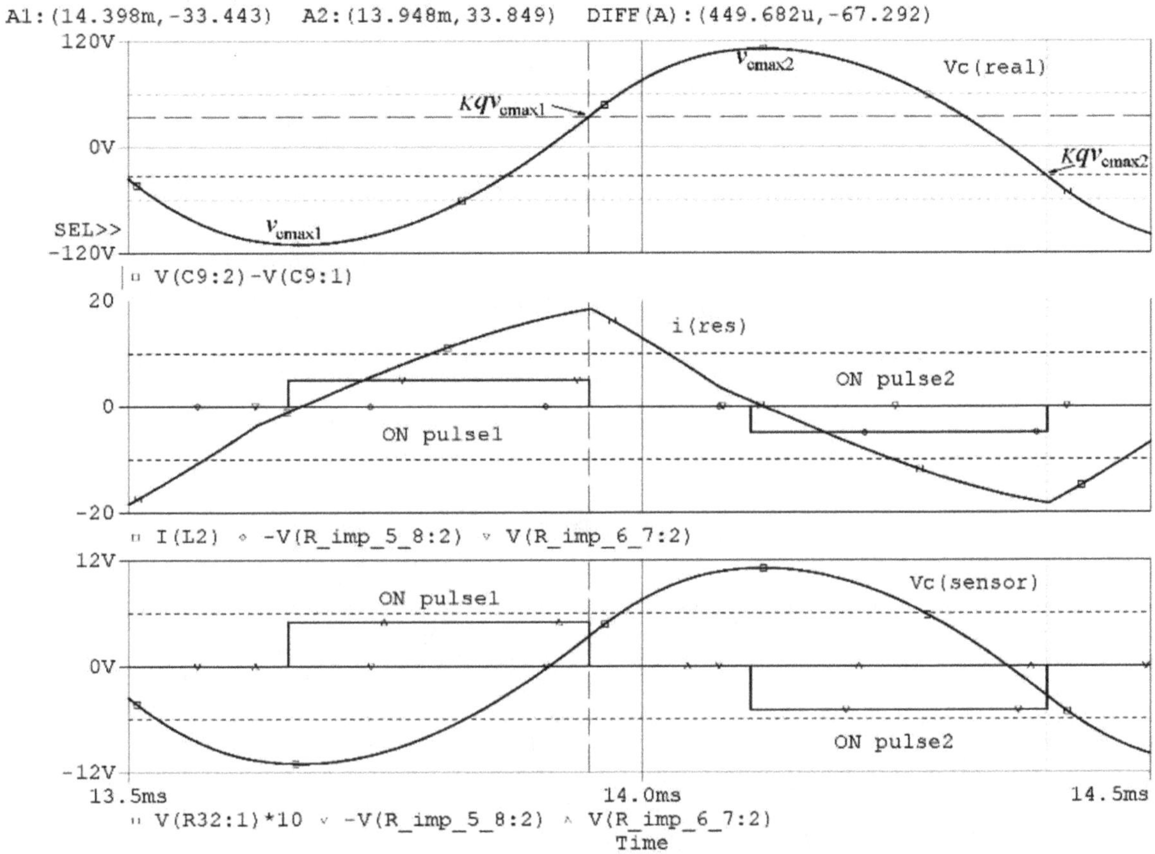

between the really applied pulses and the pulses produced by the control system is in their width. The control pulse in Figure 17 starts when the anti-parallel diode of the same transistor is still conducting, but the conducted current is close to the zero crossing, while the really applied pulse starts immediately after switching off the previous transistor diagonal. The short pulses produce the same effect, as it was pointed out earlier. They eliminate the lagging of the control reaction at the boundary between FM and PWM.

To permit the operation in PWM (at a fixed maximum frequency), the circuit of Figure 16 was provided with four latches which do not allow for the frequency to rise above a certain defined value. When the half wave of the resonant current finishes too early, the latches do not allow for the next pulse to start but force the circuit to wait until the necessary minimum width of the half period finishes. In the next circuit, to simplify this mostly analogue implementation, the latches are excluded but in the same time, the real gate pulses are provided by the control circuit (Figure 18).

To prove that the circuit of Figure 18 is operating in the same way as anticipated by the functional (block) diagram of Figure 14, its simulation results are presented in Figure 19. The plots in the figure are produced in a steady-state, reached by sufficiently long time of simulation run. The circuit operation demonstrated in Figure 19 produces graphics like the plots already shown in Figure 17, with the disconnected feedback, as in Figure 16.

Figure 18. SLSR converter instantaneous control circuit with connected feedback and removed latches

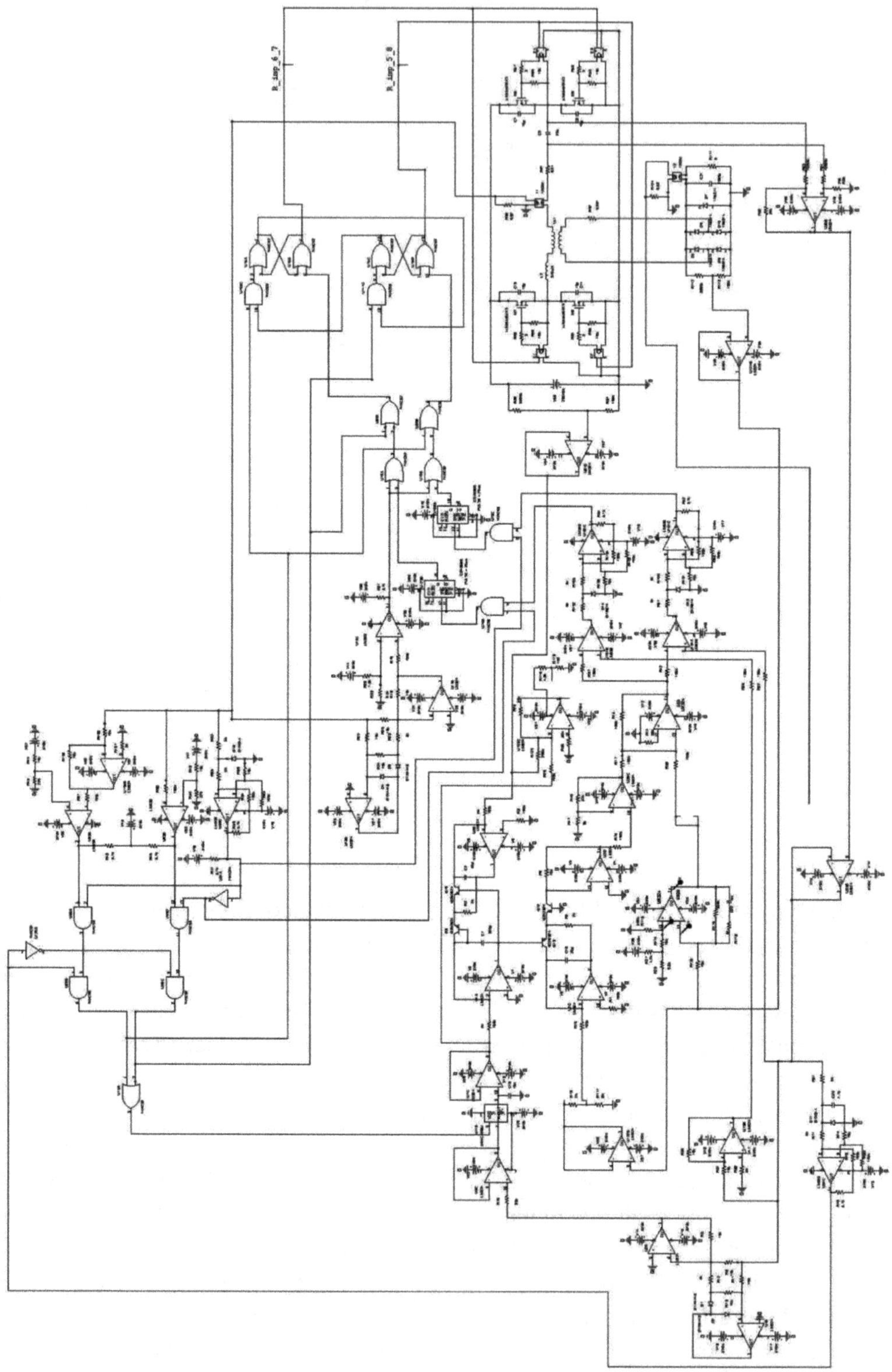

Figure 19. Simulation results of circuit in Figure 18. Upper plot: vo; Middle plot: vc/8 and i(Lr); Lower plots: control pulses to the power transistors diagonals

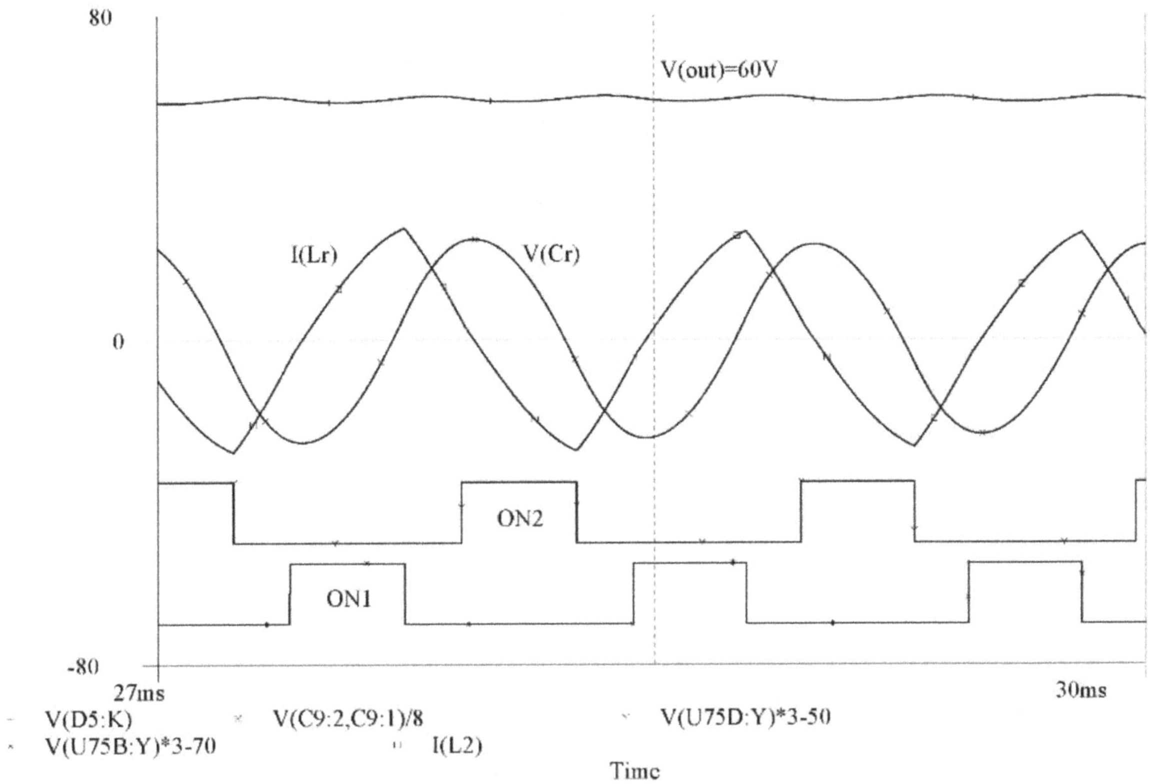

In steady-state the control is producing symmetrically displaced gate pulses, that are shown in the lower plots of Figure 19. The middle plots exhibit the waveforms of the resonant current and resonant capacitor voltage (value, reduced 8 times, for easier display). The upper plot corresponds to the output voltage, showing a visible ripple because of the relatively low capacitance (1000 μF) of the filter and because of the scaled (lowered) switching frequency.

The illustration of the decision-making procedure can be found in Figure 20. It illustrates the idea of the energy correction (i.e. by applying the scaled error voltage). The previously measured amplitude of the resonant capacitor voltage $V(C_r)_{max}$ is multiplied by the normalized output voltage q. The "Error Signal" (shown in the figure) is added to the signal "$qV(C_r)_{max}$". The result of the calculation is then inverted because the actual capacitor voltage is negative. Then the value named "$V(C_r)$-calc.", equal to the sum of "$qV(C_r)_{max}$", and the "Error Signal" is then compared to the measured voltage of the resonant capacitor "$V(C_r)$-measured".

The gate pulse in Figure 20, has its end (marked "OFF") exactly when the measured resonant capacitor voltage crosses the calculated "$V(C_r)$-calc." value. The beginning of the gate pulse (denoted "ON") appears slightly before the maximum of the resonant capacitor voltage, when the anti-parallel diode is still conducting. By this switching on of the transistor the ZVS is guaranteed. In Figure 20 the signal "Error Voltage" is changing relatively slowly, because of the heavy filtering.

Figure 20. Control pulses "ON pulse" and the resonant capacitor voltage oscillation "V(Cr)-measured" compared with the calculated value for the turn-off: "V(Cr)-calc."

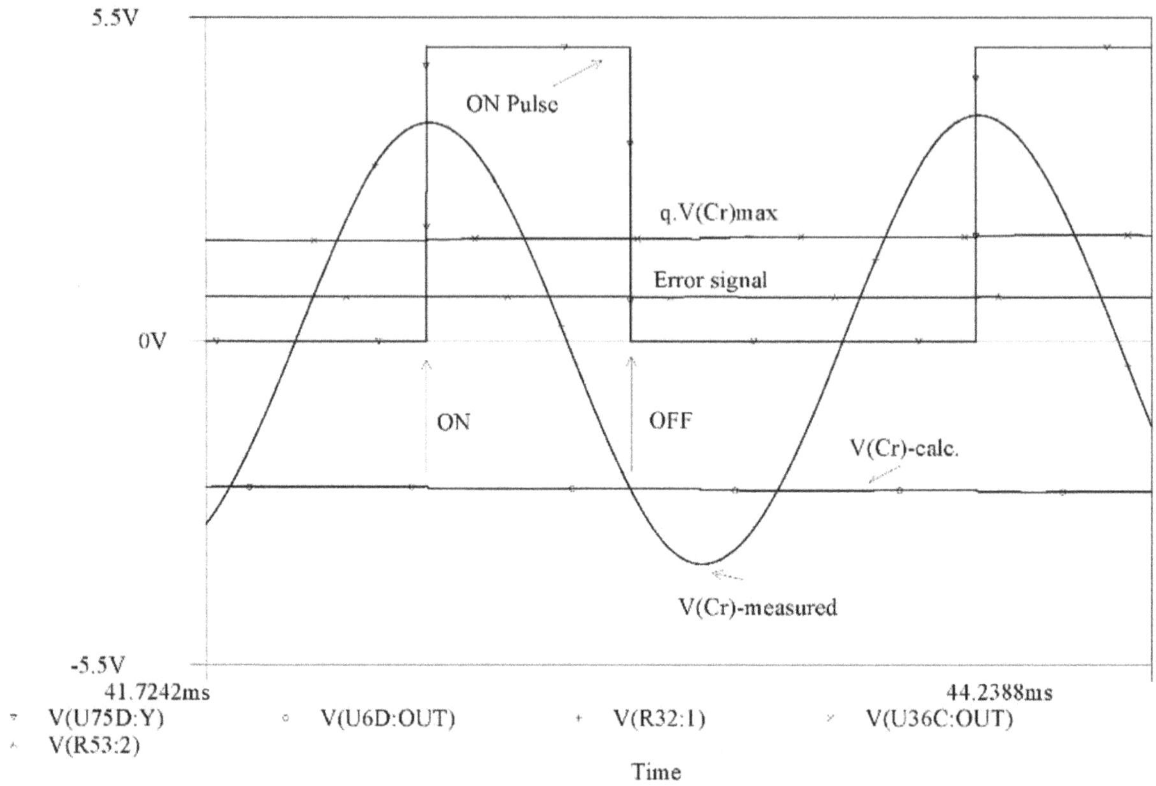

The reaction of the converter to a very fast command to increase the output voltage from zero to a defined value is shown in Figure 21. This figure represents graphically the maximum possible step up speed. In fact, it shows an almost completed sinusoidal resonant current waveforms.

The initiation of the process is based on the analogue circuits asymmetry and the first switching half period is very short (Figure 21). After this initial oscillation the resonant current waveforms represent an almost complete half-sinusoid. The output of the logical circuit "U90B:A" turns on the final latch, producing the gate command for the first diagonal of power transistors (positive half waves of the resonant current). The output of "U75B:B" is turning off the same latch. The operation of "U90A:A" and "U90B:Y" is analogous, concerning the negative half-wave diagonal.

It is seen that the stop-pulses (U75B:B and U90B:Y) sometimes appear twice. This is completely normal as the first pulse is the one calculated by equation (12). The second stop-pulse is emitted synchronously with the start pulse of the opposite diagonal. This second pulse is the guarantee of the super-resonant functioning as it will not permit the transistor to remain conducting when the resonant current waveform rear slope is approaching its zero crossing. The second stop-pulse has no influence if the transistors are already stopped, but its importance is seen during the transient, when the current waveforms are almost completed sinusoids. Another stop-pulse could possibly appear when the protection circuit switches-off the transistors prematurely.

Figure 21. Output voltage V(out), primary current I(L) and resonant capacitor voltage V(C) in transient process. Command logic: 1st diag. U90B:A-start, U75B:B-stop; 2nd diag. U90A:A-start, U90B:Y-stop

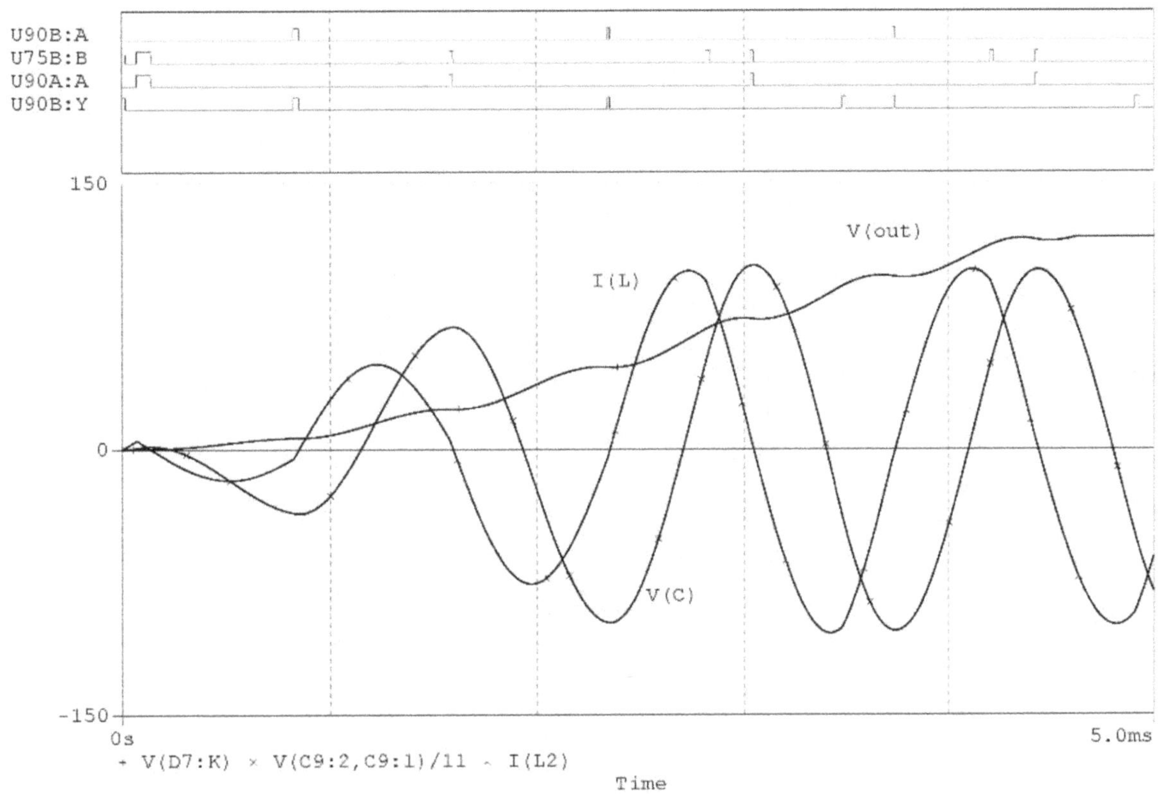

The process of fast increase of the output voltage is represented more clearly in a state variable plane (v_c, i_r) plot, as shown in Figure 22. This state variable diagram is produced from the simulation results presented in Figure 21 and includes the initial two and a half periods starting from zero. The horizontal axis in Figure 22 is scaled by multiplying it by 11.5, in order to achieve a better round-shaped state variable diagram of the resonant capacitor voltage v_c and the resonant current i_r.

In Figure 22 the first (positive) half-wave of the resonant current has a very low amplitude (roughly 5 A) and the resonant capacitor voltage is almost not oscillating (roughly v_{cmax}=2 V). This first half wave ends when the (indicated by arrow) command "T_{off1}" is produced by the logic circuit "U75B:B" (shown in Figure 21). This is the result of the first calculation of Kqv_{cmax} when v_{cmax1} is initially zero. For this first calculation, the correction value cannot be left zero and to achieve excitation, some specified low v_{cmax} value is introduced. The first oscillation in an (analogue) auto-oscillating circuit depends always on this asymmetry of the construction. A future digital control can introduce a more appropriate value for the initiation of the process.

The subsequent turn-off points in Figure 22 represent the regulation better. For example, "T_{off2}" appears after almost full 180° of the resonant half-wave. In fact, this is the maximum energy portion which the LC tank can transfer. The next half-waves (after T_{off2}) are represented by gradually shorter arcs than the full 180° (i.e. points T_{off3}, T_{off4}, T_{off5}, etc.). This reflects the error signal decrease. Approaching the

Figure 22. State variable diagram (vc and ir) of a transient process at increasing output voltage (ir is scaled through multiplying it by 11.5)

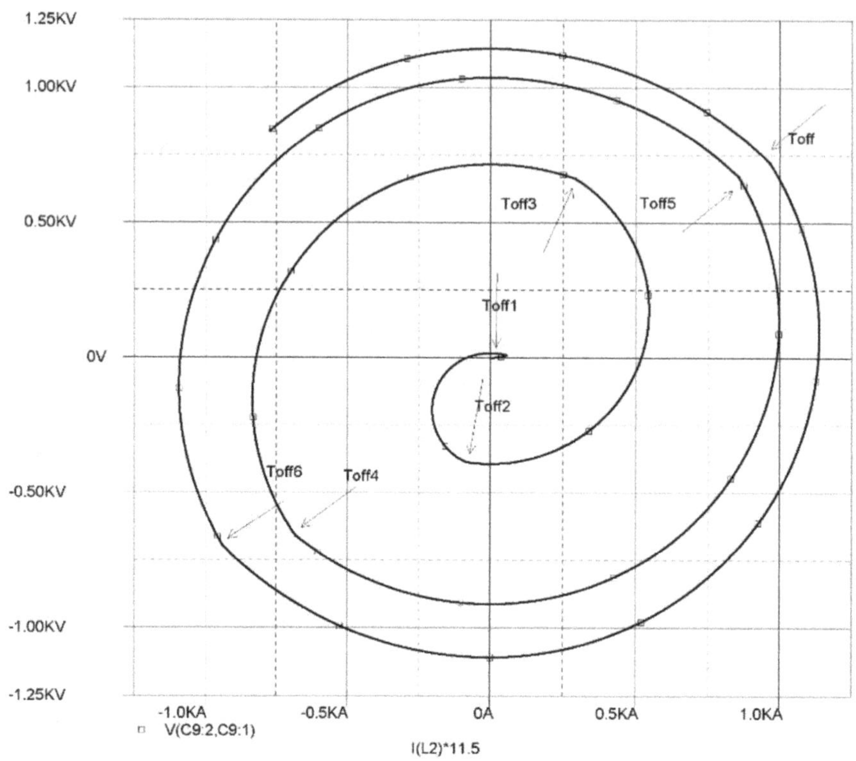

steady-state operation, the difference between the initial amplitude v_{cmax1} and the final one v_{cmax2} becomes gradually smaller, entering a steady-state operation.

The high speed of increase that the output voltage can achieve is seen in Figure 23 where the steady state is reached in 40 µs real time (i.e. 4 ms of scaled time).

When a high-speed digital circuit, i.e. FPGA, would be used the calculations will be done more precisely and the transient is expected to be even faster. The limited speed of this analogue implementation has mainly its explanation in the block indicated as "Rear Slope Level Comparator" from Figure 14 which produces the start-pulse earlier than the resonant current crosses the zero (necessary for the ZVS). In the case when the compared resonant current is low or the output voltage is low, the pulse will be produced much earlier. Unfortunately, the actual transistor stop-pulse will be produced equally earlier, thus limiting the speed of the transient process. This must be corrected by applying a proportional level of the "early than zero" detection, which is easier when the circuit is programmable (i.e. digitally implemented).

To compare the normal steady-state operation shown in Figure 23 to the emergency case of activated protection circuits, the internal current level in the circuit of Figure 18 was limited at 100 A, but the consumption of the load corresponds approximately to this limitation. This means that the resonant current is cut prematurely if more than 100 A is necessary for the given half period of the resonant current and as a result the operation is potentially unstable. The result is shown in Figure 24.

Figure 23. Transient process from zero to a fixed output voltage V(out)

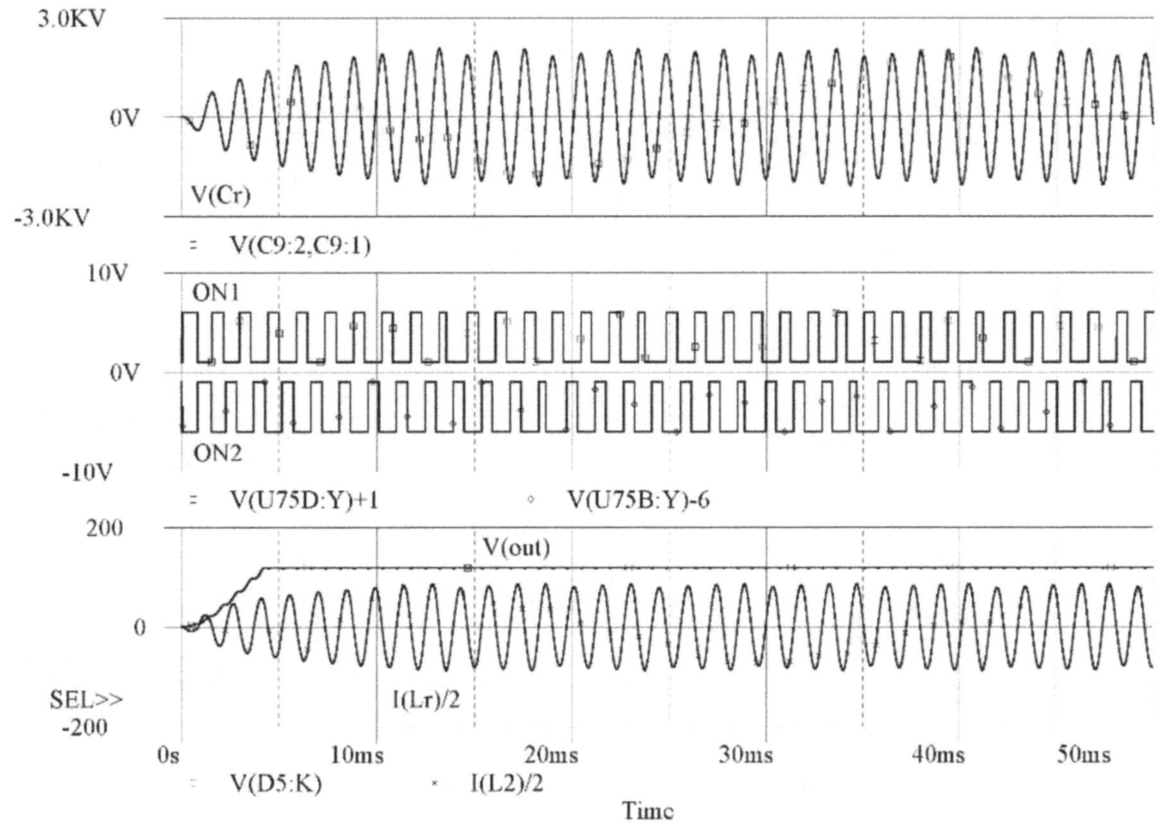

As it can be seen the disturbance is not affecting the general result of the converter operation even if the current limit is reached every 2-3 periods. Some variations are observed in the resonant capacitor voltage amplitude, but this is also observed slightly during the normal (unlimited) operation shown in Figure 23. These variations are expected also in the real circuits, as the resonant capacitor value will have slightly different initial conditions and tolerances. The circuit is correcting those frequent disturbances by its immediate reaction in correcting the width of the next pulse.

As it is shown in Figure 24 some of the control pulses have different width, and some of the resonant capacitor voltage amplitudes are also different. As each pulse width is decided individually, the switching frequency F is only approximately constant in a steady-state. In fact, the slight change of the frequency is compensating the resonant capacitor voltage variation. It is a difficult challenge to keep constant either the amplitude v_{cmax} or the frequency F (even in steady-state). A slight change in one of them needs correction in the other, but this correction can be executed only in the next half period.

An experimental set is shown in Figure 25. The maximum power during the experiments was more than 300W.

Figure 24. Transient process and maintaining the fixed output voltage V(out) in conditions of limited resonant current

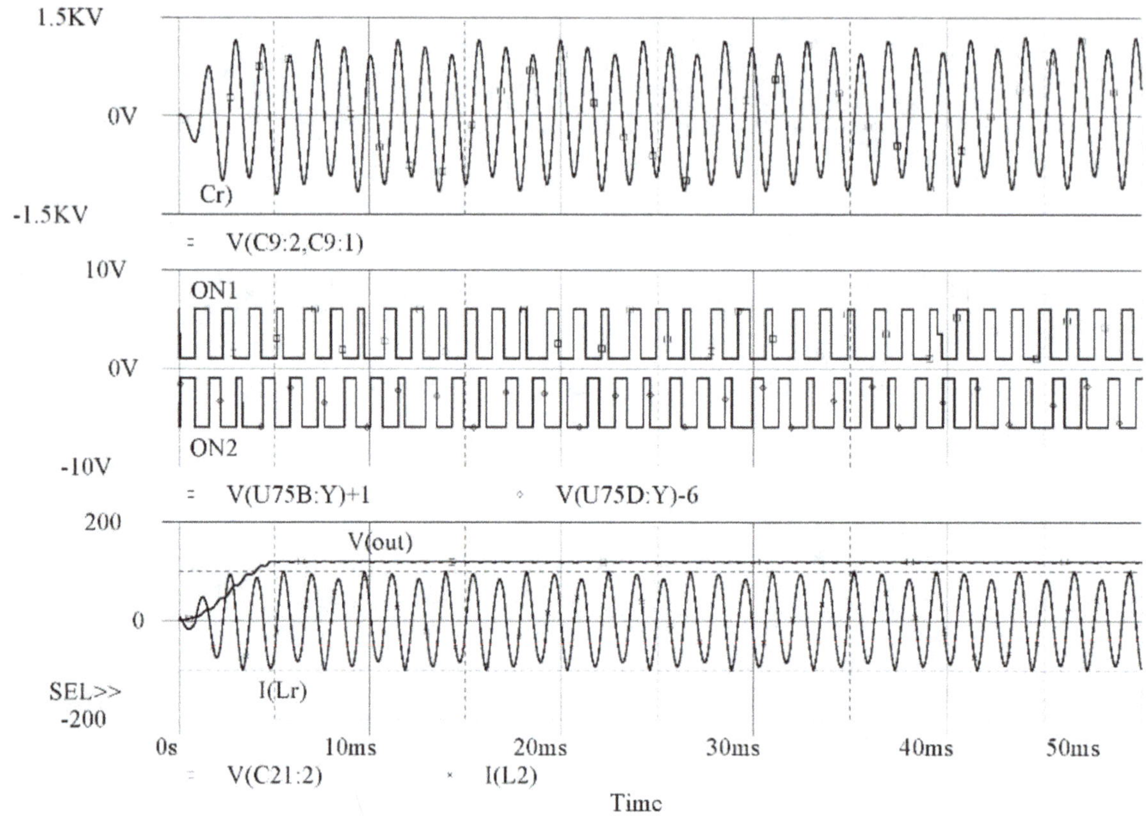

Figure 25. Experimental SLSR power converter for wireless energy transfer

7. CONCLUSION

The three control methods described in the beginning of this chapter: frequency mode (FM), pulse width mode (PWM), and their combination (FM/PWM), served as a base for several power converter constructions, also in contactless versions. The basic FM and PWM methods are presented along with their problems. The limited speed of transient reaction and limited scale of regulation of the FM method and the large scale of regulation of the PWM method are combined in an original combined FM/PWM method which shows a deep regulation for the low output power and a high efficiency for the high output power. Unfortunately, the problem of the transient speed remains in this combined regulation and another problem appears: the lagging reaction at the boundary between PWM and FM operation.

A solution to the difficulties of the combined FM/PWM regulation is presented as a fourth, totally new method for instantaneous regulation of the energy delivered to the resonant circuit. This new instantaneous approach for the regulation involves measuring the state variables and regulating each portion of supplied energy individually. The result of this regulation is similar to the effect produced by the current mode (CM) control applied to the hard-switching power converters. By properly calculated switching, the highest possible speed of the transient process is achieved.

This instantaneous control method allows to regulate the contactless energy converter by applying the value of the magnetic coupling coefficient implicitly, as a variable which is observed and considered in the total calculation, the total feedback inclusion. Inside the primary side, in the transmitter of energy, the converter is taking care for its healthy operation. The internal good operation is important to be fast and safe. The outside feed-back is including the result: the energy obtained by the receiver. This process is mainly left at the responsibility of the receiver, especially when the receivers are numerous, e.g. dynamic charging of the cars on the road. Thus, the circuit follows the changes in the magnetic coupling slowly and the operation of the contactless power converter is internally corrected rapidly, considering the resulted energy transfer accordingly, but slowly.

The presented schematic implementation in the chapter is mainly analogue and its viability is demonstrated by simulation results. The experimental converter works now at a higher frequency by applying digital circuits (FPGA). The control is and will be further improved, and it is recently becoming more flexible, functional and reliable. The time limitations (difficult to apply in the analogue version) will be further widened, thus permitting a simpler solution for the maximum frequency limitation and for the deep PWM regulation. The minimum frequency limitation is not so difficult to be made safely, to avoid the possible dangerous sub-resonant operation and to avoid the excessive power circulation in the resonant tank.

The problems now are mainly connected to the capacity of the controller for a rapid calculation at the speed that is necessary. For example, the most widely applied switching frequency for the electric vehicle charging is 110 kHz. This means that the decision will be taken in tens of nanoseconds, as the resonant capacitor voltage has a useful slope length of not more than 2 μs to 4 μs (including the necessary shortened pulses). If the calculated points are to be at least 50, the controller calculates and reacts in 10-20 ns. The solution of this problem is currently under investigation.

REFERENCES

Amghar, B., Darcherif, A.-M., Barbot, J.-P., & Boukhetala, D. (2014). Modelling and control of series resonant converter for high voltage applications. *Proceedings of IEEE Energycon 2014*, 216-221. DOI: 10.1109/ENERGYCON.2014.6850431

Dias, J.-V., Kim, H., & Jang, D. (2011). Computer model for railway inductive power supply using Valtchev model. *Proceedings of International Conference on Electrical Machines and Systems (ICEMS)*. 10.1109/ICEMS.2011.6073510

Gigov, G., Krusteva, A., & Valtchev, S. (2016). Experimental study of wireless inductive system for electric vehicles batteries charging. *Proceedings of 17th IEEE-PEMC2016 IEEE International Power Electronics and Motion Control Conference*, 286-290. 10.1109/EPEPEMC.2016.7752012

Hewlett-Packard. (1982). Alimentations modulaires 50 W pour les constructeurs OEM (revue du Hewlett-Packard). *Toute l'électronique, 471*, 43-48.

Jang, Y., Jovanović, M., & Ruiz, J., & Liu, G. (2015). Series-resonant converter with reduced frequency-range control. *Proceedings of IEEE Applied Power Electronics Conference and Exposition (APEC)*, 1453-1460. DOI: 10.1109/APEC.2015.7104539

Jovanovic, M., Liu, K.-H., Oruganti, R., & Lee, F. (1987). State-plane analysis of qasi-resonant converters. *IEEE Transactions on Power Electronics*, *2*(1), 36–44. doi:10.1109/TPEL.1987.4766330

Kim, M., Lee, D. S., & Youn, M. J. (1991). A new state feedback control of resonant converters. *IEEE Transactions on Industrial Electronics*, *38*(3), 173–179. doi:10.1109/41.87584

Klesser, H., & Klaassens, J. B. (1986). Transformer-induced low-frequency oscillations in the series-resonant converter. *Proceedings of IEEE Power Electronics Specialists Conference PESC*, 236-245. 10.1109/PESC.1986.7415567

Lyshevski, S. E. (2000). Resonant converters: Nonlinear analysis and control. *IEEE Transactions on Industrial Electronics*, *47*(4), 751–758. doi:10.1109/41.857955

Medeiros, R., Valtchev, S., & Valtchev, S. (2012). The efficient and stable charging of electric vehicle batteries: simplified instantaneous regulation. Springer Verlag Series: Technological Innovation for Value Creation. *IFIP Advances in Information and Communication Technology, 372*, 486-493.

Mohan, N., Undeland, T., & Robbins, W. (2003). *Power electronics (converters, applications and design)* (3rd ed.). New York: John Wiley & Sons.

Nguyen, V., & Dhyanchand, J. (1987). An implementation of current-mode control for a series-resonant DC-DC converter. *Proceedings of IEEE Applied Power Electronics Conference APEC*, 266-273. 10.1109/APEC.1987.7067161

Nguyen, V., Dhyanchang, J., & Thollot, P. (1986). Steady-state and small signal analysis of series-resonant converter: a novel graphical approach. *Proceedings of IEEE Industrial Application Society Conference IAS*, 692-701.

Rossetto, L. (1992). A simple control technique for series resonant converters. *Proceedings of IEEE Power Electronics Specialists Conference PESC'92*, *2*, 787-792. 10.1109/PESC.1992.254803

Souesme, B., Cheron, Y., & Metz, M. (1989). Study of a control method to gain the best dynamic performances of the series resonant converter. *Proceedings of European Power Electronics Conference EPE'89*, 1041-1047.

Valtchev, S. (1980). *Transistor generators for high-frequency (440 kHz) induction heating. Investigation Contract Report, VMEI, Dept. Power Electronics*. TU Sofia.

Valtchev, S. (1988). *High-frequency DC-DC 2.5 kW converter (a super-resonant mode voltage booster for the telephone exchanges battery rooms, with 15 output stages in parallel). Industrial Contract of TU Sofia and The Plant for Power Supplies*. Report, Pernik.

Valtchev, S. (1990). Some regulation characteristics of pulse-width modulated series resonant power conversion. *Proceedings of Power Electronics and Motor Control Conference PEMC'90*, 83-87.

Valtchev, S. (2001). *High-frequency super-resonant series resonant DC/DC converters (converters of 3 kW, 5 kW, 10 kW, 17 kW, some for contactless energy transfer)*. Industry Consultancy Contract.

Valtchev, S., Jorge, L. R., Craciunescu, A., & Brandisky, K. (2012). High frequency wireless energy for improving the EV autonomy. *Proceedings of International Symposium on Electric Vehicle and 4-th Annual Conference of Polish Society for Environment Friendly Vehicles*.

Valtchev, S., & Klaassens, J. B. (1990). Efficient resonant power conversion. *IEEE Transactions on Industrial Electronics*, *37*(6), 490–495. doi:10.1109/41.103453

Valtchev, S., Klaassens, J. B., & Wesenbeeck, M. P. N. (1995). Super-resonant converter with switched resonant inductor with PFM-PWM control. *IEEE Transactions on Power Electronics*, *10*(6), 760–765. doi:10.1109/63.471296

Valtchev, S., Medeiros, R., Valtchev, S., & Klaassens, J. B. (2011). An instantaneous regulation for the wired and wireless super-resonant converters. *Proceedings of International Telecommunication Energy Conference INTELEC*. 10.1109/INTLEC.2011.6099875

Valtchev, S., Miletiev, R., Arnaoudov, R., & Valtchev, S. (2009). Control strategy for efficient operation of super-resonant SLSR (contactless) converters. *Proceedings of ICEST'09, Veliko Turnovo*, 527-530. http://pessoa.fct.unl.pt/ssv/ICEST_2009_Valtchev.pdf

Valtchev, S., Neves-Medeiros, R., Krusteva, A., Gigov, G., & Avramov, P. (2014). A wireless energy transceiver based on induction heating equipment. *Proceedings of 16th International PEMC Conference and Exposition, Antalya, Turkey*, 1017-1023. 10.1109/EPEPEMC.2014.6980642

Valtchev, S., & Valtchev, S. (2012). Experimental study of the energy portions control in series resonant converter. *Proceedings of 13th International Conference OPTIM*. 10.1109/OPTIM.2012.6231994

Wahjudi, A. (1992). *Modelvorming en simulatie van een super-resonante omzetter met een hoogspanningstransformator* (M.Sc. thesis). Faculteit der Elektrotechniek, TU Delft.

ENDNOTES

1. The simulation is done by employing realistic circuit elements, and because of this is showing slightly deformed rear slope of the resonant current by the magnetization current.
2. For example, the measuring in a half-bridge converter with divided resonant capacitor is easier as one of the measuring points is the ground. The full bridge power stage requires the use of a differential amplifier.

Chapter 6
Electromagnetic Influence of WPT on Human's Health:
Modelling, Simulation, and Measurement

Elena Baikova
Instituto Politécnico de Setúbal, Portugal

Luis Romba
Universidade Nova de Lisboa, Portugal

Stanimir Valtchev
Universidade Nova de Lisboa, Portugal

Rui Melicio
Universidade de Lisboa, Portugal

Vitor Fernão Pires
Instituto Politécnico de Setúbal, Portugal

ABSTRACT

The focus of this chapter is the electromagnetic interference (EMI) and the electromagnetic compatibility (EMC) that the wireless power transfer (WPT) systems reveal as problems. The wireless power transfer (WPT) was introduced by Nikola Tesla more than one hundred years ago, and only recently it attracted the attention of specialists, due to the improved technical means. The WPT technology now has many applications, especially for charging of various electronic devices (i.e., mobile phones, laptops, implants, and home appliances), informatics, and electronics equipment. The high-power equipment and installations (e.g., intelligent machining systems, robots, forklift trucks, and electric cars) are also getting wireless. Moreover, much attention has been focused on the electric transportation system for improving the safe and convenient charging of electric vehicle (EV) batteries.

DOI: 10.4018/978-1-5225-5870-5.ch006

1. INTRODUCTION

The focus of this chapter is the electromagnetic interference (EMI) and the electromagnetic compatibility (EMC) that the Wireless Power Transfer (WPT) systems reveal as problems. The wireless power transfer was introduced by Nikola Tesla more than one hundred years ago, and one decade ago was attracted the attention of specialists, due to the improved technical means. The WPT technology has now many applications, especially for charging of various electronic devices, i.e., mobile phones, laptops, implants and home appliances, informatics and electronics equipment (Lu, Wang, Niyato, Kim, & Han, 2016). The high-power equipment and installations, e.g. intelligent machining systems, robots, forklift trucks and electric cars are also getting wireless. Moreover, much attention has been focused on the electric transportation system for improving the safe and convenient charging of the electric vehicles (EV) batteries (Boys, Covic, & Green, 2002; Li, & Mi, 2015; Imura, Okabe, & Hori, 2009; Baikova, Valtchev, Melicio, & Pires, 2016a).

Fundamentally, there are two different methods of WPT, defined by physical phenomena of the electromagnetic fields (EMF) propagation: near field and far field. The near field methods are based on capacitive coupling, inductive coupling and magnetic resonance and are used for wireless power transmission over relatively short distances, usually much shorter than 1 m, exceptionally reaching up to a few meters. The far field methods allow greater distances power transfer and usually involve electromagnetic (EM) energy in a form of radiation, including microwave, lasers and radio wave transmissions (Brown, 1984).

Today the inductive coupling is considered a functional and mature technology between the near field methods. The advantages of inductive coupling are the simple design, convenient operation, high efficiency in close distance (typically less than a coil diameter) and high safety (Lu et al., 2016). Despite the above advantages, there are drawbacks, namely a limited transmission range, a low yield and a requirement for accurate alignment between the transmitter and receiver coils (Boys et al., 2002; Ahn et al., 2011). The applications based on capacitive coupling are applicable in the smart card equipment or in small robots (Kline, Izyumin, Boser, & Sanders, 2011), but they are limited by the low transmitted power and the short distance. Among the existent near field methods, the best result was achieved by the group of Massachusetts Institute of Technology that uses magnetic resonance enabling longer distance efficient WPT transmission (Imura et al., 2009, Kurs et al., 2007).

The magnetic resonance technique is recognized to be the most suitable for wireless EV battery charging (Li et al., 2015; Imura et al., 2009; Sample, Meyer, & Smith, 2009) since it has more positional freedom and does not need an accurate parking position of the vehicle as in the case of generic, non-resonant inductive coupling. With the progress of magnetic resonance technology, the transmitted power and the distance between the source and the load is expected to increase. The performance of these magnetic resonance systems requires a bi-directional exchange of information between the transmitter and the receiver, i.e., the required power, the operation frequency, the vehicle identification, the payment information. The EMF, through which the energy is transferred between the transmitter coil and the receiver coil, is a high-frequency field and induces EMI into the other devices. The transmitted power for wireless EV charging is in the order from several to tens kW. So, the high-intensity EMF can have an adverse impact on other electrical and electronic equipment and on the communication channel between the transmitter and receiver. Moreover, the EMF produced by the WPT system may induce high voltages and currents in the human body (Christ et al., 2013). The stray EMF emitted by WPT systems it's far to be uniform and could exceed the reference levels set defined by international guidelines (Pinto, Lopresto, & Genovese, 2017).

The EMC issue of the WPT system requires a detailed analysis of the EM processes in the human body exposed to radiate EM energy. It is necessary to evaluate the human exposure to the WPT systems EMF emission. The "in vivo" study of the EMF distribution inside the human body is not possible. Therefore, the only opportunity to estimate the impact of the EMF on the living organisms is using modelling and simulation (Baikova et al., 2018a). On the other hand, it is possible to measure the EMF intensity produced by the WPT system in the surrounding environment without the presence of the human body. The model will indicate the expected intensity of the possibly dangerous EMF and the measured values will prove the modelling without endangering the living beings. Thus, it is necessary to compare the measured field and the reference level of the ICNIRP guideline. That comparison will be useful for evaluation of the human body exposure.

Recently, some articles were published, discussing the modelling and simulation of the EM process in wireless energy transfer systems. In Isaev, Vasileva, Budko, and Lefebvre, (2016); Bosshard, Kolar, and Wunsch, (2014); Chang, Quan, Zhu, Zong, and Zhou, (2014); Kim, Park, Kim, Kim, & Ahn,, (2014); Wen, and Huang, (2017); and Laakso, and Hirata, (2013), a computational model of the WPT system is created based on the circuit parameters calculation using the finite element method (FEM). The whole WPT system is analysed and verified by using simulation software.

In Isaev et al., (2016), the model for the wireless energy transfer system with inductively-coupled coils of irregular shape is presented. As inductances, coil resistance and self-capacitances are calculated using the EM energy assessment made by the FEM. Also, the 3-D distribution of the magnetic vector potential with varying air gaps between the coils is obtained.

In Bosshard et al., (2014) a modelling of WPT system based on inductive coupling between two coils, using the FEM tools and Ansys Maxwell software, is presented. The calculation methods to predict power loss, equivalent circuit values, and stray fields are discussed. The simulation results are compared to the measured stray field of the inductive energy transfer system.

In Chang et al., (2014) the process of coils designing and of a WPT system building using the FEM and MATLAB simulation is presented. The coupling coefficient of the coils, the efficiency and maximum power transferred via the system have been discussed based on the simulation. To validate the model and to verify the design method the simulation results are compared with the experimental ones.

In Kim et al., (2014) a concept and design methodology for a reactive shield for WPT system is proposed. The reactive shield reduces the incident magnetic field leakage from a WPT system by generating a cancelling magnetic field without consuming additional power. The efficiency of the resonant reactive shield is validated by simulations and measurements.

In Wen et al., (2017) the scenario of multiple WPT systems working in proximity is discussed. The induced electric field and specific absorption rate in a human body standing in the vicinity of the WPT system are also simulated and evaluated by FEM.

In Laakso et al., (2013) an electric field induced in a human body by magnetic field leaked from a WPT system for charging of an electric vehicle with a power value of 7 kW is evaluated. The magnetic field from the WPT system is modelled computationally using FEM and a model is validated by comparison with the field measured in a previous study.

2. RADIATION SAFETY STANDARDS

There are various international organizations which have been involved in the preparation and formulation of the radio frequencies (RF) safety standards for the human being. The international standards, as a rule, have no regulatory or binding status and become mandatory only if officially adopted within a country. The exposure guidelines are based on a critical analysis of scientific research. The limits are established according to scientific studies and are reviewed periodically, to ensure that they meet the current scientific requirements (BC Centre for Disease Control and National Collaborating Centre for Environmental Health [BCCDC], 2013; International Commission on Non-Ionizing Radiation Protection [ICNIRP], 2010; IEEE Standard C95.1., 2005].

Among the most respected standardization organizations and international experts involved in formulating radiation safety standards can be named:

- International Commission on Non-Ionizing Radiation (ICNIRP), an independent organization, which provides scientific advice and guidance on the health and environmental effects of non-ionizing radiation (NIR) to protect people and the environment from detrimental NIR exposure (ICNIRP, 2010);
- Federal Communications Commission (FCC) is an independent agency of the United States government that develops the RF exposure guidelines for USA which are derived from the recommendations of Institute of Electrical and Electronics Engineers (IEEE) and of the US National Council on Radiation Protection and Measurements (NCRP);
- Institute of Electrical and Electronics Engineers (IEEE) is a non-government organization in USA that has issued recommendations for human exposure to RF EMF. The guidelines recommended by IEEE are also applied in Canada. The IEEE safety limits are generally similar to those of the ICNIRP, with a few exceptions (IEEE, 2005);
- International Electrotechnical Commission (IEC) is a non-governmental international standards organization that prepares and publishes international standards for all electrical, electronic and related technologies;
- World Health Organization (WHO) is a United Nations agency specialized in health matters. In 1996 it established the International EMF Project to assess the scientific evidence of possible health effects of EMF in the frequency range from 0 Hz to 300 GHz to protect public health. At present the World Health Organization is working to arrange a base for international harmonization of RF/Microwaves radiation safety standards;
- European Committee for Electrotechnical Standardization (CENELEC) which cooperates in all matters relating to the health aspects and biological effects of EMF of frequencies from 0 Hz to 300 GHz with ICNIRP, WHO and relevant services of the European Commission.

The most respected organization in the safety standards development at the international level is the ICNIRP, which is formally recognized by WHO. The ICNIRP develops guidelines for exposure limits and provides the material for the WHO publications on non-ionizing radiation. The IEEE is also internationally well-recognized organization. The safety standards developed by these organizations are generally similar, with a few exceptions (Scientific Committee on Emerging and Newly Identified Health Risks [SCENIHR], 2015).

The different standards for the occupational public (workers) and for the general public exposure are applied in the most of countries (ICNIRP, 2010; IEEE, 2005). The workers are considered to be healthy adults, conscious of the risks and being exposed only during their working hours. Instead, the general public represents a diversity of the health sensitivity and it is potentially exposed 24 h daily, being unconscious of the risks. So, for general public the exposure limits are stricter.

3. ELECTROMAGNETIC FIELD IMPACT ON HUMAN HEALTH

The increasing use of the WPT technology in industrial and consumer electronic equipment brings more concerns about EMC due to EM emissions produced by WPT systems in the surrounding environment. The EMC applied to the WPT systems is associated not only with the protection of an electrical and electronic equipment from the negative impact of EM emissions that may occur in its surrounding area, but also to the protection of human health against EM pollution produced by the WPT systems.

The EM energy transferred by the WPT system and radiated into the surrounding environment is absorbed by living beings. The level of the transmitted EM power required for largescale or commercial implementation of WPT could be significant (Lin, 2013).

In the recent years, the number of scientific investigations dedicated to the impact of EMF on the living organisms has significantly increased (Redlarski et al., 2015). Some research claims that WPT technology is safe for living beings (Kurs et al., 2007), but it is impossible to predict the impact on human health of this exposure over time (International Agency for Research on Cancer [IARC], 2013; Staebler, 2017).

Numerous researchers have investigated the possible health effects of the EM radiation and have reached the conclusion that EMF is harmful and can have an adverse effect on human body. The EM radiation interacts with biological tissue and can be reflected, transmitted, refracted or dispersed throughout the body (Staebler, 2017). Reflected and scattered fields may proceed in different directions than those of the incident EMF, whereas the transmitted and refracted fields interact with the biological body in different ways. Interactions with the biological body depend on the frequency, waveform and induced field value, as well as energy deposited or absorbed in the biological system (SCENIHR, 2015; Staebler, 2017). In addition, the distribution of the fields within a biological system, namely in the human body, is affected by the distance and location of the source in relation to the body, anatomy, posture and surrounding environment (IARC, 2013; Staebler, 2017).

Exposure to EM fields has a cumulative effect, increasing with exposure time and dose value (IARC, 2013; Rifai, & Hakami, 2014). Even weak electric and magnetic fields, in the case of prolonged exposure, can produce potentially harmful effects on health. The long-term result is the health problems, namely degenerative processes in the central nervous system, leukemia, brain cancer, hormonal disorders (IARC, 2013; Staebler, 2017). Thus, further research is required on the cumulative effect of the EM fields on human health.

Therefore, it is necessary to evaluate the human exposure to the WPT system (BCCDC, 2013). This way, the measurements of EMF strength near the WPT system and the comparison between measured field and the reference level of the ICNIRP guideline can be useful for evaluation of the human body exposure.

Most of the studies are more frequently associated to simulation and measurement of the magnetic field, i.e. B-field, produced by the WPT system Thus, there is a gap regarding the study of electric fields, i.e E-field, irradiated by the WPT system. Nevertheless, the coupling mechanisms of the electric and magnetic incident-field components are different. Therefore, for a more complete characterization of human exposure, both fields must be determined distinctly (IARC, 2013).

Therefore, the simulation and measurement of the E-field become important for the overall characterization of human exposure. This research attempts to fill this gap by examining the electric field created by the WPT system and, consequently, the human exposure to that, contributing to the area of EMC in WPT systems.

The ICNIRP reference levels of E-field for occupational and general public exposure (ICNIRP, 2010) are shown in Figure 1.

According to the ICNIRP Guideline, in the frequency range of 3 kHz to 1 MHz, the limit for electric field exposure, in case of general public, is E = 83 V/m, as shown in the Figure 1. For the occupational public, the exposure limit established by the ICNIRP Guideline is E = 170 V/m.

According to the recommendations of ICNIRP (ICNIRP, 2010) in situations of simultaneous exposure to EM fields of different frequencies, these exposures are additive in their effects. Therefore, in the case of simultaneous exposure to fields of different frequencies, the following criterion called exposure coefficient (ICNIRP, 2010) should be applied:

$$\sum_{j=1Hz}^{10MHz} E_j / E_{R,j} \leq 1 \tag{1}$$

where E_j is the electric field strength at frequency j, $E_{R,j}$ is the electric field strength reference level at frequency j given by ICNIRP (ICNIRP, 2010).

Figure 1. Electric field reference levels (ICNIRP, 2010)

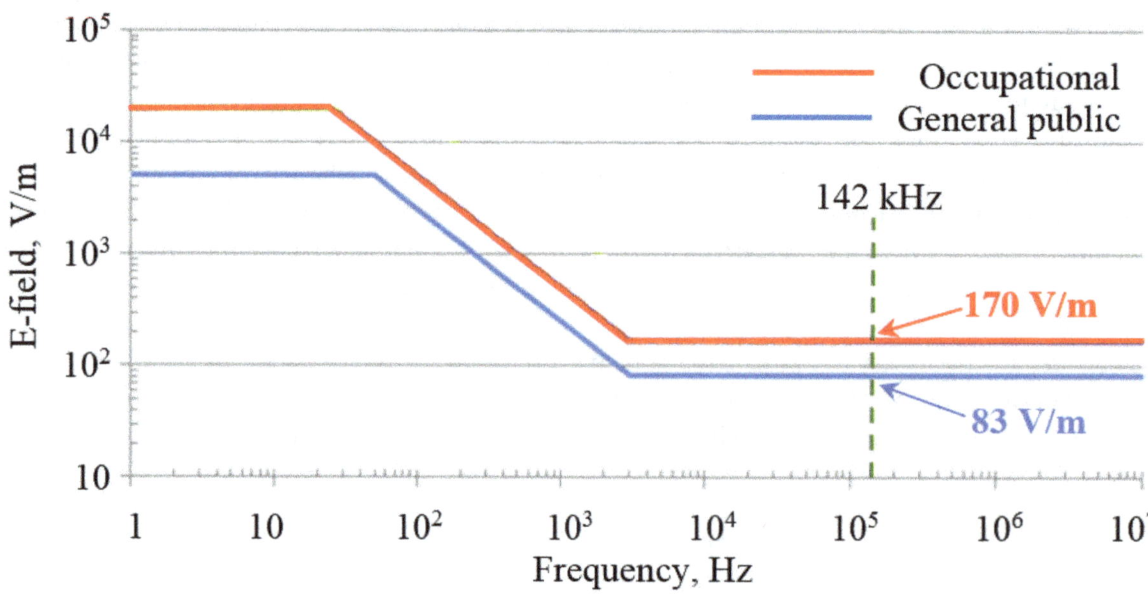

4. MODELLING

The used model of the 2-coil WPT system is the usual one which can be found in the literature (Baikova et al., 2018a). For the electromagnetic field simulation, the FEM for frequency-domain analysis is carried out.

The fundamental governing equations of electromagnetism are Maxwell's equations which describe all the classical electromagnetic phenomena and are given in their differential formulation by (Jin, 2010):

$$\nabla \times \boldsymbol{E} = -\frac{\partial \boldsymbol{B}}{\partial t} \tag{2}$$

$$\nabla \times \boldsymbol{H} = \boldsymbol{J} + \frac{\partial \boldsymbol{D}}{\partial t} \tag{3}$$

$$\nabla . \boldsymbol{D} = \rho \tag{4}$$

$$\nabla . \boldsymbol{B} = 0 \tag{5}$$

where the vectors **E** is the electric field, **H** is the magnetic field, **D** is electric flux density, **B** is magnetic flux density and **J** is conducting current density. Equations (2), (3), (4) and (5) are, respectively, Faraday's Law, Ampere's generalized law, Electric Gauss's Law and the Magnetic Gauss's Law. The four vectors, **E**, **H**, **D** and **B** together form the mathematical representation of the same physical phenomenon - the electromagnetic field.

Also, the equation of continuity can be written as:

$$\nabla . \boldsymbol{J} = -\frac{\partial \rho}{\partial t} \tag{6}$$

The second set of equations, namely, the constitutive relations, which is needed to complete the information contained in the system constructed by equations (2) to (5), is given by

$$\mathbf{B} = \rho_0 \rho_r \boldsymbol{H} \quad \mathbf{D} = \varepsilon_0 \varepsilon_r \boldsymbol{E} \quad \mathbf{J} = \sigma . \boldsymbol{E} \, . \tag{7}$$

From (2) to (7) the magnetic vector potential A can be defined by:

$$\left(\mathrm{j}\omega\sigma - \omega^2 \varepsilon_0 \varepsilon_r \right) \boldsymbol{A} + \nabla \times \boldsymbol{H} = \boldsymbol{J} \tag{8}$$

The magnetic flux density **B** can be defined in terms of the magnetic vector potential **A**, what permits to obtain the magnetic field distribution (Jin, 2010) given by:

$$\boldsymbol{B} = \nabla \times \boldsymbol{A} \tag{9}$$

Finally, the electric field is calculated regarding the Ampere's law (3).

5. PROPOSED DESIGN AND SIMULATION

5.1 Proposed Design

The proposed design for the WPT charger system consists of ferrite cores and of the transmitting and receiving circuits including a couple of coils with two respective capacitors for compensation of the inductance, so that the system can operate at resonance frequency. The coils are made of the copper tube of radius r = 3 mm with number of turns of N = 5. The air gap between the coils can vary in the range from 4 cm to 15 cm.

A Frequency Domain study using a 2D axisymmetric model was used to investigate the WPT system containing two spiral coils with a ferrite core. To the set of the spiral coils, whose geometry is not axisymmetric, a common geometric simplification was applied. This simplification comprises of the representation of a flat spiral inductor as a series of concentric circular wires. Thus, the 3D problem is reduced to a 2D axisymmetric problem, which permits significantly reducing the simulation time.

Through a 2D model, it is possible to generate a 3D model by rotating around the z-axis, as shown in Figure 2.

Figure 2. 3D model generated by rotating the 2D model around the z-axis

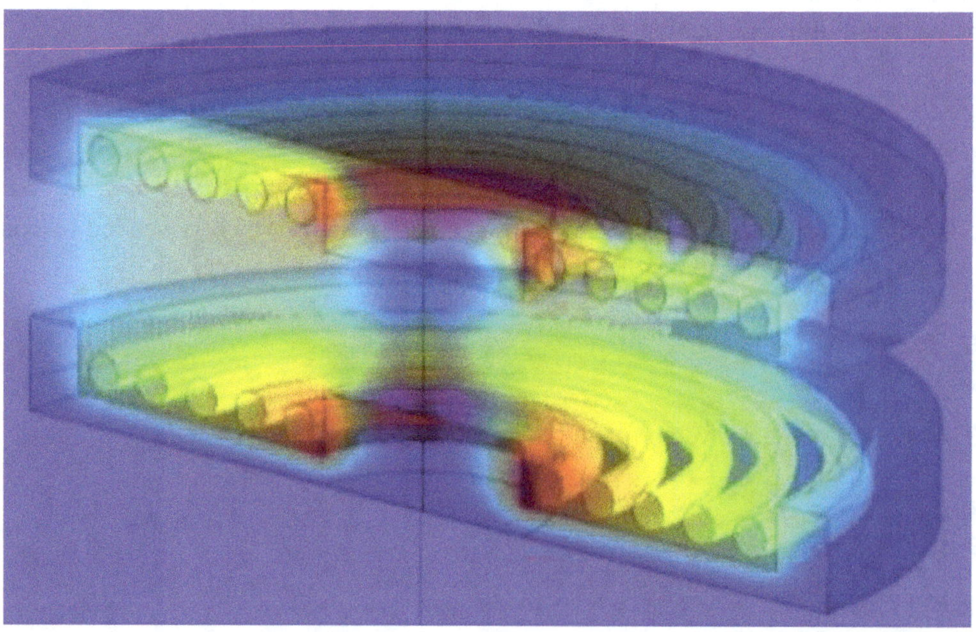

Figure 3 shows the spiral transmitter/receiver coil geometry (a) and the simplified model (b). The length of the conductor in both cases (a) and (b) is practically the same. Thus, it can be assumed that the accuracy of the solution will be minimally influenced. The approach when the spiral coil is modelled as a set of circular currents is often used in the practice of computational simulation (Rueter, 2016; Gendron, Hazel, Boudreault, Champliaud, & Pham, 2017).

The model of the WPT system, including coils, ferrite core and surrounding (air) domain, is simulated for the given setup and dimensions as indicated in Section 6. The model with dimensions in mm, used for the computational simulation is shown in Figure 4.

The mesh obtained for the studied model is presented in Figure 5. The mesh is more concentrated in the region of windings of coils and in the region of ferrite cores, where the field is maximum and has a high gradient.

The rest of the system, namely, power supply, resonance capacitors and load, has been simulated into Electrical Circuits interface. The respective components were added, connected by their node numbers, and simulated as electrical circuits.

5.2 Simulation Results

The intensity distribution of the E-field and the B-field were investigated at the designed resonant frequency of 142 kHz carrying out the FEM analysis and applying the numerical software.

Initially, the strength distribution of the E-field and the B-field with output power Pout = 800 W from WPT system were performed. The E-field distribution for the resonant frequency of 142 kHz and for the air gap between the transmitter and the receiving coils of $\Delta = 4$ cm is shown in the Figure 6. The bar on the right is demonstrating the values of E-field (V/m) in certain points of the model.

Following, the distribution of the E-field intensity at the variation of the distance between the transmitting and receiving coils was simulated. In Figure 7, the simulation results of the electric field distribution for the modelled WPT charging system for the air gap of 4 cm (a), 8 cm (b) and 15 cm (c) between the transmitting and the receiving coils, are shown.

Figure 3. (a) Flat spiral coil of the transmitter and receiver; (b) Simplified model

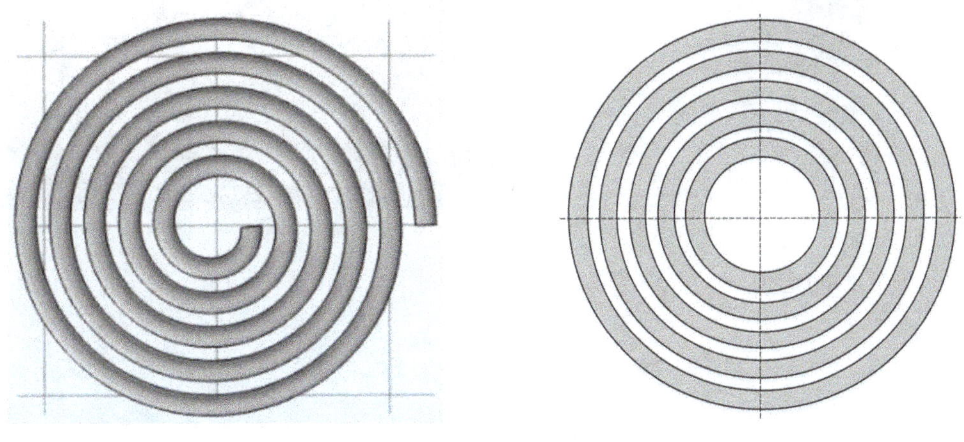

a) b)

Figure 4. Modelling set of ferrite magnetic cores and transmitter/receiver coils

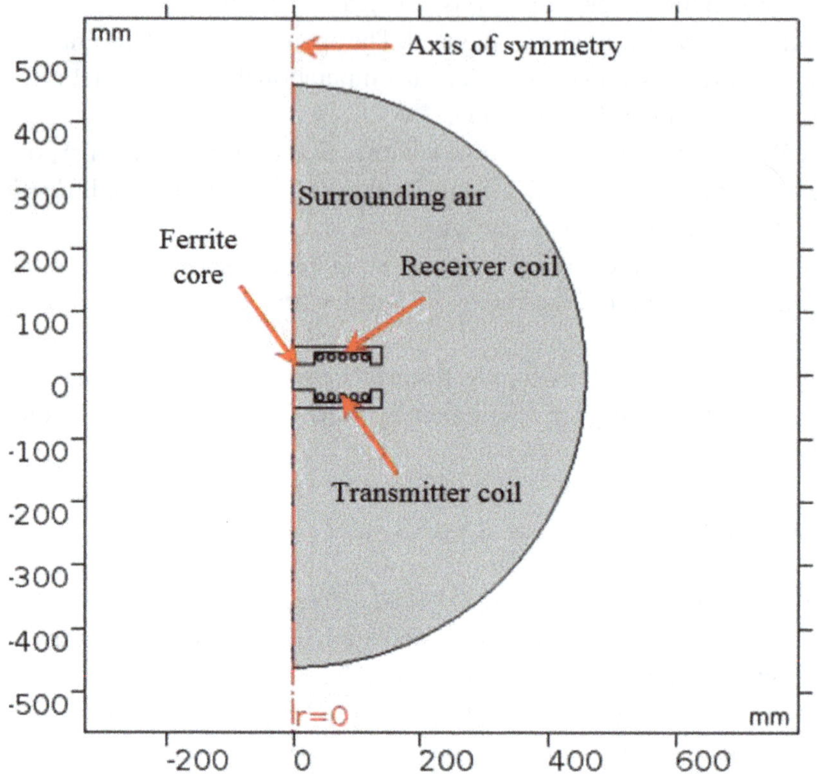

Figure 5. (a) Mesh of the model; (b) Mesh detail in the coils region

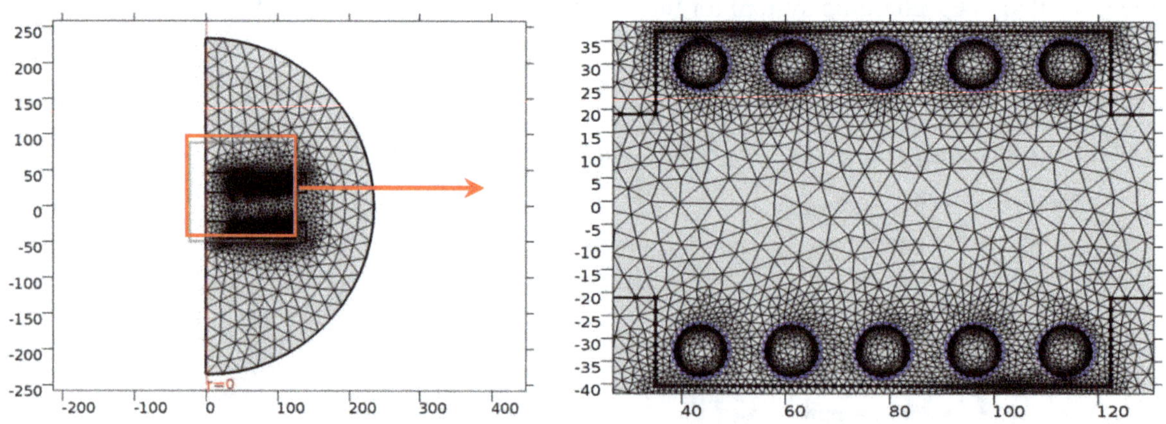

(a) (b)

Figure 6. Electric field generated by WPT system

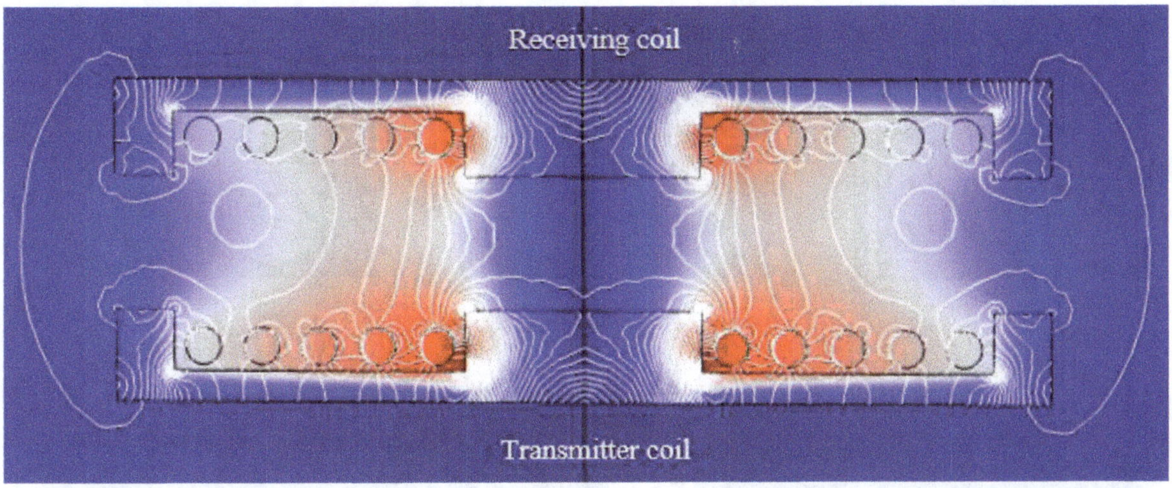

Figure 7. WPT system E-field generated at air gap of: a) 4 cm; b) 8 cm; c) 15 cm

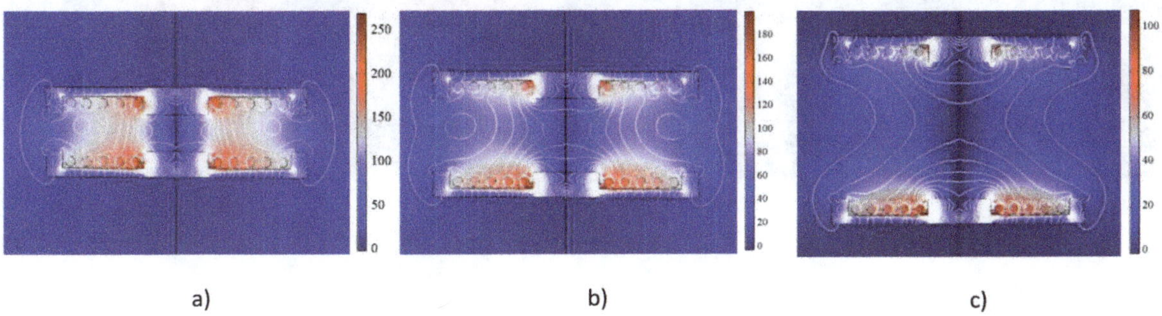

The maximum field intensity in the area restricted by the ferrite cores is around 250 V/m, which exceeds the exposure limits defined by ICNIRP (ICNIRP, 2010). Outside these regions, the field intensity is below the ICNIRP reference level. These results also illustrate that increasing the distance from the WPT system will decrease the electric field strength.

The Figure 8 introduces the magnetic field density distribution around the WPT system for air gap of 4 cm (a) and 8 cm (b) between the transmitter and the receiving coils with ferrite cores. In addition, the magnetic field density distribution for air gap of 4 cm without ferrite cores is shown in the Figure 8 (c). The bar on the right illustrates the magnitude of magnetic flux density B (T). The red arrows characterize the direction and magnitude of the magnetic flux density vector.

As it was expected, the magnetic flux density is more concentrated in ferrite core region, which as well performs the function of the effective shield reducing stray magnetic fields produced by the WPT system.

It should be mentioned, that the simulation demonstrates the function of the ferrite cores as efficient shields which permits to decrease stray magnetic fields produced by the WPT system. Due to high relative permeability ferrite cores is restricting and guiding the magnetic flux. Therefore, the distribution of EM field is restricted to a area between the transmitting and the receiving coils. The distribution of the electric field for WPT system for the air gap of 4 cm with and without ferrite cores, (a) and (b) respectively, is shown in the Figure 9.

Figure 8. WPT system B-field generated at air gap of: a) 4 cm; b) 8 cm; c) 4 cm without ferrite cores

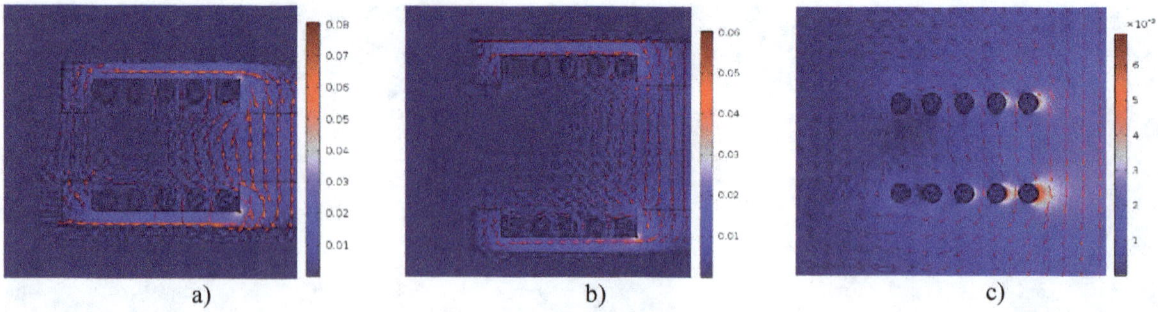

Figure 9. E-field distribution with (a) and without ferrite cores (b)

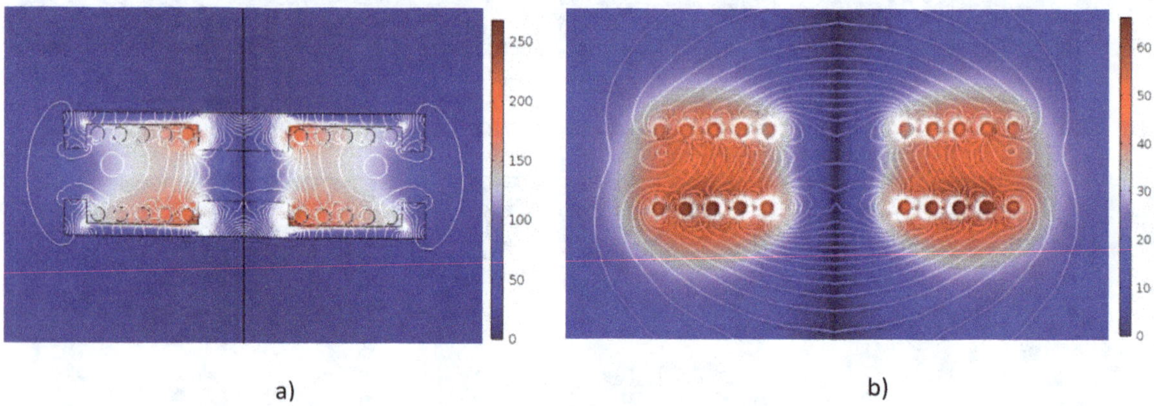

In Figure 9 (a), it can be observed that the electric field generated by WPT system with ferrite cores, is more concentrated between ferrite plates than in Figure 9 (b), i.e. without ferrite cores. Furthermore, the dispersion of electric field is much limited in Figure 9 (a). In addition, the electric field strength generated by WPT system is much higher when ferrite cores are used, as it can be verified from the bar on the right sides of the Figure 9. Thus, due to lower reluctance of the ferrite cores, the EMF is concentrated between them, protecting the biological objects, as well.

The values of the E-field were determined at radial distances of 4 cm, 20 cm and 40 cm from WPT system and are presented in Table 2.

6. EXPERIMENTAL SETUP

6.1 Prototype

For an experimental verification e validation of the model was used a prototype of WPT charger of a previous research (Baikova, Valtchev, Melicio, Krusteva, & Pires, 2016b; Baikova, Valtchev, Melicio, Krusteva, & Gigov, 2016c). The WPT system for EV charging is shown in the Figure 10.

The output power level of the WPT charger system mounted at the Power Supply Electronics Laboratory of the Technical University of Sofia is of 5 kW. The prototype parameters of the WPT charger used in this study can be found in the literature (Baikova et al., 2018a).

The operation frequency and the select distance between the transmitter and receiver coils are typical for the normal operation in charging batteries of electric vehicles. The values of inductances and capacitances provide the same self-resonant frequency of the transmitter and receiver side, which is 142 kHz.

The electric scheme of the studied WPT system for EV charging is shown in Figure 11.

The prototype of the WPT system consists of a solid ferrite core EE6527, the transmitter and receiver coils being made in a copper tube with a spiral configuration. The ferrite cores have demonstrated their vantage as magnetic shields against stray magnetic fields (Strauch, Pavlin, & Bregar, 2015) protecting the living organisms. The circuit parameters of the WPT system are shown in Table 1.

Figure 10. WPT charger system

Figure 11. Electric scheme of the WPT system for EV charging

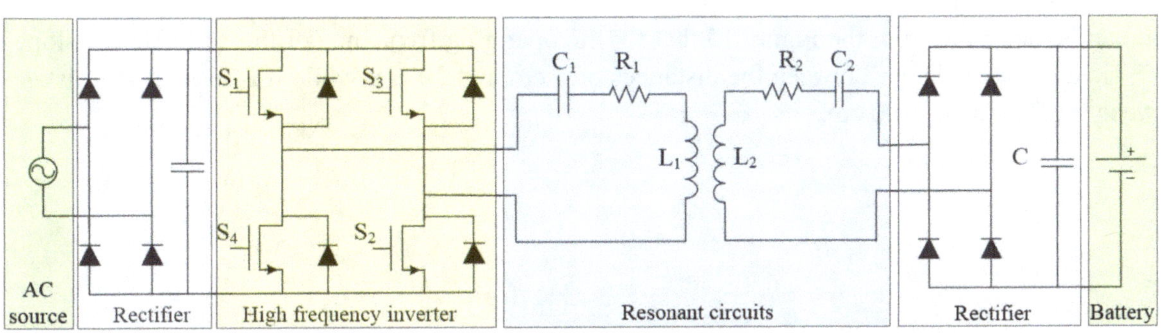

Table 1.System parameters

Parameter	Designation	Value
L_1	Transmitter side self-inductance	64 μH
L_2	Receiver side self-inductance	30 μH
$N_1=N_2$	Number of turns in transmitter and receiver coils	5
C_1	Transmitter side capacitance	19.6 nF
R_1	Transmitter coil resistance	0.276 Ω
Q_1	Transmitter coil quality factor	206
C_2	Receiver side capacitance	41.9 nF
R_2	Receiver coil resistance	0.184 Ω
Q_2	Receiver coil quality factor	145
R_L	Load resistance	32 Ω
f res	Resonant frequency	142 kHz

6.1 Prototype Results

The goal of this experiment is the measuring of electromagnetic field magnitude at different radial distances emitted by the WPT charger for EV operating at 142 kHz frequency range.

To obtain the E-field magnitude, a measuring instrument Narda SRM- 3000 Selective Radiation Meter and an antenna for the 100 kHz÷3 GHz frequency domain was used. The measurements were performed at varying distances from the prototype, namely, of 4 cm, 20 cm and 40 cm, considering the possibility that humans could be found near the EV while charging. The harmonic magnitudes of the electric field produced by the WPT system for the air gap between the transmitter and receiver of $\Delta = 4$ cm, output power Pout = 800 W and with the measurement distance variation between 4 cm and 40 cm (Baikova et al., 2018a) are shown in Figure 12.

The RMS values of the harmonics generated by the WPT system operating at a fixed frequency value of 142 kHz with the air gap between transmitter and receiver $\Delta = 4$ cm and Pout = 800 W is shown in Figure 12. For the harmonics generated by the WPT system at 142 kHz frequency, the dominant harmonics are the 2nd, the 1 st, and the 3rd order harmonics.

More measurements were performed for the different level of output power Pout at the distance from the prototype of 4 cm, 20 cm and 40 cm. The E-field RMS values for the 1st harmonic with varying distance between the transmitting and the receiving coils and for output power Pout of 120 W, 300 W and 800 W is shown in Figure 13

It can be observed from the Figure 13 that for the operating frequency of the 142 kHz the slope of the E-field curves is larger between the distances of 4 cm and 20 cm, while it is less for the distances between the 20 cm and 100 cm.

Figure 12. RMS values of the E-field harmonics generated by the WPT system with air gap Δ = 4 cm

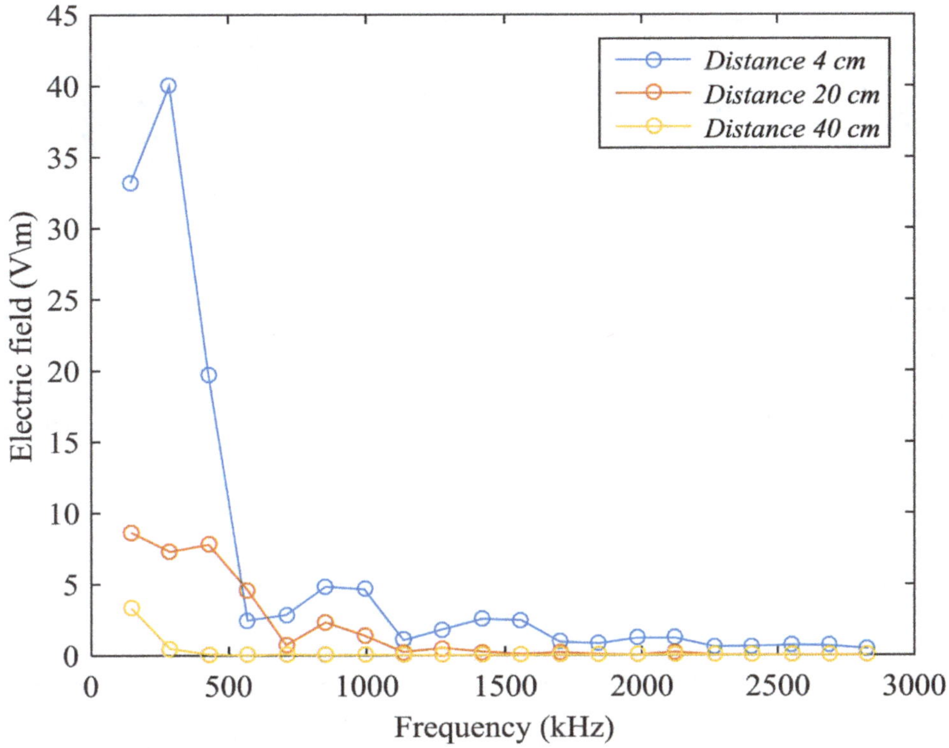

7. DISCUSSING THE RESULTS

In Table 2 the measured and simulated magnitudes of the electric field for output power Pout = 800 W from the studied WPT system are compared.

The simulation results for the electric field distribution in the developed model are in a good conformity with the measurement data of the prototype and have good correlation revealing a simulation error of 7.16%. However, the simulation results are lower than measured results due to the simplifying assumptions adopted in the model. One of the simplifications implies the approach, when a spiral coil is modeled as a set of circular filamentary currents. So, the transmitting and receiving coils of spiral geometry must be adapted for solving in a two-dimensional axisymmetric spatial dimension and were modeled as a set of circular filamentary currents. Such an approach is often used in the practice of computer modeling. One more simplification has to do with the fact that the WPT system considered in this study system is not really a linear system. The error between the simulation result and the experiment result can also be explained by the linearity assumptions adopted in computational modeling.

Considering the case of simultaneous exposure to multiple frequencies, which are additive in their effects, the criteria regarding reference levels of field strengths stipulated by ICNIRP according (1) were applied. The results of applying the multiple frequency exposure criteria and the total level of the first 20 harmonics of the electric field produced by the WPT prototype for the measuring distance of 4 cm are presented in Table 3.

Figure 13. E-field 1st harmonic generated by WPT system

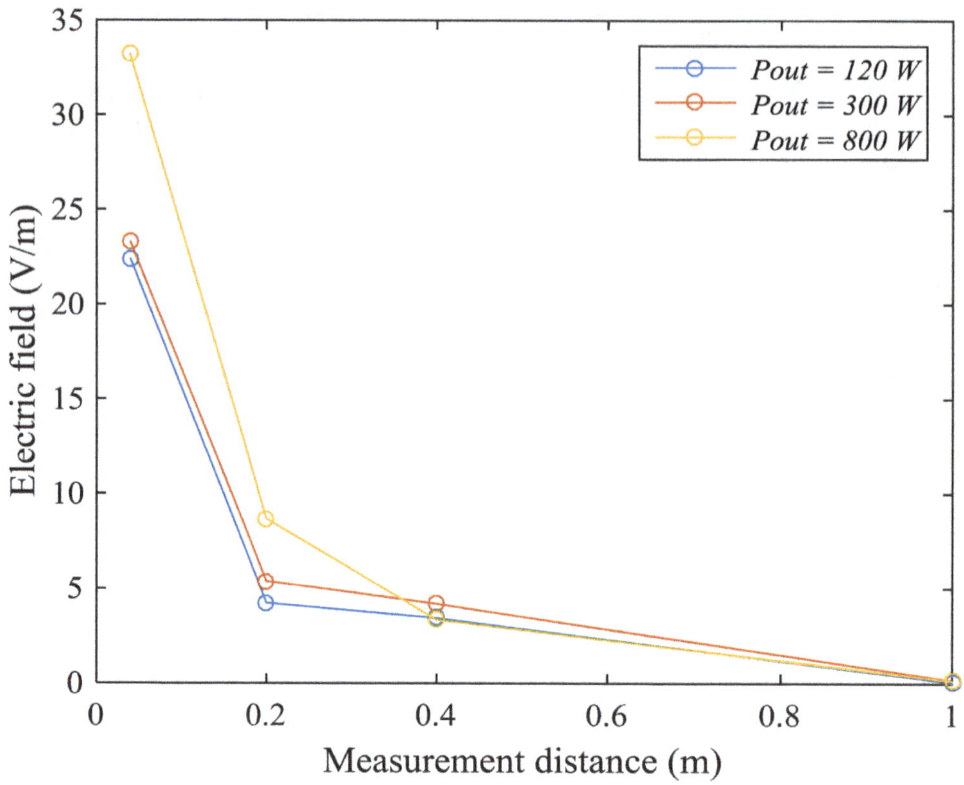

Table 2. Electric field strength

Distance of Measurement (cm)	Air gap Δ = 4 cm		
	Pout = 800 W		
	Electric Field (1st harmonic), V/m		
	Simulated	Measured	Difference
4	31.02	33.19	6.54%
20	7.98	8.65	7.75%
40	3.08	3.32	7.2%

From the simulation results, the following observations are done:

- It is observed that a strong E-field is present in the area restricted by the ferrite cores and near the transmitting coil;
- The maximum value of E-field registered in this area, for the output power of the 800 W, is 250 V/m. Nevertheless, on the outer of the ferrite regions, the electric field level is below the exposure limits determined by ICNIRP (ICNIRP, 2010);

Table 3. Electric field strength of multiple frequencies

Harmonic number	*Pout* = 800 W Air gap Δ = 4 cm E-field magnitude, RMS, V/m
1ª	33.193
2ª	40.018
3ª	19.679
4ª	2.454
5ª	2.8526
6ª	4.811
7ª	4.627
8ª	1.057
9ª	1.821
10ª	2.545
11ª	2.458
12ª	0.940
13ª	0.822
14ª	1.202
15ª	1.199
16ª	0.592
17ª	0.612
18ª	0.703
19ª	0.671
20ª	0.453
Total of harmonics $\sum E_j$, V/m	122.711
Reference level defined by ICNIRP $E_{R'j}$ V/m	83
Coefficient of Exposure $k_{exp}=\sum$ Ej/ ER	1.478

- The ferrite cores play a significant role in the WPT system of confining and guiding the magnetic flux improving coupling and reducing EMI induced in other electronic devices.

From the measured results, the following observations can be done:

- As expected, the experimentally measured magnitude of the electric field is a function of the distance to the measuring antenna from the WPT system, which can be observed in Figure 13;
- The electric field is much stronger near the WPT system, but decreases inversely proportional to the square of distance to the measuring point, i.e. $E \propto \frac{1}{d^2}$, according to the theoretical prediction (Bienkowski & Trzaska, 2012);

- Applying the multiple frequency rule determined by ICNIRP (ICNIRP, 2010) to results of Table 3 it can be concluded that the value of the multiple frequency exposure criteria is 1.478, being significantly greater than the unit. Thus, the overall level of the measured electric field considering the first 20 harmonics is above the exposure limits defined by ICNIRP (ICNIRP, 2010) at 47.8%. The fact that the measured value of the EM field parameters exceeds the reference level does not necessarily imply excessive exposure (Baikova, Romba, Melicio, & Valtchev, 2018b). In these circumstances, it will be necessary to check whether the basic restrictions are met.

8. CONCLUSION

The intensified use of WPT technology in low power and high power applications has raised concerns about possible adverse health effects for humans. Thus, the evaluation of the electromagnetic field impact on the human's health is essential for the wireless power systems future spreading. This impact is not only immediate, but it has also a cumulative effect. This means that the adverse changes in the human body do not occur instantaneously, but during timely exposure and can even affect the subsequent generations.

A WPT radiated emission assessment can be performed using modelling and numerical simulation methods. Based on the modelling and simulation it is possible to predict the probable compliance of a studied WPT to the applicable safety limits or EMC standards.

The application of computer simulation methods in correlation with physical methods of electromagnetic field strength measurements near the WPT system provides novel approaches to construct an efficient and safe WPT, for EV charging, or for another purpose.

In this study, the modelling and simulation of the WPT system operating at the frequency of 142 kHz are performed. Initially, the electric field and the magnetic flux density distribution for different air gap size between the transmitter and the receiver coils was simulated by numerical software for developed WPT model. Then measurements of the electric field, produced by the WPT prototype for EV charging, were performed. Finally, the simulated electric field was compared to the measured values to confirm the validity of the finite element model. The results are presented and compared to the measured stray electric field radiated by the prototype of WPT system.

Considering the case of simultaneous exposure to multiple frequencies, the criteria defined by ICNIRP (2010) was applied to evaluate the additive effect of the electric field of different frequencies produced by the WPT prototype. The total level of the measured electric field considering the first 20 harmonics is above the exposure limits defined by ICNIRP (ICNIRP, 2010) at 47.8%. To make the WPT more safe and acceptable to the public and to protect living beings from elevated levels of the electric field, it is proposed the use of transmitter/receiver system shielding.

ACKNOWLEDGMENT

The authors would like to thank Dra. A. Krusteva and the Power Supply Electronics Laboratory of the Technical University of Sofia for their valuable assistance. This work is funded by Portuguese Funds through the Foundation for Science and Technology-FCT under the project LAETA, reference UID/EMS/50022/2013; and by ICT project (UID/GEO/04683/2013) with the reference POCI010145FED-ER007690; Portuguese Foundation for Science and Technology (FCT) under Project UID/EEA/04131/2013.

REFERENCES

Ahn, S., Pak, J., Song, T., Lee, H., Byun, J.-G., Kang, D., & Kim, J. (2011). Low frequency electromagnetic field reduction techniques for the on-line electric vehicle (OLEV). In *Proceedings of the IEEE International Symposium on Electromagnetic Compatibility*. Fort Lauderdale, FL: IEEE.

Baikova, E., Romba, L., Melicio, R., & Valtchev, S. (2018b). Simulation and experiment on electric field emissions generated by wireless energy transfer. In L. M. Camarinha-Matos, K.O. Adu-Kankam, & M. Julashokri (Eds.), Technological Innovation for Resilient Systems (pp. 243-251). Cham, Switzerland: Springer. doi:10.1007/978-3-319-78574-5_23

Baikova, E., Valtchev, S., Melicio, R., Krusteva, A., & Gigov, G. (2016c). Study on electromagnetic emissions from wireless energy transfer. In *Proceedings of the IEEE 17th International Conference on Power Electronics and Motion Control (PEMC)*, Varna, Bulgaria: IEEE. 10.1109/EPEPEMC.2016.7752045

Baikova, E., Valtchev, S., Melicio, R., Krusteva, A., & Pires, V. F. (2016b). Study of the electromagnetic interference generated by wireless power transfer systems. *International Review of Electrical Engineering, 11*(5), 526–534.

Baikova, E. N., Romba, L., Valtchev, S. S., Melicio, R., Fernão Pires, V., Krusteva, A., & Gigov, G. (2018a). Electromagnetic field generated by a wireless energy transfer system: Comparison of simulation to measurement. *Journal of Electromagnetic Waves and Applications, 32*(5), 554–571. doi:10.1080/09205071.2017.1399832

Baikova, E. N., Valtchev, S. S., Melicio, R., & Pires, V. M. (2016a). Electromagnetic interference from a wireless power transfer system: experimental results. In *Proceedings of the International Conference on Renewable Energies and Power Quality (ICREPQ'2016)*. Madrid, Spain: ICREPQ. 10.24084/repqj14.560

Bienkowski, P., & Trzaska, H. (2012). *Electromagnetic measurements in the near field* (A. Duffy, Ed.). SciTech Publishing, Inc. doi:10.1049/SBEW042E

Bosshard, R., Kolar, J. W., & Wunsch, B. (2014). Accurate finite-element modelling and experimental verification of inductive power transfer coil design. In *Proceedings of the 29th IEEE Annual Applied Power Electronics Conference and Exposition (APEC)*. Fort Worth, TX: IEEE.

Boys, J. T., Covic, G. A., & Green, A. W. (2002). Stability and control of inductively coupled power transfer systems. *IEE Proceedings. Electric Power Applications, 147*(1), 37–43. doi:10.1049/ip-epa:20000017

Brown, W. C. (1984). The history of power transmission by radio waves. *IEEE Transactions on Microwave Theory and Techniques, 32*(9), 1230–1242. doi:10.1109/TMTT.1984.1132833

Chang, R., Quan, L., Zhu, X., Zong, Z., & Zhou, H. (2014). Design of a wireless power transfer system for EV application based on finite element analysis and MATLAB simulation. In *Proceedings of the IEEE Conference and Expo Transportation Electrification Asia-Pacific (ITEC Asia-Pacific)*. Beijing, China: IEEE. 10.1109/ITEC-AP.2014.6941225

Christ, A., Douglas, M. G., Roman, J. M., Cooper, E. B., Sample, A. P., Waters, B. H., & Kuster, N. (2013). Evaluation of wireless resonant power transfer systems with human electromagnetic exposure limits. *IEEE Transactions on Electromagnetic Compatibility, 55*(2), 265–274.

European Commission. (2015). Opinion on Potential health effects of exposure to electromagnetic fields (EMF). Luxembourg: Scientific Committee on Emerging and Newly Identified Health Risks (SCENIHR).

Gendron, M., Hazel, B., Boudreault, E., Champliaud, H., & Pham, X.-T. (2017). Modelling circular inductors coupled to a semi-infinite magnetic medium considering the proximity effect. *Journal of Electromagnetic Waves and Applications*, *31*(13), 1232–1254. doi:10.1080/09205071.2017.1336491

IEEE Standard C95.1. (2005). *IEEE Standard for Safety Levels with Respect to Human Exposure to Radio Frequency Electromagnetic Fields, 3 kHz to 300 GHz*. USA: IEEE.

Imura, T., Okabe, H., & Hori, Y. (2009). Basic experimental study on helical antennas of wireless power transfer for electric vehicles by using magnetic resonant couplings. In *Proceedings of the IEEE Vehicle Power and Propulsion Conference (VPPC '09)*. Dearborn, MI: IEEE. 10.1109/VPPC.2009.5289747

International Agency for Research on Cancer (IARC). (2013). *Non-ionizing radiation, part 2: radiofrequency electromagnetic fields*. Lyon, France: International Agency for Research on Cancer.

International Commission on Non-Ionizing Radiation Protection (ICNIRP). (2010). Guidelines for limiting exposure to time-varying electric, magnetic and electromagnetic fields (1 Hz – 100 kHz). *Health Physics*, *99*, 818–836. PMID:21068601

Isaev, Y. N., Vasileva, O. V., Budko, A. A., & Lefebvre, S. (2016). *Parameters assessment of the inductively-coupled circuit for wireless power transfer*. Paper presented at the International Conference on Mechanical Engineering, Automation and Control Systems, Moscow, Russia.

Jin, J.-M. (2010). *Theory and computation of electromagnetic fields*. John Wiley and Sons. doi:10.1002/9780470874257

Kim, S., Park, H.-H., Kim, J., Kim, K., & Ahn, S. (2014). Design and analysis of a resonant reactive shield for a wireless power electric vehicle. *IEEE Transactions on Microwave Theory and Techniques*, *62*(4), 1057–1066. doi:10.1109/TMTT.2014.2305404

Kline, M., Izyumin, I., Boser, B., & Sanders, S. (2011). Capacitive power transfer for contactless charging. In *Proceedings of the 26th IEEE Annual Applied Power Electronics Conference and Exposition*. Fort Worth, TX: IEEE.

Kurs, A., Karalis, A., Moffatt, R., Joannopoulos, J. D., Fisher, P., & Soljacic, M. (2007). Wireless power transfer via strongly coupled magnetic resonances. *Science*, *317*(5834), 83–86. doi:10.1126cience.1143254 PMID:17556549

Laakso, I., & Hirata, A. (2013). Evaluation of the induced electric field and compliance procedure for a wireless power transfer system in an electrical vehicle. *Physics in Medicine and Biology*, *58*(21), 7583–7593. doi:10.1088/0031-9155/58/21/7583 PMID:24107570

Li, S., & Mi, C. C. (2015). Wireless power transfer for electric vehicle applications. *IEEE Journal of Emerging and Selected Topics in Power Electronics*, *3*(1), 4–17. doi:10.1109/JESTPE.2014.2319453

Lin, J. C. (2013). Wireless power transfer for mobile applications, and health effects. *IEEE Antennas & Propagation Magazine*, *55*(2), 250–253. doi:10.1109/MAP.2013.6529362

Lu, X., Wang, P., Niyato, D., Kim, D. I., & Han, Z. (2016). Wireless charging technologies: Fundamentals, standards, and network applications. *IEEE Commun. Surveys Tuts.*, *18*(2), 1413–1452. doi:10.1109/COMST.2015.2499783

National Collaborating Centre for Environmental Health. (2013). *Radiofrequency toolkit for environmental health practitioners*. Vancouver, Canada: British Columbia Centre for Disease Control.

Pinto, R., Lopresto, V., & Genovese, A. (2017). *Human exposure to wireless power transfer systems: a numerical dosimetric study*. Paper presented at the 11th European Conference on Antennas and Propagation (EUCAP), Paris, France. 10.23919/EuCAP.2017.7928254

Redlarski, G., Lewczuk, B., Żak, A., Koncicki, A., Krawczuk, M., Piechocki, J., & Gradolewski, D. (2015). The influence of electromagnetic pollution on living organisms: Historical trends and forecasting changes. *BioMed Research International*, *2015*, 1–18. doi:10.1155/2015/234098 PMID:25811025

Rifai, A. B., & Hakami, M. A. (2014). Health hazards of electromagnetic radiation. *The Journal of Bioscience and Medicine*, *2*(08), 1–12. doi:10.4236/jbm.2014.28001

Rueter, D. (2016). Induction coil as a non-contacting ultrasound transmitter and detector: Modelling of magnetic fields for improving the performance. *Ultrasonics*, *65*, 200–210. doi:10.1016/j.ultras.2015.10.003 PMID:26522956

Sample, A. P., Meyer, D. A., & Smith, J. R. (2009). Analysis, experimental results, and range adaptation of magnetically coupled resonators for wireless power transfer. *IEEE Transactions on Industrial Electronics*, *58*(2), 544–554. doi:10.1109/TIE.2010.2046002

Staebler, P. (2017). *Human exposure to electromagnetic fields: from extremely low frequency (ELF) to radiofrequency*. John Wiley and Sons. doi:10.1002/9781119384533

Strauch, L., Pavlin, M., & Bregar, V. B. (2015). Optimization, design, and modelling of ferrite core geometry for inductive wireless power transfer. *International Journal of Applied Electromagnetics and Mechanics*, *49*(1), 145–155. doi:10.3233/JAE-150029

Wen, F., & Huang, X. (2017). Human exposure to electromagnetic fields from parallel wireless power transfer systems. *International Journal of Environmental Research and Public Health*, *14*(2), 1–15. doi:10.3390/ijerph14020157 PMID:28208709

Chapter 7
Rectennas for Microwave-Based Wireless Power Transfer (WPT)

Mukesh Kumar Khandelwal
Bhagwan Parshuram Institute of Technology, India

ABSTRACT

Electromagnetics has an important role in power and energy industry. In this chapter, the concept of rectenna is reviewed. The history of rectenna for wireless energy harvesting and transmission is discussed. Finally, examples are employed to illustrate some rectenna design and measurement issues such as rectenna impedance matching and its conversion efficiency. It is also shown that rectennas can harvest wireless energy efficiently under certain conditions and have the potential to become a power supplier for some special applications.

INTRODUCTION

According to current scenario, there are lots of wireless systems which are used for wireless communication like GSM, WiFi, BLE, Bluetooth etc. If one is willing to transmit the power from one point to another point wirelessly, it is still a current research challenging area for all researchers. Wired systems are usually used for power transmission from one point to another point but everyone is looking for Wireless Power Transmission (WPT) system on the account of comfortability. Number of commercial and non-commercial sectors are there where WPT systems are required including WSN networking, Remote Location based industries etc. Currently these applications are using battery system or direct wired system mostly. But the major problem associated with battery-operated-system is the life of battery.

Wireless Power Transfer (WPT) is the process of transferring the power from one circuit (place) to another circuit (place) without using any physical medium or interconnecting elements. Several approaches are there for wireless power transfer including Inductive WPT, Capacitive WPT, Laser, Microwave etc. Inductive Power Transfer (IPT) is the most popular technology among all the available WPT technologies and is being extensively studied particularly from the last two decades. The area of wireless power transmission is very interesting and it is in demand now-a-days in developing countries as most of the developing nations (i.e. China, India, Pakistan etc) are rapidly improving their standards of living. The

DOI: 10.4018/978-1-5225-5870-5.ch007

technology is in its infancy but the overall benefits from its maturation could be significant to society as a whole. These trends point to an energy demand that will grow at even a larger rate. Wireless power transmission could one day allow us to generate solar power on a satellite and beam it down to Earth, transmit power to a water treatment plant for a disaster relief operation or power a flying communication relay station from a terrestrial station.

In current stage, there are lots of researchers who are working in the field of power transmission and have developed several techniques for transmitting the electricity (power) over the long distances without wires.

Rectennas are used now-a-days for transmitting the power wirelessly over a distance. Rectenna is nothing but a combination of Rectifier and antenna (Rectifier + Antenna = Rectenna) supported by a low pass filter. Antenna is a device which used to transmit and receive the signals (Electromagnetic waves). Antenna works as a transducer which convert the electrical signal to Electromagnetic (EM) signal while using at transmitting end and EM signal to electrical signal in the case of reception. In the case of rectenna, power is transmitted through an EM radiator (antenna) and at receiving end an antenna is there to receive the EM signals. These received EM signal is converted into electrical signal which is further passed through a Low Pass Filter (LPF) to avoid the noise and unwanted signals then pass through the rectifier circuit which works as a voltage multiplier. The combination of receiving antenna, LPF, and rectifier is referred as Rectenna. Different types of load can be connected in further network. By this method, we can transmit power wirelessly which would not be harmful for human being or living things.

This chapter is organized in seven sections. Introduction and background information is given in first part of this chapter and referred as Introduction. Then, classification of power transmission is provided in second section. In next section, power transmission through microwaves and requirement of antennas are discussed followed by a discussion on energy harvesting and use of microwave for energy harvesting. Further, key parameters associated with rectennas are presented and in successive section different type of rectennas are discussed. In last section, conclusion of this chapter is presented.

CLASSIFICATION OF POWER TRANSMISSION

As we know there are lots of approaches for power transmission and we can divide them in following two major parts:

1. Wired Power Transmission
2. Wireless Power Transmission (WPT): In Wired Power Transmission System, both point (transmitting and receiving points) are physically connected via conducting media whereas in WPT power is transmitted from one point to another point wirelessly. WPT systems may further divided in following three subsections:
    a. Electromagnetic
    b. Induction
    c. Magnetic. Each subsection is further divided in following sub-parts:
        i. Microwave
        ii. Laser

iii. Capacitive
iv. Resonant
v. 5) Inductive

This division of wireless power transmission system is shown in Figure 1, this figure gives the details about the classification of WPT system. A comparison between different methods is given in Table 1.

WIRELESS POWER TRANSMISSION USING MICROWAVES

Microwave wireless power transmission is a wide range process in which long distance electric power transmission becomes possible. This process uses the microwave voltage source which emits the microwaves. The microwave source acts as a transmitting antenna and a microwave receiver is attached with the load which acts as receiving antenna. The received microwaves are then converted back in to electrical energy through which the load is driven. Different parts of the wireless power transmission through microwaves are briefed as following.

Figure 1. Classification of wireless power transmission

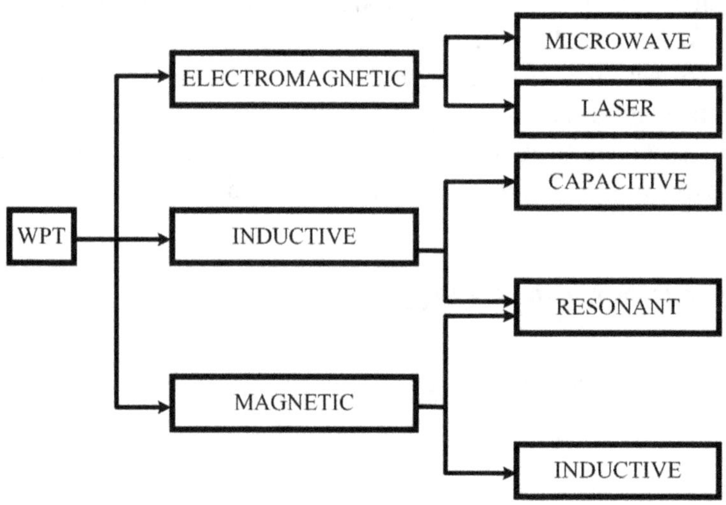

Table 1. Comparison of different methods

Magnetic Resonance Method	Microwaves Method	Lasers Method
It is economical as the equipment used is cheap and easily available	Relatively expensive as compared to other methods	Implies same economic conditions of mutual induction
Useful for implementation of the small distance applications	This method implies for long distance applications	Used for small distance but could be used for longer distances when high intensity beam is involved
It is safe from biological point of view.	Injurious for health because of high frequency rays (1GHz to1000 GHz)	The laser method is also injurious to human health

The microwave source antenna acts as transmitting antenna at the base station. It uses the microwaves of high frequency ranging from 1GHz to 1000 GHz (Summerer et al., 2009). There are many types of microwaves source antennas each of which has its own efficiency. Usually the slotted wave guide, micro strip patch and parabolic dish antennas are used for this purpose (Tomar et al., 2012). For high power applications the slotted waveguide antennas are used because of their high efficiency.

The microwave receiving antenna is mounted at the load end and due to high frequency of microwaves it could be used for large distance applications of wireless power transmission. At the load end the microwaves are received by microwaves receiving antenna and then the received microwaves are converted back into dc power. The unit which receives microwaves and then converts back to the dc power is called rectenna. The rectenna is mounted at the load end.

A typical rectifying antenna used to produce dc power from microwave energy is called rectenna. These are extensively used in microwave wireless power transmission systems. As defined in Yeoh (2010), "simple rectenna consists of a dipole antenna with an RF diode connected across the dipole elements. The diode rectifies the AC current induced in the antenna by the microwaves, to produce dc power, which powers a load connected across the diode."

RF Energy Systems

As the demand for power increases, the need for alternate energy sources has become essential. Today, there is an active research area on alternative ways to extract energy from the environment around us and convert it into electrical energy for energizing low power electronic circuits directly or store it for later use. Energy harvesting techniques thus use inexhaustible resources to provide unlimited energy for life of electronic devices. Many sources of energy, such as light, wind, temperature, vibration, radio waves and even pH have been transformed into usable energy using a variety of techniques in recent past for various applications (Shashank & Inman, 2009). One such form of energy which can be harvested is obtained from the Radio Frequency (RF) and Microwave Frequency (MWF). There is abundant energy transmitted from nearby base stations, wireless local area networks, FM/AM radio and TV broadcasting towers, especially in urban areas. These transmissions are prevalent throughout the seasons. Block diagram of WPT system is shown in Figure 2.

Among the various sources for energy harvesting, RFE is of special interest because of its ambient omnipresence irrespective of the time, day or season of the year and most importantly it provides mobility. Here, the challenge lies in utilizing or storing this petite ambient energy which can be employed as a secondary power source for battery recharging or remote power with battery backup. They can be utilized as direct power for battery free systems such as wireless sensors.

Figure 2. Block Diagram of Wireless Power Transmission System

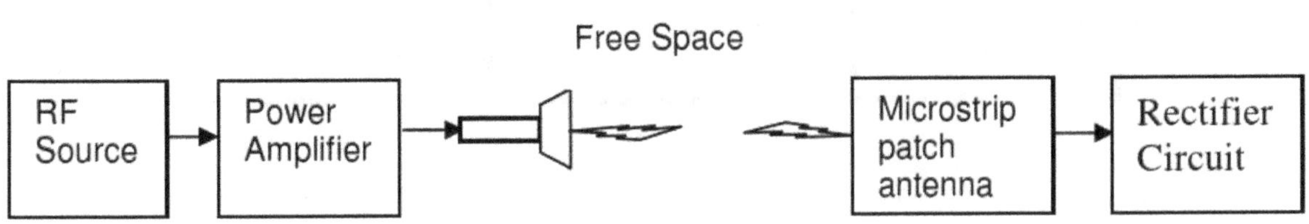

It is scalable to many nodes without a change to source and can even be embedded between walls. Hence, it is evident that RFE has an edge over other harvesting techniques. On the other hand, harvesting energy from these sources poses some challenges (Cellular Operators Association of India, n.d.); the ambient power available from ambient RFE is very low as it is an inverse function of distance from the source which calls for high gain antennas. Further, RF to Direct Current (DC) converters with low power dissipation should be employed in order to achieve high efficiency. Reflection and absorption of RFE is possible due to obstacles along the transmission path. Also, in case of multiband RF harvesting, broadband impedance matching is of utmost importance for maximum power transfer at all desired frequencies in the band.

RF Energy Sources in India

In India, the prominent RF sources which enable RF harvesting commercially are TV Broadcast: 180-220 MHz (transmitted power is few tens of kW), FM Radio Systems: 88-108 MHz (transmitted power few tens of kW), AM Transmission: 540-1600 kHz (transmitted power few hundred kW), Mobile Cellular Handsets (transmitted power 1 W to 2 W), WiFi: 2.45 and 5.8 GHz and Cell Tower Transmission of 10-20 W per carrier. Cell tower transmissions in India include Technology and Frequency bands depicted in Table 1. India, the country with the second largest number of mobile phone users (Arrawatia et al., 2011) globally is home for 7,47,917 Base Transreceiver Stations (BTS) as of 30.06.2013 (Secretariat, 2014; Department of Telecommunications, n.d.) with 864.72 million wireless connections. These cell towers transmit 24 hours every day all year-round and hence can be used as a continuous source of usable energy.

RECTENNA SYSTEM

The RF Scavenging system comprises of a receiving antenna, a pre-rectification filter, a rectifier, post-rectification filter and finally a storage element as shown in Figure 3. The antenna is designed to receive the ambient RF signals at a preselected frequency band. The expected received power for free space model is given by (1) in (Rappaport, 2001).

Figure 3. Block diagram of a generic RF scavenging system

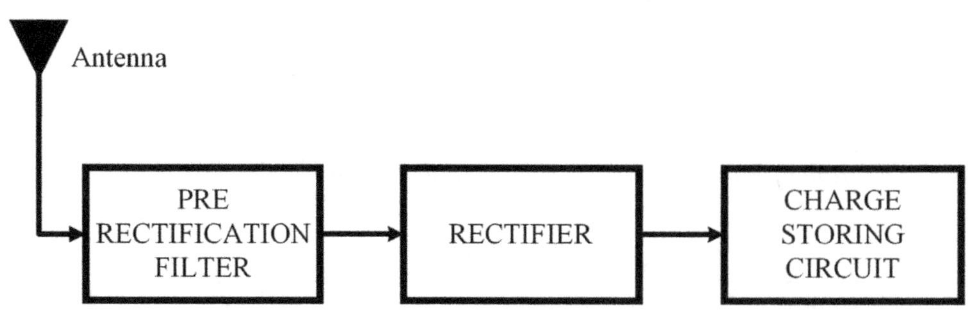

$$P_{Ow_r} = \frac{P_{Ow_t} G_{tr} G_{re} \lambda^2}{(4\pi R)^2 L} \tag{1}$$

where, Pow_r is the power received by a receiving antenna having a gain G_{re} and located at a distance R from a transmitting antenna of gain G_{tr} with a transmitting power of Pow_t while the system loss factor is L. Equation (1) holds true for line of sight propagation but at times it is not possible to have line of sight path between the transmitter and the receiver. The RF signals undergo scattering and reflection due to the obstacles in the propagation path thereby the receiving antenna collects signals from multiple paths. The received power in such cases is given by the two-ray propagation model given by (2) as in (Rappaport, 2001).

$$P_{Ow_r} = \frac{P_{Ow_t} G_{tr} G_{re} h_{tr}^2 h_{re}^2}{R^4 L} \tag{2}$$

Here, the height of the transmitting and receiving antennas is represented by h_{tr} and h_{re} respectively. A pre-rectification filter limits the frequency of the incoming RF signal to ensure that it is operating within the desired frequency band and prevents the reradiation of higher order harmonics produced by the nonlinear IV characteristics of the diode. This is then transformed into DC power by the diode based RF to DC converter forming the rectifier. The antenna-rectifier combination is termed as Rectenna. The output DC filter smoothes the output DC voltage and current by attenuating high frequency harmonics present in the RF signal or generated by the rectification process itself.

Antenna

Antenna is one of the major elements responsible for collecting the incoming RFE at various desirable frequencies. Antennas with resonance at single or multiple applicable frequencies are designed to meet different specifications. Generally, omnidirectional antennas are employed in scavenging systems so as to enable reception from all directions. Hence dipole antennas are conventionally preferred, but the most commonly used antennas include the dipole, loops, parabolic dish, horn, helical and microstrip antennas. The main factors to be considered in antenna design for RFE harvesting systems are desired frequency of operation, power handling capability, radiation efficiency of the antenna, polarization, bandwidth, weight and size of the antenna (Shreshta et al., 2013; Pavone et al., 2012). A broadband antenna may be preferred in cases where energy at various frequency bands is targeted for recycling, but in order to increase the gain, array antennas need to be considered. Also to increase efficiency of reception, low pass filters (LPF) that can suppress harmonics are included between the antenna and the rectifying device. Normally, circularly polarized antennas are preferred since polarization mismatch can be minimized. With the growing demand for compact wireless devices, the size of the antenna is also an important design aspect. Hence printed miniaturized antennas are being worked on. Though several types of antennas can be considered in a rectenna, the most popularly used is the printed microstrip patch antenna. The reason being they are light weight, low profile, aligning any shape, can be integrated with integrated circuits, design versatility and low mass production costs using printed circuit technology.

A printed broadband rectenna element and a 64-element array are extensively characterized for energy recycling of arbitrarily polarized stray fields in the microwave region to usable DC energy in (Joseph, 2003). An electromagnetically coupled square microstrip antenna which uses stacked multi-resonator configuration for broadband operation is used in the RFE harvesting system in (Arrawatia et al., 2011). A reconfigurable microstrip patch antenna allowing switching between two circular polarizations and consisting of a 3-dB hybrid coupler is proposed in (Wei et al., 2007). Here, by using only a single pole double throw switch the polarization switching is achieved and a simple design of the DC bias network is adopted. An RFE scavenging system comprising of printed differential antenna is presented in (Scorcioni et al., 2013) for wireless powering applications in the UHF band.

Filter and Rectifier

The rectifier circuit in a scavenging system rectifies the alternating current signal received through the antenna and a DC output is collected at the load/storage element (usually supercapacitors). The LPF is a precursor to the rectifier, hence called as a pre-rectification filter. This LPF not only rejects the undesired signal interference from the surroundings, but also suppresses the unwanted higher harmonics rejected by the rectifying circuit and is designed in such a way that it provides impedance matching between the antenna and the rectifier.

Hence the filter plays a critical role in determining the system's ultimate performance. The specifications of an RF scavenger depends on several factors such as power handling capability, Q factor, insertion loss, desired operating frequency and size. High Q filters such as cavity filters provide good interference rejection and improved insertion loss but at the cost of large dimensions. On the other hand, printed filters are comparatively prone to losses but can be compact, conformal and miniaturized.

A matching circuit is required between a filter and a rectifier to have impedance matching thereby leading to maximum power transfer. The possible rectifier circuits that can be implemented in a scavenging circuit are single shunt full wave rectifier, a full wave bridge rectifier or any other type of hybrid rectifier employing Schottky diodes. At RF the input power received is not enough to forward bias a diode hence the other alternative preferred is to use a Schottky diode. Compared to other diodes, Schottky diodes have low forward voltage and high switching speed (Din et al., 2012). Efficiency of the rectifier depends on Schottky diode and its power losses due to conduction and switching is low. The most commonly used circuit is a series or parallel half wave rectifier with single or multiple stage voltage doubler structure to enhance output DC voltage or a dual diode full wave rectifier to increase the alternating current to DC conversion efficiency. The main challenge in the design of rectifiers in these systems is to acquire high conversion efficiency (Shreshta et al., 2013). There are two suggested methods to meet this design impediment; firstly, to ensure that maximum received power is delivered to the rectifying diode by usage of broadband antennas, large antenna arrays and circularly polarized antennas as proposed in (Shreshta et al., 2013). Secondly, suppression of harmonics generated by the rectifying diode which is taken care of in the filter design. The important aspect of RF energy scavenging system is energy storage for which supercapacitors are most suitable (Alam et al., 2013). Supercapacitors store 10 to 100 times more energy per unit volume or mass and can deliver and accept charge much faster than batteries.

Several rectifying circuits for RF harvesting systems have been worked on from time to time. A voltage doubler is designed using Dickson Multiplier topology by Mahima et al. in (Cellular Opporators Association of India, n.d.). The RFE is converted into DC using a differential reconfigurable configuration which improves the sensitivity to -16dBm and efficiency to 60% at 840 MHz to 975 MHz is discussed in (Scorcioni et al., 2013). RFE harvesting system with a single wideband 377 E shaped patch antenna, a Pi matching network and a 7-stage voltage doubler circuit designed at the downlink radio frequency range of GSM900 band is discussed in (Reese et al., 2012). A Grainacher voltage doubler circuit, also known as single stage Cockroft Walton voltage multiplier circuit, is used in (Din et al., 2012) to double the input voltage. A 10 stage Villard voltage multiplier circuit was implemented using Manhattan Technique in (Alam et al., 2013) to achieve an output voltage approximately ten times the input voltage.

CURRENT SCENARIO

Currently, RF scavenging systems are in use in several ultra low powers, battery free applications, such as RFID tags, contactless smart cards and wireless sensor networks. For instance, RFID tags are integration of a sensor, transmitter and an antenna which are enabled to harvest adequate RFE (in W) from the RF signal transmitted by the reader in order to send the identification information.

As discussed in Publitek European Editors (2013) these systems find application in battery free wireless sensors installed in an RF ambient to wake up and power the sensor, to read data and/or send back data. This is being utilized in the good tracking (logistics), environmental monitoring and intelligent building sectors. Texas Instruments has launched a passive low frequency interface, which operates from the ISO 18000-2 RF air interface protocol. Designed for short range, battery free, two way communications, in applications such as data logging and medical biosensor devices, the device operates by scavenging the RFE transmitted from the base station. In particular, it provides reliable two-way communication in harsh environments (Publitek European Editors, 2013; Texas Instruments, 2014). Texas Instruments (2014) has also launched an ultra low power circuit that enables battery free power to applications such as wireless sensor networks, monitoring systems, smoke detectors, wearable medical devices and mobile accessories. The circuit delivers high power conversion efficiency from 10A to 50mA output currents and consumes only 350nA.

ANTENNAS FOR WPT

One of the most promising methods to harvest the wireless energy is to use a rectenna which is a combination of a rectifier and an antenna. A typical block diagram is shown in Figure 4. The wireless energy can be collected by the antenna attached to rectifying diodes through filters and matching circuit. The rectifying diodes convert the received wireless energy into DC power. The low-pass filter will match the load with the rectifier and block the high order harmonics generated by the diode to achieve high energy conversion efficiency which is the most important parameter of such a device.

There are at least two advantages for rectennas: (1) the life time of the rectenna is almost unlimited and it does not need replacement (unlike batteries). (2) It is "green" for the environment (unlike batteries, no deposition to pollute the environment).

Figure 4. Block diagram of a rectenna with a load

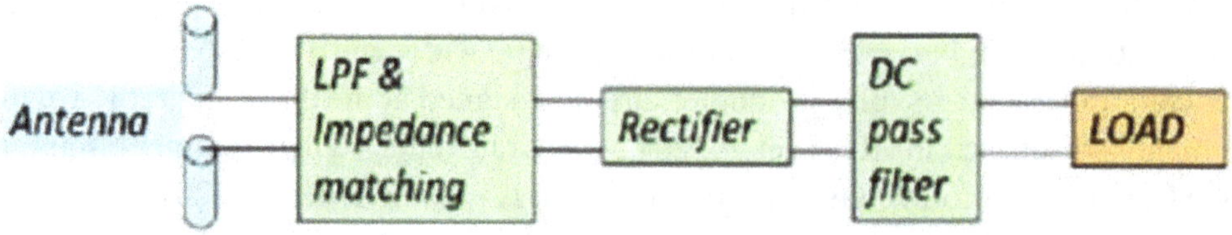

BACKGROUND OF WIRELESS POWER TRANSMISSION AND ENERGY HARVESTING

Over 100 years ago, the concept of wireless power transmission was introduced and demonstrated by Tesla (1904; 1904), he described a method of "utilizing effects transmitted through natural media". This method has been brought particularly into prominence in recent years. In fact, Tesla was unsuccessful to implement his wireless power transmission systems for commercial use, but he transmitted power from his oscillators which operated at 150 kHz to light two light bulbs. The reason for his unsuccessful attempt was that the transmitted power was radiated to all directions at 150 kHz radio wave whose wave length was 20 km and the efficiency was too low.

Based on the development of the microwave tubes during the World War II, rectification of microwave signals for supplying DC power through wireless transmission was proposed and researched in the context of high power beaming since 1950s. B. C. Brown started the modern era of wireless power transmission with the advancement of high-power microwave tube at Raytheon Company (Brown, 1984). By 1958, a 15 KW average power-band cross-field amplifying tube was developed that had a measured overall DC to RF conversion efficiency of 81%. The first receiving device for efficient reception and rectification of microwave power emerged in the early 1960's.

A rectifying antenna or rectenna was developed by Raytheon. The structure consisted of a half-wave dipole antenna with a balanced bridge or single semiconductor diode placed above a reflecting plane. The output of the rectenna element is then connected to a resistive load. 2.45 GHz is widely used as the transmitting frequency because of its advanced and efficient technology base, location at the centre of an industrial, scientific, and medical (ISM) band and its minimal attenuation through the atmosphere even in heavy rainstorms. From the 1960's to the 1970's the conversion efficiency of the rectenna was increased at this frequency (Shinohara, 2006).

Conversion efficiency is closely linked to the microwave power that is converted into DC power by a rectenna element, the greatest conversion efficiency ever recorded by a rectenna element occurred in 1977 by Brown in Raytheon Company sing a GaAsPt Schottky barrier diode, a 90.6% conversion efficiency was recorded with an input microwave-power level of 8 W. This rectenna element used aluminum bars to construct the dipole and transmission line (Raytheon Company, 1977). Later, a printed rectenna design was developed at 2.45 GHz with efficiencies around 85% (Brown & Triner, 1982). More recently, McSpadden and Chang used the rectenna as a receiving antenna attached to a rectifying circuit that efficiently converts microwave energy into DC power (McSpadden & Chang, 1994).

As an essential element of the rectenna, the antenna of rectenna can be any type such as a dipole (Brown, 1984; Shinohara, 2006; Raytheon Company, 1977; Brown & Triner; 1982; McSpadden & Chang, 1994; Mateu & Moll, 2005; Brown, 1986), Yagi-Uda antenna (Shinohara et al., 1998; Gutmann & Gworek, 1979), microstrip antenna (Ito et al., 1993; McSpadden & Chang, 1994; Fujita et al., 1996), monopole (Aoki et al., 1997), coplanar patch (Chin et al., 2005), spiral antenna (Hagerty et al., 2000), or even parabolic antenna (Fujino & Ogimura, 2004). The rectenna can also take any type of rectifying circuit such as single shunt full-wave rectifier (McSpadden & Chang, 1994; Fujita et al., 1996; Aoki et al., 1997), full-wave bridge rectifier (Brown, 1984; Shinohara et al., 1998), or other hybrid rectifiers (Gutmann & Gworek, 1979). The circuit, especially the diode, mainly determines the RF to DC conversion efficiency, rectennas with FET (Aoki et al., 1997) or HEMT (Chin et al., 2005) appeared in recent years. The world record of the RF-DC conversion efficiency among developed rectennas is approximately 90% at 8 W input of 2.45 GHz (Brown, 1984). As shown in Figure 5, the RF-DC conversion efficiency of the rectenna with a diode depends on the microwave power input intensity and the optimum connected load. When the power is small, or the load is not matched, the efficiency becomes quite low. The efficiency is also determined by the characteristic of the diode which has its own junction voltage and breakdown voltage, if the input voltage to the diode is lower than the junction voltage or is higher than the breakdown voltage the diode does not show a rectifying characteristic. As a result, the RF-DC conversion efficiency drops with a lower or higher input than the optimum.

It is worth noticing that all the recorded high conversion efficiencies were generated from high power incident level due to the reason we mentioned above. For low power incident level, a measured conversion efficiency of 21% was achieved at a power incident of 250 $\mu W/cm^2$ (Monti et al., 2011), of course, in principle a high efficiency should be achievable.

There are basically two approaches to increase the efficiency at the low microwave power density. The one is to increase the antenna aperture as shown in (Fujino & Ogimura, 2004). There are two problems for this approach. It produces a high directivity, and this is only applied for exclusive applications as SPS satellite experiment and not for low power applications like RFID or microwave energy recycling. The other approach is to develop a new rectifying circuit to increase the efficiency at a weak microwave input.

Figure 5. General relationship between microwave to DC power conversion efficiency and input power (Yoo & Chang, 1992)

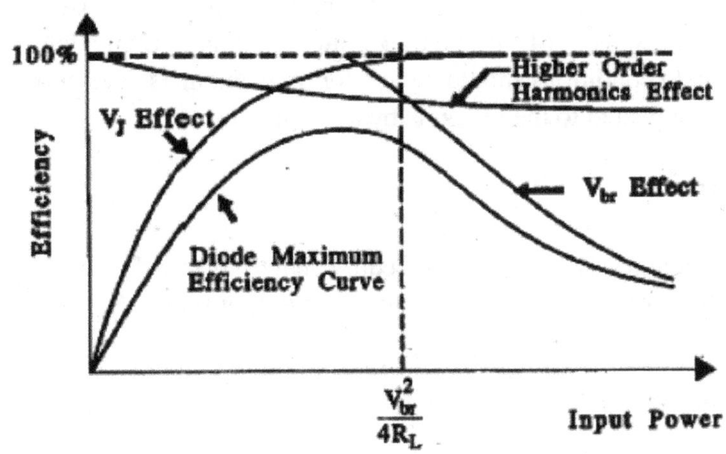

RECTENNA DESIGNS

Bandwidth of Rectenna

The rectenna can be divided into narrow-band rectennas and wide-band rectennas. Many narrow-band rectennas have been developed. For example, a dual-polarized patch rectenna at 2.4 GHz and a high conversion efficiency rectenna at 5.8 GHz were designed by McSpadden in 1994 and 1998 respectively (Mateu & Moll, 2005; McSpadden et al., 1998). However, very few broad-band rectennas have been developed which are the most desirable rectennas for wireless energy harvesting. They may collect energy from systems operating at different frequencies to maximize the output power at a given location.

Rectenna Impedance Matching

To optimize the rectenna for maximum power transfer, the antenna impedance must be matched to the impedance of rectifier diode. For example, in (McSpadden et al., 1998), Alpha SMS7630-079 Schottky diodes were used for rectification, and a source-pull simulation was used to obtain the diode input impedance. For a variable input power, the resulting DC voltage is quantified for each source impedance as shown on Figure 6 (a) (McSpadden et al., 1998). The results shown on the Smith Chart indicate that the optimum source impedance must be presented to each diode moving counter-clockwise with an increase in frequency, and closer to the centre of the Smith Chart with an increase in input power. The region of optimal source impedance is used to optimize the antenna design which needs to match the diode impedance.

The simulated input impedance for a spiral antenna is plotted on Fig 6 (b). In this rectenna structure the antenna is directly connect to a Schottky diode not a transmission line. Therefore, the spiral antenna should match to the diode input impedance. Form Figure 6 (a) and (b), a good impedance matching between the antenna elements and their respective diodes occurs in the region around 2.7 GHz for low-input powers and around 5.8-8 GHz for high-input powers.

Rectenna Testing

A typical experimental setup adopted to perform rectenna measurements is illustrated in Figure 7. To produce a variable power density, a signal generator and an amplifier are used. Measurements can be performed by varying the distance *R* from the transmitter antenna to the rectenna or the output power. For both cases, the far field conditions should be met for both antennas. The output voltage is measured from the load resistant connected to the rectenna by a voltage meter.

Rectenna Efficiency

The efficiency of the rectenna system is basically equivalent to its transfer function. The general definition of any efficiency (η) used hereafter is the ratio of the output power P_{out} over the input power Pin:

$$\eta = P_{out} / P_{in} \tag{3}$$

Figure 6. (a) Simulated range of optimal source impedance for Schottky diode from 1 to 16 GHz and with an input power from -30 dBm to 10 dBm, (b) Simulated input impedance for spiral antenna (Mc-Spadden et al., 1998)

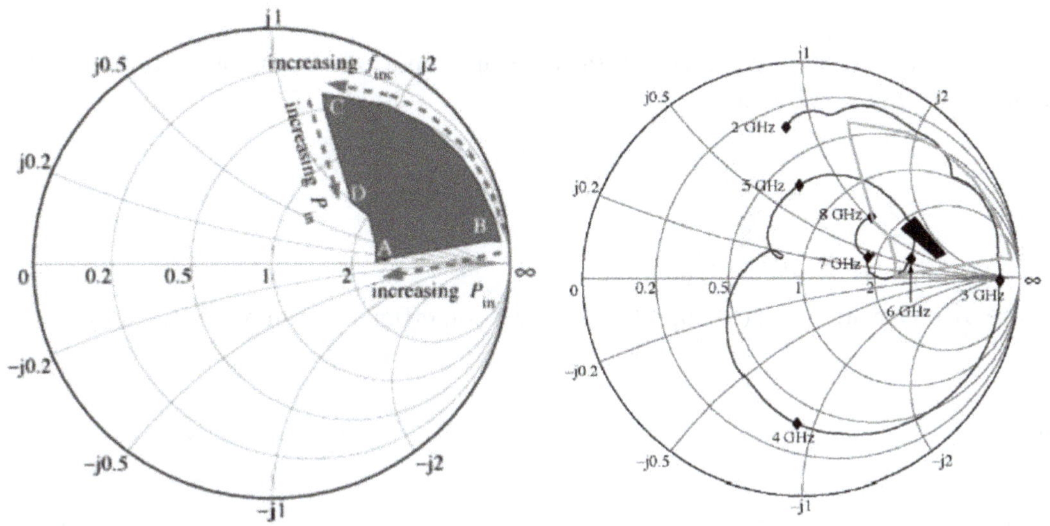

Figure 7. Experimental Set-up for rectenna characteristic

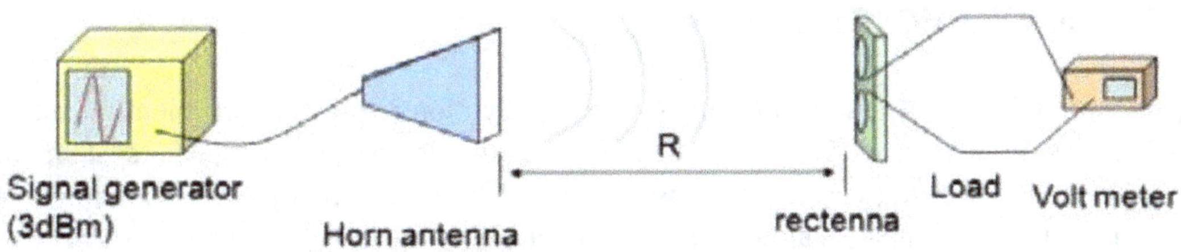

The conversion efficiency (η) of the whole system is the DC power at the receiver end over the AC input power captured by the system (antenna). This efficiency is strongly dependent on the power density (P_d) distributed across the receiver aperture. The maximum incident power density can be expressed as

$$P_d = \frac{P_t G_t}{4\pi R^2} \tag{4}$$

where P_t is the transmitted power, G_t and G_r are the gains of the transmitter and receiver antennas, and R is the distance. The effective area A_{eff} for antenna is given as

$$A_{eff} = \frac{\lambda^2 G_r}{4\pi} \tag{5}$$

Therefore, use equations (4) and (5) to obtain the power received by antenna as

$$P_{in} = P_d . A_{eff} \tag{6}$$

Because the rectenna output is DC power, thus the output power we can obtain from the output voltage generated on the load resistance by

$$P_{out} = \frac{V^2_{out,DC}}{R_{load}} \tag{7}$$

Therefore, from equation (3), (6) and (7) the conversion efficiency can be obtained by

$$\eta = \frac{V^2_{out,DC}}{R_{load}} \tag{8}$$

This definition is slightly different from the one used in the photovoltaic community and it will give higher efficiency due to the use of the effective aperture rather than the physical antenna aperture.

COMPARISON OF RECTENNAS

Recently, fast growth of wireless power transmission application requires an efficient rectenna in delivering RF energy to be rectified into DC voltage for overall wireless power transfer system. Wireless technology powered using ambient radiation which is widely used in many wireless communications such as agricultural management, healthcare management, factory monitoring and so forth. The wireless power transfer system must operate efficiently and the losses of energy during receiving and conversion of signal processes should be minimized by suppressing the unwanted signal. However, the interface between the antenna and the nonlinear circuit component such as diode or FET for active integrated antenna (AIA) or rectenna system has a harmonic suppression filter which contributes additional insertion loss. The basic idea implementation of antenna with harmonic suppression is to avoid spurious radiation that easily produced at high-order resonant frequencies of the antenna from the circuits. This weak signal then flow to the rectifier circuit for the RF-to-DC conversion and the process are repeated caused the output of the rectenna is low for wireless power transfer system. One of the important issues found in (Din et al., 2012) shows that the input impedance of the antenna design would allow the second harmonic at high-order frequency which occur at twice of its design frequency. However, higher order harmonic can be blocked as input impedance at higher order almost zero or unmatched.

Several research works conducted recently as recorded in Table 2 featuring important of antenna design at 2.45GHz (ISM Band) applications (Malaysian Communications and Multimedia Commission, 2013) which not only act as radiator but also able to suppress the unwanted harmonics produced due to spurious radiation or excitation at high-order frequency. Several studies have been conducted on antenna harmonic rejection technique in order to proposed new design with improvement in overall system performance in term of its good return loss, voltage standing wave ratio (VSWR), radiation pattern, directivity and gain. The best technique will be applied on the basic structure of radiating element to control the distribution of current flow on the patch antenna like slit, stub, defective ground structure or else. In turn, the performance of the overall system will be improved since the technique used will maximized power at the fundamental frequency and suppressed the radiated power at harmonic frequencies to achieve harmonic rejection characteristics. However, antenna harmonic suppression will have maximum power transfer at its design frequency if input impedance of antenna is 50ohm and perfectly matched with the characteristic impedance of transmission line in order to obtain maximum energy transfer between transmission feed line and an antenna. Harmonic suppression characteristic of a CPW-fed circular slot antenna using single slot on a ground conductor was demonstrated by authors Yeo & Kim (2009) and this design has multi-band rejection characteristic where the second and higher rejection bands are integer-multiple of the first band and this is generated by inserting single slot on a ground conductor of the antenna. The simulated and measured results show good agreement and the integer multiple notch bands can be adjustable by varying the length of the slot on the ground plane.

PARAMETERS DETERMINING RECTENNA PERFORMANCE

There are six parameters that determine the performance of a rectenna system and appear in the expression of the figure of merit (FOM) (Etinger et al., 2017; Marian et al., 2011; Joshi & Moddel, 2016). They can be divided into following three categories.

1. **Input Conditions:** Power at the antenna terminals (Pin) and the frequency ($f = \omega/2\pi$) of the incoming electromagnetic wave.
2. **Diode Characteristics:** Diode area (A_D), capacitance (C_D), forward resistance (R_f), and reverse resistance (R_r) in the form of the current density [$J(V)$] characteristics of the diode, determined by approximating the diode $I(V)$ by a piecewise linear characteristic.
3. **Antenna Parameters:** Radiation resistance (R_S) of the antenna, where we assume that the coupling efficiency between free-space and the antenna is unity, the antenna is in resonance (so that its impedance is purely resistive), and the propagation of the surface plasmons from the antenna to the diode is lossless.

The rectenna schematic is shown in Figure 4. As will be shown, the rectenna performance depends on the input conditions, which means that the rectenna and the diode have to be designed based on the application. The six parameters determining the *FOM* of the rectenna can be grouped into three recognizable terms.

- **Diode Reverse Leakage Power:** The reverse leakage power (P_{leak}) is the dc power lost due to the flow of finite reverse leakage current (I_{leak}) in the diode at the operating voltage (V_O) of the rectenna. Since the rectenna operates in the second quadrant of the diode $I(V)$, the diode reverse leakage current at V_O flows in the opposite direction to the dc illuminated current (I_{illum}) and, therefore, degrades rectenna performance. The P_{leak} at the V_O due to the reverse leakage current is ($P_{leak} = V_O I_{leak}$).

Table 2. Comparison of research work for rejection of Harmonics at 2.4 GHz

SN	Authors Year	Antenna Harmonic Rejection		Application
		Technique	Performance	
1	(Suh & Chang, 2002)	Open ended T-strip CPS	Efficiency: 82.7%	2.45GHz and 5.8GHz
2	(Yeo & Kim, 2009)	CPW-fed circular slot with slot on ground. Slot on A Ground Conductor	S11< -10 dB E-plane: Bidirectional H-plane: Omni-directional	2.45GHz
3	(Ong et al., 2010)	Unbalanced slots	S11: <20 dB AR: 0.15 dB Gain:10.8 dBi	2.45GHz
4	(Riviere et al., 2010)	Slot at ground	S11: -25 dB Efficiency: 34% Gain: 4.3 dB	2.45GHz
5	(Arai, 2011)	Slit and stub	Efficiency: 53.0%	2.45GHz
6	(Harouni et al., 2011)	Compact DCP patch antenna	BW: 2.1GHz 3dB(AR):705MHz Efficiency: 63% RHCP: 25.24 dB LHCP: 6.8 dB	2.45GHz
7	(Huang et al., 2012)	Four right-angles slit	S11: -45 dB Efficiency: 37.8% BW: 137 MHz 3dB(AR): 30 MHz ARmin: 0.32 dB	2.45GHz
8	(Rahim et al., 2012a)	Slit and stub	S11: -48 dB	2.45GHz
9	(Rahim et al., 2012b)	Slot and open stub.	S11: -40 dB	2.45GHz
10	(Hung et al., 2012)	T-shape slot	BW: 70 MHz AR_{min}: 0.6 dB 3dB(AR): 40 MHz RHCP: 7.98 dBic LHCP: 8 dBic	2.45GHz
11	(Rahim et al., 2013)	Circular slot defected ground structures	S11: -20 dB Dir:9.887dBi AR: 0.26 dB RHCP: 4.119 dB LHCP: 4.149dB Gain: 6.655dB	2.45GHz
12	(Sennouni et al., 2013)	Inclined slot	VSWR: 1.054 Main lobe Magnitude:9.2 dBi Direction: 35.0°	2.45GHz
13	(Zhang et al., 2014)	Planar topology	Gain: 2.05 dBi BW: 400MHz Size:122 x 122mm	2.45GHz
14	(Sim et al., 2014)	Ring slot	S11: -24 dB BW: 210 MHz	2.45GHz

A good energy harvesting diode has a leakage current such that $P_{in}/P_{peak} \gg 1$.

- **Diode Responsivity:** For efficient rectification, the diode must be asymmetric around the rectenna V_0 and have a large responsivity. The diode current responsivity, which is the dc current out per unit ac power in, can be written as

$$\beta_\Delta = \frac{1}{\Delta}\left(\frac{I(V_0+\Delta)-2I(V_0)+(V_0-\Delta)}{I(V_0+\Delta)-I(V_0-\Delta)}\right) \tag{9}$$

Here, the Δ is a step voltage that depends on the operating regime of the rectenna. Although the diode voltage amplitude (V_D) is the relevant parameter, due to its dynamic nature and dependence on V_O, we use V_S as an approximation for V_D to determine the operating regime. We approximate Δ by the source voltage (V_S) if the rectenna operates in the classical regime ($V_S \gg h\omega/q$) and is equal to $h\omega/q$ if the rectenna is in the quantum regime ($V_S < h\omega/q$). A formula for determining V_S is given as

$$V_s = \sqrt{8R_S P_{AC}} \tag{10}$$

At optical frequencies, the antenna parameters (resonance frequency, radiation resistance, etc.) do not follow classical antenna theory and might need some modifications, but (3) would still apply in estimating the *FOM*.

3) *Coupling efficiency between the antenna and the diode:* The input conditions and the diode *I*(*V*) characteristics determine the diode ac resistance (RD). For quantum operation, RD is the secant resistance of the diode because in quantum operation, the oscillating current samples the diode at only two points in its *I(V)* curve. Since calculating the RD in the classical case is not straightforward, we approximate it using the same expression as the quantum RD but with a Δ (= V_S) suitable for classical operation

$$R_D = \frac{2\Delta}{\frac{V_0+\Delta}{R_f} - \frac{V_0-\Delta}{R_r}} \tag{11}$$

The R_D in parallel with R_S determine the effective ac resistance of the system, which along with the C_D gives the R_C time constant and the cut off frequency (f_c) of the rectenna

$$f_c = \frac{1}{2\pi \frac{R_S R_D}{R_S + R_D} C_D} \tag{12}$$

The coupling efficiency is written as,

$$\eta_C = \frac{\frac{4R_S R_D}{\left(R_S + R_D\right)^2}}{1 + \left(2\pi f \frac{R_S R_D}{R_S + R_D} C_D\right)^2} \tag{13}$$

The η_C is 100% when $R_S = R_D$ and $f \ll f_c$.

CONCLUSION

In this chapter, the concept of rectennas and their history is reviewed. Some important design issues of rectennas have also been addressed. Moreover, a detailed discussion on the rectenna efficiency has been conducted. To achieve high energy conversion efficiency, many parameters must be considered. Several rectennas available in literature are compared.

REFERENCES

Aoki, Otsuka, & Idogaki Shibata. (1997). Microwave energy transmission system for microrobot. *IEICE Transactions on Electronics*, *80c*(2), 303–308.

Arai, H. (2011). Wireless Power Transfer System. *Journal of the Korean Institute of Electromagnetic Engineering and Science*, *11*(3).

Arrawatia, Maryam, & Kumar. (2011). RF Energy Harvesting System from Cell Towers in 900MHz Band. *National Conference on Communications*, 1-5.

Bin Alam, S., Ullah, M. S., & Moury, S. (2013). Design of a Low Power 2.45 GHz RF Energy Harvesting Circuit for Rectenna. *International Conference on Electronics & Vision*, 1-4.

Brown, W. C., & Triner, J. F. (1982). Experimental thin-film, etched-circuit rectenna. IEEE MTT-S Int, Microwave Symp.Dig., 185-187.

Brown. (1984). The history of power transmission by radio waves. *IEEE Trans. MTT*, *32*(9), 1230-1242.

Brown. (1986). A microwave powered, long duration, high altitude platform. *MTT- S International Microwave Symposium Digest*, *86*(1), 507-510.

Cellular Operators Association of India. (n.d.). *Annual Report, 2013-2014*. Retrieved from http://www.dot.gov.in/sites/default/files/AR%202013-14%20English%20%282%29_1.pdf

Department of Telecommunications. (n.d.). *Ministry of Communications and Information Technology, Annual Report, 2012-2013*. Retrieved from http://meity.gov.in/writereaddata/files/Annual%20Report%20 2012-13.pdf

Din, N. M., Chakrabarty, C. K., Bin Ismail, A., Devi, K. K. A., & Chen, W. Y. (2012). Design of RF Energy Harvesting System for Energizing Low Power Devices. *Progress in Electromagnetics Research, 132*, 49–69. doi:10.2528/PIER12072002

Electronic and mechanical improvement of the receiving terminal of a free-space microwave power transmission system. (1977). Raytheon Company, Wayland, MA, Tech.Rep.PT-4964, NASA REP. CR135194.

Etinger, A., Pilossof, M., Litvak, B., Hardon, D., Einat, M., Kapilevich, B., & Pinhasi, Y. (2017). Characterization of a Schottky Diode Rectenna for Millimeter Wave Power Beaming Using High Power Radiation Sources. *Acta Physica Polonica A, 131*(5), 1280–1284. doi:10.12693/APhysPolA.131.1280

Fujino & Ogimura. (2004). A rectangular parabola rectenna with elliptical beam for sps test satellite experiment. *Proc. of the Institute of Electronics, Information and Communication Engineers*, 1-10.

Fujita, Kaya, Kunimi, Ishii, Ogihata, & Kusaka Fujino. (1996). A dual polarization microwave power transmission system for microwave propelled airship experiment. *Proc. de ISAP, 96*(2), 393–396.

Gutmann, R. J., & Gworek, R. B. (1979). Yagi-uda receiving elements in microwave power transmission system rectennas. *The Journal of Microwave Power, 14*(4), 313–320. doi:10.1080/16070658.1979.11689166

Hagerty, J. A., Lopez, N. D., Popovic, B., & Popovic, Z. (2000). *Broadband rectenna arrays for randomly polarized incident waves*. IEEE. doi:10.1109/EUMA.2000.338757

Harouni, Z., Cirio, L., Osman, L., Gharsallah, A., & Picon, O. (2011). A Dual Circularly Polarized 2.45GHz Rectenna for Wireless Power Transmission. *IEEE Antennas and Wireless Propagation Letters*, 10.

Huang, F.-J., Yo, T.-C., Lee, C.-M., & Luo, C.-H. (2012). Design of Circular Polarization Antenna with Harmonic Suppression for Rectenna Application. *IEEE Antennas and Wireless Propagation Letters*, 11.

Instruments, T. (2014). *ULP Meets Energy Harvesting: A Game-Changing Combination for Design Engineers*. Available: http://www.ti.com/lsds/ti/apps/alternative_energy/harvesting

Ito, T., Fujino, Y., & Fujita, M. (1993). Fundamental experiment of a rectenna array for microwave power reception. *IEICE Trans. Commun. E (Norwalk, Conn.), 76-B*(12), 1508–1513.

Joseph, A. H. (2003). *Nonlinear Circuits and Antennas for Microwave Energy Conversion* (Ph.D. dissertation). Department of Elect. and Comp. Eng., University of Colorado.

Joshi, S., & Moddel, G. (2016, May). Simple Figure of Merit for Diodes in Optical Rectennas. *IEEE Journal of Photovoltaics, 6*(3), 668–672. doi:10.1109/JPHOTOV.2016.2541460

Jui-Hung, C., Lin, D.-B., Weng, K.-L., & Li, H.-J. (2012). Novel T-shape Slot Couple Feed Dual Circular Polarized Rectenna. *Proceedings of ISAP*.

Lok Sabha Secretariat. (2014). *Norms for the Setting Up of Telecom Towers, its Harmful Effects and Setting Up of Security Standards in Expansion of Telecom Facilities*. New Delhi: Ministry of Communications and Information Technology.

Malaysian Communications and Multimedia Commission. (2013). *Requirements for Devices using Ultra-Wideband (UWB) Technology Operating in the frequency bands of 30MHz to 960MHz, 2.17GHz to 10.6GHz, 21.65GHz to 29.5GHz and 77GHz to 81GHz", SKMM SRSP-549*. UWB.

Marian, V., Menudier, C., Thévenot, M., Vollaire, C., & Verdier, J. (2011). Efficient design of rectifying antennas for low power detection. *International Microwave Symposium*. doi:10.1109/MWSYM.2011.5973270

Mateu, L., & Moll, F. (2005). Review of energy harvesting techniques and applications for microelectronics. *Proceedings of the SPIE Microtechnologies for the New Millennium*, 359-73. 2005.

McSpadden, J. O., & Chang, K. (1994). A dual polarized circular patch rectifying antenna at 2.45 GHz for microwave power conversion and detection. *IEEE MTT-S International Microwave Symposium Digest*, 1749-52. 10.1109/MWSYM.1994.335103

McSpadden, J. O., Fan, L., & Chang, K. (1998, December). Design and experiments of a high-conversion efficiency 5.8-GHz rectenna. *IEEE Transactions on Microwave Theory and Techniques, VOL.*, *46*(12), 2053–2060. doi:10.1109/22.739282

Michaelides, E. E. (2012). *Alternative Energy Sources. In Green Energy and Technology*. Springer.

Monti, G., Terricone, L., & Spartano, M. (2011). X-band planar rectenna. *IEEE Antennas and Wireless Propagation Letters*, *10*, 1116–1118. doi:10.1109/LAWP.2011.2171029

Ong, C. S., Karim, M. F., & Ong, L. C. (2010). A Compact 2x2 Circularly Polarized Antenna Array for Energy Harvesting. *Proceedings of Asia-Pacific Microwave Conference*.

Pavone, D., Buonanno, A., D'Urso, M., & Corte, F. D. (2012). Design Considerations for Radio Frequency Energy Harvesting Devices. *Progress in Electromagnetics Research B*, *45*, 19–35. doi:10.2528/PIERB12062901

Publitek European Editors. (2013). *Tune In, Charge Up: RF Energy Harvesting Shows its Potential*. Available: http://www.digikey.com

Rahim, R. A., Hassan, S. I. S., Malek, F., & Junita, M. N. (2012a). A 2.45 GHz Circular Patch Antenna with Harmonic Suppression for Wireless Power Transmission. *IEEE Colloquium on Humanities, Science &Engineering Research (CHUSER)*.

Rahim, R. A., Hassan, S. I. S., Malek, F., Junita, M. N., Jamlos, M. F., & Azizan, M. N. (2012b). A 2.45GHz Harmonic Suppression Rectangular Patch Antenna. *International Symposium on Computer Applications and Industrial Electronics (ISCAIE)*. 10.1109/ISCAIE.2012.6482105

Rahim, R. A., Syed, I. S. H., Malek, F., Nordin, J. M., & Hassan, H. F. (2013). Harmonics Suppression Single-fed Dual-Circularly Polarized Microstrip Patch Antenna for Future Wireless Power Transmission. *IACSIT International Journal of Engineering and Technology*, *5*(5).

Rappaport, T. S. (2001). *Wireless Communications: Principle and Practice* (2nd ed.). Prentice Hall.

Reese, J., Brown, K., & Zhen, J. (2012). *RF Energy Harvesting* (Graduate Thesis). University of Cincinnati.

Riviere, S., Alicalapa, F., Douyere, A., & Lan Sun Luk, J. D. (2010). A Compact Rectenna Device at Low Power Level. *Progress in Electromagnetics Research C*, *16*, 137–146. doi:10.2528/PIERC10071604

Scorcioni, Larcher, Bertacchini, & Vincetti. (2013). An Integrated RF Energy Harvester for UHF Wireless Powering Applications. *IEEE Conf. on Wireless Power Transfer,* 92-95.

Sennouni, M. A., Zbitou, J., Tribak, A., Benaissa, A., & Latrach, M. (2013). *Circular Polarized Square Patch Antenna Array for Wireless Power Transmission*. IEEE. doi:10.1109/IRSEC.2013.6529698

Shashank & Inman. (2009). *Energy Harvesting Technologies* (Vol. 21). New York: Springer.

Shinohara, Kunimi, Miura, Matsumoto, & Fujiwara. (1998). Open experiment of microwave power experiment with automatically target chasing system (Japanese). *IEICE Trans. B-II, J81-B-II,* (6), 657-661.

Shinohara, N. (2006). *Wireless power transmission for solar power satellite*. Retrieved from http://www.sspi.gatech.edu/wptshinohara.pdf

Shreshta, Noh, & Choi. (2013). Comparative Study of Antenna Designs for RF Energy Harvesting. *International Journal of Antennas and Propagat.*

Sim, C.-Y.-D., Chang, M.-H., & Chen, B.-Y. (2014). Microstrip-Fed Ring8 Slot Antenna Design with Wideband Harmonic Suppression. *IEEE Transactions on Antennas and Propagation*, *62*(9), 4828–4832. doi:10.1109/TAP.2014.2330595

Summerer, L., & Purcell, O. (2009). *Concepts for wireless energy transmission via laser. Europeans Space Agency (ESA)*. Advanced Concepts Team.

Tesla, N. (1904). The transmission of electric energy without wires. *The Thirteenth Anniversary Number of the Electrical World and Engineer.*

Tesla, N. (1904). *Experiments with Alternate Current of High Potential and High Frequency*. McGraw.

Tomar & Gupta. (2012). Wireless power Transmission: Applications and Components. *International Journal of Engineering, 1*(5).

Wei, W. B., Liu, Q. Z., Yin, Y. Z., & Zhou, H. J. (2007). *Reconfigurable Microstrip Patch Antenna with Switchable Polarization* (Vol. 75). Progress in Electromagnetics Research.

Xue Chin, Q. (2005). Design of a 5.8-Ghz. rectenna incorporating a new patch antenna. *IEEE Antennas and Wireless Propagation Letters*, *4*, 175–178. doi:10.1109/LAWP.2005.846434

Yeo, J., & Kim, D. (2009). Harmonic Suppression Characteristic of a CPW-Fed Circular Slot Antenna Using Single Slot on a Ground Conductor. *Progress in Electromagnetics Research Letters*, *11*, 11–19. doi:10.2528/PIERL09072108

Yeoh, W (2010). *Wireless power transmission (WPT) application at 2.4 GHz in common network.* Academic Press.

Yoo & Chang. (1992). Theoretical and experimental development of 10 and 35 GHz rectennas. *IEEE Transactions on Microwave and Techniques, 40*(6).

Young-Ho, S., & Chang, K. (2002). A High Efficiency Dual-Frequency Rectenna for 2.45GHz and 5.8GHz Wireless Power Transmission. *IEEE Transactions on Microwave Theory and Techniques, 50*(7).

Zhang, Liu, Meng, Wu, Lee, Xu, … Kim. (2014). Design of a Compact Planar Rectenna for Wireless Power Transfer in the ISM Band. *International Journal of Antennas and Propagation*.

Chapter 8
Radio-Frequency Identification (RFID)

Mukesh Kumar Khandelwal
Bhagwan Parshuram Institute of Technology, India

ABSTRACT

RFID technology has been widely adapted in industries for uses in logistical tracking, highway tolling, building access, and transportation ticketing. These applications have generally been limited to the original intended use of RFID, that is identification and a replacement for bar codes. Research in this area focuses on increasing the read rate, range, and reliability of their RF tags. The WPT is the enabling technology for realizing a true internet of things. Broad sensor networks capable of monitoring environmental pollutants, health-related biological data, and building utility usage are just a small fraction of the applications which are part of an ever-evolving ubiquitous lifestyle. Realizing these systems requires a means of powering their electronics sans batteries. Removing the batteries from the trillions of these envisioned devices not only reduces their size and lowers their cost, but also avoids an ecological catastrophe. This chapter discusses new theoretical models of RFID, communication standards, radio channel characteristics, RFID readers and tags.

INTRODUCTION

Ambient backscatter uses existing radio frequency signals, such as radio, television and mobile telephony, to transmit data without a battery or power grid connection. Each such device uses an antenna to pick up an existing signal and convert it into tens to hundreds of microwatts of electricity. It uses that power to modify and reflect the signal with encoded data. Antennas on other devices, in turn, detect that signal and can respond accordingly. Initial implementations can communicate over several feet of distance, even with transmission towers up to 10.5 kilometres (6.5 mi) away. Transmission rates were 1k bits per second between devices situated 0.45 metres (1 ft 6 in) apart inside and 0.75 metres (2 ft 6 in) apart outside, sufficient to handle text messages or other small data sets. Circuit sizes can be as small as 1 sq. mm. This approach would let mobile and other devices communicate without being turned on. It would also allow unpowered sensors to communicate, allowing them to function in places where

DOI: 10.4018/978-1-5225-5870-5.ch008

external power cannot be conveniently supplied. With the advancements in Wireless Power Transfer (WPT), researchers are looking for backscatter communications for low-cost tracking. One of the first instances of the use of backscatter occurred in World War II. With the advent of RADAR, both sides could tell when an aircraft was approaching. However, there was no method to identify whether that aircraft was friendly or an enemy. The German Air Force is credited with discovering that they could identify their aircraft by having their pilots manoeuvre their planes in a certain manner which changed their Radar Cross-Section (RCS), and therefore the response on the radar screen (Dobkin, 2008). This method is a precursor to how backscatter communications work and will be discussed in this Chapter. Few years later in 1937, the British debuted an identify friend or foe (IFF) transponder system called the XAE, and later Mark I, as a more secure method of identifying their aircraft. This system, when hit with a RADAR pulse, would send out a signal which identified the aircraft. It is accredited with being the first active RFID (Radio Frequency Identification) system.

As early as 1945, Soviet Scientist Leon Theremin had designed a passive method of listening to conversations. This device, now known as the `Great Seal Bug,' allowed the Soviets to spy on the American embassy in Moscow for seven years until its discovery (Finkenzeller, 2003). It used a resonator which, when sound wave vibrated a membrane, modulated any sound in the room onto an external RF (Radio Frequency) signal that Soviet Spies pointed at the American Embassy. The Americans were unable to discover this device because it had no active components which to detect. It is accredited with being one of the first passive RFID systems despite not being officially disclosed until much later.

During the time that the Great Seal Bug was in use, Harry Stockman officially referenced the first method of communication by reflected radio waves in a paper published in 1948 (Stockman, 1948). While much of the necessary technology for performing true passive, backscattered communication did not exist, Stockman nevertheless foresaw the benefits of reflected radio waves for passive communication. It wasn't until a few years later in 1952 that D.B. Harris filed a U.S. Patent referencing the first relative of the modern RFID tag (Harris, 1952). While Harris's power transfer mechanisms were not as advanced as Brown's, the path for integration was set.

The 1960s and 1970s represented a huge leap in the progress of RFID tags thanks to advancements in semiconductor technology and interest by the community. Practical applications such as Sensormatic and Checkpoint electronic article surveillance (EAS) devices for retailers (Landt, 2005), the Ratheon Raytag (Freedman, 1973), and Los Alamos National Laboratory nuclear material and cattle tracking (Koelle et al., 1975; Landt, 2001) increased the interest in RFID technology and led to further investment and development by the government and industry alike. The work out of Los Alamos led to the development of passive UHF RFID which is directly linked to the logistical tracking and road tolling systems currently in use.

The success of RFID has attracted the attention of the WSN (Wireless Sensor Network) community (Roy et al., 2010). The desire for an internet of things (IOTs), to `sense everything' in current society, has been increasing. Furthermore, `sensing everything' demonstrates a method to more efficiently manage resources. Numerous applications in ecological, military, and healthcare applications have been suggested. Wireless, distributed sensing provides an effective means to implement these ideas with RFID providing a particularly attractive solution because of their low cost and small size.

Most current trends in RFID research mention sensing to at least some degree. The incorporation of sensing in current RFID technology puts additional constraints on the system design as the sensors must also be powered. Thus, researchers have begun examining some of the work of the SSP (Sensor Signal Processing) community. While the SSP community typically focuses on high-power (greater than

20 dBm) energy conversion, their work on semiconductors and rectennas can be directly applied to the very low-power (less than 0 dBm) situations where WSN applications occur.

Some of the First work in this RFID-enabled sensing area has come from Zoya Popovic at the University of Colorado at Boulder (Paing et al., 2007) and Joshua Smith at the University of Washington (Smith et al., 2006). The Intel Wireless Identification and Sensing Platform (WISP), which Smith did most of his work with, is a passive UHF RFID with a programmable microcontroller that can communicate via the EPC Global Generation (Gen) 2 protocol using of the-shelf RFID readers. This research tool has proven very effective as the Gen 2 protocol is one of the most used and the project code, schematics, and WISP samples have been made available to researchers and hobbyists alike. It has spawned a great deal of interest and work in the sensors (Chaudhuri et al., 2008; Buettner et al., 2009), computation (Chae et al., 2007; Ransford et al., 2008), protocol (Ko et al., 2007), and security (Chae et al., 2007; Holcomb et al., 2007) areas. Much of the other work in the RFID space has involved other methods of energy harvesting (Sample et al., 2011; Roundy et al., 2003; Sample & Smith, 2009), moving to higher frequencies (Durgin, 2008), novel antenna designs (Trotter et al., 2012), flexible substrates (Yang et al., 2007), and improved circuit designs (Curty et al., 2005; Liu et al., 2011; Mbombolo & Park, 2011) among many others.

Both SSP and RFID applications have numerous advantages in moving to microwave frequencies. Microwave antennas are substantially smaller than some of their UHF and VHF counterparts, and this smaller size helps to reduce the antenna aperture required and to focus the antenna beams more easily. Since the antenna size is the dominating component in the size of a rectenna or RFID-enabled sensor, space can be used more efficiently. This reduction in size (and therefore weight) can help SSP applications reduce the amount of land required for ground-based stations and help RFID applications minimize the profiles of their tags and sensors. The world-wide availability of the unlicensed 5.8 GHz microwave band, an ISM frequency band in the United States, has led research in microwave-power transfer to focus on this frequency. In fact, several other benefits of operating at 5.8 GHz for RFID have been discussed apart from the reduced footprint including the following: greater bandwidth availability, reduced on-object penalty (Durgin, 2008), spatial signalling techniques (Grigin & Durgin, 2009; Grifin, 2009), greater material sensing capability (Hasan et al., 2011a; Hasan et al., 2011b), and plasma penetration ability (Valenta et al., 2010a; Valenta et al., 2010b) among others. Another key technological development in RFID and WPT is the power-optimized waveform (POW) (Trotter et al., 2009; Trotter & Durgin, 2010) or multi-sine excitation (Boaventura & Carvalha, 2011; Fernandes et al., 2011). A POW is a waveform whose spectrum is shaped to increase the efficiency of RF energy harvesting circuits, especially at low powers. By increasing this efficiency, the range and reliability or power available to the RF tag will also be increased. A multi-sine accomplishes this increase in efficiency via a large peak-to-average power ratio (PAPR) which helps to overcome the energy harvesting circuit diode-voltage threshold limitation. When incident signals are not greater than this threshold, the circuit efficiency suffers. Trotter and Durgin (2010) and Boaventura and Carvalha (2011) have independently shown that POWs do increase the efficiency of charge-pump circuits. However, their work has not explored in detail the limitations of this increase in energy harvesting efficiency or attempted to design a waveform and energy harvesting circuit in tandem.

Radio-frequency Identification (RFID) is used for Automatic Identification and Data Capture (AIDC) as the replacement of the bar codes. There are two major parts used in this technology; (i) RFID Reader, and (ii) RFID Tag. RFID reader tracks the tags and automatically identifies the objects by electromagnetic fields. RFID Tags are of two types Passive and Active which contains electronically stored information.

Passive Tags collect energy from a nearby RFID reader's radio waves while Active tags need a local power source such as battery and may be tracked from hundreds of meters distance from RFID reader. By this technology number of RFID tags can be tracked simultaneously without any physical contact. There is a need of line of sight contact in barcodes which is not required for RFID. RFID decreases the cost of system and increases the reliability of data associated with objects. RFID has various applications in the field of logistical tracking, toll collection, employee identification, ticketing etc (Finkenzeller, 2003). Cost of RFID hardware is reducing as the demand of RFID systems is increasing throughout the world. Advanced VLSI technologies have allowed the possibilities of integration of passive RFID-enabled sensors with Wireless Sensor Networks (WSN) (Smith et al., 2006). Passive/battery-less RFID-enabled sensors are cheap maintenance-free. Due to backscattered radio waves these advantages are possible. RFID has known as variety of technologies now-a-days. RFID is an electronic means of identification and may be used with RF sensors or RF tags. This new terminology may also involve hardware which communicates via backscatter radio, the same method which semi-passive and passive RFID tags use. Therefore, RFID are in use for identification of objects as well as for new sensing or computation devices which are being researched now-a-days.

RFID tags use two methods for communication; (i) Magnetic or Inductive Coupling, and (ii) Backscatter Radio System. Magnetically coupled systems typically operate at LF (Low Frequency) and HF (High Frequency) at 135 kHz and 13.56 MHz, respectively. These systems use the same terminology as backscatter RFID systems but have a shorter range as magnetic coupling is a near field phenomenon. Usually, this range is limited to the longest dimension of the transmit antenna (Finkenzeller, 2003). Backscatter radio systems; rely on far field propagating electromagnetic waves at frequencies greater than 400 MHz thus have a greater range.

STANDARDS IN RFID TECHNOLOGY

RFID systems are comprised of two parts: the reader (also known as an interrogator or base station) and the tag (Dobkin, 2008). RFID systems differ from conventional radio systems in that they use backscattered radio waves to communicate. While traditional radio systems have active transmitters on both the mobile station and the base station, RFID systems only have one active transmitter at the reader. This reader typically broadcasts a continuous wave (CW) signal which travels to the mobile station (RFID tag) as shown in Figure 1. The tag modulates its data onto this signal by changing its radar cross-section (RCS) by means of switching its antenna load. This modulated signal then returns to the reader where it is demodulated and decoded.

Frequency Bands

In the United States, RFID tags typically operate in Federal Communication Commission (FCC) designated, unlicensed Industrial, Scientific, and Medical (ISM) communication bands at 433/915 MHz, 2.4 GHz, and more recently 5.8 GHz (Dobkin, 2008). Table 1 lists some common parameters for the ISM bands in the United States including maximum power transmission and effective isotropic radiated power (EIRP). Similar frequency bands exist elsewhere in the world, but exact regulations tend to vary by country. Note that inductively-coupled tags operating at 135 kHz and 13.56 MHz are absent from this list.

Figure 1. A typical semi-active/passive RFID system communicates via backscatter radio; a reader broadcasts a CW signal to the environment where RF tags then backscatter this signal back to the reader; the reader then must process the received signal and demodulate the tag data

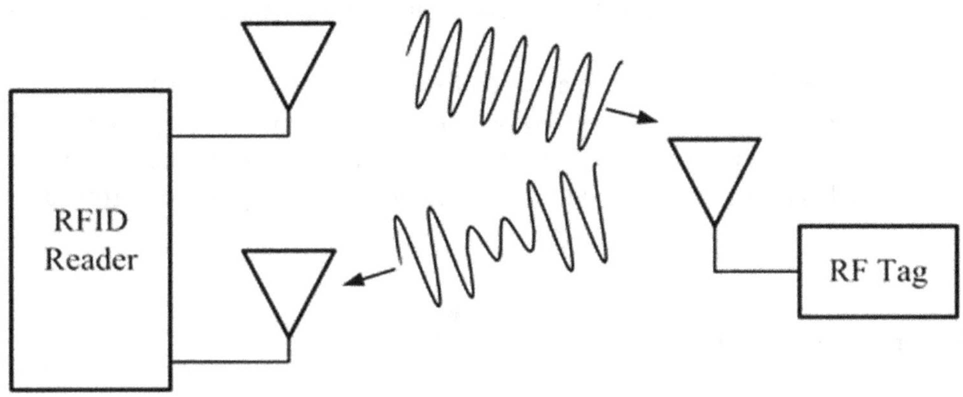

Table 1. Common parameters of RFID bands in the United States (EPCglobal, 2008).

Centre Frequency (GHz)	Bandwidth (MHz)	Max Power (W)	Max EIRP (W)
0.915	26	1	4
2.45	83.5	1*	4*
5.8	150	1	4**

*Antenna gains greater than 6 dBi can be used if the radiated power is reduced by 1 dB for every 3 dB the antenna exceeds 6 dBi.
**Antenna gains greater than 6 dBi can be used without reducing the radiated power.

915 MHz UHF RFID systems offer the largest range for the lowest cost of any passive RFID system. Therefore, these systems are among the most widely commercially adopted. However, many additional benefits at higher frequencies exist including smaller footprints, additional bandwidth, and a larger permissible EIRP, among others (Grifin, 2009). Thus, it is possible to meet and even exceed the limitations of the additional path loss at higher frequencies such as 5.8 GHz.

In this present chapter, RFID is discussed in detail. Communication standards and channel characteristics are discussed. Further, RFID readers, RFID tags and sensors and their workings are described. Then power management and other factors related to RFID technology are presented.

Communication Standards

Besides the frequency and transmit power standards which are set by regulatory agencies, RFID systems follow a set of communication standards which set various protocols and modulations. The most well-known of these standards is the EPC Class-1 Generation-2 UHF RFID (also simply known as `Gen2') (EPCglobal, 2008). There are other `low-power' standards that are designed for WSNs including IEEE 802.15.4, Bluetooth, ZigBee, and many other proprietary standards. However, the Gen2 standard is used by most of UHF RFID tags, which is the de facto standard for inventory and logistics tags. While the specifics of this standard are outside the scope of this discussion, it is important to note that the communication protocol affects the power consumption of the tag. For example, when two-way communication between the tag and the reader is desired (in send-and-acknowledge schemes such as in the EPC standard), the tag must use an ADC so it may sample and decode the incoming reader command. Not only must the tag power an additional peripheral, it must also remain powered during the entirety of the command.

Protocols are designed in part to avoid packet collisions, to increase security, and to ensure maximum spectral efficiency. Removing or simplifying them may not always be a possibility. Therefore, a trade-off between communication efficiency and power availability must be made when the protocol is selected. In some applications, such as building temperature monitoring, packet error rate may not be as important as low-cost sensors.

CHANNEL CHARACTERISTICS

Unlike typical radio communication systems, in RFID systems there are two different radio channels. The forward channel expresses the radio environment from the reader to the tag while the backscatter channel describes the environment from the tag back to the reader. While each channel environment may be modelled identically, the statistics of the channel fading will change since the radio wave will pass through two unique channels, not necessarily the same one twice. These channels can be modelled according to the Friis Transmission formula. However, in backscatter, each channel will include other terms not typically used in conventional wireless communication.

The forward channel or `power-up' link budget is expressed in (1). This log-scale link budget is called power-up because this equation models the amount of energy which is available at the output of the tag antenna. For battery-free tags, this is the maximum power available to power-up the tag.

$$P_t = P_T + G_T + G_t - 20log\left(\frac{4\pi r_f}{\lambda}\right) + X_f - \theta - B_f - F_P \quad (1)$$

where P_t is the received power at the RF tag, P_T is the transmitted power by the base station, G_T and G_t are the gains of the reader transmitter and tag antennas, respectively, r_f is the forward-link distance between the RFID transmitter and the RF tag, λ is the free space wavelength, X_f is the polarization mismatch between the RFID transmitter and the RF tag antennas, θ is the RF-tag antenna on-object gain penalty, B_f is the forward-link path blockage loss, and F_P is the power-up fade margin, typically chosen as Rayleigh or Rician.

On the backscatter link, a similar link budget is used, but several factors are counted twice and others are added. These additional terms are largely due to the fact that the forward and backscatter signal paths do not need to be symmetrical. The backscatter link budget is introduced in (2).

$$P_R = P_T + G_T + 2G_t + G_R - 20log\left(\frac{4\pi r_f}{\lambda}\right) - 20log\left(\frac{4\pi r_b}{\lambda}\right) + X_f + X_b + M - 2\theta - B_f - B_b - F_b \quad (2)$$

.where P_R is the received power at the RFID reader, G_R is the gain of the RFID reader antenna, r_b is the backscatter-link distance between the transmitter and receiver, X_b is the polarization mismatch between the RF tag and RFID reader receiver antennas, M is the modulation factor, B_b is the backscatter-link path blockage loss, and F_b is the backscatter fade margin, typically chosen as a product Rayleigh or product Rician. An in-depth description of these link budgets is discussed by Griffin in (Grifin & Durgin, 2013).

RFID READERS

RFID readers typically come in two forms. In mono-static systems, the reader has a single antenna which is used for both transmitting and receiving. It accomplishes this configuration via an RF circulator. However, poor isolation in the circulator can lead to increased interference in the receiver because of transmit leakage. Bi-static systems eliminate the need for a circulator and better isolate the transmit and receive paths through the use of two antennas. By having independent antennas, the isolation, and therefore sensitivity, can be improved which can help to increase the signal-to-interference-plus-noise ratio (SINR).

RFID readers must demodulate backscattered signals from tags in the presence of a variety of noise and interference sources. Received backscatter signals typically have very low amplitudes as their path loss increases with the fourth-power of distance. In addition to common thermal and flicker noise sources, RFID readers have the added disadvantage of unmodulated backscatter interference. These sources can include self-interference due to transmitted energy leaking directly into the receive chain as well as signals which reflect off of background clutter. These interferers can be many orders of magnitude greater than the signal level of the backscatter signals. Thus, readers typically use direct down-conversion (homodyne) receiver architecture with AC coupling as shown in Figure 2. The use of this architecture allows the use of a single RF oscillator to down convert and effectively eliminate the CW power in the RF carrier, including the high-power unmodulated backscatter and self-interference. This fully coherent receiver can also be used to make very sensitive phase measurements. This topology also allows the reader to transmit and receive continuously in full-duplex mode.

As part of the homodyne receiver architecture with AC coupling, capacitors are required before the signal(s) can be sampled by an ADC to eliminate any DC component as a result of the demodulation of the previously mentioned CW interference power. The capacitors form a high-pass, RC filter with a time constant controlled by the value of the capacitor. This value must be large enough to remove any strong, unmodulated signals, but small enough to keep the data signal itself from decaying. To improve the performance of RFID systems, which employ direct down-conversion receivers, the RF tag modulation may be modified via tag hardware and/or coding/software modifications (Durgin et al., 2013).

RFID TAGS AND RFID-ENABLED SENSORS

RF tags can be active, semi-passive, or passive. Active tags require an on-board battery which powers the tag electronics and their active transmitter. They do not communicate via backscatter and operate like a traditional radio communication system. Due to their active transmitter, active RFID tags act like a transponder, and therefore typically have the largest ranges and fastest data rates. However, these advantages come at the cost of increased price, size, and complexity due to the necessity of a battery and

Figure 2. (a) A simplified 5.8 GHz direct down-conversion (homodyne) transceiver with AC coupling; the transmit chain amplifies the 5.8 GHz carrier wave; the receive chain mixes the received signal directly to baseband in-phase and quadrature outputs which are then sampled by an ADC (a USRP SDR in this case); (b) RFID Module.

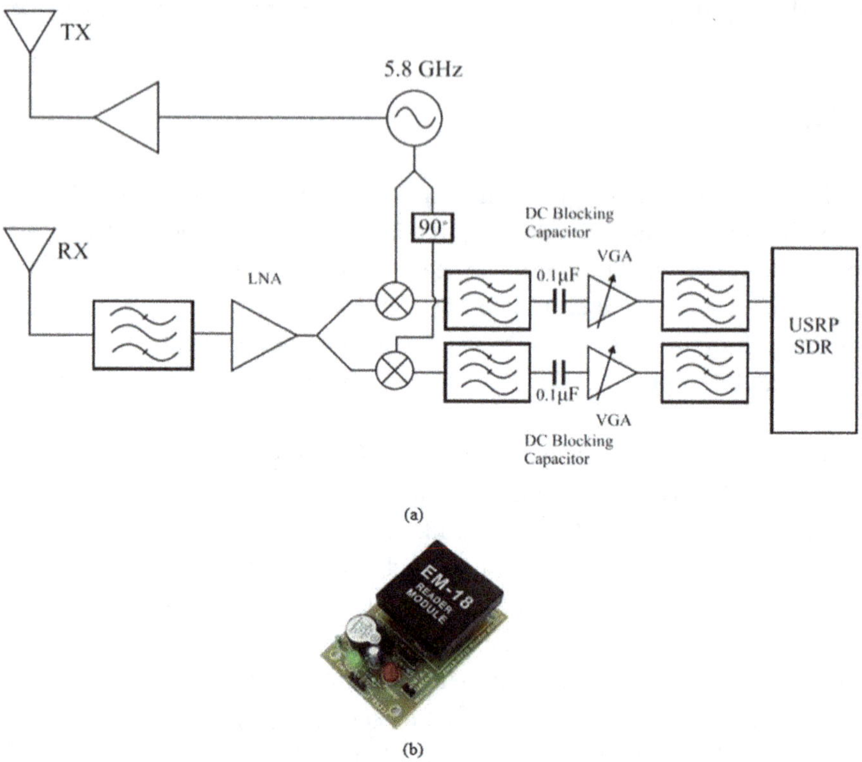

on-board RF hardware. Semi-passive tags also use a battery, but only to power the tag electronics when within reader range. All communication is done via passive backscatter, which eliminates the necessity of complex RF hardware. Passive tags use no battery and harvest all their energy from signals from the reader or some other source. The reader-delivered signal is used for both backscatter communication and energy harvesting. Due to the need to harvest their energy, passive tags typically have the shortest ranges and lowest data rates. However, their simplicity, and thereby low cost, makes them ideal for a variety of applications.

Active and semi-passive tags are generally backscatter-link limited. This means that their maximum range is governed by the sensitivity of the receiver. It must be able to detect the extremely weak backscattered signal in the presence of non-linear reflections, self-interference, and modulated and un-modulated backscattered signals. On the other hand, passive RF tags are generally forward-link or power-up link limited. This description means that the maximum range at which a tag can operate is limited by the voltage available at the tag antenna. When the voltage fails to reach a specified value, the tag does not have sufficient energy to power-up. This failure occurs because of a limitation in the electronic circuit utilized for energy harvesting. Unless the incident signal produces a voltage level greater than the threshold voltage of the diodes in the energy harvesting circuit, current will not be able to low, and the RF tag will not have any energy for processing or communication.

PASSIVE RF TAGS

Passive or battery-free RFID tags are an attractive option for WSNs and RFID applications because of their minimum- to zero-maintenance requirements. This benefit is principally due to the fact that passive RFIDs do not utilize a battery for power storage. Instead, passive tags must harvest energy from some other source. Options can include solar (Sample et al., 2011), motion or vibration (Roundy et al., 2003), ambient RF (Sample & Smith, 2009), or most commonly, an RF signal generated by the RFID reader. RFID tags typically rely on their readers for energy as the reader presence usually signifies a desire to read the tag, therefore a need for energy. However, in some WSN applications, additional methods of energy harvesting could be required if the tag is expected to be removed from the presence of a reader for a long period of time. Passive RFID tags are unique communication devices in that they do not have complex RF hardware such as mixers, oscillators, or amplifiers. The use of backscatter permits a much simpler RF front end and no battery while allowing the tags to be small, low-cost, and power efficient. Figure 3 shows the main components of a semi-passive/passive RFID tag.

Several components make up a semi-passive/passive RFID tag. The tag is primarily controlled by a piece of digital logic which stores the tag identification information, receives and processes reader commands, interrogates sensors (if applicable), and backscatters information back to the reader. While commercial solutions utilize custom ASICs for this digital logic, low-yield applications in the research field tend to use microcontrollers (Smith et al., 2006), complex programmable logic devices (CPLDs), or field-programmable gate arrays (FPGAs) (Roy et al., 2010). Many of these devices tend to have integrated oscillators, timers, comparators, ADCs, DACs, and digital communication buses, thereby allowing for a single integrated circuit to complete many of the tag tasks (Texas Instruments, 2008).

One or more antennas are used by the tag for backscatter communication, receiving reader commands, and for energy harvesting. This/These antenna(s) can be linearly or circularly polarized depending on their application and are often designed to be as omnidirectional as possible. This uniform coverage allows the tag to communicate with the reader regardless of its orientation. However, if the tag position can be controlled or is known a priori, a directional antenna with much greater gain can be used. If a single antenna or antenna array is used, it is simultaneously connected to a load modulator, energy harvester, and envelope detector which all affect the load on the antenna1, an important consideration for communication and energy harvesting.

In backscatter communication, a load modulator changes the impedance on the antenna so that the tag can communicate with the reader. It is controlled by digital logic which sends the encoded data to the modulator. Typically, this load modulator is a simple transistor switch which changes between two different impedances. This choice limits the tag to two-state ASK or PSK modulation schemes. While higher-order modulation schemes have been demonstrated by (Koo, 2010; Thomas & Reynolds, 2010), low-order modulations are typically used because of the low SNR at the receiver. In the event that a single antenna is used, one impedance state may be the energy harvester input impedance while the other state would be a short or open circuit. When this occurs, the modulation index of the tag is reduced which decreases the differential backscattered power to the reader (Nikitin & Rao, 2008). However, a separate antenna could be used for both the modulator and energy harvester to maximize the differential backscattered power levels.

An energy harvester and power management circuitry are responsible for collecting sufficient energy to power the tag and any associated sensors. As previously mentioned this energy can come from a variety of sources but is typically reader-delivered RF power.

Figure 3. (a) This block diagram shows the main components of a semi-passive/passive RFID tag. Some form of digital logic samples sensors, receives reader commands, and controls a load modulator to backscatter data back to a reader. An energy harvester provides power for all of these devices (Trotter, 2011), (b) RFID Tags.

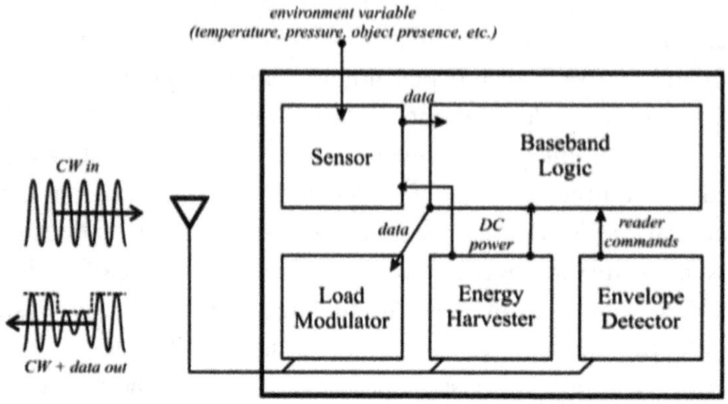

(a)

(b)

An envelope detector consists of a rectifier or voltage doubler and allows an ADC to sample the received waveform from the reader. This waveform can then be decoded into a reader command, and the tag can operate accordingly. It is important that the capacitors in the envelope detector be sized appropriately to the reader command chip rate so that an accurate waveform can be sampled.

Associated sensors in the tag are powered by the energy harvester and interrogated by the digital logic. These sensors may measure quantities such as temperature, pressure, or gas concentration among others. Several researchers have also integrated sensing into the load modulator by using a variation in impedance due to changing material properties as the transducer (Hasan et al., 2011b; Lee et al., 2010). Various researches have been reported in these days in the same field (Almada et al., 2012; Valenta & Durgin, 2013; Sutardja & Rabaey, 2018; Bao et al., 2018; Gope et al., 2018; Shi et al., 2018; Fahim et al., 2018; Chakaravarthi et al., 2018; Mbacke et al., 2018; Xiao et al., 2018; Yang et al., 2018a; Colella & Catarinucci, 2018; Caccami & Morrocco, 2018; Zhang et al., 2018; Kim & Vinck, 2018; Guidi et al., 2018; Yang et al., 2018b).

POWERING A PASSIVE RFID TAG

Passive RFID tags typically rely solely on RF power from their readers for all of their energy. Energy harvesting circuits such as rectennas or charge pumps convert RF energy to DC energy that is able to power the tag electronics. However, these circuits suffer from inefficiencies and behave non-linearly. Equation (3) follows from (1) and gives the DC power available from the output of the energy harvesting circuit.

$$P_{tagDC} = P_T + G_T + G_t - 20log\left(\frac{4\pi r_f}{\lambda}\right) + X_f - \theta - B_f - F_P + L_{PC} \tag{3}$$

where P_{tagDC} is the DC power at the output of the energy harvester, L_{PC} is the energy harvester RF to-DC conversion gain which is a function of the tag received RF power P_t as defined in (4) (Almada et al., 2012; Valenta & Durgin, 2013). All other terms follow from (1).

$$L_{PC} = 10log_{10}\left(\eta_{RF-DC}\right) = 10log_{10}\left(\frac{P_{tagDC}}{P_t}\right) \tag{4}$$

where $\eta_{RF\text{-}DC}$ is the RF-to-DC conversion efficiency of the energy harvesting circuit. The largest value of L_{PC} will be zero. This number will occur only if the energy-harvesting circuit is 100% efficient.

Operating Zones

In the absence of multi-path interference, as the tag moves farther from the reader, the amount of RF power decreases. By (3), the amount of DC power also decreases. At close distances when $P_{tagDC} \geq P_{task}$, where P_{task} is the power requirement to complete a task on-board a tag, the data throughput is said to be protocol limited (Roy et al., 2010). The tag can continuously power all on-board sensors, receive commands from the reader, and perform necessary communications. The only tag limitation is the amount of data it can deliver to the reader as determined by the protocol. However, many times the power available is insufficient for continuous operation. In this case when $P_{tagDC} < P_{task}$, the tag is said to be power limited. Energy must be stored in an energy storage device, typically a capacitor, until enough energy exists to complete a task. The tag communication rate is thereby reduced. Under these circumstances, a capacitor or battery must charge until the stored energy E_{stored} is greater than or equal to the task energy E_{task} according to (5) and (6).

$$E_{stored} = \frac{1}{2}C\left(V_{tagDC}^{\ 2} - V_{dd}^{\ 2}\right) \tag{5}$$

$$E_{task} = \int_{T_0}^{T_0+T_{task}} P_{task} dt = V_{dd} \int_{T_0}^{T_0+T_{task}} I_{task} dt \quad (6)$$

where C is the value of the energy harvesting circuit output capacitor, V_{dd} is the voltage level at which the circuit logic operates, T_{task} is the time required for the task, and I_{task} is the time varying current draw of the logic during the task. Equation (5) describes the usable tag energy stored for the circuit to operate. Since the logic requires a minimum voltage V_{dd} to function, the voltage level must be greater than this value. Only voltage overhead, the available voltage above V_{dd}, can be used for tag operations. Equation (6) gives the energy level required for a tag to complete a specific task. This equation is simply the power consumption of the tag multiplied by the time it takes to finish its task. Rearranging (5) and solving for V_{tagDC} gives the minimum voltage required at the output of the energy harvester to execute a task on a tag as shown in (7).

$$V_{tagDCMIN} = \sqrt{\frac{2E_{task}}{C} + V_{dd}^{\ 2}} \quad (7)$$

When the tag moves further away from the reader and $V_{tagDCMIN} < V_{tagDC}$, it is said to be voltage limited and will not power on. Some energy harvesting circuits will step up the voltage that is received by the tag. In this case, $V_{tagDCMIN}$ may fail to be large enough because received power levels are too low for the energy harvester to charge at all or power levels are too intermittent and leakage currents deplete available charge before large enough levels can be reached. Therefore, $V_{tagDCMIN}$ sets the maximum operating range for a given task energy of E_{task}.

Power Management

Since the tag power level can vary with distance from the reader and multi-path interference, the tag must have a mechanism to detect whether or not the current energy level is sufficient for operation. This mechanism must be able to detect the energy level and be able to wake up the digital logic when sufficient energy is present. Most often when waiting for energy accumulation, the tag remains in an ultra-low power state with its logic circuitry and the majority of its peripherals powered down. In this state, the tag waits for the energy harvester to charge with enough energy so that it can power on and complete its task.

Voltage supervisor circuits are used to monitor the charge accumulation on the energy harvesting circuit output capacitor by measuring the voltage on the capacitor V_{tagDC}. When this voltage reaches $V_{tagDCMIN}$, this implementation sends the main logic a wakeup signal to tell it that there is sufficient energy stored to complete its task. An ADC may also be used to measure this voltage, but this method requires additional logic and time and therefore power. Thus, the voltage supervisor circuit is a preferable method.

While digital logic can operate over a range of voltages, there is typically a preferred voltage level from a power consumption point of view. Furthermore, input voltage swings can lead to variation in on-chip oscillators and affect the device timing and stability. Thus, voltage regulators are used to supply a steady voltage level for the RF tag. It is important to note that as the voltage regulator together with the voltage supervisor is the circuits from which the energy harvester load and consequently conversion efficiency are determined.

Figure 4 shows an example of the power management circuitry for a passive RF tag. Output current from the energy harvesting circuitry builds up charge on an output capacitor and supplies power to a voltage supervisor and voltage regulator. Simultaneously, the voltage supervisor monitors the output capacitor voltage. The regulator provides power to the tag logic and associated peripherals, which are powered down or in a low power state while the output capacitor is charging. Finally, a zener diode acts as overvoltage protection for the supervisor and regulator.

Power Consumption

The limited power harvested by a passive RF tag must be used for a variety of tasks both in hardware and software. Although the specific implementations and architectures of RF tags can vary, some of the basic components which must have their power consumption taken into account are listed below. Care must be taken to minimize the current draw of each component and maximize the efficiency of each software subroutine.

Hardware element power consumption considerations

- Baseband logic/Microcontroller
    - CPU
    - Oscillator(s)
    - Analog-to-digital converter
    - SPI/I2C digital communications
    - Timers
- RF switch
- Voltage regulator
- Power Management
- Sensor(s)

Figure 4. The power management circuitry for an RF tag must regulate and measure the energy harvesting circuit output voltage. When this voltage reaches a predetermined level, it will wake up the tag logic. Furthermore, this circuitry sets the load conditions for the energy harvesting circuit which affects its RF-to-DC conversion efficiency.

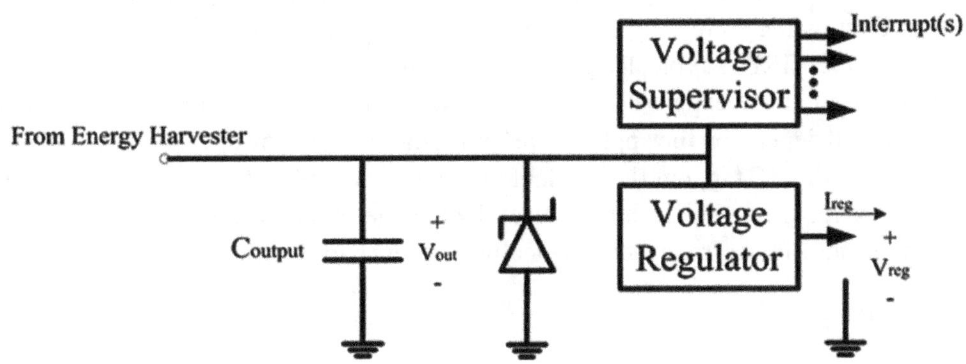

As modern microcontrollers or other digital logic devices implement a variety of protocols and tag control, it is important to consider the power consumption of the software implementations. By modifying the tag behaviour or implementation of a tag operation, the power consumption can drastically vary. Some major tag operations are listed below.

Software implementation power consumption considerations

- Initialization
- Reception of reader commands
- Sensor interrogation
- Packet formation
- CRC calculation
- Coding/Modulation
- Backscattering

RANGE AND RELIABILITY OF RF TAGS

Passive RFID tags and RFID-enabled sensors are typically forward-link limited due to the lack of power available to the tag. When this minimal power level is not met, the tag will fail to power up and communication cannot occur. Several items can be addressed to extend this forward-link limit and thereby increase the range and reliability of RF tags. Many of these will be application specific and have impacts on the system performance.

- Power-up link budget equation
- Decreased duty cycle of operation
- Simplified communication protocol, modulation, and encoding
- ASIC/More efficient semiconductor/packaging
- Optimized energy harvesting circuit design
- Improved excitation waveform

Power-Up Link Budget Equation

Items in the power-up link budget from (1) can be adjusted to increase the tag received power P_t and therefore the available DC power P_{tagDC}. If these levels are increased, the tag maximum operating range can be extended. First, while increasing the transmit power P_T and the reader transmit antenna gain G_T will increase range, the EIRP is typically government regulated. Second, the tag antenna gain G_t can potentially be increased at the expense of coverage. However, tag orientation may be uncontrollable. Thus, this increase will depend on the application. Third, moving to a lower frequency will also increase the power received by the RF tag, but the available frequencies are limited by regulating agencies.

The antenna sizes required to efficiently collect the RF energy at these low frequencies may also be impractical. Furthermore, lower frequencies typically have less bandwidth available and more congestion which may lead to communication challenges. Finally, if the tag location is controllable, the polarization mismatch X_f, on-object gain penalty θ, and path blockage B_f can be reduced to zero. If it is not, then appropriate statistical models must be made to ensure the tag will function within an accepted probability of error. This analysis could be similar to the one used to estimate the power-up fade margin, which is typically a function of the environment.

Decreased Duty Cycle

The duty cycle of the tag in its on state could be reduced. This change could come in the form of lowering the number of samples taken by a sensor or lowering the number of times a tag communicates with the reader. By reducing the need to perform calculations and/or sample sensors, RF tag task energy E_{task} could be lowered. Following from (7), if E_{task} is lowered, $V_{tagDCMIN}$ is reduced which would increase range.

Simplified Communication Protocol, Modulation, and Encoding

Like decreased duty cycle, if the communication protocol is simplified, the amount of energy required by the tag is reduced. Complex modulations and encoding all require an increased number of clock cycles on the digital logic. By minimizing the number of clock cycles, the power requirements of the tag can be reduced. However, the application may not tolerate the lack of security or high probability of error associated with eliminating some of these techniques, so these additions and subtractions must be carefully weighed.

ASIC/More Efficient Semiconductor/Packaging

It is well known that ASICs offer superior efficiency and lower operating voltages compared to discrete designs. Discrete semiconductors tend to have high losses associated with their packaging or implementations, especially at microwave frequencies. Additional research into improving these methods would help to decrease these losses and improve the overall efficiency. That said, an RFID-enabled sensing solution implemented on an ASIC would be expected to have a larger range. However, the excessive cost and inexibility with an ASIC design may not suit all applications.

Optimized Energy Harvesting Circuit Design

Choosing the most efficient energy harvesting circuit design is critical to obtaining the highest RF-to-DC conversion efficiency possible. The previously mentioned techniques all focus task on obtaining the highest amount of RF energy available at the antenna output or reducing the tag power consumption. However, this technique focuses on efficiently converting the received power to usable DC energy. There are a variety of energy harvesting circuit designs in literature, but few attempts at comparing them. However, there are certainly trade-offs in the number of stages in charge pumps, the diodes chosen, and the values of the capacitors and load.

Improved Excitation Waveform

Changing the reader delivered waveform that powers the tag from the standard continuous waveform (CW) to a high peak-to-average power ratio (PAPR) waveform such as a POW or multi-sine can increase the power available to the RFID tag. POWs and multi-sines can deliver a higher instantaneous power/ voltage to an RF tag than a CW waveform can. This high instantaneous power helps to increase the RF-to-DC conversion efficiency of energy harvesting circuits.

CONCLUSION

In this chapter, Radio Frequency Identification (RFID) technology is discussed in detail. A brief history of RFID is provided along with the developments in technology in the introduction part. RFID standards, required frequency bands and channel characteristics are described with the advantages of RFID over traditional technologies. Components of RFID i.e. RFID readers, RFID tags active and passive are discussed. Various parameters associated with RFID technology are discussed including operating zones, power management, power consumption, power link budget etc.

REFERENCES

Almada, M., Blanca-Pimentel, L., Block, J., Gonzalez, J., Valenta, C., & Durgin, G. (2012). Power Conversion Gain as a design metric for RFID Systems. *2012 IEEE Antennas and Propagation Society International Symposium*, 1-2. 10.1109/APS.2012.6349367

Bao, D., Zou, Z., Nejad, M. B., Qin, Y., & Zheng, L.-R. (2018). *A Wirelessly Powered UWB RFID Sensor Tag with Time-Domain Analog-to-Information Interface. IEEE Journal of Solid-State Circuits.*

Boaventura, A., & Carvalha, N. (2011). Maximizing DC power in energy harvesting circuits using multisine excitation. *IEEE MTT-S International Microwave Symposium Digest*, 1 - 4. 10.1109/MWSYM.2011.5972612

Buettner, M., Prasad, R., Philipose, M., & Wetherall, D. (2009). Recognizing Daily Activities with RFID-Based Sensors. *11th International Conference on Ubiquitous Computing*.

Caccami & Marrocco. (2018). Electromagnetic Modeling of Self-tuning RFID Sensor Antennas in Linear and Nonlinear Regimes. *IEEE Transactions on Antennas and Propagation*, 1-1.

Chae, H., Yeager, D., Smith, J., & Fu, K. (2007). Maximalist Cryptography and Computation on the WISP UHF RFID Tag. *Proceedings of the Conference on RFID Security*.

Chakaravarthi, Prasath, Philip, Jayaganthan, Velmurugan, & Arunachalam. (2018). Reusable Passive Wireless RFID Sensor for Strain Measurement on Metals. *IEEE Sensors Journal*, 1-1.

Chaudhri, R., Lester, J., & Borriello, G. (2008). An RFID Based System for Monitoring Free Weight Exercises. *The 6th ACM Conference on Embedded Networked Sensor Systems*. 10.1145/1460412.1460488

Colella & Catarinucci. (2018). Electromagnetic Design of UHF RFID Tags Enabling a Novel Method to Retrieve Sensor Data. *IEEE Journal of Radio Frequency Identification*, 1-1.

Curty, J., Joehl, N., Krummenacher, F., Dehollain, C., & Declercq, M. (2005). A model for µ-power rectifier analysis and design. *Circuits and Systems, IEEE Transactions on*, *52*(12), 2771–2779. doi:10.1109/TCSI.2005.854294

Dobkin, D. M. (2008). *The RF in RFID. Passive UHF RFID in Practice*. Elsevier.

Durgin. (2008). *The Hidden Benefits of Backscatter Radio and RFID at 5.8 GHz*. URSI 2008.

Durgin, G., Valenta, C., Akbar, M., Morys, M., Marshall, B., & Lu, Y. (2013). Modulation and Sensitivity Limits for Backscatter Receivers. *IEEE International Conference on RFID*, 124 - 130. 10.1109/RFID.2013.6548145

EPCglobal. (2008). *EPC RF Identity Protocols Class-1 Generation-2 UHF RFID Protocol for Communications Version 1.2.1*. Author.

Fahim, ElBatt, Mohamed, & Al-Ali. (2018). Towards Extended Bit Tracking for Scalable and Robust RFID Tag Identification Systems. *IEEE Access*, 1-1.

Fernandes, R., Boaventura, A. S., Carvalho, N., & Matos, J. (2011). Increasing the range of wireless passive sensor nodes using multisines. *IEEE International Conference on RFID-Technologies and Applications*, 549 - 553. 10.1109/RFID-TA.2011.6068599

Finkenzeller, K. (2003). *RFID Handbook: Fundamentals and Applications in Contactless Smart Cards and Identification*. John Wiley and Sons. doi:10.1002/0470868023

Freedman, N. (1973). Raytag, an electronic remote data readout system. *Proceedings Carnahan Conference Electronic Crime Countermeasures*, 104 - 107.

Gope, Lee, & Quek. (2018). Lightweight and Practical Anonymous Authentication Protocol for RFID-Systems Using Physically Unclonable Functions. *IEEE Transactions on Information Forensics and Security*, 1-1.

Grifin, J. D. (2009). *High-frequency modulated-backscatter communication using multiple antennas* (Ph.D. dissertation). Georgia Institute of Technology.

Grifin, J. D., & Durgin, G. D. (2009, April). Complete link budgets for backscatter radio and RFID systems. *IEEE Antennas & Propagation Magazine*, *51*(2), 11–25. doi:10.1109/MAP.2009.5162013

Grifin, J. D., & Durgin, G. D. (2009). Multipath Fading Measurements for Multi-Antenna Backscatter RFID at 5.8 GHz. *IEEE International Conference on RFID*, 322 - 329. 10.1109/RFID.2009.4911197

Guidi, F., Decarli, N., Dardari, D., Mani, F., & D'Errico, R. (2018). *Millimeter-Wave Beamsteering for Passive RFID Tag Localization. IEEE Journal of Radio Frequency Identification.*

Harris, D. (1952). *Radio transmission systems with modulatable passive responder*. U.S. Patent 2 927 321.

Hasan, A., Peterson, A., & Durgin, G. (2011a). Feasibility of Passive Wireless Sensors Based on Reected Electro-Material Signatures. *ACES Journal of Applied Electromagnetics*, *25*(6), 552–560.

Hasan, A., Peterson, A. F., & Durgin, G. D. (2011b). Reected electro-material signatures for self-sensing passive RFID sensors. *IEEE International Conference on RFID*, 62 - 69.

Holcomb, D., Burleson, W., & Fu, K. (2007). Initial SRAM State as a Fingerprint and Source of True Random Numbers for RFID Tags. *Proceedings of the Conference on RFID Security*.

Instruments, T. (2008). *Mixed Signal Microcontroller*. Retrieved from http://www.ti.com/lit/ds/slas632/slas632.pdf

Kim, Y. G., & Vinck, H. (2018). *Anti-Collision Algorithms for FM0 Code and Miller Subcarrier Sequence in RFID Applications. IEEE Transactions on Vehicular Technology*.

Ko, Roy, Smith, Lee, & Cho. (2007). An Enhanced RFID Multiple Access Protocol for Fast Inventory. *IEEE Globecom 2007*, 4575 - 4579.

Koelle, A., Depp, S., & Freyman, R. (1975, August). Short-range radio-telemetry for electronic identification using modulated RF backscatter. *Proceedings of the IEEE*, *63*(8), 1260–1261. doi:10.1109/PROC.1975.9928

Koo, G. (2010). *Signal constellations of a retrodirective array phase modulator (RAPM)* (Master's thesis). Georgia Institute of Technology.

Landt, J. (2005, October/November). The History of RFID. *IEEE Potentials*, *24*(4), 8–11. doi:10.1109/MP.2005.1549751

Landt, J. (2001). *Shrouds of Time - The History of RFID*. Retrieved from http://www.transcore.com/pdf/AIM%20shrouds_of_time.pdf

Lee, H., Naishadham, K., Tentzeris, M., & Shaker, G. (2010). A Novel Highly Sensitive Antenna-Based Smart Skin Gas Sensor Utilizing Carbon Nanotubes and Inkjet Printing. *IEEE Antennas and Propagation Symposium*, 1593-1596.

Liu, D., Li, F., Zou, X., Liu, Y., Hui, X., & Tao, X. (2011). New analysis and design of a RF rectifier for RFID and implantable devices. *Sensors (Basel)*, *11*(June), 6494–6508. doi:10.3390110706494 PMID:22163968

Mbacke, Mitton, & Rivano. (2018). A survey of RFID readers anticollision protocols. *IEEE Journal of Radio Frequency Identification*, 1-1.

Mbombolo, S., & Park, C. (2011). An improved detector topology for a rectenna. *Proceedings of the IMWS-IWPT*, 23-26. 10.1109/IMWS.2011.5877083

Nikitin, P., & Rao, K. V. S. (2008). Antennas and Propagation in UHF RFID Systems. *IEEE International Conference on RFID*, 277 - 288. 10.1109/RFID.2008.4519368

Paing, Morroni, Dolgov, Shin, Brannan, Zane, & Popovic. (2007). Wirelessly-powered wireless sensor platform. *Proceedings of the 37th European Microwave Conference*, 999-1002.

Ransford, B., Clark, S., Salajegheh, M., & Fu, K. (2008). Getting things done on computational rfids with energy-aware checkpointing and voltage-aware scheduling. *USENIX Workshop on Power Aware Computing and Systems*.

Roundy, S., Wright, P. K., & Rabaey, J. M. (2003, July). A study of low level vibrations as a power source for wireless sensor nodes. *Computer Communications*, *26*(11), 1131–1144. doi:10.1016/S0140-3664(02)00248-7

Roy, S., Jandhyala, V., Smith, J., Wetherall, D., Otis, B., Chakraborty, R., ... Sample, A. (2010, September). RFID: From Supply Chains to Sensor Nets. *Proceedings of the IEEE*, *98*(9), 1583–1592. doi:10.1109/JPROC.2010.2053690

Sample, A., Braun, J., Parks, A., & Smith, J. (2011). Photovoltaic enhanced UHF RFID tag Antennas for dual purpose energy harvesting. *IEEE International Conference on RFID*, 146 - 153. 10.1109/RFID.2011.5764615

Sample, A., & Smith, J. (2009). Experimental results with two wireless power transfer systems. *IEEE Radio and Wireless Symposium*, 16 - 18. 10.1109/RWS.2009.4957273

Shi, Guo, Yan, Yu, Luo, & Li. (2018). Optimizing Directional Reader Antennas Deployment in UHF RFID Localization System by Using a MPCSO Algorithm. *IEEE Sensors Journal*, 1-1.

Smith, J., Sample, A., Powledge, P., Mamishev, A., & Roy, S. (2006). A wirelessly powered platform for sensing and computation. *8th International Conference on Ubiquitous Computing (Ubicomp)*, 495 - 506. 10.1007/11853565_29

Stockman, H. (1948, October). Communication by means of reected power. *Proceedings of the IRE*, *36*(10), 1196–1204. doi:10.1109/JRPROC.1948.226245

Sutardja, C., & Rabaey, J. M. (2018). Isolator-Less Near-Field RFID Reader for Sub-Cranial Powering/Data Link of Millimeter-Sized Implants. *IEEE Journal of Solid-State Circuits*, 1-11.

Thomas, S., & Reynolds, M. (2010). QAM backscatter for passive UHF RFID tags. *IEEE International Conference on RFID*, 210 - 214.

Trotter, M., Valenta, C., Koo, G., Marshall, B., & Durgin, G. (2012). Multi-Antenna Technologies for Enabling Passive RFID Tags and Sensors at Microwave Frequencies. *IEEE International Conference on RFID*, 1 - 7.

Trotter, M. S. (2011). *Range finding in passive wireless sensor networks using power optimized waveforms* (Ph.D. dissertation). Georgia Institute of Technology.

Trotter, M. S., & Durgin, G. D. (2010). Survey of range improvement of commercial rfid tags with power optimized waveforms. *IEEE International Conference on RFID*, 195 - 202. 10.1109/RFID.2010.5467265

Trotter, M. S., Grifin, J. D., & Durgin, G. D. (2009). Power-optimized waveforms for improving the range and reliability of RFID systems. *IEEE International Conference on RFID*, 80 - 87. 10.1109/RFID.2009.4911196

Valenta, C., & Durgin, G. (2013). Link Budgets for Backscatter Radio and RFID Systems Using Power-optimized Waveforms. *IEEE International Symposium on Antennas & Propagation*, 237-244. 10.1109/APS.2013.6711817

Valenta, C., Graf, P., Trotter, M., Koo, G., Durgin, G., Daly, W., & Schafer, B. (2010b). Transient Backscatter Channel Measurements at 5.8 GHz Across High-Voltage Insulation Gaps. *AMTA 2010 Symposium*.

Valenta, C., Graf, P., Trotter, M., Koo, G., Durgin, G., & Schafer, B. (2010a). Backscatter Channel Measurements at 5.8 GHz Across High-Voltage Corona. *IEEE Sensors Conference*, 2400. 10.1109/ICSENS.2010.5690964

Xiao, Aflaki, Lang, Zhang, Tao, Py … Chang. (2018). Printed UHF RFID Reader Antennas for Potential Retail Applications. *IEEE Journal of Radio Frequency Identification*, 1-1.

Yang, Crisp, Penty, & White. (2018a). RFID Enabled Health Monitoring System for Aircraft Landing Gear. *IEEE Journal of Radio Frequency Identification*, 1-1.

Yang, L., Rida, A., Vyas, R., & Tentzeris, M. (2007). RFID Tag and RF Structures on a Paper Substrate Using Inkjet-Printing Technology. *IEEE Transactions on Microwave Theory and Techniques*, *55*(12), 2894–2901. doi:10.1109/TMTT.2007.909886

Yang, T., Xiang, J., Wang, Y., Tan, X., Wang, J., Yan, N., ... Min, H. (2018b). *An Active Tag Using Carrier Recovery Circuit for EPC Gen2 Passive UHF RFID Systems. IEEE Transactions on Industrial Electronics*.

Zhang, W.-H., Cheong, P., Lu, W.-J., & Tam, K.-W. (2018). *Planar Endfire Circularly Polarized Antenna for Low Profile Handheld RFID Reader. IEEE Journal of Radio Frequency Identification*.

Chapter 9
Wireless Power Transfer to Implantable Medical Devices With Multi-Layer Planar Spiral Coils

N. Sertac Artan
New York Institute of Technology, USA

Reza K. Amineh
New York Institute of Technology, USA

ABSTRACT

Implantable medical devices such as pacemakers, implantable cardioverter defibrillators, deep brain stimulators, retinal and cochlear implants are gaining significant attraction and growth due to their capability to monitor the health condition in real time, diagnose a particular disease, or provide treatment for a particular disease. In order to charge these devices, wireless power transfer technology is considered as a powerful means. This eliminates the need for extra surgery to replace the battery. In this chapter, some of the major implanted medical devices are reviewed. Then, various wireless power transfer configurations are reviewed briefly for charging such devices. The chapter continues with reviewing wireless power transfer configurations based on the multi-layer printed or non-printed planar spiral coils. At the end, some of the recent works related to using multi-layer planar spiral coils for safe and efficient powering of IMDs will be discussed.

INTRODUCTION

Today, a variety of implantable medical devices (IMDs) such as pacemakers, implantable cardioverter defibrillators (ICD)[1], deep brain stimulators (DBS), retinal[2] and cochlear implants[3] are vital for many patients. These devices fulfill various objectives such as: monitoring the health condition in real-time, diagnosing a particular disease, or providing treatment for a particular disease.

DOI: 10.4018/978-1-5225-5870-5.ch009

Despite the advantages, there are still many challenges hindering the efficient and widespread use of IMDs. One major issue is that these devices typically require high power to deliver the promised functionality but, on the other hand, providing safe and continuous energy supply for these devices is not technically straight-forward. Conventional IMDs get extra power supply by battery or a wired charging system which not only increases the burden on the patient but also increases the pain, risk of infection, and requirement for frequent surgeries. Wireless power transmission (WPT) to IMDs resolves this problem because of providing sustainable power without the need for extra surgeries or using wires passing through the skin. That is why they have become the most important and innovative technology to solve the problems regarding powering of the IMDs. Also, referred as transcutaneous energy transfer systems (TETS), WPT to IMDs has been successfully used for powering few implanted medical devices so far such as those for neural recordings and drug delivery.

Despite the advantages described above, there are still numerous challenges along the way to safely charge IMDs employing WPT. For example, delivering high power to implants without damaging the tissue due to heating while keeping the implant size small is still a topic under intensive investigation.

In this chapter, we first briefly introduce the application areas of WPT for implantable medical devices, and then study various WPT configurations. We will continue by studying WPT configurations based on the multi-layer printed or non-printed planar spiral coils. At the end, we describe some of the recent works related to using multi-layer planar spiral coils for safe and efficient powering of IMDs.

APPLICATIONS OF WPT IN IMPLANTABLE MEDICAL DEVICES

The design of WPT systems for an implantable device relies heavily on the application characteristics. For instance, recharging a battery, or powering devices, which uses current stimulation of the tissue as a way of therapy such as DBS require high instantaneous power. Such high power can stress the tissue if it causes temperature increases beyond the safety limits. In other applications such as cochlear implants, where the primary coil stays on continuously, albeit providing only a fraction of the current, the sizes of both primary and secondary coils are of primary concerns (Zeng, Rebscher,Harrison, Sun, & Feng, 2008). Additionally, the properties of the underlying tissue should also be carefully analyzed and taken into account when designing the WPT. In many cases, the WPT is used in a dual role as a telemetry system, or it may need to coexist with a telemetry system. The co-design of these two systems adds additional limits on the system design on parameters such as bandwidth and quality factor. Finally, various regulatory issues must be cleared to have an implantable biomedical device to be used in a clinical setting; imposing further constraints on the device design.

In the following paragraphs, exemplary IMDs are discussed to better assess the requirements and limitations of IMDs for different applications. Note that these examples are by no means an exhaustive list of IMDs, and such a discussion is beyond the scope of this chapter. Interested readers can refer to Agarwal, Jegadeesan, Guo, and Thakor (2017) for an up-to-date and extensive review of WPTs for IMDs.

Cochlear Implants

Cochlear implants partially restore hearing by stimulating the auditory nerve via electrical currents. As of 2016, over half a million cochlear implants (CIs) have been implanted, and this number is expected to grow to one million by 2020 (Wilson, 2017). The device consists of an external unit, which receives the audio signals and transmits them to the implanted unit, which generates the current pulses to stimulate the auditory nerve. The implant has no power source of its own, and rely on the external unit to provide power via a wireless link. The stimulation is applied via electrodes. There are typically between 8 and 24 electrodes in a CI, and each electrode receives stimulation current amplitudes ranges between 10 μA and 2 mA. The wireless power link budget is expected to be in the range of 20–40 mW (Agarwal et al., 2017). The RF link efficiency is around 40% with a coil distance of 4–10 mm over skin (Zeng et al., 2008).

There is a 1 to 1,000 ratio between the auditory neuron size and electrode size in cochlear implants (Zeng, 2017). Recently, optical stimulation of the auditory neurons rather than electrical stimulation to provide superior spatial resolution has been gaining attention (Hernandez et al., 2014). However, the power consumption of optical stimulation is much higher due to the high power consumption of the LEDs used in stimulation (e.g., 3.5 mW/LED) (Goßler et al., 2014).

Cortical Implants

Monitoring of electrophysiological activity at a large scale via non-invasive scalp Electroencephalography (sEEG) is a routine diagnostic tool for neurological diseases and disorders such as epilepsy. However, sEEG has limited spatial and spectral resolution and only shows events spanning millions of neurons. On the other hand, monitoring and stimulating of individual neurons or a small set of neurons via action potentials, and local field potentials can be achieved by implanting penetrating electrodes into the brain tissue for applications such as Brain-Computer Interfaces (BCI). The penetrating electrodes can provide superior spatial and spectral resolution, however, their span is limited, and there are challenges in using these electrodes reliably for extended periods of time. An approach that has capabilities and limitations in between the sEEG and penetrating electrodes is recording from the surface of the brain, i.e., Electrocorticography (ECoG). ECoG has superior spatial and spectral resolution compared to the sEEG, while it is more reliable than using penetrating electrodes. With the recent developments in electrode technology (Viventi et al., 2011), ECoG electrodes can offer very high resolution. Processing, and transmitting (Kim, Artan, Viventi, & Chao, 2012; Trumpis et al., 2017) of such a large amount of data demand large power to be delivered to these devices.

Sodagar et al. demonstrated an implantable system capable of recording from 64-electrodes while consuming an overall power of 14.4 mW (Sodagar et al., 2009). The power is transferred to the IMD via inductive coupling over coils wound over E-shape ferrite cores. The system is tested with a 1 cm gap between the primary and secondary coils. Ha reported Encapsulated Neural Interfacing Acquisition Chip (ENIAC); a single chip integrating the electrodes and a coil on-chip for WPT along with all the necessary circuitry (Ha, 2016). The on-chip coil has an inductance of 23.7 nH and quality factor of 11.2. The optimal power transfer is achieved at the resonant frequency of 190 MHz, delivering 2 mW to the implant over a distance of 5 mm.

Cardiac Implants

Irregular or abnormal heartbeats known as heart rhythm disorders or arrhythmias are very common and can be fatal. Various types of devices has been developed to help regulate the heartbeat such as pacemakers, and Implantable Cardioverter Defibrillators (ICDs), as well as to for chronically monitoring the heart rhythm such as the Implantable Cardiac Loop Recorder. The typical power consumption for a pacemaker is 8 μW. Although, current pacemaker batteries can last 7-13 years, and current ICD batteries can last for 5-10 years, the battery replacement surgeries can be completely eliminated (except for device malfunction cases), if the batteries of these devices can be charged remotely. However, it is a challenge to transmit enough power to an implant deep inside the body without causing any damage due to the tissue via heating. Kim et al. showed that 51.2 μW power can be delivered at 1 GHz or 1.7 GHz to an implanted 1 mm coil at a distance of 5 cm for an implantable device placed on the heart (Kim, Ho, Chen, & Poon, 2012). Ho et al. used a source with a patterned metal structure with 4 ports, and excited at a controlled phase at each port at 1.6 GHz (Ho et al., 2014). This source is reported to power an implantable device via a 2-mm receiver coil, which needs at least 10 μW to work. The distance between the coils is also reported as 5 cm (1 cm air gap and 4 cm heterogeneous tissue).

Most recently, Abiri took a system approach for designing a cardiac pacemaker (Abiri et al., 2017), where they divided their system into two parts; a stimulator implanted to the anterior cardiac vein, and a controller, which is implanted in the thoracic cavity. The stimulator has no battery and only has a tank circuit, an AC/DC converter, and a DC/DC converter, and it is fully powered and controlled by the controller unit, so that the stimulator size can be minimized. The receive coil consists of 4 circular and 3 meandering turns with a height of 15 mm, and width of 3 mm. This approach allowed an intermittent power transfer rather than continuous power transfer thus improving safety. A 2 V amplitude at the receiver antenna with a load of 500 Ω is achieved at a distance of 3 cm without misalignment. Even with various 45° displacement and angular misalignments same voltage can be achieved at 2 cm distance.

Retinal Implants

The ultimate goal of visual prosthesis is to restore vision to a functional level for the blind via electrical stimulation (Lewis et al., 2016). Retinal implants can help restore vision by stimulating the retina if the blindness is due to age-related macular degeneration (AMD) and retinitis pigmentosa (RP). One such device, which received regulatory approvals in countries including Canada, EU and USA is the ARGUS II ® retinal prosthesis system (Zhou, Dorn, & Greenberg, 2013). The ARGUS II device is powered by a wireless link with a receiver coil implanted around the eye, and a transmitter coil which is placed on glasses worn by the subject. The device is reported to be powered at around 45 mW (Agarwal et al., 2017).

REGULATORY ISSUES AND SAFETY

Safety is the primary concern for any medical device, but especially for implantable medical devices. Specific Absorption Rate (SAR) is the metric commonly used for measuring energy absorption in tissue. SAR is given in terms of Watts per Kilogram and is usually defined at two levels; the average rate for the whole body and the rate for a tissue weighing 1 to 10 g. The latter can also be specific to a particular tissue or organ. The SAR at the surface of the tissue induced by a primary circular coil of a WPT can be given as Christ,Douglas, Nadakuduti, and Kuster (2013):

$$SNR = \frac{\rho_t}{\sigma_t}\left(\frac{NM\omega I}{2\pi R_t}\right)^2 \qquad (1)$$

where the coil is specified with the number of its turns, N and its current, I with angular frequency, ω. ρ_t and σ_t are the tissue's conductivity and density, R_t is the radius of the current loop in the body, and M is the mutual inductance between the primary coil and the tissue. The SAR due to WPTs decays with the square of distance, and so it is lower for tissues deeper in the body. In Table 1, we showed the SAR limits defined by IEEE, International Commission on Non-Ionizing Radiation Protection (ICNIRP), and U.S. Federal Communications Commission (FCC).

Various aspects of WPT systems should be taken into account for safety, such as the biocompatibility and resilience of materials used, the amount of energy transferred and lost via the WPT, interference to/from the WPT or IMD electronics, in general. More recently, the security of these devices has also become a concern as more attacks are being identified that can take control of these systems and can cause serious, sometimes fatal damage.

Typically, IMDs are enclosed in a hermetically sealed metallic casing made from biocompatible materials such as titanium. However, metal casing is not an option for encapsulating the coils, which themselves are made from non-biocompatible material (e.g., copper). Instead, the coils are encapsulated into polymeric biocompatible materials such as Polydimethylsiloxane (PDMS) (Agarwal et al., 2017). When modeling the medium in between the primary and secondary coils, the dielectric properties of the polymeric material should also be considered. Often, for reducing the complexity of the implantation surgeries, flexibility of these materials are also desired. With the flexible encasing, the coil can move or bend along with the encasing as the patient moves. This introduces additional challenges as the shape of the implanted coil along with its distance and its alignment to the primary coil will not be fixed and should be taken account to keep efficiency high.

WPT TECHNOLOGIES TO MEDICAL IMPLANTS

At present, three main technologies are being studied and developed for WPT to IMDs: ultrasound technique, electromagnetic induction technique, and magnetic resonance technique. Below, we briefly describe pros and cons of ultrasound technique followed by a more detailed discussion about magnetic induction and resonance techniques which are the main topics focused in this chapter.

WPT via Ultrasound

In order to efficiently power small IMDs (in the order of mm) at large distances, recently, the ultrasound technique has attracted attention due to the following advantages: 1) the attenuation of acoustic waves

Table 1. SAR limits defined by IEEE, International Commission on Non-Ionizing Radiation Protection (ICNIRP), and U.S. Federal Communications Commission (FCC)

Standard	Whole Body SAR [W/kg]	Local SAR [W/kg]
IEEE Std C95.1-2005(Chou et al., 2005)	0.4	10
ICNIRP (Ahlbom et al., 1998)	0.08	2 (head) or 4 (limbs)
U.S. FCC (Cleveland & Ulcek, 1999)		1.6 over 1 g cube of tissue

inside the tissue is much smaller than the attenuation of electromagnetic waves, 2) operation at lower frequencies is possible due to the low acoustic velocity inside the tissue, and 3) higher intensities of ultrasonic can be applied safely (with less health concerns) compared to the electromagnetic waves (Seo, Carmena, Rabaey, Alon, & Maharbiz, 2013; Meng, Ibrahim, & Kiani, 2015). Despite these advantages, optimization of ultrasonic WPT systems is at early stages and, so far, only few functional devices have been demonstrated.

WPT via Electromagnetic Induction Coupling

WPT via electromagnetic induction has been the most common method over the past few decades to either directly power up the IMDs or recharge their batteries wirelessly (Yin, Borton, Aceros, Patterson, & Nurmikko, 2013; Zhou & Greenbaum, 2009). So far, the typical size of IMDs that are charged wirelessly and operate with powers in the order of tens of mWs is in the order of centimeters (cm) (Yin, Borton, Aceros, Patterson, & Nurmikko, 2013). On the other hand, attempts have been made to reduce the size of IMDs to the millimeter (mm) order. Such small devices will be less invasive and pave the way toward using them in applications such as biosensing, multisite neural recording or stimulation, and even optogenetics. For these applications the power requirements are in the order of hundreds of μW to several mWs (Biederman et al., 2013;Yeh et al., 2013).

To have an efficient WPT link, two important factors need to be maximized: 1) the power delivered to the load (PDL) and 2) the power transmission efficiency (PTE). In a magnetic induction system, the PTE and PDL of inductive links highly depend on the geometries of transmitter and receiver coils and their coupling distance, (Jow & Ghovanloo, 2007;Kiani, Jow, & Ghovanloo, 2011). Since the transmitter coil is outside the body, there is less constraint on its overall size. But the receiver coil size is often limited by the IMD application.

To maximize both PTE and PDL, the size of the receiver coil has to be increased as much as the application allows (Jow & Ghovanloo, 2007). Also, to increase PDL, theoretically, the transmitted power can be increased. However, in WPT to IMDs, another important factor that plays a role in the design of the WPT link is the extent of the tissue exposure to the electromagnetic field. This exposure needs to be maintained within safety limits. In other words, the relevant regulatory requirements need to be satisfied (Lazzi, 2005). Thus, compared to conventional cm-sized IMDs, efficient and safe WPT to mm-sized IMDs becomes quite challenging.

Increasing the operation frequency to hundreds of MHz or even GHz has been proposed recently in some studies to improve the PTE for WPT to mm-sized IMDs placed several mm to tens of mm inside the tissue, (for example see Poon, O'Driscoll, & Meng, 2010;Ho et al., 2014;Bjorninen et al., 2012;Ahn & Ghovanloo, 2016). This is in contrast to tens of MHz and below that have often been used for powering cm-sized IMDs, for example in Jow and Ghovanloo, (2007) (in these works, in the first step, the PTE of the WPT link is optimized and then the maximum allowable PDL (within safety limits) is found afterwards). Unlike cm-sized IMDs in which PTE optimization is the key (sufficient PDL can be safely achieved with a cm-sized receiver coil), in high-performance mm-sized IMDs, achieving high PDL within safety limits is challenging. This challenge exists, for example, for IMDs employed for electrical stimulation and optogenetics. In Ibrahim and Kiani (2016), a new figure-of-merit (FoM) has been proposed for optimal design of inductive WPT to mm-sized IMDs. The proposed FoM combines both PTE and maximum allowable PDL under safety constraints. This FoM can be employed to find the optimal geometry of the transmitter and receiver coils as well as operating frequency. An inductive link, designed accordingly, offers several advantages compared to previously proposed WPT links including: 1) operating at frequencies in the order of tens of MHz. This increases the PDL values in mW-range, and 2) transmitter/receiver coil geometries are optimized simultaneously. This is in contrast to the previously proposed works in which operating frequency has been mainly restricted by the receiver coil size, and 3) the optimal gap between transmitter coil and the tissue is estimated employing the FoM parameter. This is in contrast to the previously proposed works in which this gap had been chosen arbitrarily within the range of 5–10 mm to reduce safety concerns.

WPT via Electromagnetic Resonance Coupling

WPT via strongly coupled magnetic resonances was first reported in 2007 by a group of MIT researchers (Kurs et al., 2007). This technique extends the transmission distance to more than three times of the diameter of the transmitting coil. Figure 1 shows a typical schematic diagram of the two-coil resonant WPT system. The coils operate in resonant state. This leads to a significant increase in the energy transfer efficiently. The WPT efficiency decreases with increasing the transmitter-receiver distance (Lee, Zhong, & Hui, 2012). Among WPT technologies, magnetic resonance technology offers the advantages of: 1) increased distance between the power supply and the receiver, 2) increased efficiency, 3) low electromagnetic radiation, and 4) satisfying the requirements of electromagnetic compatibility. These features greatly broadens the applications of WPT technology including those in the field of biomedical engineering applications. As a critical part of the WPT system using this method, the resonant coils must be cautiously designed for the specific resonant frequency and high-quality factor (Q). One drawback for such systems is that the resonance frequency of the resonators changes as the transmitter-receiver distance (gap between the resonators) changes. This is in particular important when this technology is supposed to operate in a specific frequency band, for example, in the industrial, scientific, and medical (ISM) band. Later, we introduce some of the approaches that have been employed to mitigate this drawback.

As discussed above, resonance-based WPT is an efficient technique when the transmitter-receiver distance has to be relatively long. In this technique, employing four-coils instead of two-coils (that is common in conventional inductive links) offers some advantages. In a four-coil system, the adverse effects of distance on the WPT efficiency (low coupling coefficient) between transmitter and receiver coils are compensated to some extent by using high quality factor (Q) coils between the transmitter side and receiver side. This significantly boosts the performance of the system in terms of the energy transfer efficiency. In other words, the effect of the low Q and low coupling between the source and load coils that causes low efficiency in a two-coil system can be compensated by using high Q coils. In such systems, unlike a two coil system, the variation of efficiency versus transmitter-receiver distance is no longer a monotonically decreasing function. Also, the WPT efficiency is less sensitive to changes in the transmitter-receiver distance between the primary (transmitter) and secondary (receiver) coils. In a four-coil energy transfer system, the configuration and dimensions can be optimized such that maximum efficiency is obtained at a certain operating distance.

Figure 2 illustrates a four-coil resonant WPT system. In this system, the sending (primary) coil and receiving (secondary) coil are the main coils that perform the WPT between the transmitter side and receiver side. These two coils have the same self-resonant frequency. It is important, for the best functionality, to make sure that they remain in the resonant state so that the energy is transferred efficiently.

Figure 1. (a) A pair of transmitter-receiver coils operating in resonant mode (LC resonators) for WPT, (b) Typical variation (exponential decay) of the efficiency versus transmitter-receiver distance d in this WPT system (Lee, Zhong, & Hui, 2012)

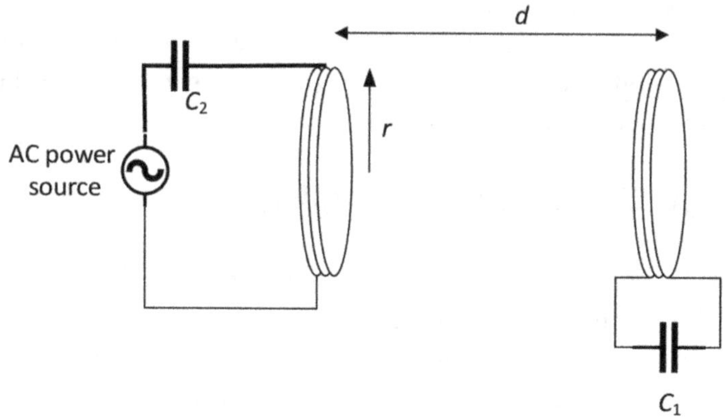

(a)

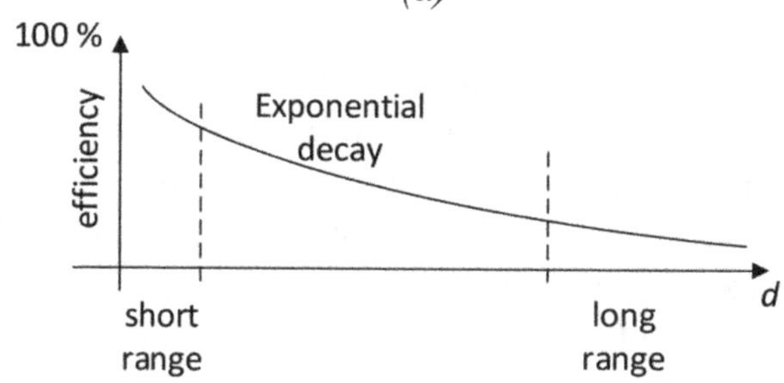

(b)

As discussed earlier, for a reasonable design, normally the transmission distance can reach one to more than three times the diameter of the resonant coils.

Unlike the induction systems for which the coupling coefficient denoted by K_{12} between the sending and receiving coils is typically much smaller than 1, resonantly coupled coils operate in a strongly coupled regime. In this regime, another coefficient $\Delta = K_{12}Q_1Q_2$ (where Q_1 and Q_2 are quality factors of the sending and receiving coils, respectively) is larger than 1 indicating a large coupling between the resonant coils. Thus, high transmission efficiency can be achieved (Chen, Chu, Lin, & Jou, 2010). This emphasizes that increasing the quality factors of the resonant coils (sending and receiving coils) is an important factor to achieve higher efficiency and transmitting energy over longer distances. For example, the results presented in Senjuti (2013) demonstrate the relative immunity of the four–coil WPT system for a wide range of transmitter-receiver distances.

As illustrated in Figure 2, the source and load are connected to coils that are inductively coupled to, but distinct from the coils that are terminated by lumped capacitances C_t and C_r. Having large Q for the sending coil and receiving coil when they operate in their resonant frequency boosts the PTE significantly. The resonant frequency for the sending coil and receiving coil can be determined by adjusting the lumped capacitances C_t and C_r.

Figure 2. Typical schematic diagram of the four-coil resonant WPT system

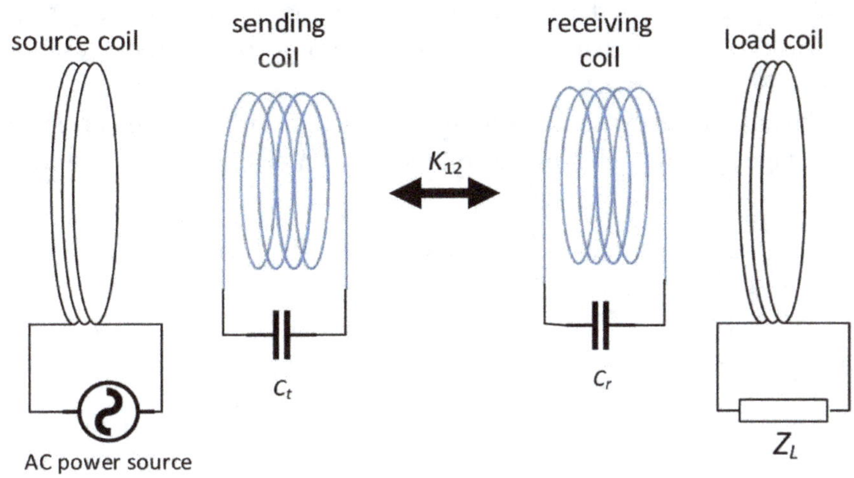

WPT Efficiency for a Two-Coil System

As discussed earlier, in a two-coil WPT system, the PTE denoted by η highly depends on the values of the Q for the primary coil (Q_p) and the secondary coil (Q_S). Another important factor that highly affects η is the mutual coupling factor (k) between the coils. This coupling factor depends on the alignment between the coils and the distance between them. Efficiency of a two-coil-based WPT system is given by Catrysse, Hermans, and Puers (2003), Rindorf, Lading, and Breinbjerg (2008), and Kumar, Mirabbasi, and Chiao (2009):

$$\eta = \frac{k^2 Q_p Q_s}{1 + k^2 Q_p Q_s} \tag{2}$$

As discussed earlier, a commonly used performance metric for comparing different designs is η versus the normalized transmitter-receiver distance R. Normalized distance is defined as the ratio of the transmitter-receiver distance d and the geometric mean of the primary and secondary coil's radii $r_m = \sqrt{r_p r_s}$. In a two-coil resonant system, as shown in Figure 1, η is decreasing drastically versus transmitter-receiver distance. This is due to the low Q and low magnetic coupling between the coils. The reason for low Q is the source and load resistances. In such systems, efficiency is typically less than 40% for $d \geq r_m$ (Wang et al., 2006;Catrysse, Hermans, & Puers, 2003;Yan, Ye, Zan, Wang, & Ma, 2007) and generally it drops quickly with distance ($\propto 1/d^3$) for $d < r_m$.

WPT Efficiency for a Four-Coil System

Couple-mode theory (Haus, Huang, 1991) has been employed in Kurs et al. (2007) and Karalis, Joannopoulos, and Soljacic, (2007) to describe four-coil resonance-based coupling. A simple circuit-based model for these systems has been presented in Kumar, Mirabbasi, and Chiao, (2009) to simplify the analysis and provide expressions for the efficiency. Here, for the sake of simplicity of analysis, the four coils are referred to as driver (or source), primary (or sensing), secondary (or receiving), and load coils and are denoted as coils 1 to 4, respectively.

By employing the circuit theory for this system, the relationship between currents passing through each coil and the voltage applied to the source coil (coil 1) can be expressed as follows:

$$\begin{bmatrix} I_1 \\ \vdots \\ I_4 \end{bmatrix} = \begin{bmatrix} Z_{11} & \cdots & Z_{14} \\ \vdots & \ddots & \vdots \\ Z_{41} & \cdots & Z_{44} \end{bmatrix}^{-1} \begin{bmatrix} E \\ \vdots \\ 0 \end{bmatrix} \tag{3}$$

where

$$Z_{mn} = \begin{cases} R_n + j\omega L_n + \dfrac{1}{j\omega C_n} & \text{for } m = n \\ j\omega M_{mn} & \text{for } m \neq n \end{cases} \tag{4}$$

E is the amplitude of voltage source applied to the driver coil, and R_n, L_n, and C_n are the effective resistance, inductance, and capacitance of the coil n. M_{mn} is the mutual inductance between coil m and n

$$M_{mn} = k_{mn}\sqrt{L_m L_n} \tag{5}$$

where k_{mn} is the coupling factor between coil m and coil n.

When all coils operate at the same resonance frequency, we have $Z_{mn} = R_n$ (for $m = n$ and $n \in \{1, 2, 3, 4\}$). For small driver and load coil inductance and relatively large distances between coils 1 and 4, coils 1 and 3, and coils 2 and 4, coupling coefficients k_{14}, k_{13}, and k_{24} can be neglected. From (3), at resonance, the current in load coil can be calculated as

$$I_4 = \frac{k_{12}k_{23}k_{34}\sqrt{Q_1Q_2}\sqrt{Q_2Q_3}\sqrt{Q_3Q_4}}{\sqrt{R_1R_4}[(1 + k_{12}{}^2Q_1Q_2)(1 + k_{34}{}^2Q_3Q_4) + k_{23}{}^2Q_2Q_3}E \tag{6}$$

where Q_n is the loaded quality factor of coil n at the frequency of operation. Then, the power-transfer efficiency can be computed as

$$\eta = \frac{\left(k_{12}^2 Q_1 Q_2\right)\left(k_{23}^2 Q_2 Q_3\right)\left(k_{34}^2 Q_3 Q_4\right)}{\left[\left(1 + k_{12}^2 Q_1 Q_2\right)\left(1 + k_{34}^2 Q_3 Q_4\right) + k_{23}^2 Q_2 Q_3\right]\left[1 + k_{23}^2 Q_2 Q_3 + k_{34}^2 Q_3 Q_4\right]} \tag{7}$$

Tuning the Resonance Frequency

Although performing much better than the induction systems, it is still well-known that longer transmitter-receiver distances in a resonant-based WPT system leads to lower coupling coefficient between the coils. To incraese the coupling coefficient, it is best to lower the transmitter-receiver distance between the coils. This, however, makes the resonant frequencies more sensitive to the mutual coupling (mutual position) between the coils increasing the probability of having different resonant frequencies for them. This degrades the efficiency. Therefore, a system to maintain the resonance frequencies somewhat unchanged versus the change in transmitter-receiver gap is required to ensure high efficiency at all mutual positions of the coils.

To solve the problem of having low efficiency due to the change in the transmitter-receiver distance, WPT systems achieve maximum PTE via various configurations and approaches. Some configurations have been proposed that control impedance matching circuits for the purpose of tuning the resonance frequency, for example see Beh, Kato, Imura, Oh, and Hori, (2013). A tunable impedance matching circuit was also implemented through varactor diodes in Won, Kim, and Jang (2010). However, the use of automatic frequency tuning techniques complicates the system structure because the impedance matching circuit should be tuned according to the transmitted power and the reflected power. Therefore, in Lee, Son, Oh, and Yu (2013) a WPT system has been proposed that uses antiparallel (forward and reverse) resonant loops (coils) without the need for further components to minimize the transfer efficiency variation versus the distance between the coils. In that work, the optimal design of these coils for achiving this objective (uniform WPT efficiency versus transmitter-receiver distance) has been presented. The proposed technique achieves frequency variation that is one-sixth of conventional unidirectional coils. In some other approaches to achieve high PTE, the transmitted power and reflected power, which are functions of the frequency, are continuously measured (Sample, Meyer, & Smith, 2011; Beh, Imura, Kato, & Hori, 2010). Since the system forms a transmission line, the power not consumed by the load or lost to parasitic resistance is reflected back to the source. Thus, an automatic frequency tuning is used to find the frequency that will result in the highest possible PTE. Yet another study suggests a WPT system that decreases resonance frequency difference between the primary and secondary coils by changing or tilting the coil configuration (Tak, Park, & Nam, 2009). Also, PTE can be improved by a frequency tracking control component that consists of a current transformer, a differential amplifier, a phase compensator, and a phase-locked loop (Fu, Zhang, & Qiu, 2009).

Realization of Magnetic Resonance-Based WPT

To implement a magnetic resonance-based WPT system, various coil configurations have been proposed. In Kurs et al. (2007), Seung-Hwan and Lorenz (2011), RamRakhyani, Mirabbasi, and Chiao (2011), Sample, Meyer, and Smith (2011), and Beh, Kato, Imura, Oh, and Hori, (2013) the helical or spiral coil made by copper wires have been employed with the parasitic capacitance between turns as the resonant capacitor.

In Lee, Zhong, and Hui, (2012), Cannon, Hoburg, Stancil, and Goldstein (2009),Imura and Hori, (2011), Dukju and Songcheol (2013), and Pantic and Lukic (2012) the tightly wound coil with external lumped capacitor has been used while the value of external capacitance plays a significant role in determining the resonant frequency.

In Hyeon-Chang, JinWook, Do-Hyeon, Kwan-Ho, and Young-Jin (2011). a self-resonant coil with a flexible coaxial-like capacitor and a helical coil has been proposed. The coaxial-like capacitor is used to provide a particular resonant frequency when combined with the parasitic capacitor and the inductance of the coil. It also helps to prevent variation of the resonant frequency due to external objects. The capacitor can be readily integrated with the helical coil. This configuration leads to high Q which is crucial to obtain high PTE.

Other common coil configurations in WPT systems are planar spiral coils. These coils can also be made by simply by using wires or using printed circuit board (PCB) technology, in this case they are called printed spiral coil (PSC) (Lee, Son, Oh, & Yu, 2011.; Wei & Qiang, 2011; Hao, Suxia, Junmin, Shyshenq, & Shahnasser, 2011; Hung-Yu, Jia-You, & Tsung-Wen, 2011; Raju, Wu, Chan, & Yue, 2014). For many years, PSCs have been used in other fields such as the inductors in RFICs and Rogowski coils. But those coil designs are not suitable for WPT systems because the requirements of coil characteristics and power carrying capacity are quite different.

In recent years, PSCs have been used in some WPT prototypes as well. Even though the power capacity is limited for these coil configurations, they are a good choice for the small and medium power transfer applications, such as portable electronic devices, robots, and IMDs. In fact, among various coil configurations, planar spiral coils, in particular PSCs, are most suitable to be employed as secondary coils for powering IMDs due to their compact size and stability. In general, PSCs have several advantages such as high precision, high stability, and ease of design and manufacture. In addition, because of the planar structure, PSCs are very suitable for miniaturization and integration with other circuits.

For most of the PSC designs, so far, wires are only placed on one side of the substrate (Lee, Son, Oh, & Yu, 2011; Wei & Qiang, 2011; Hao, Suxia, Junmin, Shyshenq, & Shahnasser, 2011; Raju, Wu, Chan, & Yue, 2014), thus the inductance is limited and the resonant frequency is very high (usually over 10 MHz) because of the small parasitic capacitance. In Casanova, Low, and Lin (2009), both layers of PCB are used to place wires and to increase the inductance but no analysis has been presented to provide an estimation of the values for the parameters.

As we know, in the resonant WPT systems, the transfer performance is very sensitive to the characteristics of resonant coils (dimensions, mechanical properties, and electrical parameters of the coils). Two important issues that need to be addressed while designing the PSC are: (1) how to design the PSC and emulate its characteristic for a precise resonant frequency and (2) how to increase the Q of PSC as much as possible.

Two options exist for designing planar spiral coils and achieve large inductances which is crucial to maximize Q and ultimately the efficiency of WPT system: (1) increasing the number of turns and (2) adding layers. The first option occupies more area on the PCB which is not desired in IMD applications. Thus, the second option is the best way to accommodate a large inductor when there is limited PCB area. The coils can be fabricated as multi-layer PSC exploiting the recent developments in the PCB manufacturing technology. In the rest of this chapter we focus more on the characterization and use of muti-layer PSCs for powering IMDs.

MULTI-LAYER PLANAR SPIRAL COILS FOR WPT

Multilayer planar spiral coils have stable inductance and require a less complex manufacturing process compared to other large inductors. The other advantage of using multi-layer planar spiral coils is the increase in the parasitic capacitance and the inductance of the coil which leads to lower resonant frequency. Although high resonant frequency (typically higher than 10 MHz) is used in some of the recent inductive WPT techniques for powering small IMDs as discussed earlier, in general, the lower frequencies, which is only several megahertz or even lower than 1 MHz, are recommended. This is due to the fact that the transmission system with lower operating frequency has less electromagnetic interferences (EMI) from the environment and also it reduces the manufacturing difficulties of the power source with high efficiency. For a simple single-layer planar spiral coil, the parasitic capacitance is typically too small and this leads to a quite high resonant frequency. Some methods to reduce the resonant frequency have been discussed in Zhang, Ho, and Fu, (2011). In Chen and Zhao (2013) a double-layer PSC has been proposed to resolve this problem and also to provide other advantages discussed earlier for multi-layer PSCs. Different from the previously proposed single-layer PSCs, this design uses both the top layer and the bottom layer of the PCB. The coils on both layers are connected in series by through-holes. Using this approach, the limited space of PCB is fully used and provides a larger value of inductance. Besides, the resonant capacitance is formed by the parasitic capacitance between various turns, especially the turns that are facing each other but on different layers. Due to that, the equivalent resonant capacitance is increased significantly. This can be employed to adjust a specific resonant frequency by tuning the wire width and thickness and permittivity of the substrate, placing the wires on both layers in staggered arrangement, etc.

It is worth mentioning that for the multi-layer PSCs, there is a limit for increasing the layers. Above a certain number of layers, the effect of increase in coil's resistance (loss) imposes a reverse effect on the power transfer efficiency of a WPT system as will be discussed in the last section.

Characterization of Multi-Layer Planar Spiral Coils

As discussed above, there are several advantages to use planar spiral inductors in IMDs. On the other hand, accuracy in designing a spiral inductor is important because it is difficult to modify the inductor once it is built on the PCB. Some formulas are available for calculating the spiral inductor for RFIC applications with inductance of less than 100 nH on a single-layer design. These formulas are not suitable for designing PSCs for WPT applications which typically require much larger inductance value. For WPT applications, some works have been proposed to design and optimize the single layer PSCs based on the analytical expressions for IMD (Jow & Ghovanloo, 2007, 2009). But providing analytical expressions for highly accurate calculation of the multi-layer spiral inductors is very challenging. There has been little analysis of the multi-layer PSC coil design with consideration of parasitic parameters. Some descriptions of PSC modeling are given by Wei and Qiang (2011) and Hao, Suxia, Junmin, Shyshenq, and Shahnasser, (2011), which are not detailed enough for an accurate design.

In multi-layer PSC, mutual inductances and capacitances exist between the layers. Calculations of these mutual parameters are very complicated. This complexity becomes more pronounced when the number of layers increases. The coupling between the layers depends on the number of turns of the coils, diameter or width of the conductor, spacing between the turns, and the distance between the two layers.

The most accurate method to characterize the performance of multi-layer coils in a WPT system is via the use of full-wave electromagnetic simulations. However, full-wave simulations are expensive to evaluate in terms of memory and time. On the other hand, empirical equations can be employed for the estimation of the electrical parameters of multi-layer planar spiral coils. In a design with more than two layers, there are more coupling factors between any two layers. Similar empirical formulas can be used to obtain each coupling factor and then calculate the total factor although the accuracy will drop drastically.

Another technique that has been proposed in Chen and Zhao (2013) for characterization of double-layer PSCs is based on simplified partial element equivalent circuit (PEEC) method and finite element method (FEM). In this technique, the circuit model with consideration of parasitic parameters and high-frequency losses is built and the impedance characteristic of the coil is simulated, which coincides well with the measurement results. Below, we describe this method in more details.

To obtain the impedance characteristic of the double-layer PSC, the PEEC method is employed in which the geometry is modeled with approximate lump circuit elements (Ruehli, 1974). In the first step, the coil geometry is divided into several parts. Figure 3 illustrates the partitioning of the square spiral coil, which is similar with that implemented in Zhang, Ho, and Fu, (2011), or even more accurate. The thick solid lines are split lines for inductance elements. They are placed on the corners of the coil. Thus, each side of the rectangle is considered as an inductor and the mutual inductance between different sides should be considered as well. The thick dashed lines are split lines for capacitance elements, which should be staggered for half element so that the equivalent capacitors can be connected between the inductors in the equivalent circuit. Besides, actually parasitic capacitances exist between every two elements, although they are all not marked in Figure 3 for figure simplicity. For a double-layer PSC with a considerable number of turns, the equivalent circuit based on this partition method will, however, be quite complicated. Therefore, in Chen and Zhao (2013), a simplified partition and equivalent method is used which offer acceptable accuracy.

Figure 3. Schematic diagram of partitioning for square spiral coil based on PEEC (solid line: split line for inductance element, dotted line: split line for capacitance element)

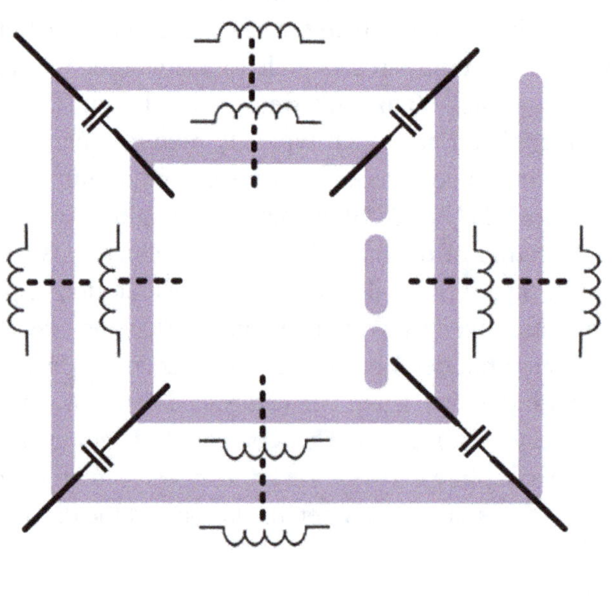

The simplified partitioning of the proposed PSC is shown in Figure 4. Some points that are worth mention in this figure are: (1) each turn is treated as only one element, and the same section is used for both inductance and capacitance elements, 2) Figure 4 only shows the top layer of the PSC, while the bottom layer and its separation is similar and in mirror of the top layer, labeled as parts 5–8. In addition, the connection of the coils on the two layers is also a split point.

In the first step, we ignore the losses and consider the self-inductance of each element, mutual inductance between elements, and parasitic capacitance between elements. Figure 5 shows the equivalent circuit. In this figure the mutual inductance is not shown for simplicity.

It is worth mentioning that due to the same partitioning of the inductance elements and capacitance elements, according to the actual geometry, the parasitic capacitances should be connected to the middle of the inductances. But this makes it difficult to evaluate the mutual inductances. Thus, as another simplification, all the parasitic capacitances will be shifted for half an element and connected between inductances in the final equivalent circuit model being described later. This process does change the circuit characteristics significantly. Because of the symmetrical structure of the coil, the simulation values of corresponding parasitic capacitances are, however, quite similar between the two splitting methods: capacitance partition staggered for half element according to the inductance partition and using the inductance partition as the capacitance partition directly. So the parasitic capacitances can be shifted half an element in the equivalent circuit, pretending their values are simulated by the former method mentioned above, and the error is acceptable for engineering design.

Figure 4. Partitioning of the top layer for the square planar spiral coil; the bottom layer of the coil has similar configuration and is a mirror of this scheme

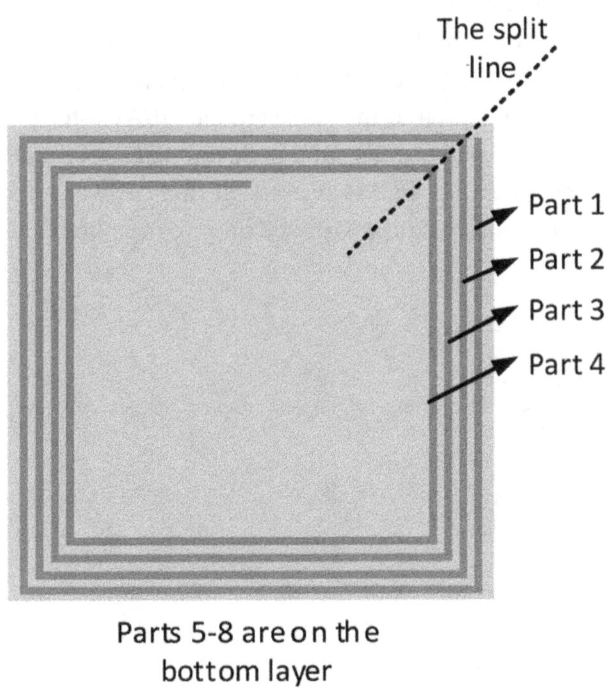

Figure 5. Original lossless equivalent circuit of the double-layer PSC

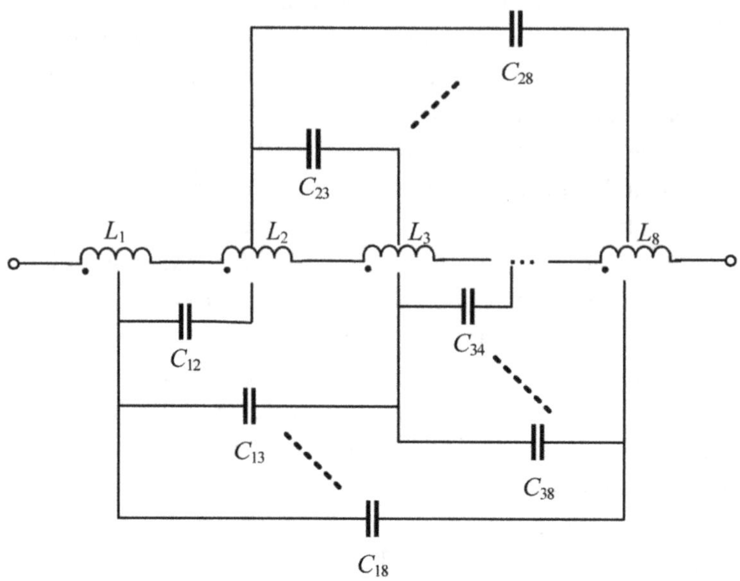

Now, modifications of the model will be discussed when considering the losses in the PSC. On the high frequency excitation, there are three types of losses existing in PSC: the radiation loss, the ohmic loss, and the dielectric loss. For the WPT via strongly coupled magnetic resonances, the radiation loss can be ignored (Kurs et al., 2007). The ohmic loss is caused by wire resistance with the impact of skin effect and proximity effect, which means that there should be a resistance added into each inductance in series. In addition, as the dielectric of parasitic capacitances, the substrate will produce a considerable dielectric loss at high frequency. This implied that each parasitic capacitance should be connected with an equivalent series resistance as well. The equivalent circuit including these losses is shown in Figure 6. To calculate the parameters in this model, the software called Ansoft Q3D Extractor (the predecessor of ANSYS Q3D Extractor) has been used, which is developed specifically for parasitic component extraction using method of moments and FEMs. Based on this software, matrices of self-inductance, mutual inductance, parasitic capacitance, and resistance of wires can be obtained in the following form:

$$\mathbf{L} = \begin{bmatrix} L_1 & M_{12} & M_{13} & & M_{18} \\ M_{21} & L_2 & M_{23} & \cdots & M_{28} \\ M_{31} & M_{32} & L_3 & & M_{38} \\ & \vdots & & \ddots & \vdots \\ M_{81} & M_{82} & M_{83} & \cdots & L_8 \\ & & & & \end{bmatrix} \tag{8}$$

Figure 6. Final equivalent circuit model for double-layer PSC

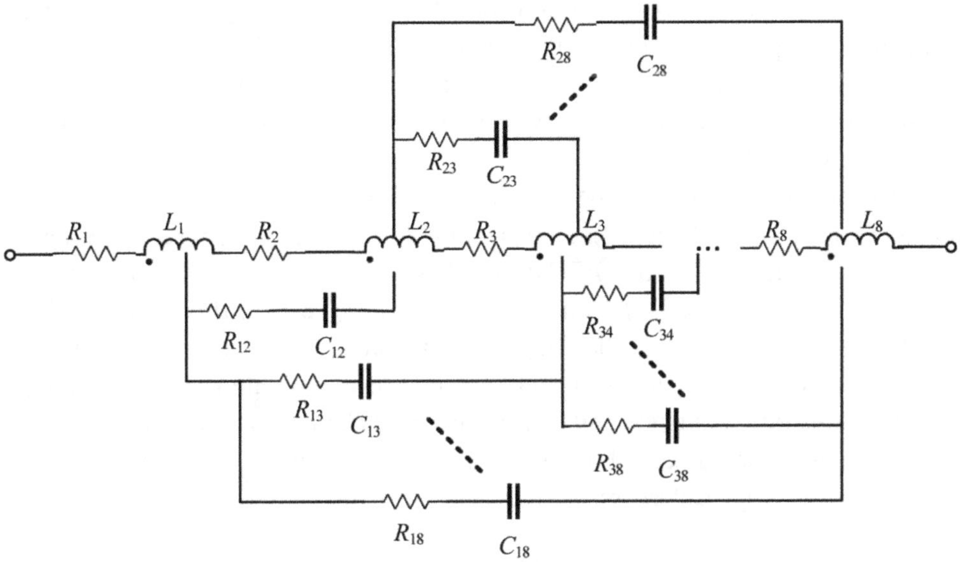

$$\mathbf{C} = \begin{bmatrix} - & C_{12} & C_{13} & & C_{18} \\ C_{21} & - & C_{23} & \cdots & C_{28} \\ C_{31} & C_{32} & - & & C_{38} \\ & \vdots & & \ddots & \vdots \\ C_{81} & C_{82} & C_{83} & \cdots & - \end{bmatrix} \tag{9}$$

$$\mathbf{R}_{\text{wire}} = \begin{bmatrix} R_1 & & \\ & \ddots & \\ & & R_8 \end{bmatrix} \tag{10}$$

The equivalent resistances caused by dielectric loss can be calculated by

$$R_{xy}\Big|_{x,y=1,2,\ldots,8} = \frac{\tan\delta}{\omega C_{xy}} \tag{11}$$

where ω is the operating angular frequency of the PSC in rad/s, $\tan\delta$ is the dielectric loss tangent of the material of substrate, R_{xy} and C_{xy} represent the equivalent resistance and parasitic capacitance between part x and part y, respectively.

Based on the circuit model shown in Figure 6, the impedance characteristic of the PSC can be simulated and the characteristic parameters, such as resonant frequency and Q , can be obtained. This could serve as a means to accomplish a coil design that satisfies particular values for the parasitic parameters, resonant frequency, and Q. In addition, as an extension, the multilayer PCB, such as four-layer or six-layer, can be used to manufacture the PSC as well, while the basic structure and the circuit model presented above are still available.

As the last approach, the analysis of the resonant frequency and Q of a multi-layer PSC based on the finite element simulations and deriving empirical models has been proposed in Gong et al., (2016). In this work, a WPT system with 4-layer PSC design has been presented. The coil is fabricated with 4 layers and epoxy resin glass fabrics layer are set between the layers. In this 4-layer coil, the parasitic capacitance and self-inductance are high leading to low resonant frequency which is desired as we discussed earlier. A parametric study has been implemented to investigate the influence of various coil dimensions on the resonance frequency and Q . Empirical formulas have been derived to expedite the parametric study process. The numerical and experimental data have been employed to derive the unknown coefficients of the empirical expressions.

Multi-Layer PSCs for Biomedical WPT Applications

Multi-layer planar spiral coils have been proposed in Artan et al. (2011) for WPT to medical implants and to increase the PTE when used as secondary coils. It has been shown that these configurations allow the delivery of the required power safely for a longer period. In other words, using these configurations, a 5 times increase in the duration of safe power delivery (not increasing the skin temperature more than $2°C$. has been achieved using multi-layer coils as the secondary coil compared to using single-layer coils even when there is a 50% misalignment between primary and secondary coils. In the same paper, it has been shown that the increase in duration of the safe power transfer is even 16 times more when the coils are aligned. The improvement in the duration of safe power transfer is achieved without any drastic change in the dimensions of the coil and only with a tiny increase in the thickness when using multi-layer planar spiral coils compared to single layers.

Also, in Artan, Patel, Ning, and Chao, (2012), a hybrid multi-layer coil configuration has been proposed as secondary coil. It consists of layers of full and hollow coils. This coil configuration can deliver power to an implant for longer durations without increasing the skin temperature over $1°C$. By introducing the hybrid multi-layer coils higher values of quality factor is obtained compared to full multi-layer coils discussed above. However, there is a limit in the number of layers for which the hybrid coil provides higher Q compared to full coil.

An example use case for multi-layer coils is for powering implantable biosensors. Biosensors have very strict size constraints, and thus may not even have a battery. Furthermore, due to size constraints such devices can only have a limited area, so having multi-layer coils as the secondardy coil reduces their overall size. Laskovski et al. simulated and tested 15 mm x 15 mm spiral coils with 7 turns and reported receiving -27 dBm power at 25 mm distance from a transmission of 6 dBm (Laskovski, Yuce, & Dissanayake, 2011). In Carrara et al. (2012) an inductive power link is used to power a subcutaneous biochip, which is aimed at detecting various metabolites in the human body. The device is designed to be small enough to be easily implanted subcutaneously. A single-layer coil is not practical for such a miniaturized system, thus a multilayer coil is used as the receiver coil. This coil is built by stacking multiple PCBs, where each PCB has an identical inductor drawn on two of its layers. This coil is not only used for receiving power but also for communication, where the data is transmitted from the implanted device via backscattering, and to the implanted device via Amplitude Shift Keying (ASK). The system can send 1.17 mW power through a 17 mm tissue, and has a maximum link efficiency of 13%. The coil has a surface area of 30 mm^2 and thickness of 816 µm. In Cavellini et al. (2015), it is pointed out that the copper metallization of the coil is cytotoxic and a conformal coating of 3 µm parylene C is used to prevent severe cytotoxicity and corrosion.

Another example use case for multi-layer coils is for neural interfaces, in particular for neurostimulation, which requires high power to be delivered by the implantable device. Kassiri et al. presented a 64-channel neurostimulator, which is powered by an 8-layer flexible receiver coil (Kassiri et al., 2014). The coils are made of 4 20 mm × 20 mm flexible two-layer PCBs. The power is transmitted via a two layer of 16 planar high-Q coils, which is designed to be placed under a rat cage, and power the implantable device for stimulating brain of a rat moving freely in the cage. The flexible design allows the coil to conform to the shape of the skull of the animal. The authors reported a Q of 24 for the receiver coil, and a power transfer efficiency of 40% at 12 cm distance.

CONCLUSION AND RECOMMENDATIONS

In this chapter, we discussed various configurations for delivering of the power wirelessly from source to the consumer side. To ensure transferring of the power over longer distances, resonance-based WPT is preferred provided that the sending and the receiving coils on both sides of the WPT system can be controlled well to operate in the resonance mode. Furthermore, to ensure the optimal use of the available space for WPT to medical implants, multi-layer PSCs are recommended. As discussed, these coils can increase the PTE when used as secondary coils compared to single layer coils. Also, it has been shown that these configurations allow the delivery of the required power safely for a longer period. The design of multi-layer PSCs, however, is difficult and this has been demonstrated in this chapter by reviewing the design process for a double layer PSC. In general, the design of these coils needs to be performed via the use of full-wave electromagnetic simulation software. For example, in Baua, Amineh, and Artan, (2018) the multi-layer PSCs with all-hollow, all-full, and hybrid configurations have been compared in terms of quality factor via the use of FEKO software which is a high frequency simulator. It has been observed that all-hollow multi-layer PSCs, with the number of turns for each layer being almost half of the number of turns for the corresponding full coil, can provide the highest quality factor. Further research need to be performed to fully investigate the capabilities and limitations of the multi-layer planar spiral coils for biomedical WPT applications.

REFERENCES

Abiri, P., Abiri, A., Packard, R. R. S., Ding, Y., Yousefi, A., Ma, J., ... Hsiai, T. K. (2017). Inductively powered wireless pacing via a miniature pacemaker and remote stimulation control system. *Scientific Reports*, *7*(1), 6180. doi:10.103841598-017-06493-5 PMID:28733677

Agarwal, K., Jegadeesan, R., Guo, Y. X., & Thakor, N. V. (2017). Wireless power transfer strategies for implantable bioelectronics: Methodological review. *IEEE Reviews in Biomedical Engineering.*

Ahlbom, A., Bergqvist, U., & Bernhardt, J. H. (1998). Guidelines for limiting exposure to time-varying electric, magnetic, and electromagnetic fields (up to 300 GHz). *Health Physics*, *74*(4), 494–522. PMID:9525427

Ahn, D., & Ghovanloo, M. (2016). Optimal design of wireless power transmission links for millimeter-sized biomedical implants. *IEEE Transactions on Biomedical Circuits and Systems*, *10*(1), 125–137. doi:10.1109/TBCAS.2014.2370794 PMID:25616074

Artan, N. S., Li, X., Patel, R., Ning, C., Ludvig, N., & Chao, H. J. (2011). Multi-layer coils for efficient transcutaneous power transfer. *Proceedings of Annual International Conference of the IEEE Engineering in Medicine and Biology Society*, 3031-3034. 10.1109/IEMBS.2011.6090829

Artan, N. S., Patel, R. C., Ning, C., & Chao, H. J. (2012). High-efficiency wireless power delivery for medical implants using hybrid coils. *Proceedings of Annual International Conference of the IEEE Engineering in Medicine and Biology Society*, 1683-1686. 10.1109/EMBC.2012.6346271

Baua, S. K., Amineh, R., & Artan, N. S. (2018). Wireless power transfer with multi-layer planar spiral coils as secondary coils. *Proceedings of IEEE Wireless Power Transfer Conference (WPTC).*

Beh, T. C., Imura, T., Kato, M., & Hori, Y. (2010). Basic study of improving efficiency of wireless power transfer via magnetic resonance coupling based on impedance matching. *Proceedings of IEEE ISIE*, 2011-2016.

Beh, T. C., Kato, M., Imura, T., Oh, S., & Hori, Y. (2013). Automated impedance matching system for robust wireless power transfer via magnetic resonance coupling. *IEEE Transactions on Industrial Electronics*, *60*(9), 3689–3698. doi:10.1109/TIE.2012.2206337

Biederman, W., Yeager, D., Narevsky, N., Koralek, A., Carmena, J., Alon, E., & Rabaey, J. (2013). A fully-integrated, miniaturized (0.125 mm^2) 10.5 μW wireless neural sensor. *IEEE Journal of Solid-State Circuits*, *48*(4), 960–970. doi:10.1109/JSSC.2013.2238994

Bjorninen, T., Muller, R., Ledochowitsch, P., Sydanheimo, L., Ukkonen, L., Maharbiz, M., & Rabaey, J. M. (2012). Design of wireless links to implanted brain-machine interface systems. *IEEE Antennas and Wireless Propagation Letters*, *11*, 1663–1666. doi:10.1109/LAWP.2013.2239252

Cannon, B. L., Hoburg, J. F., Stancil, D. D., & Goldstein, S. C. (2009). Magnetic resonant coupling as a potential means for wireless power transfer to multiple small receivers. *IEEE Transactions on Power Electronics*, *24*(7), 1819–1825. doi:10.1109/TPEL.2009.2017195

Carrara, S., Cavallini, A., Ghoreishizadeh, S., Olivo, J., & De Micheli, G. (2012). Developing highly-integrated subcutaneous biochips for remote monitoring of human metabolism. *IEEE Sensors*, 1-4.

Casanova, J. J., Low, Z. N., & Lin, J. (2009). A loosely coupled planar wireless power system for multiple receivers. *IEEE Transactions on Industrial Electronics*, *56*(8), 3060–3068. doi:10.1109/TIE.2009.2023633

Catrysse, M., Hermans, B., & Puers, R. (2003). An inductive power system with integrated bidirectional data transmission. *Proceedings of XVII Eurosensors*, 843-846.

Cavallini, A., Jost, T. R., Ghoreishizadeh, S. S., Olivo, J., de Beeck, M. O., Gorissen, B., ... Carrara, S. (2015). A Subcutaneous Biochip for Remote Monitoring of Human Metabolism: Packaging and biocompatibility assessment. *IEEE Sensors Journal*, *15*(1), 417–424. doi:10.1109/JSEN.2014.2339638

Chen, C. J., Chu, T. H., Lin, C. L., & Jou, Z. C. (2010). A study of loosely coupled coils for wireless power transfer. *IEEE Transactions on Circuits and Wystems. II, Express Briefs*, *57*(7), 536–540. doi:10.1109/TCSII.2010.2048403

Chen, K., & Zhao, Z. (2013). Analysis of the double-layer printed spiral coil for wireless power transfer. *IEEE Journal of Emerging and Selected Topics in Power Electronics*, *1*(2), 114–121. doi:10.1109/JESTPE.2013.2272696

Chou, C.-K., D'Andrea, J., & Petersen, R. (2005). *IEEE Standard for Safety Levels with Respect to Human Exposure to Radio Frequency Electromagnetic Fields, 3 kHz to 300 GHz*. New York, NY: Institute of Electrical and Electronics Engineers.

Christ, A., Douglas, M., Nadakuduti, J., & Kuster, N. (2013). Assessing human exposure to electromagnetic field from wireless power transmission systems. *Proceedings of the IEEE*, *101*(6), 1482–1493. doi:10.1109/JPROC.2013.2245851

Dukju, A., & Songcheol, H. (2013). A study on magnetic field repeater in wireless power transfer. *IEEE Transactions on Industrial Electronics*, *60*(1), 360–371. doi:10.1109/TIE.2012.2188254

Fu, W., Zhang, B., & Qiu, D. (2009). Study on frequency-tracking wireless power transfer system by resonant coupling. *Proceedings of IEEE 6th International Conference on Power Electronics and Motion Control*, 2658-2663.

Gong, F., Cong, Y., Wei, Z., Yin, B., Chi, H., & Luan, X. (2016). Analysis of the multi-layer printed spiral coil for wireless power transfer system used in medical implants. *7th Asia Pacific International Symposium on Electromagnetic Compatibility*, 460-462.

Goßler, C., Bierbrauer, C., Moser, R., Kunzer, M., Holc, K., Pletschen, W., ... Schwarz, U. T. (2014). GaN-based micro-LED arrays on flexible substrates for optical cochlear implants. *Journal of Physics. D, Applied Physics*, *47*(20), 205401. doi:10.1088/0022-3727/47/20/205401

Ha, S. (2016). *Silicon integrated high-density electrocortical and retinal neural interfaces* (PhD Dissertation). University of California, San Diego, CA.

Hao, J., Suxia, L., Junmin, Z., Shyshenq, L., & Shahnasser, H. (2011) Design and optimization of printed spiral coils in wireless power transfer for biomedical implants. *Proceedings of 4th International Conference on Biomedical Engineering and Informatics*, 1064-1067.

Haus, H., & Huang, W. (1991). Coupled-mode theory. *Proceedings of the IEEE*, *79*(10), 1505–1518. doi:10.1109/5.104225

Hernandez, V. H., Gehrt, A., Reuter, K., Jing, Z., Jeschke, M., Schulz, A. M., & ... . (2014). Optogenetic stimulation of the auditory pathway. *The Journal of Clinical Investigation*, *124*(3), 1114–1129. doi:10.1172/JCI69050 PMID:24509078

Ho, J., Yeh, A., Neofytou, E., Kim, S., Tanabe, Y., Patlolla, B., ... Poon, A. (2014). Wireless power transfer to deep-tissue microimplants. *Proceedings of the National Academy of Sciences of the United States of America, 111*(22), 7974–7979. doi:10.1073/pnas.1403002111 PMID:24843161

Hung-Yu, S., Jia-You, L., & Tsung-Wen, C. (2011). Study of contactless inductive charging platform with core array structure for portable products. *Proceedings on International Conference on Consumer Electronics, Communications, and Networks,* 756-759.

Hyeon-Chang, S., JinWook, K., Do-Hyeon, K., Kwan-Ho, K., & Young-Jin, P. (2011). Self-resonant coil with coaxial-like capacitor for wireless power transfer. *Proceedings of Asia-Pacific Microwave Conference*, 90-93.

Ibrahim, A., & Kiani, M. (2016). A figure-of-merit for design and optimization of inductive power transmission links for millimeter-sized biomedical implants. *IEEE Transactions on Biomedical Circuits and Systems, 10*(6), 1100–1111. doi:10.1109/TBCAS.2016.2515541 PMID:28055825

Imura, T., & Hori, Y. (2011). Maximizing air gap and efficiency of magnetic resonant coupling for wireless power transfer using equivalent circuit and neumann formula. *IEEE Transactions on Industrial Electronics, 58*(10), 4746–4752. doi:10.1109/TIE.2011.2112317

Jow, U., & Ghovanloo, M. (2007). Design and optimization of printed spiral coils for efficient transcutaneous inductive power transmission. *IEEE Transactions on Biomedical Circuits and Systems, 1*(3), 193–202. doi:10.1109/TBCAS.2007.913130 PMID:23852413

Jow, U., & Ghovanloo, M. (2009). Modeling and optimization of printed spiral coils in air, saline, and muscle tissue environments. *IEEE Transactions on Biomedical Circuits and Systems, 3*(5), 339–347. doi:10.1109/TBCAS.2009.2025366 PMID:20948991

Jr. Cleveland, R. F., & Ulcek, J. L. (1999). *Questions and Answers about Biological Effects and Potential Hazards of Radiofrequency Electromagnetic Fields*. Washington, DC: Federal Communications Commission.

Karalis, A., Joannopoulos, J., & Soljacic, M. (2007). Efficient wireless nonradiative mid-range energy transfer. *Annals of Physics, 323*(1), 34–48. doi:10.1016/j.aop.2007.04.017

Kassiri, H., Bagheri, A., Soltani, N., Abdelhalim, K., Jafari, H. M., Salam, M. T., ... Genov, R. (2014, September). Inductively-powered direct-coupled 64-channel chopper-stabilized epilepsy-responsive neurostimulator with digital offset cancellation and tri-band radio. *40th EEE European Solid State Circuits Conference (ESSCIRC)*, 95-98. 10.1109/ESSCIRC.2014.6942030

Kiani, M., Jow, U., & Ghovanloo, M. (2011). Design and optimization of a 3-coil inductive link for efficient wireless power transmission. *IEEE Transactions on Biomedical Circuits and Systems, 5*(6), 579–591. doi:10.1109/TBCAS.2011.2158431 PMID:21922034

Kim, S., Ho, J. S., Chen, L. Y., & Poon, A. S. Y. (2012). Wireless power transfer to a cardiac implant. *Applied Physics Letters, 101*(7), 073701. doi:10.1063/1.4745600

Kim, T., Artan, N. S., Viventi, J., & Chao, H. J. (2012). Spatiotemporal compression for efficient storage and transmission of high-resolution electrocorticography data. *Proceedings* of *2012 Annual International Conference of the IEEE*, 1012-1015.

Kumar, A., Mirabbasi, S., & Chiao, M. (2009). Resonance-based wireless power delivery for implantable devices. *Proceedings of IEEE Biomedical Circuits and Systems Conference*, 25-28. 10.1109/BIOCAS.2009.5372092

Kurs, A., Karalis, A., Moffatt, R., Joannopoulos, J. D., Fisher, P., & Soljacic, M. (2007). Wireless power transfer via strongly coupled magnetic resonances. *Science*, *317*(5834), 83–86. doi:10.1126cience.1143254 PMID:17556549

Laskovski, A. N., Yuce, M. R., & Dissanayake, T. (2011). Stacked spirals for biosensor telemetry. *IEEE Sensors Journal*, *11*(6), 1484–1490. doi:10.1109/JSEN.2010.2091123

Lazzi, G. (2005). Thermal effects bioimplants. *IEEE Engineering in Medicine and Biology Magazine*, *24*(5), 75–81. doi:10.1109/MEMB.2005.1511503 PMID:16248120

Lee, C. K., Zhong, W. X., & Hui, S. Y. R. (2012). Effects of magnetic coupling of nonadjacent resonators on wireless power domino-resonator systems. *IEEE Transactions on Power Electronics*, *27*(4), 1905–1916. doi:10.1109/TPEL.2011.2169460

Lee, W. S., Son, W. I., Oh, K. S., & Yu, J. W. (2011). Contactless energy transfer systems using antiparallel resonant loops. *IEEE Transactions on Industrial Electronics*, *60*(1), 350–359. doi:10.1109/TIE.2011.2177611

Lewis, P. M., Ayton, L. N., Guymer, R. H., Lowery, A. J., Blamey, P. J., Allen, P. J., ... Rosenfeld, J. V. (2016). Advances in implantable bionic devices for blindness: A review. *ANZ Journal of Surgery*, *86*(9), 654–659. doi:10.1111/ans.13616 PMID:27301783

Maghami, M. H., Sodagar, A. M., Lashay, A., Esfahani, H. R., & Riazi-Esfahani, M. (2014). Visual prostheses: The enabling technology to give sight to the blind. *Journal of Ophthalmic & Vision Research*, *9*(4), 494. doi:10.4103/2008-322X.150830 PMID:25709777

Meng, M., Ibrahim, A., & Kiani, M. (2015). Design considerations for ultrasonic power transmission to millimeter-sized implantable microelectronic devices. *Proceedings of IEEE Biomedical Circuits and Systems Conference*. 10.1109/BioCAS.2015.7348295

Pantic, Z., & Lukic, S. M. (2012). Framework and topology for active tuning of parallel compensated receivers in power transfer systems. *IEEE Transactions on Power Electronics*, *27*(11), 4503–4513. doi:10.1109/TPEL.2012.2196055

Poon, A., O'Driscoll, S., & Meng, T. (2010). Optimal frequency for wireless power transmission into dispersive tissue. *IEEE Transactions on Antennas and Propagation*, *58*(5), 1739–1750. doi:10.1109/TAP.2010.2044310

Raju, S., Wu, R., Chan, M., & Yue, C. (2014). Modeling of mutual coupling between planar inductors in wireless power applications. *IEEE Transactions on Power Electronics*, *1*, 114–121.

RamRakhyani, A. K., Mirabbasi, S., & Chiao, M. (2011). Design and optimization of resonance-based efficient wireless power delivery systems for biomedical implants. *IEEE Transactions on Biomedical Circuits and Systems*, *5*(1), 48–63. doi:10.1109/TBCAS.2010.2072782 PMID:23850978

Rindorf, L., Lading, L., & Breinbjerg, O. (2008). Resonantly coupled antennas for passive sensors. *Proceedings of IEEE Sensors*, 1611–1614.

Ruehli, A. E. (1974). Equivalent circuit models for three-dimensional multiconductor systems. *IEEE Transactions on Microwave Theory and Techniques*, *22*(3), 216–221. doi:10.1109/TMTT.1974.1128204

Sample, A. P., Meyer, D. A., & Smith, J. R. (2011). Analysis, experimental results, and range adaptation of magnetically coupled resonators for wireless power transfer. *IEEE Transactions on Industrial Electronics*, *58*(2), 544–554. doi:10.1109/TIE.2010.2046002

Senjuti, S. (2013). *Design and optimization of efficient wireless power transfer links for implantable biotelemetry systems*. University of Western Ontario.

Seo, D., Carmena, J. M., Rabaey, J. M., Alon, E., & Maharbiz, M. M. (2013). *Neural dust: an ultrasonic, low power solution for chronic brain-machine interfaces*. arXiv preprint arXiv:1307.2196

Seung-Hwan, L., & Lorenz, R. D. (2011). Development and validation of model for 95%-efficiency 220-W wireless power transfer over a 30-cm air gap. *IEEE Transactions on Industry Applications*, *47*(6), 2495–2504. doi:10.1109/TIA.2011.2168555

Sodagar, A. M., Wise, K. D., & Najafi, K. (2009). A wireless implantable microsystem for multi- channel neural recording. *IEEE Transactions on Microwave Theory and Techniques*, *57*(10), 2565–2573. doi:10.1109/TMTT.2009.2029957

Stronks, H. C., & Dagnelie, G. (2014). The functional performance of the ARGUS II retinal prosthesis. *Expert Review of Medical Devices*, *11*(1), 23–30. doi:10.1586/17434440.2014.862494 PMID:24308734

Tak, Y., Park, J., & Nam, S. (2009). Mode-based analysis of resonant characteristics for near-field coupled small antennas. *IEEE Antennas and Wireless Propagation Letters*, *8*, 1238–1241. doi:10.1109/LAWP.2009.2036133

Trumpis, M., Insanally, M., Zou, J., Elsharif, A., Ghomashchi, A., Artan, N. S., ... Viventi, J. (2017). A low-cost, scalable, current-sensing digital headstage for high channel count μECOG. *Journal of Neural Engineering*, *14*(2), 026009. doi:10.1088/1741-2552/aa5a82 PMID:28102827

Viventi, J., Kim, D.-H., Vigeland, L., Frechette, E. S., Blanco, J. A., Kim, Y.-S., ... Litt, B. (2011, December). Flexible, foldable, actively multiplexed, high-density electrode array for mapping brain activity in vivo. *Nature Neuroscience*, *14*(12), 1599–1605. doi:10.1038/nn.2973 PMID:22081157

Wang, G., Liu, W., Sivaprakasam, M., Zhou, M., Weiland, J. D., & Humayun, M. S. (2006). A dual band wireless power and data telemetry for retinal prosthesis. *Proceedings of IEEE EMBS Confeferce*, 28-38.

Wei, W., & Qiang, F. (2011). Design and simulation of printed spiral coil used in wireless power transmission systems for implant medical devices. *Proceedings of Annual International Conference of the IEEE Engineering in Medicine and Biology Society*, 4018-4021. 10.1109/IEMBS.2011.6090998

Wilson, B. S. (2017). The modern cochlear implant: A triumph of biomedical engineering and the first substantial restoration of human sense using a medical intervention. *IEEE Pulse*, *8*(2), 29–32. doi:10.1109/MPUL.2017.2649039 PMID:28328495

Won, D. H., Kim, H. S., & Jang, B. J. (2010). 13.56 MHz wireless power transfer system using loop antennas with tunable impedance matching circuit. *Journal of Korean Electromagnetic Engineering Society*, *21*(5), 519–527. doi:10.5515/KJKIEES.2010.21.5.519

Yan, G., Ye, D., Zan, P., Wang, K., & Ma, G. (2007). Micro-robot for endoscope based on wireless power transfer. *Proceedings of International Conference in Mechatronics Automation*, 3577-3581. 10.1109/ICMA.2007.4304140

Yeh, A., Ho, J., Tanabe, Y., Neofytou, E., Beygui, R., & Poon, A. (2013). Wirelessly powering miniature implants for optogenetic stimulation. *Applied Physics Letters*, 103.

Yin, M., Borton, D., Aceros, J., Patterson, W., & Nurmikko, A. (2013). A100-channel hermetically sealed implantable device for chronic wireless neurosensing applications. *IEEE Transactions on Biomedical Circuits and Systems*, *7*(2), 115–128. doi:10.1109/TBCAS.2013.2255874 PMID:23853294

Zeng, F. G. (2017). Challenges in improving cochlear implant performance and accessibility. *IEEE Transactions on Biomedical Engineering*, *64*(8), 1662–1664. doi:10.1109/TBME.2017.2718939 PMID:28650803

Zeng, F. G., Rebscher, S., Harrison, W., Sun, X., & Feng, H. (2008). Cochlear implants: System design, integration, and evaluation. *IEEE Reviews in Biomedical Engineering*, *1*, 115–142. doi:10.1109/RBME.2008.2008250 PMID:19946565

Zhang, X., Ho, S. L., & Fu, W. N. (2011). Quantitative analysis of a wireless power transfer cell with planar spiral structures. *IEEE Transactions on Magnetics*, *47*(10), 3200–3203. doi:10.1109/TMAG.2011.2147768

Zhou, D., & Greenbaum, E. (2009). *Implantable Neural Prostheses 1*. New York: Springer.

Zhou, D. D., Dorn, J. D., & Greenberg, R. J. (2013). *The ARGUS II retinal prosthesis system: An overview*. Paper presented at the 2013 IEEE International Conference on Multimedia and Expo Workshops (ICMEW), San Jose, CA. 10.1109/ICMEW.2013.6618428

ENDNOTES

1. Medtronics Implantable Cardioverter Defibrillators: http://www.medtronic.com/us-en/patients/treatments-therapies/icd-devices/models-dimensions-descriptions.html
2. Argus II Retinal Prosthesis: http://www.secondsight.com/g-the-argus-ii-prosthesis-system-pf-en.html
3. Cochlear Nucleus Implant: http://www.cochlear.com/wps/wcm/connect/us/home/treatment-options-for-hearing-loss/cochlear-implants

Chapter 10
Dynamic Pivoting Antenna-Module-Based Flexural Wireless Power Charger for Various Wireless Sensor

Ming-Shen Jian
National Formosa University, Taiwan

Chen Yen-Lung
National Formosa University, Taiwan

ABSTRACT

IoT is one of the popular topics for the past few years. The study achieves the data in the internet collecting and sharing by data transmission wired or wireless from information carriers. In most of IoT application, the main information devices are embedded systems and micro sensors. Due to the hardware limitations, most of information devices use the wireless transmission to do the data exchange like Wi-Fi, NFC, and Bluetooth. However, most of these devices are powered by batteries. In the other words, there are some problems that the user has to face: time limit, hard to replace batteries, and different specifications. In order to solve these problems, a dynamic antenna-based wireless power charger based on electromagnetic coupling of coil is proposed to replace traditional batteries in this chapter. Flexible printed circuit can allow the user changes in the circuit into a suitable shape by cutting, reversing, and coiling. In this way, the user can improve the electricity supply effect by adjusting the power of wireless transmission.

INTRODUCTION

Currently, the Internet Of Things (IOT) is the popular implementation and research topic. Most IOT devices are integrated with sensor components, network component, and power component. Each IOT device obtains the information of environment by the sensors. However, these sensors need the power. Considering the physical implementation, connecting with the wire power line for all IOT devices is impossible. To exchange the battery which as the power source of the IOT device increases the cost of

DOI: 10.4018/978-1-5225-5870-5.ch010

time and human resource. Therefore, to transmit the power wirelessly is the possible solution for the IOT devices.

Originally, most sensors are powered by wire line connection. These sensors can work with stable power. Considering the real implementation and environment, most sensors are movable or not located at only one location. Therefore, how to provide the power for the sensors or embedded systems becomes the research topic.

Today, most movable sensors or embedded systems use the battery as the power supply. In other words, using the battery as the power should reserve the space for the battery. Hence, the size of sensor or embedded system will be bigger for adopting the battery as the power provider. However, the battery should be exchanged to keep the stable power supply. In other words, the system users have to exchange the battery frequently. It will cost time and money.

Based on the concept: electromagnetic coupling of coil, many application were proposed. Most popular application is radio frequency identification (RFID) system. Radio Frequency Identification (RFID) is the popular wireless induction system used in IOT, personal payment, or customized individual services (Jian, 2009a, 2009b; Wu et al., 2011; Jian, 2012). When an independent RFID tag approaches the RFID antenna, the induction between RFID tag and antenna happens. At the same time, due to the electromagnetic coupling of coil, the RFID tag can receive the power from the antenna of RFID reader and active the chip embedded on RFID tag. Finally, the signal of data and be replied to RFID reader via wireless electromagnetic wave. Hence, the wireless services between the RFID tag user and the RFID reader (which includes the services) can be provided. In other words, the power can be provided to the RFID tag wirelessly. For example, the Taiwan Far Eastern Electronic Toll Collection Co., Ltd used the RFID based Electronic Toll Collection for the nationwide freeways (FETC, 2012).

Since the power can be transmitted from one device to another via wireless electromagnetic wave, no wire connection will be needed. It also means that the device can be powered based on electromagnetic wave without the battery (Ostaffe, 2015; Wu et al., 2015). Generally used wireless power transmission is implemented based on the single on demand designed coil or antenna. Some power transmission transmitters use the antenna (coil) arrays which are consisted of multiple small antenna or coil. However, due to the oscillator circuit designing of the transmitter, most of the single or arrays of the antenna (or coil) are designed with fixed size (area), fixed shape, or fixed turns of the coil. The antenna or coil is not extendable or dynamically changeable.

Unfortunately, the performance of induction between RFID tag and antenna depends on 1. the induction direction (angle) of the coil and electromagnetic field, 2. the total amount of coil, and 3. the area of the coil. If the power is not strong enough, the induction cannot be happened. Or, the chip embedded in the tag will not be acted. In addition, if the total amount of coil or the area of the coil is too small, the inducted electromagnetic field will not be enough. In addition, the induction range or distance will also affect the performance of the power transmission.

In other words, how to dynamically change and construct the antenna, circuit, and the sensor system is important for power transmission (Cheng, 2003; Jian, 2017; Smith, 2011; Lee, 2012). Figure 1 presents the structure of the RFID tag: Near-Field Loop-Antenna for High Frequency (HF) RFID tag, and Far-Field Dipole antenna for Ultra High Frequency (UHF) RFID tag.

According to the size of the current physical IOT devices, the space for the battery may be different. Therefore, to design the adaptive power transmission module for the various spaces instead of the battery is important. In this research, the wireless power charger with dynamic pivoting antenna module consists of cuttable soft-substrate power charger circuit and the pivoting antenna module is proposed.

Figure 1. The structure of different types of RFID tags

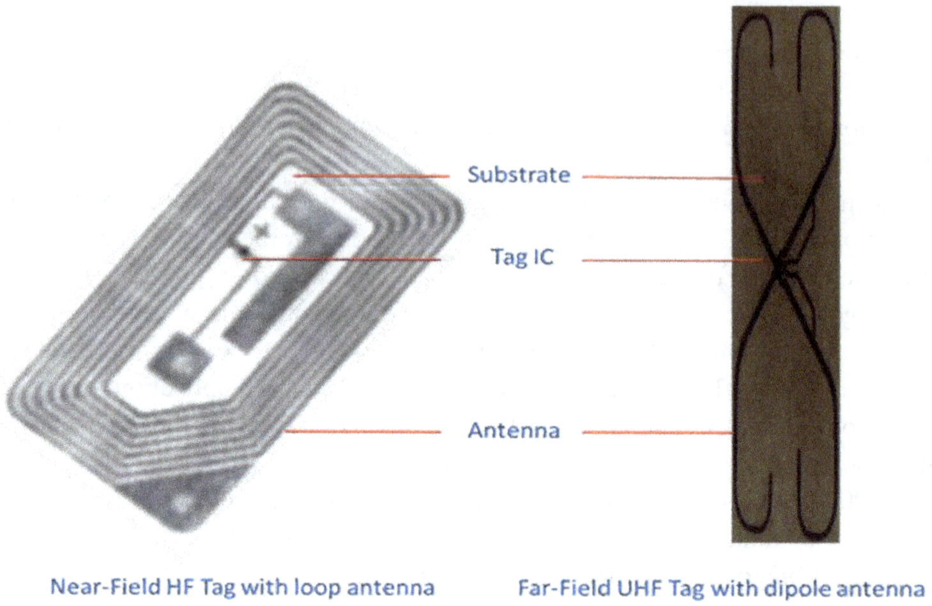

The proposed module can 1) reduce the cost of designing the sensor system, 2) reduce the space of power supply, 3) adaptively re-construct for the real implementation environment. System designer can adaptively extend the induction antenna and change the shape or size of the antenna. In other words, the various wireless sensor systems can be wireless powered based on the proposed module. The wireless sensors may be powered without the need of battery.

BACKGROUND

IOT systems and devices are generally used and implemented in normal life. There are many sensors which can create the services inside the house, in the vehicle, or deployed in the public space. However, all these IOT devices or wireless sensors need power. If these sensors are working without wired power, it means that these IOT devices or wireless sensors need batteries or build-in power generators as the power source.

In IOT, the data mining is an important part of the big data analysis, which collects data through the database, wireless sensors or embedded system. However, considering the real implementation of these IOT devices or wireless sensors, even the mobile phones, to provide the power via fixed wire is too difficult, especially for the movable devices. Hence, the wireless recharger system for power providing is proposed and used.

Based on electromagnetic coupling, the power can be transmitted between the power sender and receiver via wireless. According to the Faraday's law of induction, a magnetic field will interact with an electric circuit to produce an electromotive force (EMF), also called electromagnetic induction.

The EMF is also given by the rate of change of the magnetic flux:

Figure 2. Inductive Coupling between RFID Reader Antenna and RFID Tag Coil

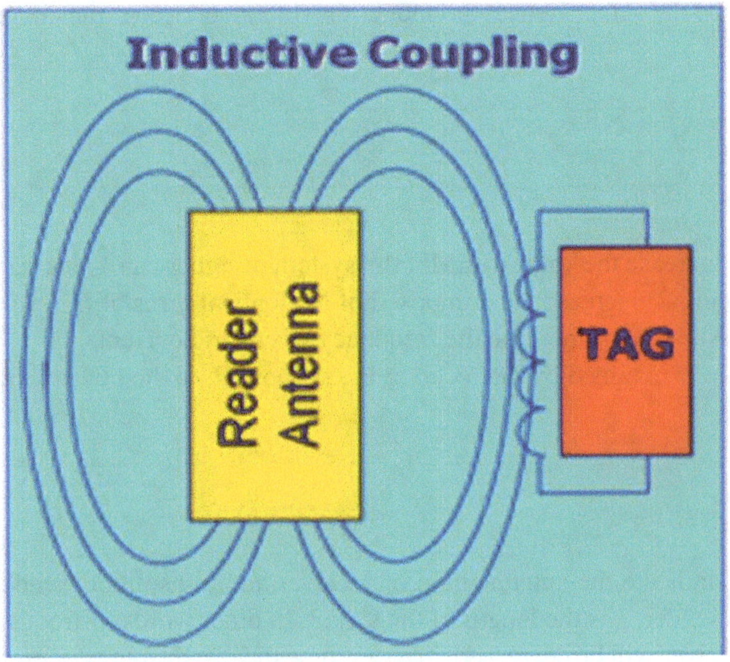

$$\mathcal{E} = -\frac{\Delta\varphi B}{\Delta t} \tag{1}$$

where Ɛ indicates the EMF and ΦB is the magnetic flux. For a tightly wound coil of wire, composed of N identical turns, each with the same ΦB, than the EMF can be presented as:

$$\mathcal{E} = -N\frac{\Delta\varphi B}{\Delta t} \tag{2}$$

If B is the magnetic field, and A is the area of coil, then the EMF can be also presented as:

$$\mathcal{E} = -N\frac{\Delta BA\cos\theta}{\Delta t} \tag{3}$$

where θ is the angle between the normal vector of area and the magnetic field.

Similar to the EMF of transformer:

$$\mathcal{E} = 4.44\times N\times\varphi\times f = 4.44\times N\times\left(B\times A\right)\times f \tag{4}$$

where f is the frequency of electric current.

Furthermore, based on the ideal situation, the power between induction antenna of sender and the coil of the receiver could be the same. Hence, the induction voltage between transformer, Vs, and receiver, Vr, can be defined as:

$$V_r \times I_r = V_s \times I_s \text{ and } \frac{V_r}{V_s} = \frac{N_r}{N_s} \tag{5}$$

In other words, to increase the value of EMF, the system designer can 1) increase the value A which indicates the area of coil, 2) increase the composed of N identical turns of the coil, 3) try to reduce the θ angle as zero as possible, or 4) increase the frequency of electric current.

To evaluate the power obtained by the receiver P_R, suppose P_T is the power from the sender. Then

$$P_R = P_T \frac{G_T G_R \lambda^2}{16\pi^2 D^2 L} \tag{6}$$

Where G_R and G_T indicate the antenna gain of receiver and transmitter (sender), D is the distance between receiver and sender, λ is the length of the signal. In other words, if frequency f is too high, the power obtained by receiver will be decreased. To overcome the disadvantage, increasing the power from the transmitter or decrease the distance between the power transformer and receiver is needed.

Therefore, in this research, the dynamic and adaptive antenna module for 1) increasing the composed of N identical turns, 2) increasing the value A which indicates the area of coil, and 3) maintaining the θ which indicates the angle between the normal vector of area and the magnetic field, is very important.

Nowadays, there are two wireless recharge standards: Wireless Power Consortium (Qi) and Alliance for Wireless Power / Power Matters Alliance (AirFuel). The sensor system or mobile devices, such as Samsung, iPhone which follows the standard can enhance the wireless power recharge. Most wireless recharge systems equip a battery to save the power. Therefore, most sensor systems should reserve the space for power storage.

However, considering the real implementation environment, the pre-designed antenna or coils for these sensor systems is made and used. Based on the final size of the product, the pre-designed antenna or coils is embedded such as Figure 3.

Considering the proposed Taiwan patent (no.I431887), the wireless power recharging system, there is the fixed size shell, induction coil (antenna), control circuits, isolation liner, and at least one electrical connection. By using the induction coil, the electromagnetic energy can be transferred as the electric energy. Based on the control circuits and the electrical connection, the embedded battery can be recharged. The concept of the patent can be shown as Figure 4. No. 110 indicates the induction coil (antenna). No. 120 is the control circuits with the electrical connection. No. 150 presents the fixed size shell. This patent shows that the power recharging system based on the wireless electromagnetic energy is possible. However, this patent doesn't present the possibility or the feasibility of the resizable battery. In other words, only the traditional fixed size battery can be used. Furthermore, this patent is proposed to recharge the energy of the battery. It is "not" proposed to replace the battery which used in the embedded system or IOT devices.

Figure 3. Pre-designed antenna or coils is embedded into trade items

Figure 4. The concept of Taiwan patent (no.I431887)

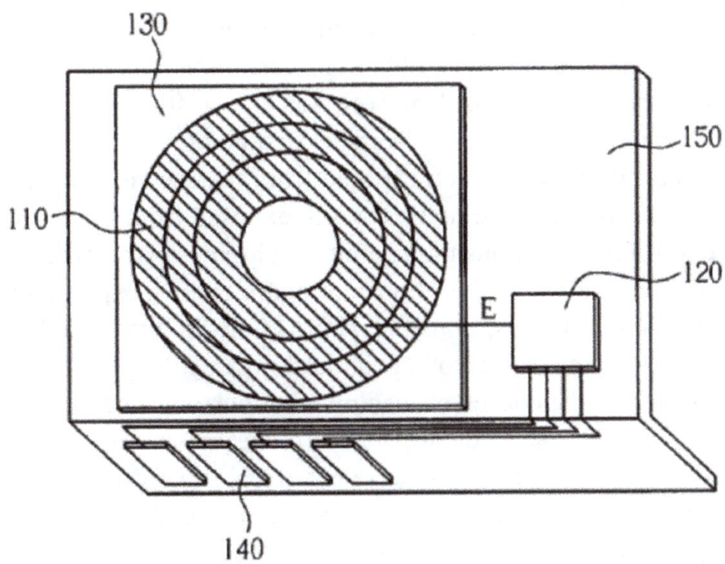

Another patent (No. M466535, Taiwan) proposed in the past tried to re-designed the wireless power transmission battery instead of the mercury battery. The traditional button size of the mercury battery is replaced. The antenna is re-designed and embedded into the mercury battery shown as the Figure 5. In figure 5, the antenna is designed as the coil for receiving the wireless electromagnetic energy. Then, the wireless power received can be provided to the embedded system and IOT device. In other words, the proposed patent can be as small as the buttons used on clothes. This patent also presents the possibility of battery recharging by receiving the wireless power. However, considering the traditional size of mercury battery, the total amount of the coil turns will be limited. To dynamically increase the size or turns of the coil is difficult. If the storage space of battery is changed, or the size of the IOT device is changed, the fixed size wireless power recharging battery cannot satisfy the new implementation requirements and limitation conditions.

Not only the induction coil of the power receiving devices but also the coil or antenna of the power transmitters should be considered. Oscillating circuit is generally used in the wireless power or energy transmission. The circuit can be shown as follows.

In traditional coils, it has a problem that the performance of wireless power or energy transmission which is related to the shape of antennas. With fixed shape, users can not adjustment the induction frequency easily. Therefore, dynamically extend the antenna for increasing the area of coil and the composed of N identical turns, or to adaptively tune the antenna for better angle between the normal vector of area and the magnetic field, changeable power charger circuit and antenna will be needed. In this way, the antenna which install on the different wireless recharging circuit will be more suitable.

Furthermore, when the telescope antenna is lengthened but could not be adjusted with the intensity of the signal, the efficiency of the antenna still is limited. For solving the said problem, the telescope antenna with a pivoting portion for adjusting the direction to receive the signal is provided.

However, the length of the said telescope antenna is fixed, so that the telescope antenna cannot receive the wider range of the signal. The pivoting portion of the telescope antenna only can be connected to the base or external device for adjusting the direction due to the restricted length of the antenna, so that the adjusted direction of the antenna is limited. Therefore, if the antenna has the flexible length an angle, and the shape or size can be adjusted easily, the efficiency and convenience can be enhanced.

According to the cited references, Son et al. (2014) disclose “a first member 710 coupled to a second member 720 by joint 740” shown as follows. However, Son et al. (2014)failed to teach the features “the end of the first transmitting unit which exposed from the first pivoting portion is a concave depression, the end of the second transmitting unit which exposed from the second pivoting portion is a convex protrusion, the end of the second transmitting unit which exposed from the third pivoting portion is a concave depression, the end of the third transmitting unit which exposed from the second pivoting portion is a convex protrusion, and the end of the third transmitting unit which exposed from the third pivoting portion is a concave depression”. That is, Son et al. (2014) failed to teach double pivoting connection structure. Figure 7 presents the proposed pivoting connection structure.

Further, both of the features “the convex-shaped pivoting portions and the corresponding concave-shaped pivoting portions” and “the concave depression end of the transmitting units and the corresponding convex protrusion end of the transmitting units” were not given. According to one aspect of the present disclosure, an antenna structure includes a base, a first extending member and a second extending member. The base has a first pivoting portion and an outputting end. The base includes a first transmitting unit located in the base, wherein two ends of the first transmitting unit are electrically connected to the outputting end and the first pivoting portion of the base respectively, and the end which connected to the first pivoting portion is exposed from the first pivoting portion. The first extending member has a second pivoting portion and a third pivoting portion, wherein the second pivoting portion of the first extending member is detachably and pivotally connected to the first pivoting portion of the base. The first extending member includes a second transmitting unit located in the first extending member, wherein two ends of the second transmitting unit are electrically connected to and exposed from the second pivoting portion and a third pivoting portion of the first extending member respectively. The second extending

Figure 5. The mercury battery size rechargeable battery with wireless energy transmission based on small coil (antenna)

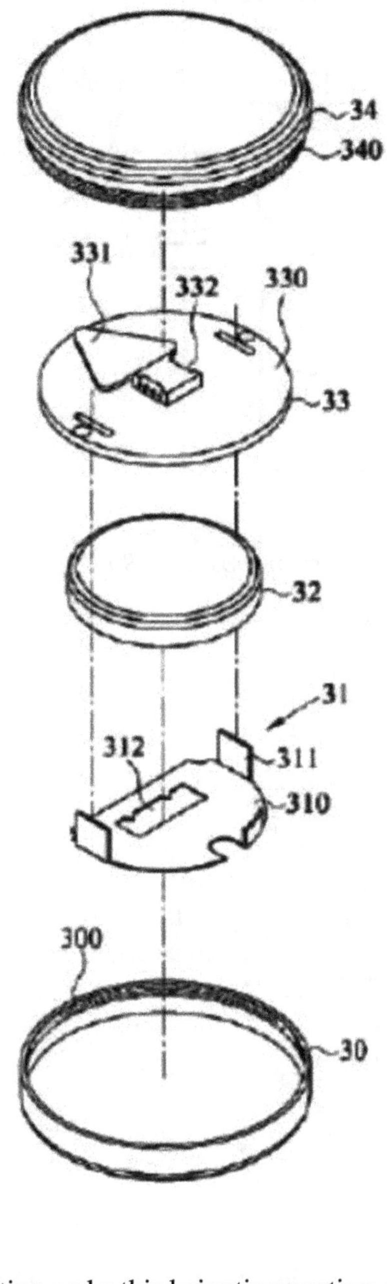

member has a second pivoting portion and a third pivoting portion, wherein the second pivoting portion of the second extending member is detachably and pivotally connected to the third pivoting portion of the first extending member.

The second extending member includes a third transmitting unit located in the second extending member, wherein two ends of the third transmitting unit are electrically connected to and exposed from the second pivoting portion and a third pivoting portion of the second extending member respectively. According to another aspect of the present disclosure, an antenna structure includes a plastic base, a first

Figure 6. Oscillating circuit for wireless electromagnetic induction

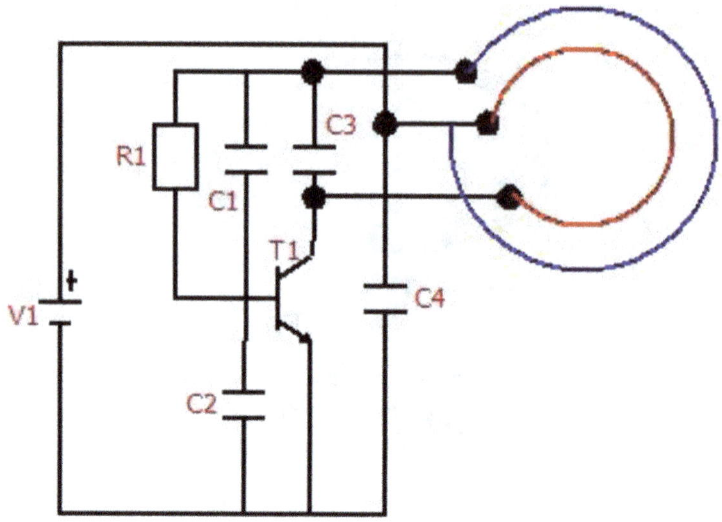

Figure 7. The pivoting connection structure proposed by Son, et al.(2014)

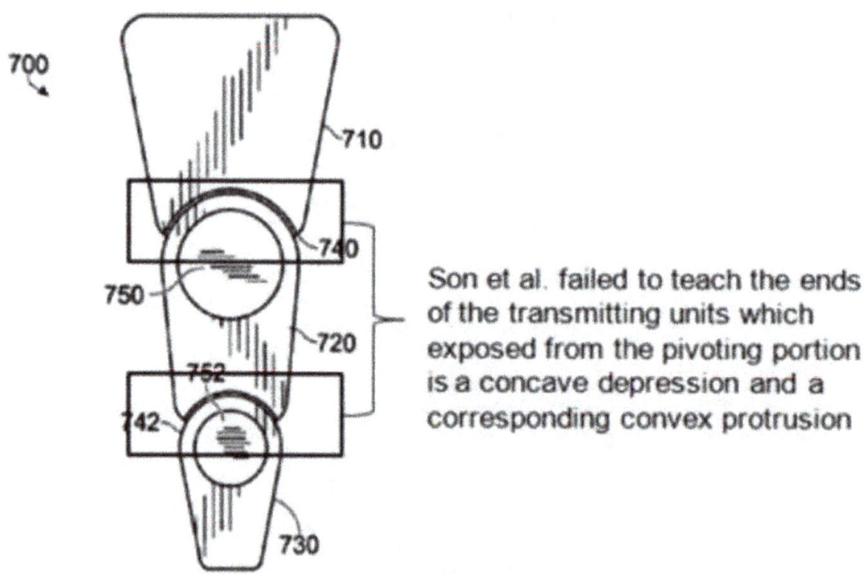

plastic extending member and a second plastic extending member. The plastic base has a first connecting portion and an outputting end. The plastic base includes a first transmitting unit located in the plastic base, wherein two ends of the first transmitting unit are electrically connected to the outputting end and the first connecting portion of the plastic base respectively, and the end which connected to the first connecting portion is exposed from the first connecting portion. The first plastic extending member has a second connecting portion and a third connecting portion, wherein the second connecting portion of the first plastic extending member is detachably connected to the first connecting portion of the plastic base.

The first plastic extending member includes a second transmitting unit located in the first plastic extending member, wherein two ends of the second transmitting unit are electrically connected to and exposed from the second connecting portion and a third connecting portion of the first plastic extending member respectively. The second plastic extending member has a second connecting portion and a third connecting portion, wherein the second connecting portion of the second plastic extending member is detachably connected to the third connecting portion of the first plastic extending member. The second plastic extending member includes a third transmitting unit located in the second plastic extending member, wherein two ends of the third transmitting unit are electrically connected to and exposed from the second connecting portion and a third connecting portion of the second plastic extending member respectively.

MAIN FOCUS OF THE CHAPTER

Cuttable-Flexural Soft-Substrate Power Charger Circuit

By using the foil, polyimide film, and adhesives, the circuit and be made as the soft-substrate. By designing the circuit, the electronic components can be embedded into a small circle. The size of the circle circuit is the same size as the mercury batteries. In other words, within the circle circuit, all the sensors or devices can be powered. In Figure 8, number-130 is the main circuit with electronic components. The size of the circle is the same as the most mercury batteries used in IOT devices. In addition, there are two connection reserved for the pivoting antenna or coils as number-150 shown. Another two connections, number-151, are used as the power output. Considering the real implementation environment, the size of power charger circuit should be dynamically changed. Therefore, number-141 indicates the cuttable location or point for the soft-substrate size changing. In other words, any size of charger circuit can be made by cutting or twisting the soft-substrate charger circuit. By connecting the number-151, the sensors or devices can be powered by the cuttable soft-substrate power charger circuit. In other words, corresponding to various spaces of the IOT devices, the proposed Cuttable-Flexural Soft-substrate Power Charger Circuit can be adaptively used by the IOT devices.

In addition, since the proposed Power Charger Circuit is made based on the Soft-substrate, the whole proposed Power Charger Circuit can be flexible. The induction coil or antenna is connected as number-150. The connection direction can be vertically connected to the proposed Power Charger Circuit. The coil is not constructed based on the Soft-substrate. In other words, the coil (antenna) is independent from the Power Charger Circuit Soft-substrate. Therefore, the flexion of the Power Charger Circuit Soft-substrate will not affect the designing of the pivoting coil (antenna). Furthermore, the power output connections shown as number-151 are also vertically connected to the proposed Power Charger Circuit. Therefore, the flexion of the Power Charger Circuit Soft-substrate can be various according to the requirements or limitations of the storage space in IOT devices. Then, the proposed Cuttable-Flexural Soft-substrate Power Charger Circuit can be used to replace the original power battery. Figure 9 presents the concept of the flexion of the Cuttable-Flexural Soft-substrate Power Charger Circuit.

To provide the power, which includes voltage and current, for the sensors, the simple power charger circuit can be designed as Figure 10. By inducting the antenna, the electromagnetic wave will create the current. By using diodes D1 and D2, the current will flow one way. If the sensor equips the capacitance, the capacitance can be used to instead of the sensor. Then, the resistance will be given to protect the circuit.

Figure 8. Cuttable-flexural soft-substrate Power Charger Circuit

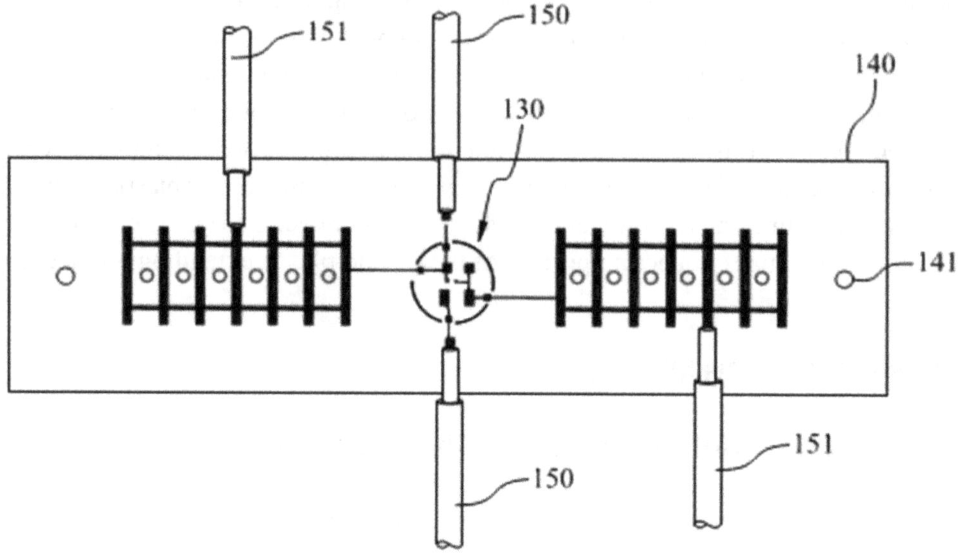

Figure 9. The flexion of the cuttable-flexural soft-substrate Power Charger Circuit

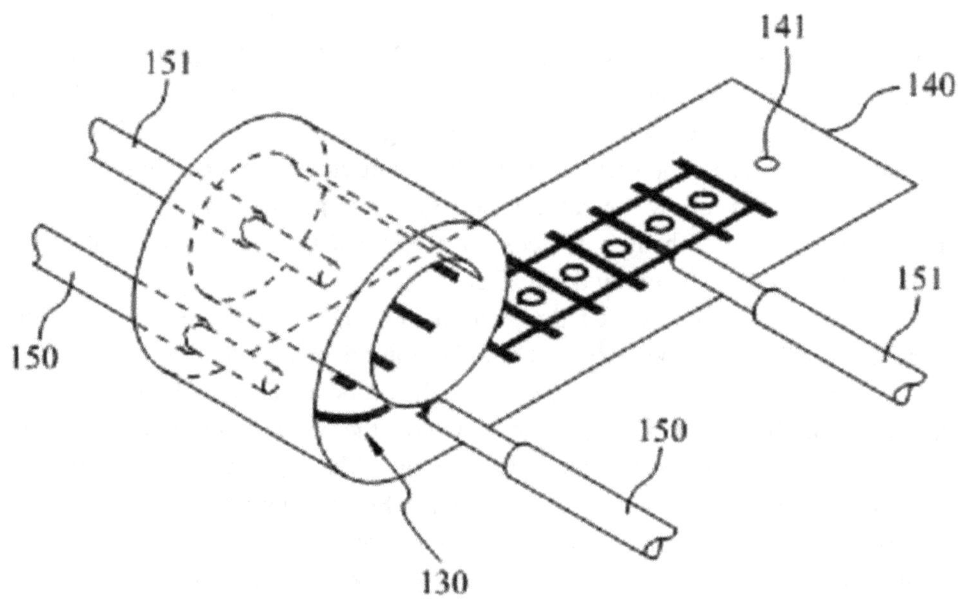

Via suitable cutting, the proposed Cuttable Soft-substrate Power Charger Circuit can be embedded into any sensor system. In other words, the sensor system does not need to exchange the battery. Also, it can be apply to several kinds of sensor system with different standards. If the system can provide the power supply source such as green energy like solar system or remote wireless induction antenna, the sensor can work continuously.

Pivoting Antenna Module

Originally, the antenna to receive or send the electromagnetic wave is on demand designed. In other words, the antenna used for devices or sensors are fixed. Due to the implementation limitation, the antenna cannot be dynamically changed. In addition to the on demand limitation of sensor system size, the power transmission is based on the electromagnetic wave. To induct or be inducted by the electromagnetic wave, the frequency of the electromagnetic wave may affect the size of shape of the antenna (Dobkin, 2007). Also, the performance will be affected.

Considering the induction based on the frequency, the limited antenna size should follow the function as the follow shows.

$$1 = \frac{300000}{\mathbf{F}} \times 0.96 \times 0.25 \tag{7}$$

where l is the length of antenna, F indicates the frequency of electromagnetic wave (MHz). If the frequency is higher, the antenna size can be decreased. However, according the function (4) and (6), although higher frequency can increase the EMF, higher frequency also decrease the power obtained by the receiver P_R very quickly that due to $C = \lambda \times F$. In other words, to adaptively extend the antenna for better performance corresponding to the system implementation environment is difficult and becomes an important issue.

Furthermore, according to the different frequency of the electromagnetic wave, the performance corresponding to the system implementation environment will be also different. For example, the higher frequency can have better performance when near the metal environment with smaller antenna size. In opposition, the low frequency can have better performance when near the liquid or moister environment. Table 1 shows the comparison.

Hence, to adopt the fixed size of antenna for different frequency, different environment or implementation is difficult. In order to provide the dynamic design for different implementation, the pivoting antenna module which can be dynamically changed as different size and different shape will be needed (Jian, 2016).

Figure 10. Power Charger Circuit

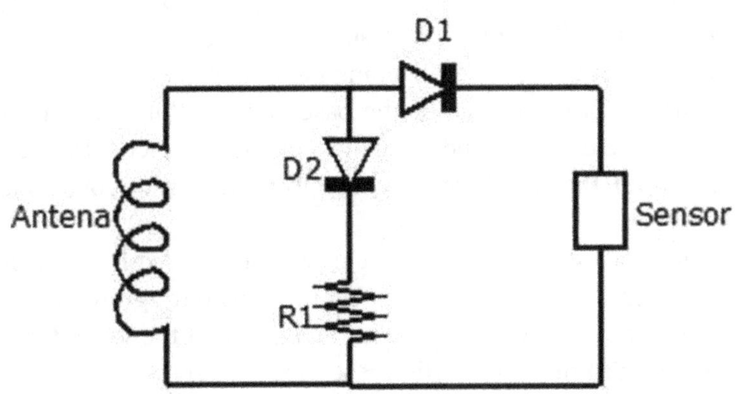

Table 1. The characteristics of different frequency of electromagnetic wave

	Low Frequency	High Frequency	Ultra High Frequency
Antenna Size	Large------------------------ > Small		
Performance Near Metal / Liquids	Better when near Liquids	--------- >	Better when near Metal

In this research, the authors proposed the extendable module for extending the length, changing the size and shape, and increasing the composed of N identical turns of coil. First, according to the frequency of electromagnetic wave used for power transmission as function 7, the corresponding antenna length, size, or shape, should be designed. Second, considering the direction of electromagnetic wave, the angle between the normal vector of area and the magnetic field can be as small as possible based on the shape of extended pivoting antenna. Third, to increasing the EMF, the total number of composed of coil can be increased by adding the module of the pivoting antenna. In other words, via suitable design, the pivoting antenna consists of jumper with the composed of coil can be designed as different modules.

The proposed antenna structure is comprising:

- A base having a first pivoting portion and an outputting end, the base comprising:
- A first transmitting unit located in the base, wherein two ends of the first transmitting unit are electrically connected to the outputting end and the first pivoting portion of the base respectively, and the end which connected to the first pivoting portion is exposed from the first pivoting portion;
- A first extending member having a second pivoting portion and a third pivoting portion, wherein the second pivoting portion of the first extending member is detachably and pivotally connected to the first pivoting portion of the base, and the first extending member comprises:
- A second transmitting unit located in the first extending member, wherein two ends of the second transmitting unit are electrically connected to and exposed from the second pivoting portion and the third pivoting portion of the first extending member respectively; and
- A second extending member having a second pivoting portion and a third pivoting portion, wherein the second pivoting portion of the second extending member is detachably and pivotally connected to the third pivoting portion of the first extending member, and the second extending member comprises:
- A third transmitting unit located in the second extending member, wherein two ends of the third transmitting unit are electrically connected to and exposed from the second pivoting portion and the third pivoting portion of the second extending member respectively;
    - Wherein the first pivoting portion of the base and the third pivoting portions of the first extending member and the second extending member are convex-shaped, and the second pivoting portions of the first extending member and the second extending member are convex-shaped;
    - Wherein the end of the first transmitting unit which exposed from the first pivoting portion is a concave depression, the end of the second transmitting unit which exposed from the second pivoting portion is a convex protrusion, the end of the second transmitting unit which exposed from the third pivoting portion is a concave depression, the end of the third trans-

mitting unit which exposed from the second pivoting portion is a convex protrusion, and the end of the third transmitting unit which exposed from the third pivoting portion is a concave depression.

In addition, the ends of the second transmitting unit and the third transmitting unit are exposed from virtual axes of the second pivoting portion and the third pivoting portion of the first extending member and the second extending member. A wireless transmitter is detachably and pivotally connected to the outputting end of the base which further comprises a controlling unit. To connect different extending members, the first extending member and the second extending member are geometric-shaped.

Following figure show three-dimensional view of an antenna structure according to others embodiments of the present disclosure. The antenna structure includes a first extending member 120 and a second extending member 130. The first extending member 120 has a second pivoting portion 121 and a third pivoting portion 122, and includes a second transmitting unit 123. Two ends 126, 127 of the second transmitting unit 123 are electrically connected to and exposed from the second pivoting portion 121 and the third pivoting portion 122 of the first extending member 120 respectively. The second extending member 130 has a second pivoting portion 121 and a third pivoting portion 122, and includes a third transmitting unit 131. Two ends 126, 127 of the third transmitting unit 131 are electrically connected to and exposed from the second pivoting portion 121 and the third pivoting portion 122 of the second extending member 130 respectively. When the second pivoting portion 121 of the second extending member 130 is pivotally connected to the third pivoting portion 122 of the first extending member 120, the exposed end 126 of the third transmitting unit 131 can be electrically connected to the exposed end 127 of the second transmitting unit 123.

The ends 126, 127 of the second transmitting unit 123 and the third transmitting unit 131 are exposed from the virtual axes X of the second pivoting portion 121 and the third pivoting portion 122 of each of the first extending member 120 and the second extending member 130. In detail, the end 127 of the second transmitting unit 123 which exposed from the second pivoting portion 121 of the first extending member 120 is a convex protrusion. The end 127 of the second transmitting unit 123 which exposed

Figure 11. 3D view of an antenna structure according to others' embodiments of the present disclosure

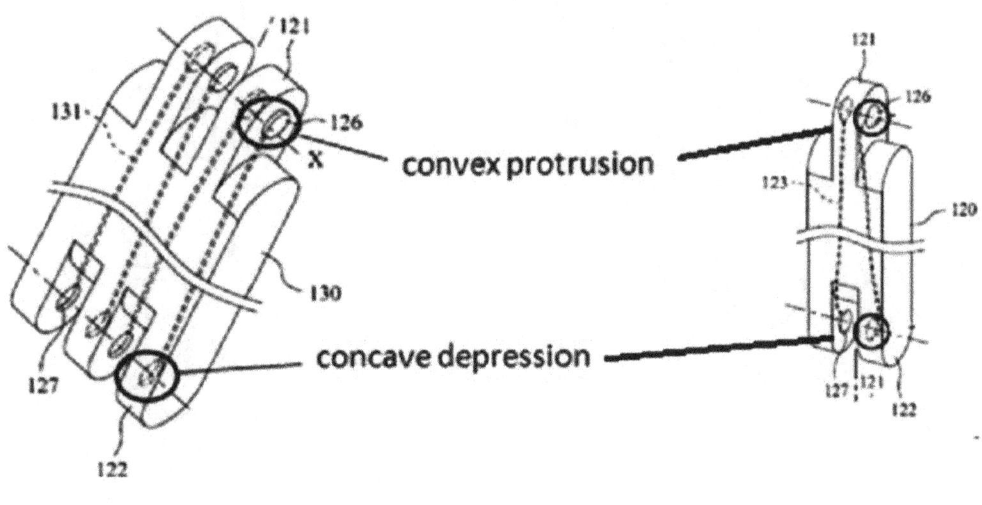

from the third pivoting portion 122 of the first extending member 120 is a concave depression. The end 126 of the third transmitting unit 131 which exposed from the second pivoting portion 121 of the second extending member 130 is a convex protrusion. The end 127 of the third transmitting unit 131 which exposed from the third pivoting portion 122 of the second extending member 130 is a concave depression. The foregoing convex protrusion can be a cylinder or a hemisphere, and the foregoing concave depression is corresponding to the convex protrusion.

When the first extending member is pivotally connected to the second extending member, the end (such as a convex protrusion) is pivotally connected to the end (such as a concave depression). Therefore, the angle of antenna structure can be varied by the pivoting connection between the first extending member and the second extending member. The antenna structure includes a plastic base, a first plastic extending member and a second plastic extending member. The plastic base has a first connecting portion and an outputting end, and includes a first transmitting unit. The first transmitting unit is located in the plastic base, wherein two ends of the first transmitting unit are electrically connected to the outputting end and the first connecting portion of the plastic base respectively, and the end which connected to the first connecting portion is exposed from the first connecting portion.

Figure 12. 3D view of antenna structure extension

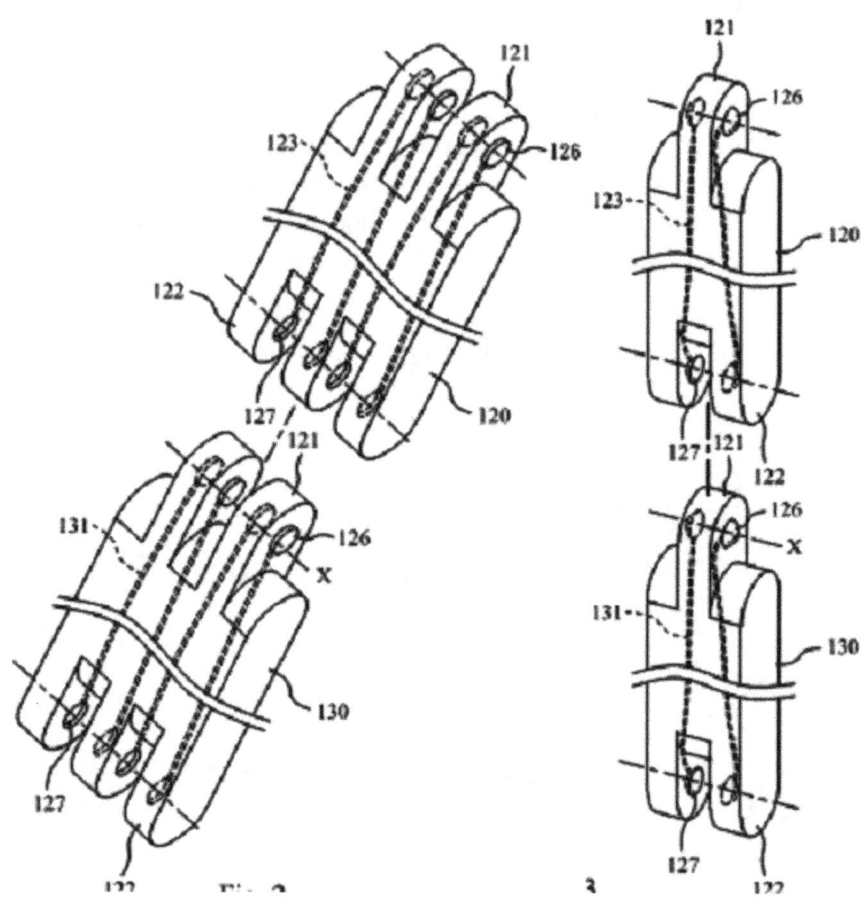

Figure 13 shows the example. There are module number 210, 220, and 230. The number 211, 222, and 221 indicate the pivoting latch for different module. Hence, the antenna designer can extend the pivoting antenna as any size and any shape for better power transmission performance. The antenna structure includes a plastic base, a plurality of first plastic extending members and a plurality of second plastic extending members, wherein the first connecting portion of the plastic base is detachably connected to the second connecting portion of one of the first plastic extending members which adjacent to the plastic base the third connecting portions of the first plastic extending members are detachably connected to the second connecting portions of the second plastic extending member, and the third connecting portion of one of the second plastic extending members which between the first plastic extending members is detachably connected to the first connecting portion of one of the first plastic extending members. When the plastic base, the first plastic extending members and the second plastic extending members are assembled, the first transmitting unit, the second transmitting unit and the third transmitting unit can be electrically connected together. When the third transmitting unit in the second plastic extending members receives the external signal, the external signal can be transmitted from the third transmitting unit through the second transmitting unit to the first transmitting unit, and then, the external signal can be outputted through the outputting end for signal processing. In detail, the plastic base, the first plastic extending members and the second plastic extending members can be made of plastic polymeric material. Therefore, the antenna structure is flexible for changing the angle thereof. Also, the antenna structure can be assembled with more first plastic extending members and more second plastic extending members on demand.

Figure 13. Extended Pivoting Antenna Module

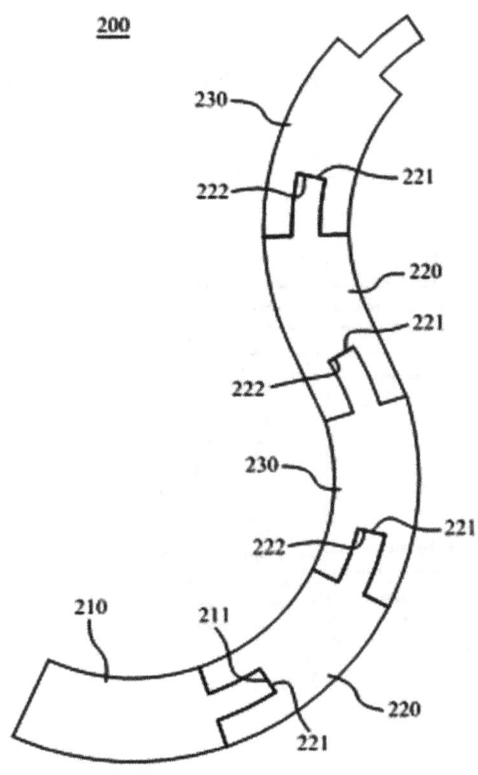

In other words, by using the module number 220 and 230, system designer can extend the pivoting antenna as any length needed, any total compose composed of N identical turns, and any area needed. Considering the connection of the Cuttable-Flexural Soft-substrate Power Charger Circuit, how to dynamically change the size or the composed of N identical turns of the coil should be considered (Jian, 2016). Suppose that there are three individual partitions: part-1) connection part for coil and power charger circuit, part-2) and extension part for increasing the area or turns of the coil. Figure 14 presents the concept of the pivoting antenna module. There are two connection reserved for the power charger circuit as number-150 shown. The connection between different parts can be done by the pair of 111A and 111B. By using the part-2 the area and shape of the antenna (or coil) can be dynamically changed. In addition, if the total amount of the coil turns will be changed or increased, the loop can be increased directly by connecting to part-3 instead of part-1 which can be indicated shown as red line in the figure. As the counterclockwise connection such as part-1➔ part-2 ➔ part-2 ➔ part-2 ➔ part-2, …, and finally connected to the part-1, the total amount of coil turns can be increased.

The usage state of an antenna structure according to still another embodiment is presented as follows. The plastic base, the first plastic extending members and the second plastic extending members of the antenna structure are assembled into an arch-shaped sensor, so that the antenna structure can be used in the airport for checking the luggage or the container. Moreover, when the antenna structure would be moved, the antenna structure can be easily folded and carried.

Therefore, by using the Extended Pivoting Antenna Module to increase the value of EMF, the system designer can 1) increase the area of coil by extending the coil via pivoting antenna module. By connecting the pivoting antenna module, the coil can be longer. In addition, the system designer can 2) increase the composed of N identical turns via using the coil jumper module. Considering the real system implementation environment, the electromagnetic wave direction from the power sender may be different. Hence,

Figure 14. Connection concept of the Extended Pivoting Antenna Module

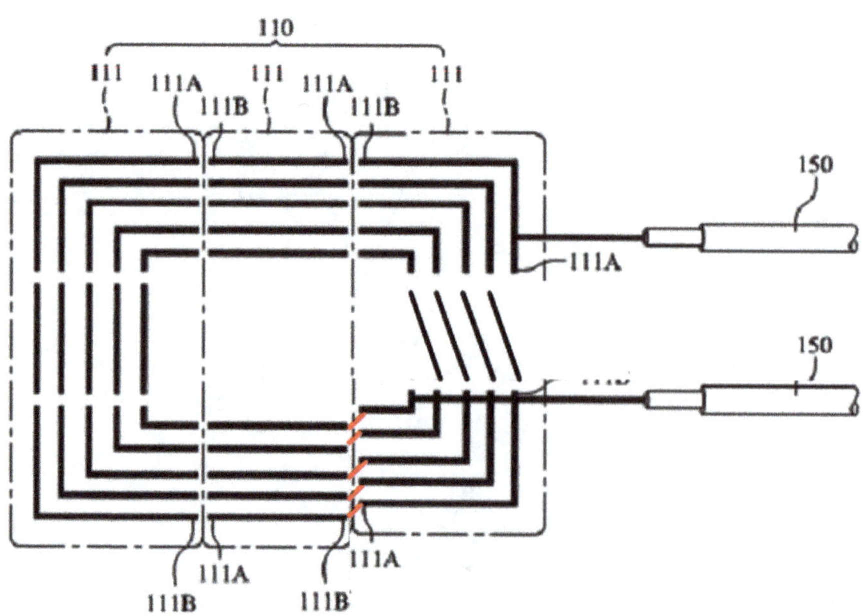

Figure 15. Example of the pivoting antenna structure for wireless induction

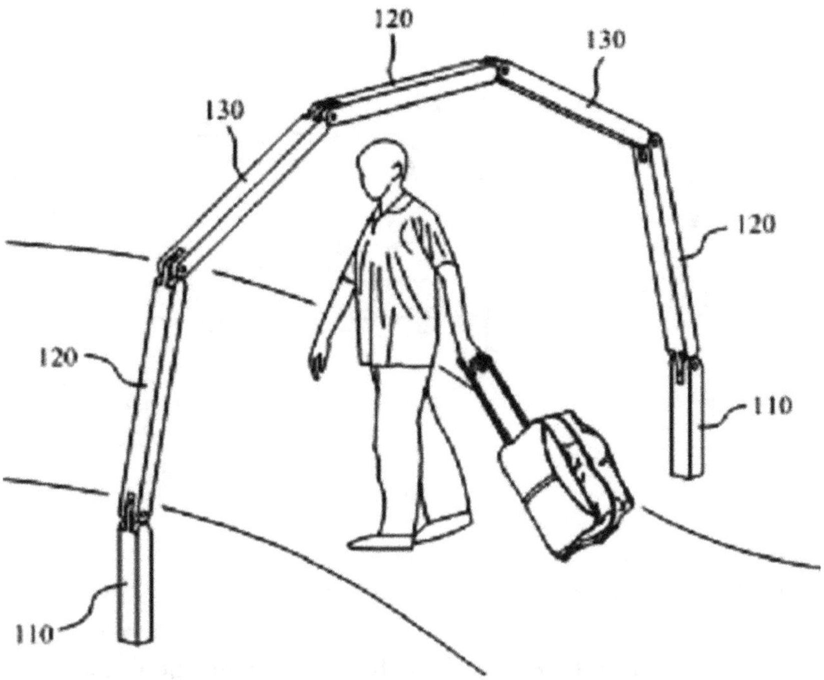

by extending the antenna, system designer can 3) try to adaptively change the normal vector of area and the magnetic field. Then, the system designer can reduce the θ angle as zero as possible. Furthermore, the proposed Extended Pivoting Antenna Module can be dynamically connected based on the on demand designed partitions. In other words, since the whole coil antenna can be dynamically changed according to the requirements, the cost of designing and developing the antenna can be reduced.

According to the proposed system, the pivoting antenna (coil) is connected to the recharging devices. The pivoting antenna (coil) 110 can be integrated and connected by multiple pivoting antenna (coil) modules shown as number-111. Based on the designing of circuit, the AC power energy can be provided to the power charger circuit shown as number-130. Then, the power charger circuit can provide the DC power energy to the IOT devices from the connection number-122. In other words, the power charger circuit can be designed to replace the original battery such as number-120 and the space number-121. Finally, the IOT device or sensor presented as number-200 can be powered.

Considering the proposed Cuttable-Flexural Soft-substrate Power Charger Circuit, the flexed substrate number-140 can be used and put inside the space number-121. According to the proposed pivoting antenna, the power charger circuit number-130 is printed based on the proposed Cuttable-Flexural Soft-substrate number-140. Corresponding to the space number-121, the multiple fixed potion holes number-141 can be used to adjusting length of the Cuttable-Flexural Soft-substrate. In other words, various spaces number-121 can be still satisfied by adjusting the suitable length of the substrate number-140. In addition, due to the connection method of the Cuttable-Flexural Soft-substrate Power Charger Circuit, the connection between the power charger circuit and the pivoting antenna (coil) will "NOT" be affected. Therefore, the space number-121 for the battery number-120 can be reduced. The IOT devices or sensors can be used

Figure 16. The concept of pivoting antenna (coil) which is connected to the recharging devices

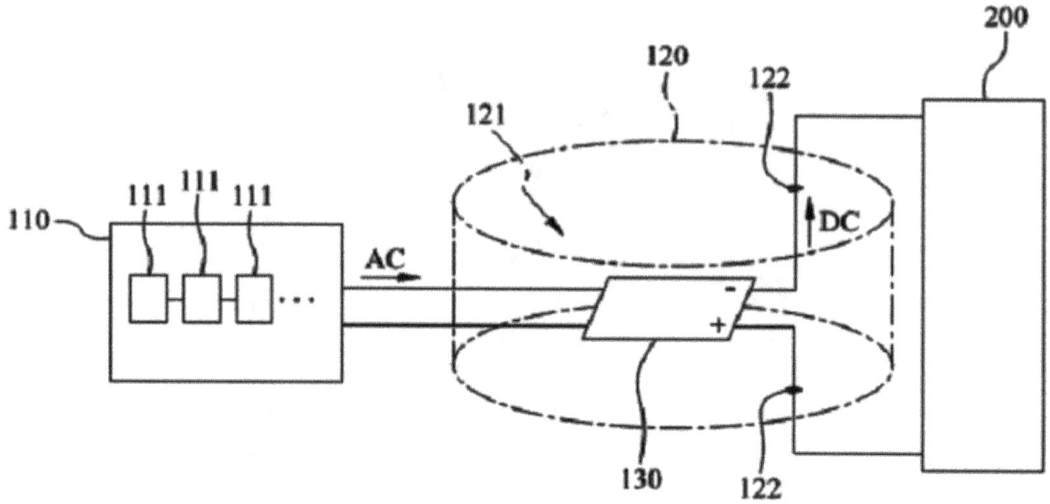

and put inside the tiny space. According to the proposed Cuttable-Flexural Soft-substrate Power Charger Circuit, the connections number-150 are connected between the power charger circuit number-130 and the pivoting antenna (coil) module number-111. In addition, the connections number-150 are connected from the power charger circuit number-130 and used as the power supply (output) source connection number-122 for the load such as IOT devices or sensors shown as number-200.

SOLUTIONS AND RECOMMENDATIONS

In this chapter, the wireless power charger with dynamic pivoting antenna module consists of cuttable soft-substrate power charger circuit and the pivoting antenna module is presented and really implemented.

By using the soft copper foil substrate (FPC soft substrate) which made by copper foil, polyimide film PI, and the adhesive, the Power Charger Circuit can be printed on the Cuttable-Flexural Soft-substrate. Figure 8 presents the physical implementation of the proposed Cuttable-Flexural Soft-substrate Power Charger Circuit. The power charger circuit is printed inside the center (circle). The connection points at the upper and bottom of the center power charger circuit circle are used to connect the pivoting antenna (coil). The left and right sides both are the power output and supply for the IOT devices or sensors. Considering the space for the power charger circuit, there are multiple holes for adjusting the total length of the soft-substrate.

To adopt the Cuttable-Flexural Soft-substrate Power Charger Circuit for the existed system, 1) the designer should cut the soft-substrate power charger circuit as the suitable size. Then, 2) considering the wireless power transmission, the antenna module should be connected to the power charger circuit. Finally, 3) the power charger circuit can be used instead of the original battery.

By using the proposed Cuttable-Flexural Soft-substrate Power Charger Circuit, anyone can use scissors to cut the substrate for any shape and any size. Or, the user can make the Cuttable-Flexural Soft-substrate Power Charger Circuit be flexural. In Figure 18, (a) the flexural and (b) cut Cuttable-Flexural Soft-substrate Power Charger Circuit is presented.

Figure 17. Real implementation of Cuttable-Flexural Soft-substrate Power Charger Circuit

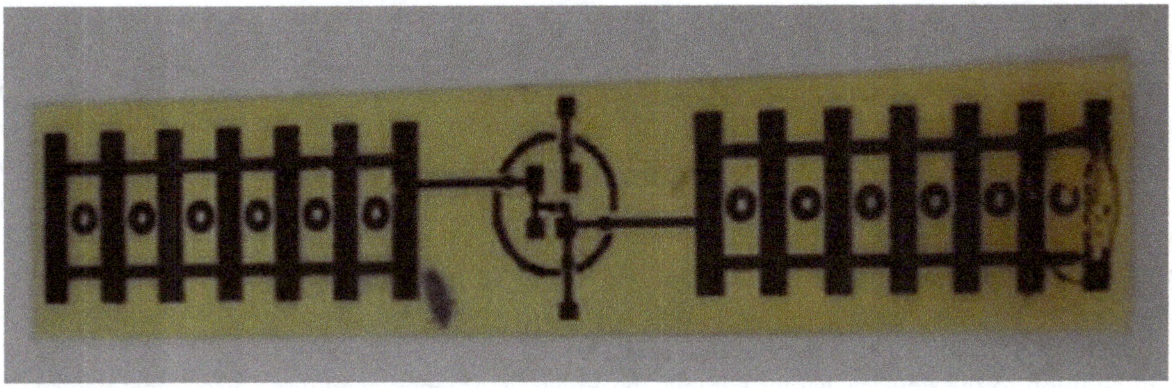

Figure 18. The (a) flexural and (b) cut Soft-substrate Power Charger Circuit

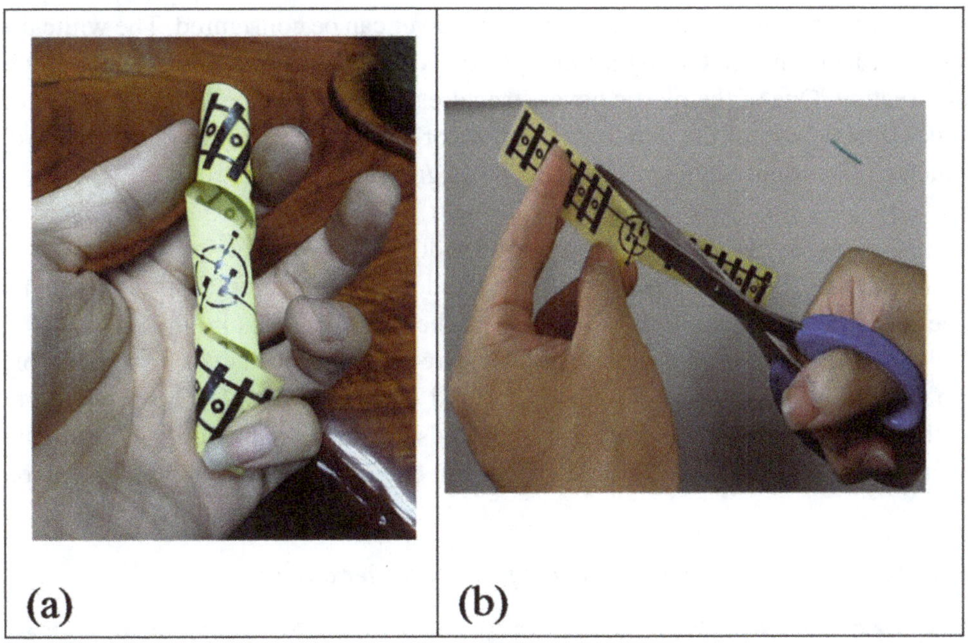

In other words, the proposed Cuttable-Flexural Soft-substrate Power Charger Circuit users can dynamically change the size, the shape of the circuit. By using the flexural type of circuit, no matter what kind of the original battery belongs to, the user can always put the flexural Power Charger Circuit instead of the original battery.

In the real implementation, the sensor system is the existed tire pressure sensor. Originally, the sensor is powered by mercury battery. For installing the wireless recharger system, the soft-substrate power charger circuit is cut as the same size as the mercury battery shown as Figure 18 (b).

In next, the pivoting antennas can receive the electromagnetic signals from the wireless power charger in which installed in elsewhere. By using the proposed dynamic pivoting antenna module based wireless power charger, the existed system can be continuously powered without changing the battery. For example, originally, the tire air-pressure sensor requires the mercury battery as the power source. The weight and the space will be increased due to the size of the battery. By cutting the soft-substrate, the center power charger circuit can be used to replace the original mercury battery. The implementation shows that the tire air pressure sensor can work after using the wireless power transmission based on the pivoting antenna (coil) and the Soft-substrate Power Charger Circuit. In other words, by cutting the Soft-substrate Power Charger Circuit, the original mercury battery can be replaced. The IOT devices can be powered. Furthermore, since the power can be transmitted wirelessly, the cost of mercury battery exchanging and manual operation can be reduced.

In addition to the Power Charger Circuit, the Pivoting Antenna Module is also implemented. To connect the partitions of the pivoting antenna (coil) more easily, each partition is printed by the 3D printer. Figure 20 presents the basic structure of the pivoting antenna (coil). White lines indicate the coil structure inside the antenna module. By using these two extension module (part) and connecting to the extra connection part, the coil of the transmitter and receiver can be configured. The white lines indicate the copper wire structure inside the 3D printed plastic cover. There are multiple copper wires in each plastic cover structure. One of the plastic cover structures will provide the loop wire structure. In other words, by suitable designing of the plug or socket, the original coil circle structure can be divided into partitions which can be connected or plugged as the original single coil structure.

In other words, via connecting to the extension part, the size or shape of the coil or antenna can be extended. For example, by horizontal extension module (part) connecting, the area of the coil can be also extended. Figure 21 shows the horizontal extended coil (antenna). The connection part for the connection between the extendable coil (antenna) and the power charge circuit can be connected at the right hand side. The white lines indicate the copper wire structure inside the 3D printed plastic cover. According to the same designing of the plastic covered extension module, the coil can be horizontal extended by the designing plug or socket. In other words, the area of the coil can be extended and increased. Furthermore, the transmitter or the receiver can be dynamically changed. The pivoted antenna

Figure 19. The existed sensor system and the cut power charger circuit

Figure 20. The basic structure of the pivoting antenna (coil)

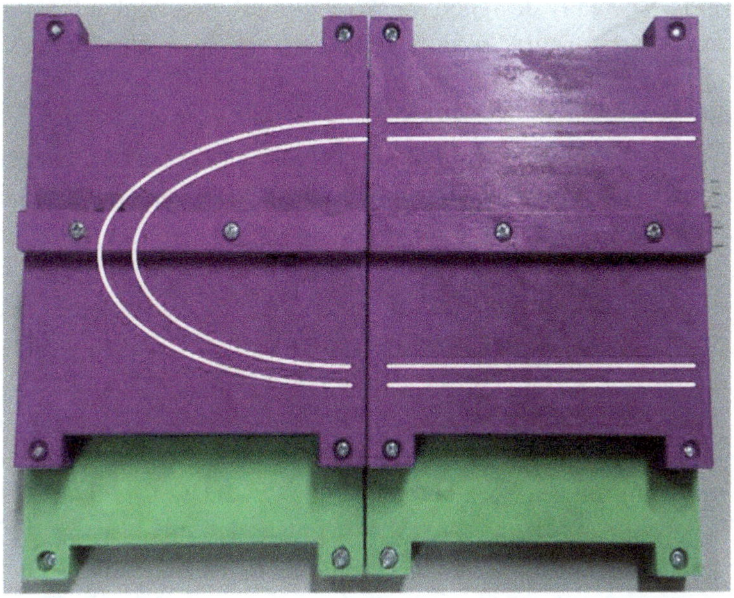

(or coil) is still integrated as the single one antenna (coil). Only one oscillating circuit is needed for the single pivoted antenna (or coil). Hence, according to the pivoting structure, the size of the transmitter and the receiver can be dynamic changed.

In addition, by using the vertical extension parts, two horizontal extended structures can be connected and integrated as the vertical extension coil (antenna). The area of the coil is increased. Furthermore, by using the Y-axis extension parts, horizontal extended structures can be connected and overlapped.

Figure 21. The horizontal extended structure of the pivoting antenna (coil)

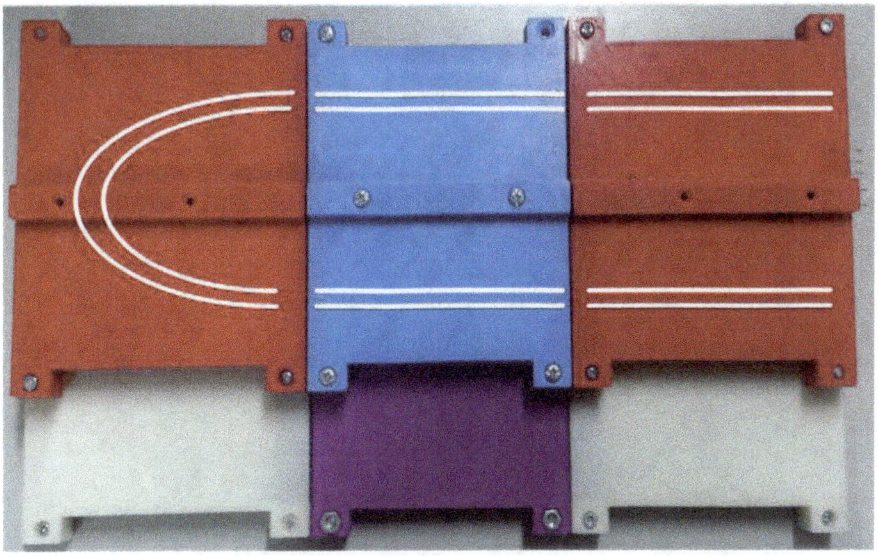

Figure 22 presents the vertical extension coil (antenna) by using two horizontal extended structures. The 3D printed plastic cover is especially designed for connecting the vertical extension antenna (coil). The connection direction of copper wire line will be changed from the vertical direction to the horizontal direction. The shape of the coil or antenna is also changed from 'O' shape loop to 'S' shape loop. In other words, the area of the coil (antenna) can be extended and increased.

In addition to the Y-axis extension parts, the Z-axis extension parts can be connected and overlapped. In other words, the total amount of coil turns can be increased dynamically. Finally, according to the requirements, the coil (antenna) can be dynamically changed. Not only the area of the coil (antenna) but also the total turns of the coil can be increased.

Therefore, the wireless pivoting antenna (coil) can be dynamically changed and extended. In other words, corresponding to the implementation environment and applications, the required coil or antenna can be pivoted. For example, the wireless power transmission can be implemented by the proposed pivoting antenna or coil. Due to the improvement of the technology, the Unmanned Aerial Vehicle (UAV) is also provided and used in many researches and applications. However, the power of the UAV is the main problem which may limit the services and the flying distance of the UAV, especially the electric power source UAV. Too huge battery with more electric power will increase the total weight of UAV and cause the consumption of power increased. Hence, to frequently recharge the battery power can light the weight of UAV. However, most recharging methods depend on wired power line and need the manual operation assistance. Therefore, by using the proposed pivoting antenna (coil), the power energy can be transmitted wirelessly. In other words, the power recharging could be done without manual operation. The power charging station for UAV can independently work. In opposition, the UAV will also equip the pivoting antenna (coil) to receive the power energy from the power charging station. When the UAV arrives the charging station, the power transmission between station and UAV can be automatically started. The proposed cuttable-flexural soft-substrate power charger circuit is light and used to provide

Figure 22. The vertical extension coil (antenna) by using two horizontal extended structures

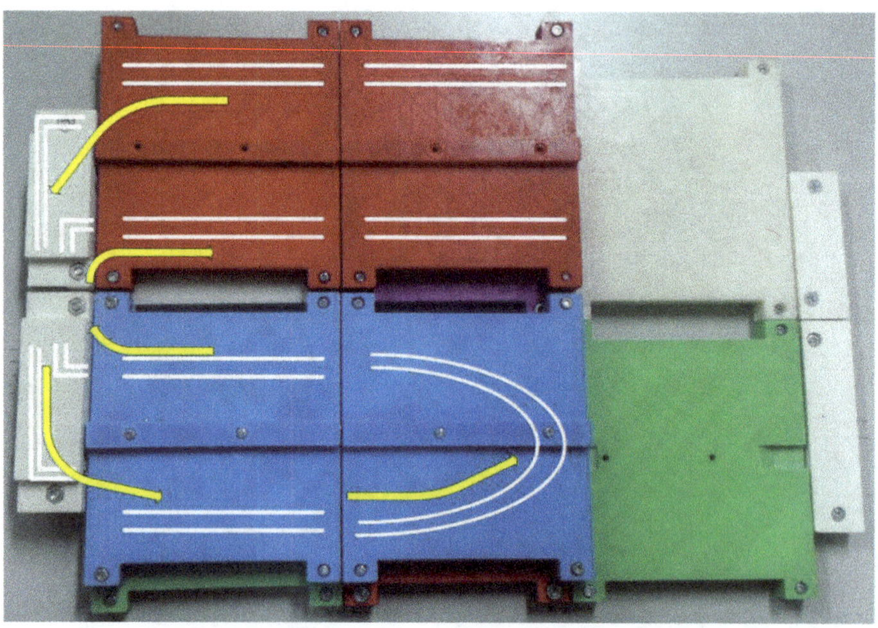

the power energy for the UAV and the battery. Since the UAV can be recharged without manual operation, the battery capacity of the UAV can be reduced.

By using the proposed pivoting antenna module, 1)Wired power line is needless. The implementation cost can be reduced. In addition, manual operation is needless. The man power can be reduced. 2) Different types of UAV can be served with the same electromagnetic coupling charging module. In other words, various types of UAV can be power recharged by the same pivoting antenna (coil) module. 3) The extra battery can be also used at the charging station to storage the power. By using the green energy sources, the power charging station can obtain the power energy independently. The charging station can be self-recharged by the green energy with more environmental protection.

In addition to the battery charging, the power generation from the green energy can be also implemented by the proposed pivoting antenna (coil) module. Currently, most green power energy resources are wind power and solar power. The Distributed Power Generation (DG) system is implemented based on multiple energy generators that can provide energy rather than centralized energy providing system. By integrating the local area power generators, batteries, and grid, the power microgrid of local area can be established (Luo, 2014) and adjusted (Talapko, 2014; Gujar, 2013). However, different natural environments will cause the performance of the various green energy resources different. Hence, based on the proposed pivoting antenna (coil) module, different green energy sources can be embedded and connected. All the green energy power generators will be connected by parallel circuit independently.

Figure 23. Wireless power transmission between power charging station and UAV based on the proposed pivoting antenna (coil) module

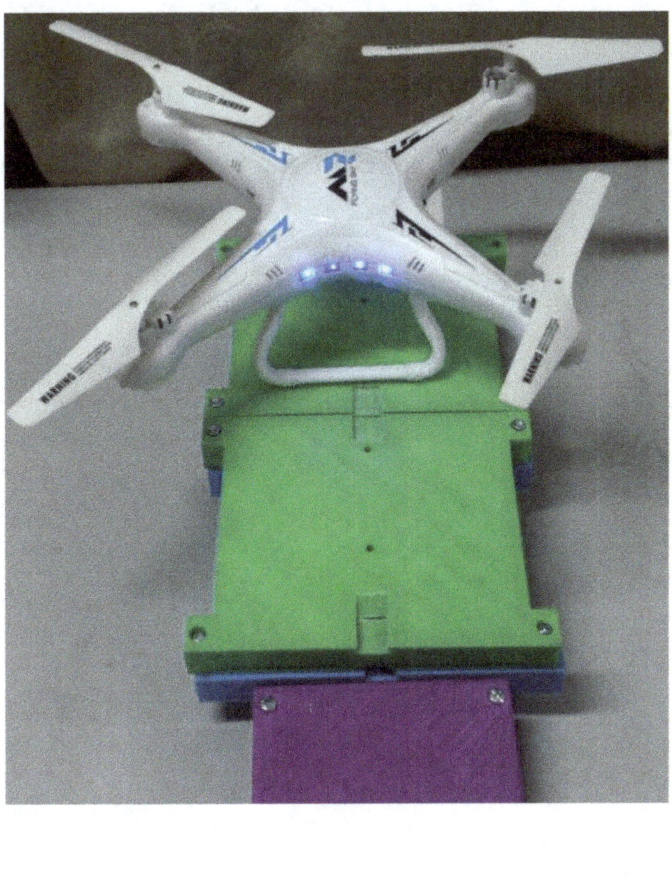

Therefore, various green power energy resources can be used. The power generated will be sent to the regulator circuit. Finally, the power can be transmitted by the pivoting antenna (coil) module. In other words, the power from different green energy source can be sent wirelessly. Figure 24 shows the concept of the green power energy transmission.

According to the proposed pivoting antenna (coil) module, the power generated from the various green energy resources can be transmitted wirelessly by the same antenna (coil). In other words, corresponding to the area of the green energy power generators deployed, the "single" antenna (coil) implemented based on the multiple pivoting antenna (coil) modules can be as the same area size as the deployment. Figure 16 presents the possible structure of the wireless power transmission based on the proposed pivoting antenna (coil) module. Each pivoting antenna (coil) module for antenna or coil area extension can be connected with two green energy power generators. The solar panel and the wind power generator can be dynamically exchanged. In figure 16, the structure is the basic structure of the pivoting antenna (coil). Each module can be connected with two green energy power resources. Therefore, in figure 16, most four solar panels or four wind power generators (or other various combinations of green energy) can be connected. By using the basic structure of the pivoting antenna (coil), the power can be transmitted.

CONCLUSION

By using the proposed dynamic pivoting antenna module based wireless power charger, the power can be continuously provided for the sensor system without battery exchanging. Comparing with the induction antenna used today, the proposed pivoting antenna module can be extended corresponding to the real system implementation environment. Furthermore, the proposed cuttable-flexural soft-substrate power charger circuit can be adaptively used for any size and shape needed. Considering the deployed environment of IOT devices, the cuttable-flexural soft-substrate power charger circuit can reduce the volume or space of the IOT devices. The physical verification shows that the development and deployment of the cuttable-flexural soft-substrate power charger circuit can be used instead of the mercury battery inside the IOT devices.

Figure 24. The concept of the green power energy transmission

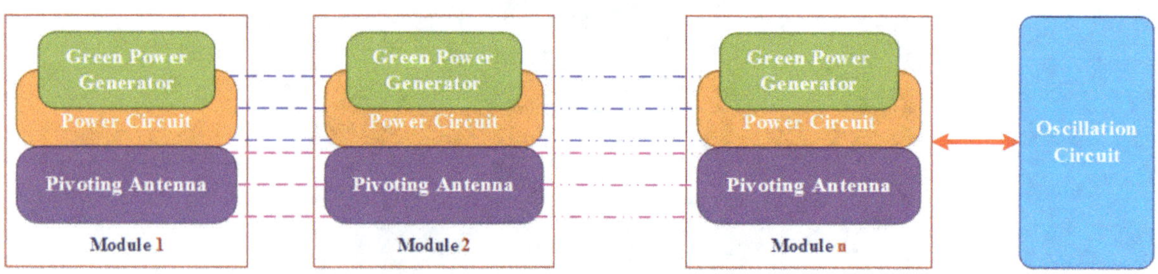

Figure 25. Green energy power harvesting based on the proposed basic structure of the pivoting antenna (coil) module

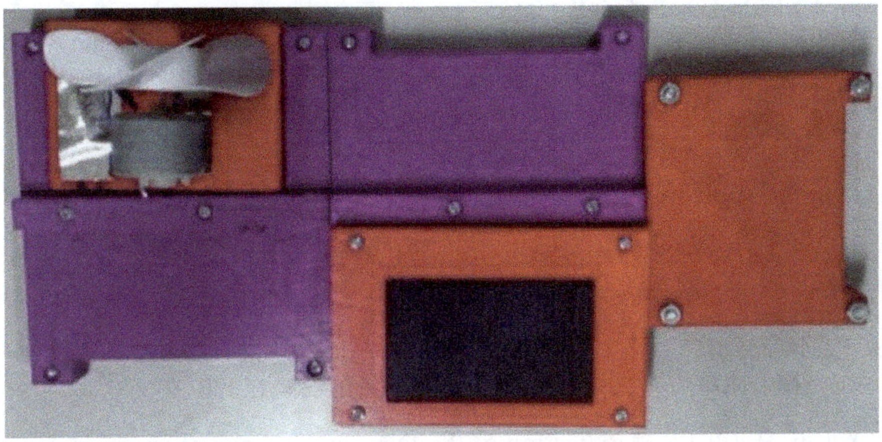

By connecting with the proposed pivoting antenna (coil) module, the power transmission can be done wirelessly. In addition, corresponding to the physical environment and requirements, the proposed pivoting antenna module can be dynamically extended for better power transmission performance. The real verification presents that the proposed dynamic pivoting antenna module based wireless power charger is adaptive for the different and existed sensor system. Even different green energy power resources can be integrated as the single power provider via the embedded single pivoted antenna (coil). Comparing with the system with multiple antennas or antenna array, the proposed pivoting antenna (coil) module can be extended more easily with simple circuit designing. The extra signal synchronization of the system between different antennas is needless.

REFERENCES

Cheng, K. W. E., & Lu, Y. (2003). Development of a contactless power converter. *IEEE ICIT*, *2*, 786–791.

Dobkin, D. M. (2007). *The RF in RFID: Passive UHF RFID in Practice*. New York: Elsevier.

Far Eastern Electronic Toll Collection Co. Ltd. (n.d.). Retrieved from https://www.fetc.net.tw/en/

Gujar, M., Datta, A., & Mohanty, P. (2013). Smart Mini Grid: An Innovative Distributed Generation based Energy System. *IEEE ISGT*, 1-5.

Jian, M. S., Chen, Y. L., & Li, S. J. (2015). Wireless Power Charger with Dynamic Pivoting Antenna Module for Various Wireless Sensor. *13th EHAC'15*, 110-113.

Jian, M. S., Chen, Y. L., Tong R. W., Lin, Y. H., Cheng C. (2017). Various Green Energy Sources in Smart Grid Power Network based on Cloud. *ICACT*, 660-663.

Jian, M. S., Chou, T. Y., & Hsu, S. H. (2009). RFID Encryption/Decryption Technology Aided Multimedia and Data Intellectual Property Protection. *WSEAS Trans. on Communications*, *8*(7), 734–743.

Jian, M. S., Chou, T. Y., & Hsu, S. H. (2012). Mobility corresponding location-aware information services based on embedded RFID platform. *International Journal of Ad Hoc and Ubiquitous Computing*, *9*(2), 122–131. doi:10.1504/IJAHUC.2012.045553

Jian, M. S., Fang, Y. C., Tong R. W., Lin, Y. H., (2016) Wireless Green Energy Power Transmission System Based on Assembly Method and Pivoting Antenna Module. *ICASI*, 1-4.

Jian, M. S., & Hsu, S. H. (2009). Aware Public/Personal Diversity of Information Services based on embedded RFID Platform. *Proc. of ICACT*, 1145-1150.

Jian, M. S., & Wu, J. S. (2013). *RFID Applications and Challenges (Chapter) of Radio Frequency Identification from System to Applications*. In-Tech.

Lee, D. S., Liu, Y. H., & Lin, C. R. (2012). A Wireless Sensor Enabled by Wireless Power. *Sensors (Basel)*, *12*(12), 16116–16143. doi:10.3390121216116 PMID:23443370

Luo, Y., He, J., & Liu, H. (2014). Application of the Distributed Generation, Micro and Smart Power Grid in the Urban Planning. *IEEE-Cyber*, 634-637.

Ostaffe, H. (2012). Retrieved August 16, 2012, from http://www.powercastco.com/uhf-rfid-sensing-passive-rfid-wireless-sensor-tags/

Smith, J. R. (2011). *WISP (Wireless Identification and Sensing Platform)*. Retrieved 2011, from https://sensor.cs.washington.edu/WISP.html

Son, S. H., & Jones, G. G. (2014). *Electronic device stand*. US Patent 8737064.

Talapko, D., & Tesnjak, S. (2014). Influence of Distributed Power Generation Sources on Improvement of Power Supply Availability in Telecom Infrastructure. *IEEE DPSP*, 12-21.

Wu, C. C., Jian, M. S., & Chou, T. Y. (2011). Environmental Affection Based RFID Potential Bio-Disease Tracking/Tracing System, WSEAS Transactions on Systems. *WSEAS Transactions on Systems*, *10*(2), 38–48.

Wu, Y., Liu, P., Qiu, J., & Dai, G. (2015). Enable sustainable sensor networks with non-contact charging: Efficient deployment of energy hubs. *Int. J. of Sensor Networks*, *18*(3/4), 172–181. doi:10.1504/IJSNET.2015.070406

Chapter 11
Research on Modelling Inductive Power Transfer for Electric Vehicles

Michela Longo
Politecnico di Milano, Italy

Morris Brenna
Politecnico di Milano, Italy

Federica Foiadelli
Politecnico di Milano, Italy

ABSTRACT

The environmental pollution caused by fossil fuels is a hot issue around the world in recent years. The gases lead to poor air quality, in particular in large cities, and the global warming that can cause ecological calamity such as tropical cyclones, heatwaves, drought, and extreme tides. International Energy Agency clearly states that the current energy trend is not sustainable environmentally, economically, and socially. Therefore, it must devise solutions to achieve the future economic growth without adverse environmental effects. The increasing diffusion of electric vehicles is driving academic and institutional research towards exploring different possible ways of charging vehicles in a fast, reliable, and safe way. For this reason, wireless power transfer systems have recently been receiving a lot of attention in the academic literature. This chapter reviews the main analytic and computational tools that are typically used to perform analyses in the context of inductive power transfer systems (IPTSs).

INTRODUCTION

In recent years, the world has experienced a greatly renovated interest in the topic of Electric Vehicles (EVs), mainly because of the increasing environmental awareness of the general public and the equally growing perception that fossil fuels are a finite resource. Thus, academic and industrial research on EVs is currently flourishing, and the global electric car stock has been steadily increasing since 2010, surpassing the 2-million-vehicles mark in 2016 (International Energy Agency, 2017).

DOI: 10.4018/978-1-5225-5870-5.ch011

However, EVs are not exempt from problems that slow down their diffusion. The first main issue is the cost of this kind of vehicles, mainly related to the on-board batteries. Moreover, as well as being expensive, they do not guarantee an energy density comparable to that of fossil fuels, nor a similar charging time. The actual energy density of a Li-ion battery is only around 100–180 Wh/kg (Catenacci et al., 2013), and improvements are slow because they do not follow Moore's law due to the fact that batteries are governed by chemistry and not electronics (Vilathgamuwa & Sampath, 2015; Rim, 2018). In the context of overcoming problems related to battery size and charging time, Wireless Power Transfer (WPT) is now the subject of intensive research, both by academic institutions and automotive industry players. Wireless charging systems provide various advantages compared to conductive systems, firstly because they do not require to manually plug-in a connector, which can be dangerous if weather conditions are not optimal (Rim, 2018). Additionally, in WPT systems the size of on-board batteries can be reduced, because the infrastructure can be designed in order to maximize charging opportunities. For example, electric buses can be recharged by a wireless system while loading and unloading passengers at stops along their route (Bi et al., 2016).

There are different types of WPT, based on their principle of operation that can be categorized as electromagnetic radiation-based (applicable for long-distance power transfer), magnetic (also known as inductive or resonant, used for short-distance power transfer), and capacitive, also used over short distances. Currently, Inductive Power Transfer Systems (IPTSs) are experiencing the highest amount of academic interest, because they generate a less harmful field compared to the capacitive case (Rim, 2018; Bi et al., 2016).

STATIC AND DYNAMIC CHARGING

Wireless power transfer systems can be classified, based on their working modes, as either *static* (charging while the vehicle is not in motion) or *dynamic* (charging while the vehicle is moving along the road) (Bi et al., 2016).

Static wireless charging can be achieved when the vehicle reaches dedicated charging areas, which can be easily implemented in urban contexts like parking lots (for private vehicles) and bus stops (for public transit vehicles) (Vilathgamuwa & Sampath, 2015). The operating principle of static WPT is to use a high frequency inverter in order to convert the power from grid frequency high frequency AC, which allows energy to be transferred from a transmitting resonating coil to a similar receiver, by means of inductive coupling. The power received by the second resonator is converted to DC quantities in order to charge the battery pack (Vilathgamuwa & Sampath, 2015). Even though the vehicle needs to be stationary, static WPT for electric vehicles appears to be more accepted by the market, due to the lower implementation cost compared to the dynamic case. Furthermore, this kind of charging system allows to shift the complexity towards the transmitting side, keeping the vehicle components as simple and light as possible (Bi et al., 2016). Obviously, the simplicity of static WPT compared to the dynamic case has a price: such systems need a bigger on-board battery in order to provide a reasonable driving range between charging sessions. While a static WPT system needs vehicles to be stationary, dynamic WPT allows the EVs to be continuously charged while in motion, ensuring a much longer driving range. For this reason, dynamically charged EVs are able to compete with Internal Combustion Engine Vehicles (ICEVs) in almost all aspects of performance, except for the cost: they are quieter, cleaner and more efficient, and with the appropriate road infrastructure, they could have unlimited range (Covic & Talbot

Boys, 2013; Park et al., 2015). Dynamically charged EVs are also called Roadway-Powered Electric Vehicles (RPEVs) because they are supplied by rails buried under the road (Park et al., 2015). Dynamic WPT can be further categorized in two types that mainly differ in the transmitter array design. The first one is called stretched track and is constituted by a coil whose dimension in the EV moving direction is greater than the receiver coil one. The second array design, instead, is the lumped transmitter coil array, composed by multiple coils with the dimension of the receiver coil and supplied by a single high frequency power source (Buja et al., 2016). Despite the advantages of unlimited range, this kind of charging system also suffers from some drawbacks: mainly, infrastructure cost for this type of WPT increases with the length of the track, limiting opportunities for development.

Another relevant problem for this type of charging system is the low coupling factor, due to the smaller transmitting region covered by the receiver coil, which makes it hard to improve the general, efficiency of such systems. Typically, ferromagnetic materials are used to guide the magnetic flux and increase the efficiency, counteracting the low coupling factor. Moreover, electromagnetic emissions in the uncoupled region are another relevant issue, due to the danger that electromagnetic radiation poses for people. The lumped track design has been found to eliminate field exposure for people while also improving the coupling factor Moreover, electromagnetic emissions in the uncoupled region are another relevant issue, due to the danger that electromagnetic radiation poses for people. The lumped track design has been found to eliminate field exposure for people while also improving the coupling factor. However, this kind of array design introduces other problems: the power source must be continuously switched from one coil to another while the load is moving along the track, and the distance between coils must be optimized. An excessive distance could make the power transfer not continuous, while on the other hand an excessively small distance could introduce a negative mutual inductance between each pair of coils, which would in turn generate parasitic currents. Of course, the distance must be optimized keeping in mind that each additional coil adds to total cost of the infrastructure (Vilathgamuwa & Sampath, 2015).

Current research on dynamic wireless charging has led to the development of an OLEV (On-Line EV) prototype by KAIST that is characterized by a high power transfer capability and an efficiency of 85%. Infrastructure cost remains a big issue for this kind of vehicle, due to the wide excavation of the road needed to install the power supply rails. Prototyping on the lumped track design has instead been initiated by a research group at the University of Auckland (Park et al., 2015; Buja et al., 2016).

ANALYTIC MODELS FOR INDUCTIVE POWER TRANSFER

Three of the main analytic models used in designing IPT systems are explored in detail, in particular the coupled coils model, the gyrator model and the magnetic mirror model. The selection of these three among all the existing models and methodologies was driven mainly by a combined evaluation of how useful they are as design tools, as well as how they correlate to the specific process of optimizing an IPT system. This is the reason why, for instance, phasor transformation models for AC converters were not considered in this report: while they are widely used in the literature, their application is not really specific to wireless power transfer systems. Among the chosen models, two – the gyrator and the magnetic mirror – have been just recently introduced in the literature, and for this reason they were considered particularly interesting.

Coupled Coils Model

A very simple, yet powerful theoretical tool for modelling inductive power transfer is the coupled coils (or coupled inductors) model (Rim & Mi, 2017). Resting upon basic circuit theory and various assumptions, it allows researchers to focus on the main differences between circuit topologies for WPT (Whang et al., 2005; Swain et al., 2012; Omori et al., 2012). Because of its simplicity, it has found widespread use in the relevant literature (Zhang & Chau, 2015; Raju et al., 2014; Shin et al., 2014; Li et al. 2016). First of all, let us consider two coupled coils, each driven by a different current source, as shown in Figure 1 (Kuzey et al., 2017; Barmada et al., 2017; Mohamed et al., 2017; Maddalena & Godoy, 2017).

It is assumed that both coils are linear, free from parasitic capacitance and that the whole circuit is electrically small (Campi et al., 2017; Paul, 2006). Furthermore, the wires connecting each coil to the respective source are assumed to be free of parasitic inductance, which could however be considered by simply inserting an additional lumped inductance in series with the rest of the circuit. According to Ampere's law, the current passing through the inductors causes magnetic fluxes, which intersect the coils. According to Faraday's law, voltages v_1 and v_2 can be obtained by derivation of these fluxes, in the following way:

$$\begin{cases} v_1 = N_1 \dfrac{d}{dt} + \left(\phi_{11} + \phi_{12} - \phi_{21}\right) \\ v_2 = N_2 \dfrac{d}{dt} + \left(\phi_{22} + \phi_{21} - \phi_{12}\right) \end{cases} \tag{1}$$

where it is assumed for simplicity that all flux lines intersect all coils. Considering the canonical definition of the inductance

$$L = \frac{N\phi}{i} \tag{2}$$

The relations above can be rewritten as

Figure 1. Two coupled coils driven by arbitrary independent current sources

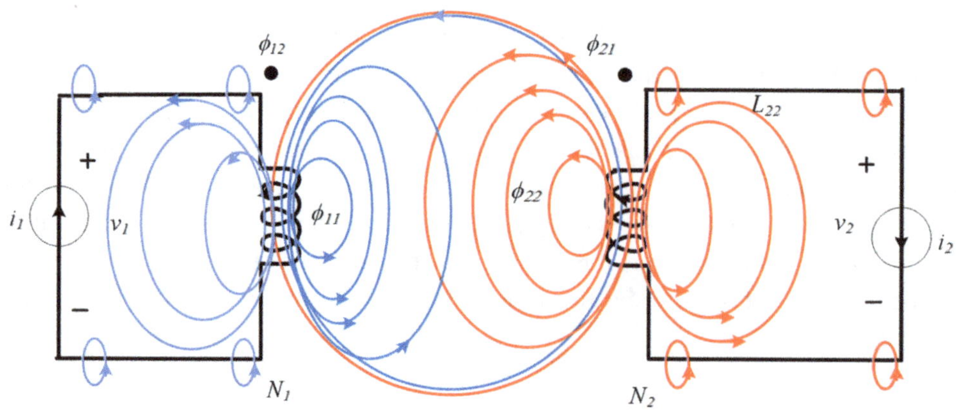

$$\begin{cases} v_1 = L_{11}\dfrac{di_1}{dt} + \left(L_{12}\dfrac{di_1}{dt} - L_{21}\dfrac{d}{dt}\dfrac{i_2}{n}\right) \\ v_2 = -L_{12}\dfrac{di_2}{dt} - \left(L_{21}\dfrac{di_2}{dt} - nL_{21}\dfrac{di_1}{dt}\right) \end{cases} \tag{3}$$

where the inductances and turn ratio are

$$L_{11} = \frac{N_1\phi_1}{i_1},\ L_{12} = \frac{N_1\phi_{12}}{i_1},\ L_{21} = \frac{N_2\phi_{21}}{i_2},\ L_{22} = \frac{N_2\phi_2}{i_2} \tag{4}$$

$$n = \frac{N_2}{N_1} \tag{5}$$

From this, a lumped-parameter circuit model can be drawn, as depicted in Figure 2.

Between the coupled coils there is now no leakage magnetic flux, therefore the co-sharing flux ϕ_m can be defined as:

$$\phi_m = \phi_{12} - \phi_{21} \tag{6}$$

In order to be usable from the technical point of view, the magnetic model depicted in Figure 2 must be updated with a suitable equivalent circuit that removes the variables of magnetic flux. This can be done using one of three different models, as shown in Figure 3:

Figure 2. Simplified lumped-parameter circuit model for the two coupled coils

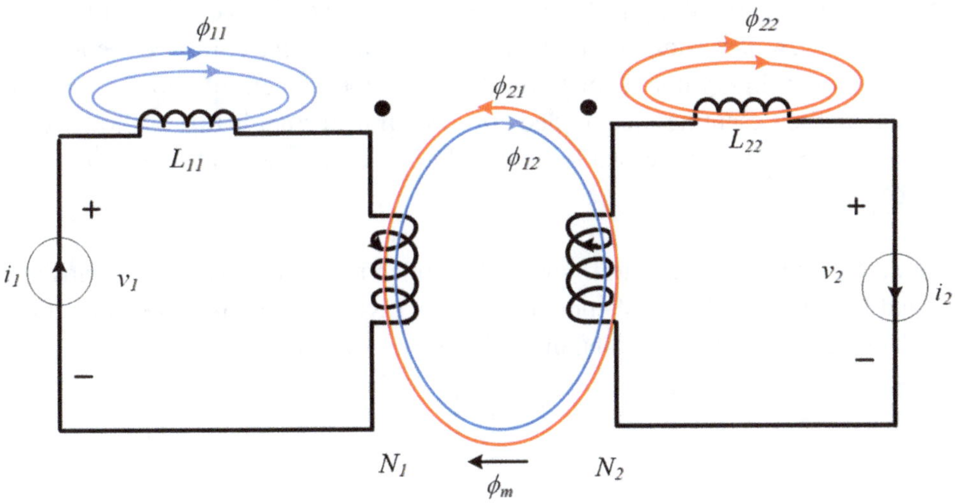

- **Transformer Model:** A first equivalent circuit model, depicted in Figure 3a, is the classic transformer, which is convenient for designing parameters when explicit circuit variables are sought for (Rim, 2018). The relationship between L_{12} and L_{21} can be found starting from (5) and (6), in the following way

$$L_{12} = \frac{N_1^2}{R_m} \tag{7}$$

$$L_{21} = \frac{N_2^2}{R_{...}} \tag{8}$$

Where R_m is the magnetic resistance of the coumpling coil. Thus

$$L_{21} = n^2 L_{12} = n^2 L_m \tag{9}$$

Which, applied to (3), yields

$$\begin{cases} v_1 = L_1 \dfrac{di_1}{dt} + v_m \\ v_2 = nv_m - L_{22} \dfrac{di_2}{dt} \end{cases} \tag{10}$$

where $v_m = i_m L_m$ and $i_m = i_1 - ni_2$.

The equivalent circuit of this system is indeed the classic transformer. In any case, this model presents some major differences with respect to the typical circuit representations of transformers: first of all, leakage inductances L_{11} and L_{22} are usually around 1% of the magnetizing inductance in conventional power transformers, while in WPT applications these values typically range between 100% and 10000%. Furthermore, when the resonant compensation circuit is attached to the coils, the secondary current can be quadratic against the primary one, something that does not happen in a conventional transformer. The main advantage of representing the coupled coils as a transformer is that it allows to explicitly determine the physical properties of the two leakage inductances, as well as the magnetizing inductance and ideal transformer.

- **M-Model:** A second model that has found wide application in wireless power transfer is the mutually coupled inductors model, or M-model, shown in Figure 3b. In this case, only three parameters are used to describe the coupled coils, in the following way

$$\begin{cases} v_1 = L_1 \frac{di_1}{dt} - M \frac{di_2}{dt} \\ v_2 = M \frac{di_1}{dt} - L_2 \frac{di_2}{dt} \end{cases} \tag{11}$$

where $M = nL_m$, $L_1 = L_{11} + L_m$, $L_2 = L_{22} + n^2 L_m$, all of which can be easily obtained from the transformer model formulation. While this representation is simpler than the transformer model, the presence of controlled voltage sources does not allow direct circuit-based analysis. In this model, L_1 and L_2 are the inductances seen from the primary and secondary sides, respectively, while M is the mutual inductance. For this reason, in (Rim & Mi, 2017) this formulation is defined as an implicit transformer model, where the physical parameters do not explicitly appear. A value for the mutual inductance M can be obtained analytically considering system geometry using methods based on Neumann's formula, as detailed in (Raju et al., 2014). In many practical cases, however, solving the problem of determining M involves the use of the Finite Element Method (FEM) or other similar 3D simulation methods, which are very reliable although they require an appropriate amount of computing power. Computational methods used in the context of IPT will be better described in "*Computational models for inductive power transfer*". Going back to the *M*-model, a relevant observation can be made about how the transferred power is handled by this analytic tool. Considering the instantaneous input and output powers:

$$\begin{cases} p_1(t) = -i_1 \cdot M \frac{di_2}{dt} \\ p_2(t) = i_2 \cdot M \frac{di_1}{dt} \end{cases} \tag{12}$$

It can be noted that they are different: for arbitrary currents, the instantaneous power is not preserved between the two controlled voltage sources. However, considering sinusoidal, single-frequency, steady-state inputs, and the static power in the phasor domain can be written as:

$$\begin{cases} P_1 = -Re\left(I_1^* \cdot Mj\omega_2 I_2\right) \\ P_2 = -Re\left(I_2^* \cdot Mj\omega_2 I_1\right) \end{cases} \tag{13}$$

and it can be proven that $P_1 = P_2$. Therefore, the power produced by a steady-state, sinusoidal source is transferred through the controlled voltage sources of the *M*-model. A measure of the amount of coupling between the coils can be obtained in the form of a very important and widely used parameter: the coupling factor *k*. This can be computed by multiplying the primary- and secondary-side open-circuit voltage gains:

$$k^2 = \left.\frac{V_2}{V_1}\right|_{I_2=0} \cdot \left.\frac{V_1}{V_2}\right|_{I_1=0} \tag{14}$$

which, after some manipulation, yields:

$$k = \frac{M}{\sqrt{L_1 L_2}} = \frac{nL_m}{\sqrt{\left(L_{11} + L_m\right)\left(L_{22} + n^2 L_m\right)}} \tag{15}$$

In inductive power transfer applications, usually $k \ll 1$, which makes M typically very small. The *M-model* finds application mainly in equation-based analysis. For purely circuit-based studies, the transformer model is preferred.

- **T-Shape Model:** A final equivalent circuit, depicted in Figure 3c, is the T-shape model or T-model for short. It can be derived with a simple rearrangement of the terms in (11), which gives

$$\begin{cases} v_1 = \left(L_1 - M\right)\dfrac{di_1}{dt} + M\dfrac{d}{dt}\left(i_1 - i_2\right) \\ v_2 = -\left(L_2 - M\right)\dfrac{di_2}{dt} + M\dfrac{d}{dt}\left(i_1 - i_2\right) \end{cases} \tag{16}$$

In this case, all the inductors are decoupled, and the resulting circuit is very simple. However, this representation does not take into account the galvanic separation between the two coils, although it is still electrically equivalent to the other two models detailed above. Furthermore, important circuit parameters such as the coupling factor and turn ratio do not appear explicitly in this model. Like the transformer model, the T-shape model finds application in circuit-based analyses.

All three models are valid both in steady-state and transient conditions, and they accept arbitrary current or voltage sources as well as loads.

The models presented so far are a great tool to analyze the behavior of two coupled coils, but a real WPT system based only on the circuits shown above would be close to useless, due to the presence of severe leakage flux effects deriving from the weak coupling between the coils. In practice, inductive power transfer systems need additional elements in order to compensate for these effects, and *LC* resonant circuits are the usual choice (Kalmar et al., 2015). A general representation of a complete ICPT system, including compensation, power supply and load is provided in Figures 4 and 5. The possible compensation circuit topologies can be classified by the number of compensating capacitors and inductors, the configuration of these elements, and the type of sources. The most common strategy – also known as basic compensation circuit, by way of its simplicity – makes use of only two capacitors, and is therefore also the least expensive way of obtaining resonant compensation.

Based on the two possible sources (voltage and current), and the four possible topologies for the capacitor elements, eight different compensation schemes can be obtained, as shown in Figures 4 and 5. In (Rim, 2018), three main conditions are identified in order to evaluate the practical applicability of each compensation topology. First, the output current or voltage should be independent from the nature of the load. Second, the compensating action should be k-independent, which means that the angular frequency at which the topology achieves maximum efficiency should not vary with k. Third, the condition of no magnetic coupling ($k = 0$) must be allowed.

Figure 3. Main equivalent circuit models for a pair of inductively coupled coils (a) Transformer model, (b) Mutual coupling model and (c) T-shape model

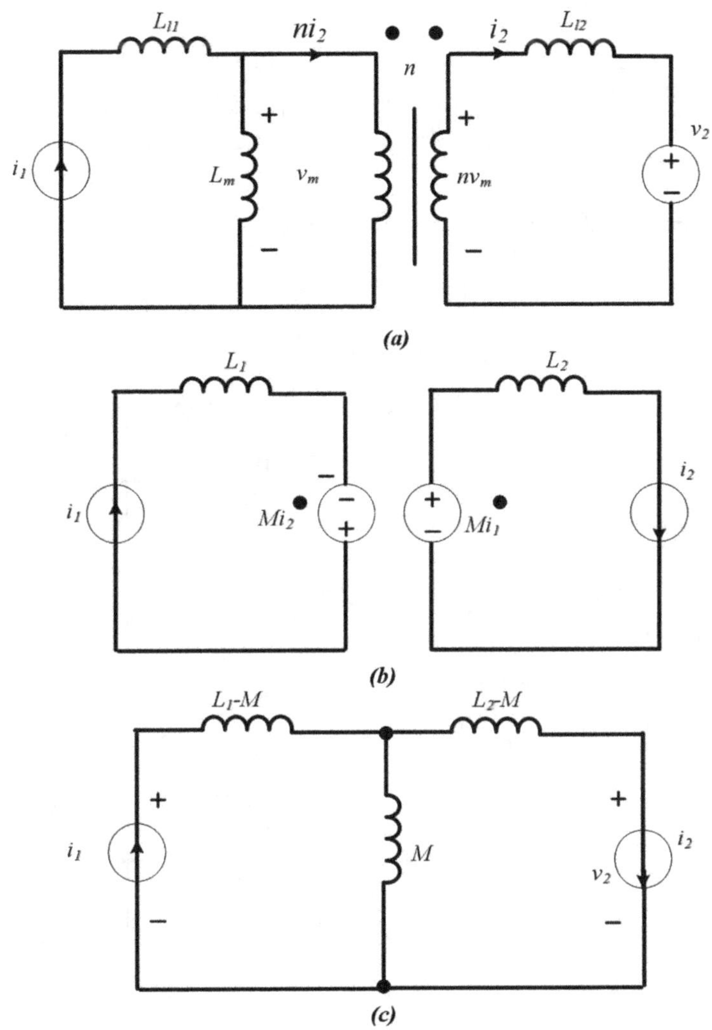

It can be verified that only I–SS and I–SP compensation strategies comply with all these three conditions simultaneously, and the substantial similarity between the two topologies – with respect to efficiency, power transfer characteristics and ratings of the reactive components – implies that the choice of one over the other depends only on practical design objectives. Specifically, the I–SS topology has a voltage-source-type output characteristic, which makes it more suitable when large output currents are needed, whereas the I-SP topology has a voltage-source-type characteristic, which makes it ideal for high-voltage applications. Current research (Pantic et al., 2011; Li Siqi et al., 2015; Li et al., 2015; Kan et al., 2017; Deng et al., 2017) is focused on designing and optimizing more advanced compensation circuits, which usually include more components and are capable of achieving higher efficiency and better power quality compared to the simple – yet effective – schemes detailed above.

Figure 4. (a) General configuration of an IPTS; Current-sources type primary series-secondary (b) series (I-SS) and (c) parallel (I-SP); Current-sources type primary parallel-secondary (d) series (I-PS) and (e) parallel (I-SP)

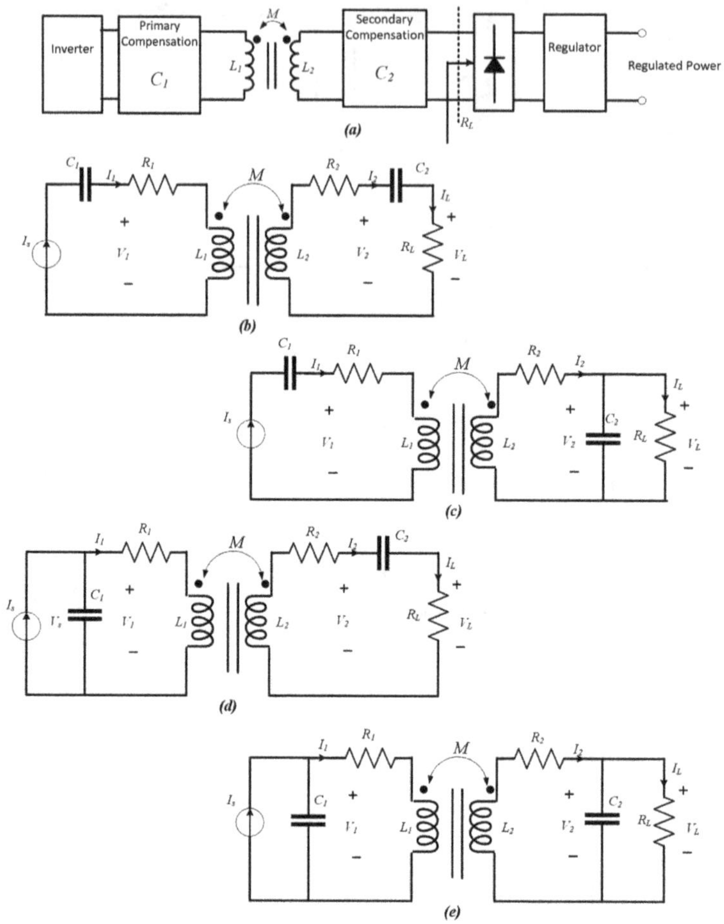

The coupled coils model, as discussed so far, has only been applied in the context of static charging. It is, in fact, quite simple to extend it to the dynamic case. A basic circuit topology for a generic dynamic WPT system modelled using the coupled coils representation where many couplings are highlighted: the main difference between the static and dynamic case is the presence of multiple coils on the transmitting side, which are individually coupled both with the secondary coil and among each other.

In (Buja et al., 2016), a simple model is given for a receiver coil moving along a fixed track where multiple transmitting coils are placed. Considering the system depicted in Figure 6a, a few assumptions are made. First, each track coil is supplied with the same sinusoidal current it. Furthermore, the distance *D* between the fixed coils is long enough to allow the pickup coil to be coupled with only one track coil at a time, and the constant speed *U* of the coil is such that during a period of the supply current the coil travels a small distance in comparison to *D*. Finally, it is assumed that the moving coil passes through the centres of the track coils. This last assumption is not really necessary, and the following procedure can be extended to a case where the pick-up coil moves along *x* with an offset along *y*. Figure 6a shows

Figure 5. Voltage-sources type primary series-secondary (a) series (V-SS) and (b) parallel (V-SP); Voltage-sources type primary parallel-secondary (c) series (V-PS) and (d) parallel (V-SP)

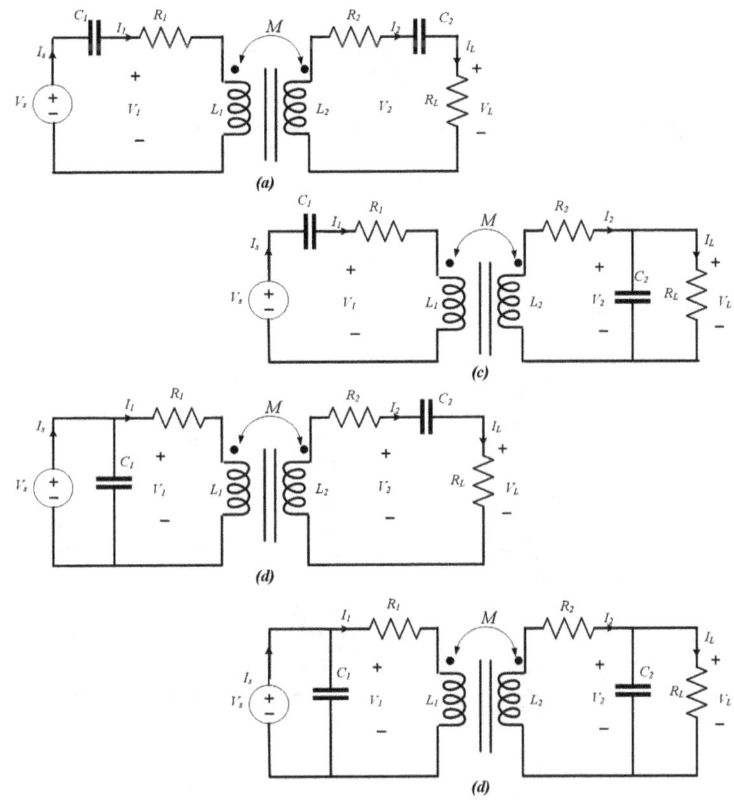

the equivalent circuit of the configuration described so far, modelled with the coupled coils method, where the pick-up coil is compensated with a series capacitor.

Let ϕ_a be the flux linkage that derives from track coil *a*, and ϕ_b the one that derives from track coil *b*. The voltage induced in the pick-up coil will have the expression:

$$v_p = -\frac{d\phi_a}{dt} - \frac{d\phi_b}{dt} \tag{17}$$

which can also be expressed as:

$$v_p = -\left[M_a\left(x\right) + M_b\left(x\right)\right]\frac{di_t}{dt} - \frac{d\left[M_a\left(x\right) + M_b\left(x\right)\right]}{dt} i_t \tag{18}$$

where $M_a(x)$ and $M_b(x)$ are the mutual inductances between the pick-up coil and the two respective track coils, which clearly are a function of *x*. The first term of (18) is completely analogous to what has been found in (11), and rep- resents the static component of the voltage. Instead, the second term is exclusive to dynamic systems, and is the motional voltage that depends specifically on the relative movement of the

Figure 6. Physical and electrical schematic configuration of a generic dynamic WPT system composed of several fixed track coils and a moving pick-up coil (a) Top view and (b) equivalent circuit

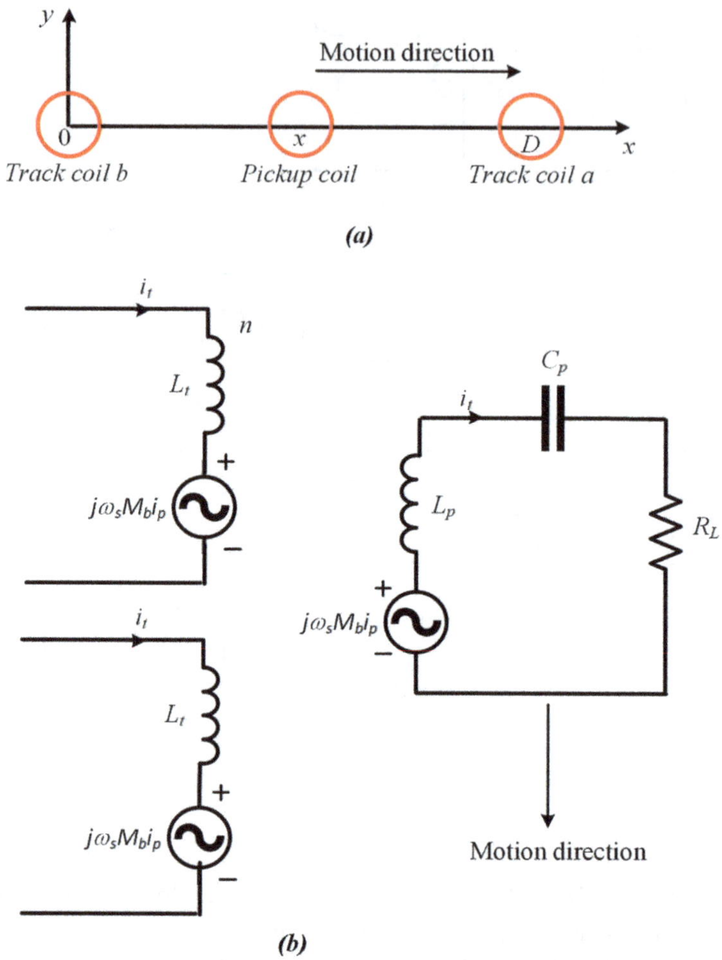

two coils. In practice, however, it can be demonstrated that the motional voltage is negligible compared to the static component (Buja et al., 2016).

By way of some additional assumptions, important considerations can be derived about the power and energy transfer in a dynamic WPT system. First of all, self-inductances are assumed constant with respect to variations in x, something that is not necessarily true in reality. Moreover, the resonating condition of the series L_pC_p branch is assumed to be independent from x as well. This means that the resonating branch can be modelled as a short circuit for any position of the pick-up coil. On the basis of what has been derived so far, the magnitude of the sinusoidal current in the pick-up coil can be expressed as:

$$I_p = \frac{V_I}{R_L} = \frac{\omega_s I_t \left[M_a\left(x\right) + M_b\left(x\right) \right]}{R_L} \tag{19}$$

where V_I is the magnitude of the static term in (18), and R_L is the load attached to the pick-up coil. The power transmitted to R_L can be written as:

$$P_L = \frac{1}{2} I_p^2 R_L = \frac{1}{2} \frac{\omega_s^2 I_p^2 \left[M_a(x) + M_b(x)\right]^2}{R_L} \tag{20}$$

which, integrated over the time T_D during which the coil covers a distance equal to D, gives the energy transmitted to the load during that period:

$$E = \int_0^{T_D} P_L(t)\, dt \tag{21}$$

since the coil moves at constant speed, the following relationship holds:

$$dt = \frac{1}{U} dx \tag{22}$$

and, after some manipulation (Buja et al., 2016), the expression for E becomes:

$$E = \frac{1}{U} \frac{\omega_s^2 I_p^2}{2R_L} \int_0^{D} \left[M_a(x) + M_b(x)\right]^2 dx \tag{23}$$

which establishes that the energy transferred from the track to the pick-up coil is inversely proportional to the speed. However, the number of track coils that are connected to the pick-up coil during T_D increases with the speed. Conclusively, it can be said that E does not depend on the coil speed. The coupled coils model has therefore proven to be a reliable, versatile and simple analytic tool which allows to analyze some of the crucial design aspects in IPT systems, both in the static and dynamic cases.

Gyrator Model

As detailed in "*Compensation Strategies*", compensation circuits used in inductive power transfer systems typically include both capacitors and inductors to increase load power through the compensation of the reactive power (Figure 7). Every compensation circuit has its own specific characteristics at a given resonance frequency, such as power factor, output resistance and current/voltage gain.

It is obvious how important it is to choose the right circuit for each IPT application, thus the preliminary analysis at the beginning of the design process must be carried out with the aim to evaluate multiple strategies and select the best fitting one. Usually, this kind of analyses are performed using analytic approaches that include matrix manipulation, from which the optimal circuit characteristics can be derived. Despite this approaches being extremely general, unified and definite, they can become quite complicated, and often require a lot of time to be solved. This is especially true in case the compensa-

Figure 7. General scheme for an IPT system, including compensation

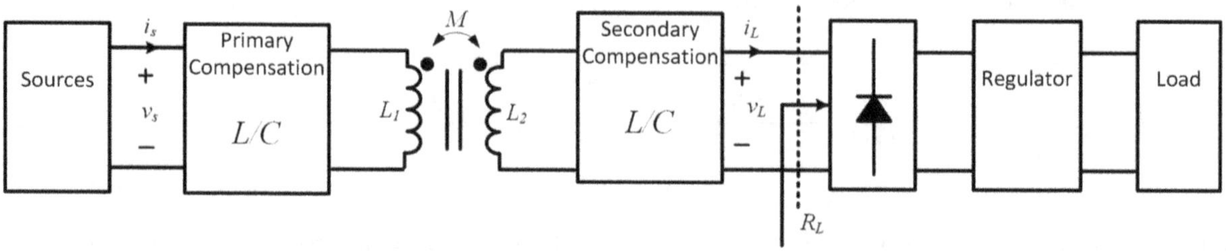

tion circuit includes more than four reactive components (Rim, 2018). An alternative graphic approach, based on the interpretation of all IPT circuits as intrinsic gyrators, has been proposed in (Sohn et al., 2015), and it allows to remarkably simplify the study of compensation circuit.

The gyrator, first introduced in (Tellegen, 1948) as one of the basic circuit elements, is complementary to the ideal transformer, and its principal property is that it inverts the voltage-current characteristic of an electrical component or an electrical network. Through the gyrator approach, it can study compensation circuits not only at resonance frequency, but also at any other frequency, due to its simplicity and systematic quality: it is simple because it does not require the solution of possibly complicated systems of equations, and it is systematic because it can be applied in the same way to different compensation schemes (Sohn et al., 2015).

In order to understand the advantages of the gyrator approach, the main properties of the basic circuit element have to be explained first. A gyrator (Figure 7) is a linear, time-invariant, and lossless two-port network that can convert the current of one port to the current of another port (Rim & Mi, 2017). The voltage-current characteristic is the following:

$$\begin{cases} I_2 = G \cdot V_1 \\ I_1 = G^* \cdot V_2 \end{cases} \tag{24}$$

where the apex * indicates the complex conjugate, and G and G^* are respectively the forward and reverse gyrator gains. From the voltage-current characteristic, a first important property can be derived as follows:

$$P_1 \equiv Re\left(V_1 I_1^*\right) = Re\left(G V_1 V_2^*\right) = Re\left[\left(G^* V_1^* V_2^{**}\right)\right] = Re\left(G^* V_1^* V_2\right) = Re\left(V_2 I_2^*\right) \equiv P_2 \tag{25}$$

which means the real power on the primary side is equal to the real power on the secondary side. As for the reactive power

$$Q_1 \equiv Im\left(V_1 I_1^*\right) = Im\left(G V_1 V_2^*\right) = Im\left[\left(G^* V_1^* V_2^{**}\right)\right] = -Im\left(G^* V_1^* V_2\right) = -Im\left(V_2 I_2^*\right) \equiv -Q_2 \tag{26}$$

which means the reactive power on the primary side is the opposite of the one at the secondary side. A gyrator can be realized using only passive components, it consists of impedances Z and $-Z$ and its equivalent circuit can be derived in both T-shape (Figure 8c) and Π-shape (Figure 8d) models.

The relationship between voltage and current in these circuits can be derived through easy calculations. Applying Kirchhoff's law to the previous networks, the following relation can be obtained:

$$\begin{cases} I_2 = \dfrac{1}{Z} V_1 \\ I_1 = -\dfrac{1}{Z} V_2 \end{cases} \tag{27}$$

which, combined with (24), gives that the sufficient condition for the circuits in Figures 8c and 8d to be gyrators is:

$$-\frac{1}{Z} = \left(\frac{1}{Z}\right)^* \leftrightarrow Re(Z) = 0 \tag{28}$$

which leads to the representation depicted in Figure 8b. In compensation circuits that resonate at frequency ω_0, purely imaginary impedances can be easily obtained using reactive components, such as:

Figure 8. (a) Circuit symbol for the gyrator and (b) containing only real impedences; (c) T-scape and (d) Π-schape equivalent circuit of a gyrator with only purely imaginary impedences

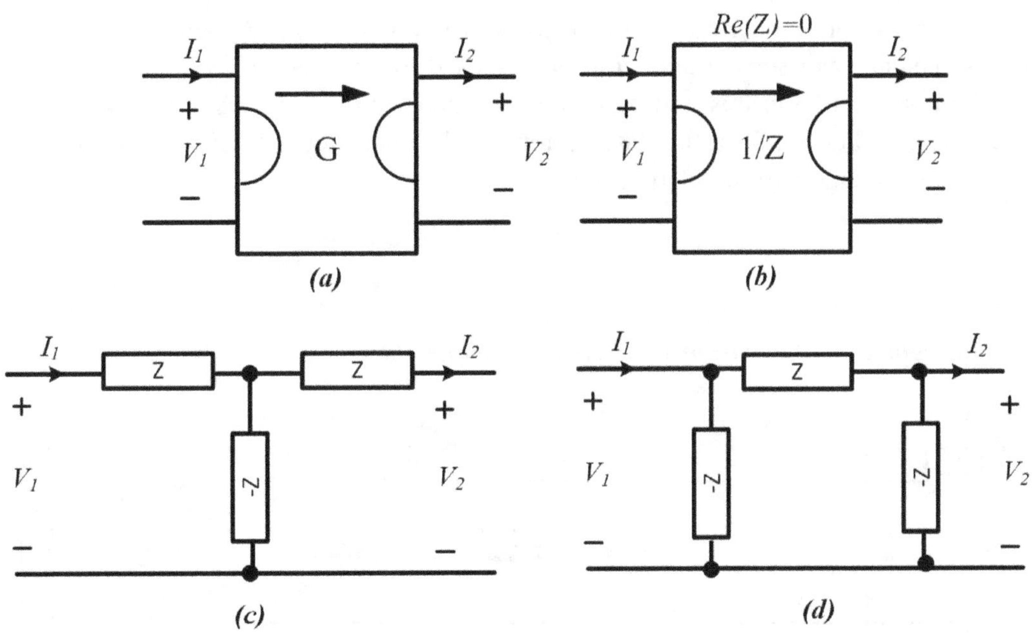

$$X_0 = \sqrt{\frac{L}{C}} = \omega_0 L = \frac{1}{\omega_0 C} \tag{29}$$

where ω_0 can be expressed using the Eq. (30):

$$\omega_0 = \frac{1}{\sqrt{LC}} \tag{30}$$

In the circuit depicted in Figure 9, L and C can be interchanged without loss of generality.

This resonant circuit can be found in many compensation schemes. In case the T- or Π-models are not explicitly expressed, some manipulation can be done by adding redundant series-series passive elements, which are effectively a short circuit, or redundant parallel-parallel elements, equivalent to an open circuit, in order to obtain the desired network model. An example of this method can be seen in the representation of a parallel-series resonant circuit in Figure 10.

It has been shown that symmetric T- and Π-model networks in resonance conditions can be reduced to an equivalent gyrator with imaginary gain. Due to this property, coupled inductors used for the exchange of power in IPTSs can be expressed as gyrators as well. Coupled inductors implicitly contain a gyrator in the T-model, in which the negative mutual inductance $-M$ can be considered as a capacitance in the following way:

$$\frac{1}{j\omega C} = -j\omega M \rightarrow C = \frac{1}{\omega^2 M} \tag{31}$$

leading to the equivalent circuits in Figure 11, which show the intrinsic nature of the coupled inductor's gyrator properties.

Now that the behavior as a gyrator of simple compensation circuits and coupled inductors has been highlighted, some characteristics of a gyrator that are used for the study of compensation circuits can be analyzed in more detail. For this study, a purely imaginary gyrator is considered (Sohn et al., 2015) with a gain $G = jX$. Starting from this assumption, all the characteristics of interest can be obtained from the following voltage-current relationships:

Figure 9. Implementation of a gyrator with reactive components

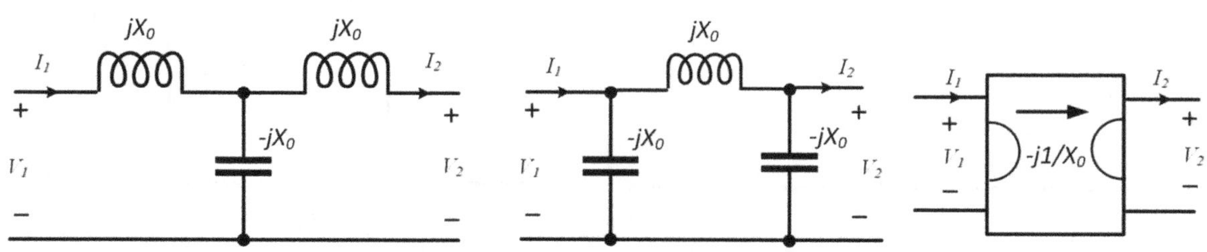

Figure 10. Representation of a parallel-series resonant circuit through gyrator

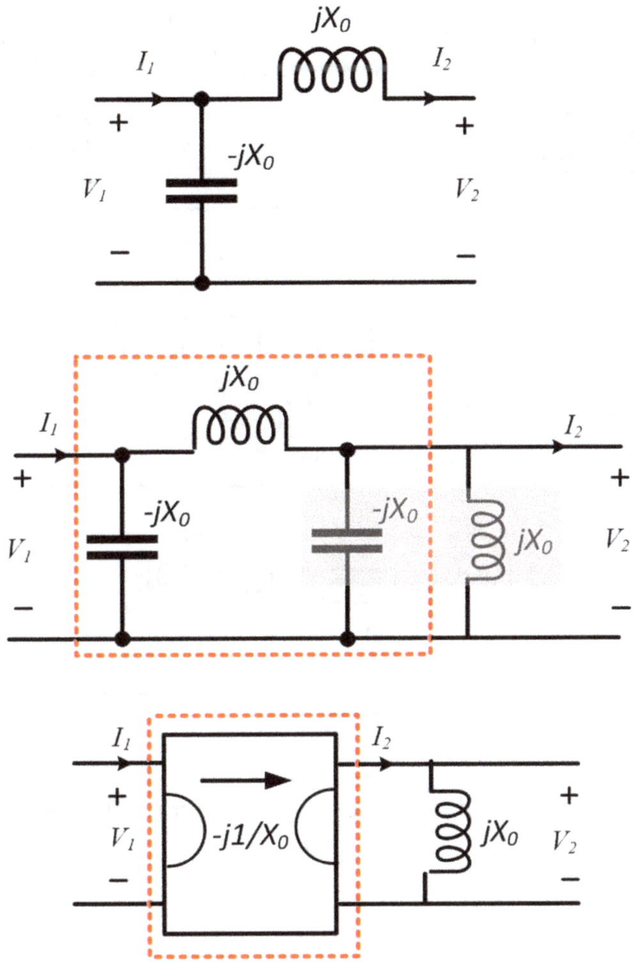

$$\begin{cases} I_2 = jX * V_1 \\ I_1 = -jX * V_2 \end{cases} \tag{32}$$

In case the second port of the gyrator terminates with an impedance Z_2, the inverted Z_2 at the primary port Z_1 can be derived from the voltage-current characteristic:

$$Z_1 = \frac{V_1}{I_1} = \frac{I_2/jX}{-jXV_2} = \frac{I_2}{X^2 V_2} = \frac{1}{X^2 Z_2} \tag{33}$$

This rule is called *Terminated Impedance Inversion Rule*, and four important features can by found from its applications:

Figure 11. Representation of coupled inductors with a gyrator

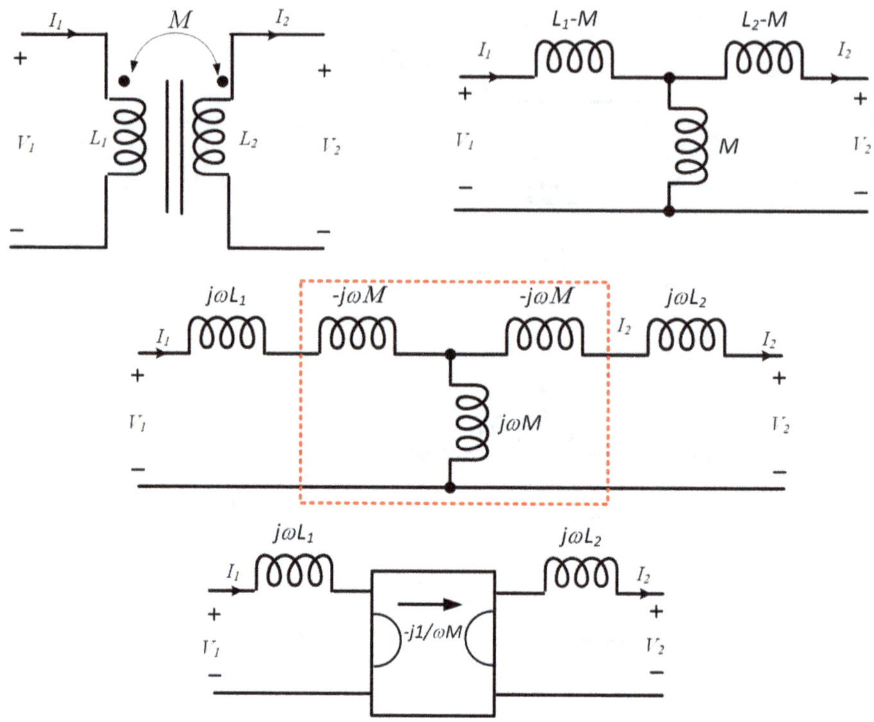

1. A short circuit at the secondary side becomes an open circuit at the primary side, and vice-versa

$$\begin{cases} \left|Z_1\right|_{\left|Z_2\right|=0} = \infty \\ \left|Z_1\right|_{\left|Z_2\right|=0} = 0 \end{cases} \tag{34}$$

2. The magnitude of the impedance can be described in the following way

$$\left|Z_1\right| = \left|Z_2\right| \cdot \frac{1}{X^2 \left|Z_2\right|^2} \tag{35}$$

3. The power factor of impedance Z_1 is equal to the power factor of the impedance Z_2

$$cos\left(Z_1\right) = \frac{R_1}{\sqrt{R_1^2 + X_1^2}} = \frac{R_2}{\sqrt{R_2^2 + X_2^2}} = cos\left(< Z_2\right) \tag{36}$$

4. An inductive impedance Z_2 becomes a capacitive impedance Z_1, and vice-versa

$$\begin{cases} Im\left(Z_1\right) > 0 \rightarrow Im\left(Z_1\right) < 0 \\ Im\left(Z_2\right)\left\langle 0 \rightarrow Im\left(Z_1\right)\right\rangle 0 \end{cases} \tag{37}$$

The rule detailed in (37) can be extended at unterminated impedance at the secondary side without affecting the external voltage-current characteristic, as it has been shown in (Rim and Mi, 2017). This new extended rule can therefore be called *Unterminated Impedance Inversion Rule*.

It will now be detailed how a series connections of multiple gyrators can be merged into a single one. If *n* gyrators are connected in series, the behavior of the equivalent gyrator can change depending on whether *n* is even or odd. For an even value of n, the conversion between voltage and current occurs an even amount of times, and the gyrators can be merged into a single transformer whose turn ratio is a real number and is equal to:

$$N = \prod_{i=1}^{n/2} \left(-\frac{X_{2i-1}}{X_{2i}} \right) \tag{38}$$

In the same way, for an odd value of *n*, voltages and currents are converted an odd number of times and the series of gyrators can be merged into a single gyrator with the following purely imaginary forward gain:

$$jX = jX_n \prod_{i=1}^{n-1/2} \left(-\frac{X_{2i-1}}{X_{2i}} \right) \tag{39}$$

So far, a general characterization of the gyrator has been provided, and it has been demonstrated that both compensation circuits and coupled inductors can be described as gyrators. In order to clarify the usefulness of this innovative method, an example of its application to a simple circuit is provided below.

The circuit chosen for this example is composed by two compensating capacitors and coupled inductors, and is supplied by a current source (*I–SP* model). As it can be seen in Figure 12, the coupled inductors can be replaced by their gyrator equivalent.

The analysis is carried out in two steps: first, all the elements of the compensation circuit are transformed into their gyrator equivalents, while the source and load remain unchanged. Then, source and load are analyzed using the properties of gyrators detailed above. After that, the series-series resonant circuit L_1C_1 becomes a short circuit while the parallel- series circuit L_2C_2 at the secondary has to be converted in its gyrator model, as shown in the top part of Figure 13.

The power factor can be easily calculated through (36), which yields:

$$PF = \frac{R_L}{\sqrt{R_L^2 + \left(\frac{1}{\omega_0 C_2}\right)^2}} \tag{40}$$

Figure 12. I-SP circuit model and Its equivalent with coupled inductor replaced by a gyrator

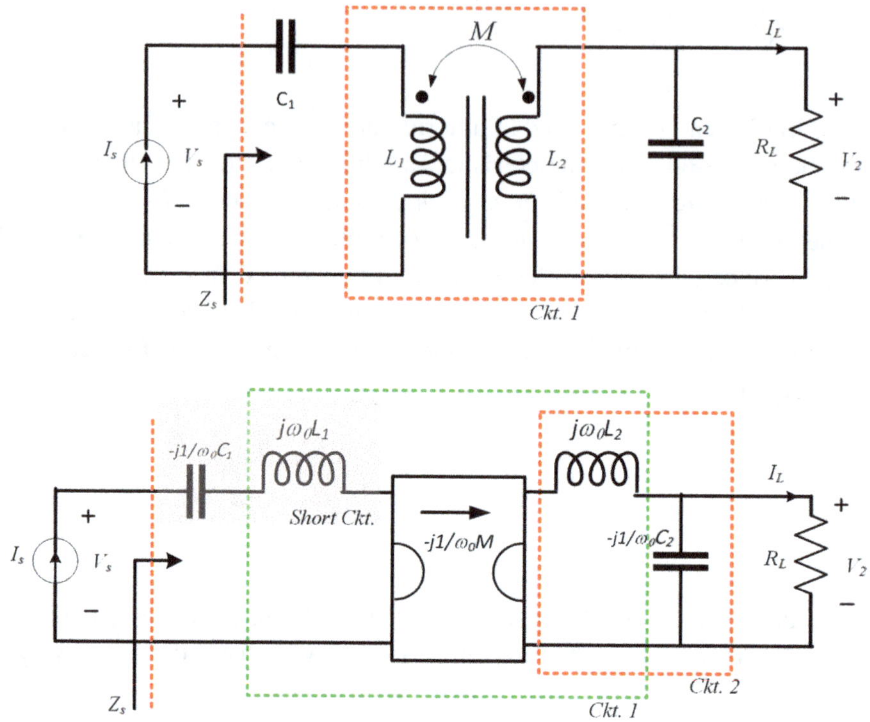

Figure 13. Further simplification of an I-SP compensation circuit through gyrator analysis

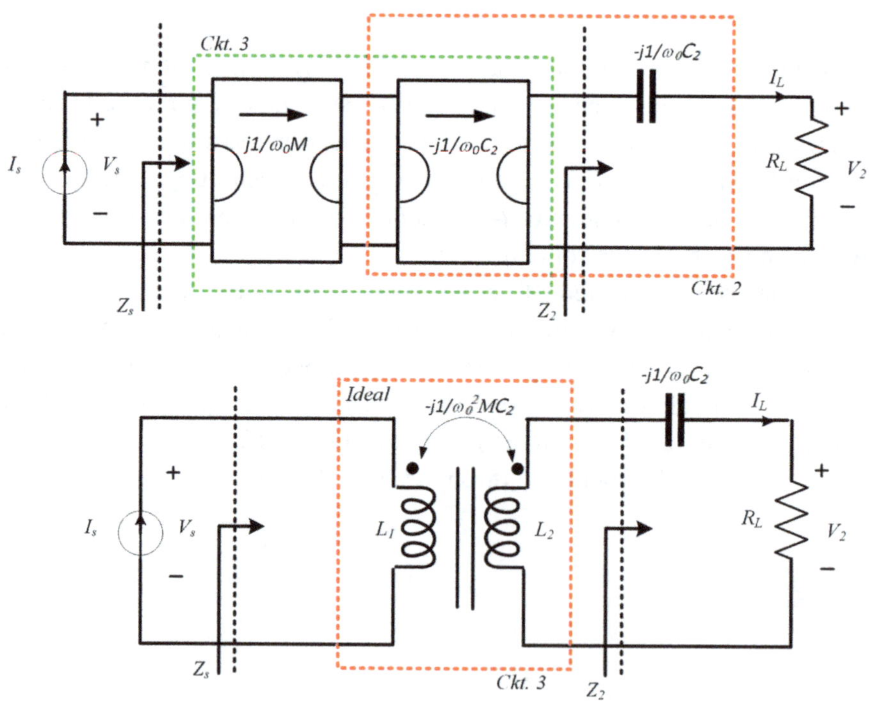

This relation provides us with the parameters that determine the reactive power rating at the source, and it was derived in a simple way by exploiting the properties of the gyrator. It should also be noted that the final transformer in Figure 13 was obtained by merging a series connection of an even number of gyrators so that the capacitive impedance Z_2 can be inverted into the capacitive impedance Z_s at the primary side. Furthermore, due to its transformer behavior, in case the load is short-circuited, the source voltage V_s becomes zero, while for $R_L = \infty$ the source voltage becomes infinity, and it results as an unsafe operation. All the characteristics of an I–SP compensation circuit obtained through gyrator analysis agree with those found in the relevant literature. The transformation of any compensation circuit into a gyrator equivalent can be carried out exactly as it was done in this example. This graphical approach can be used to design circuits with the specific characteristics required.

Magnetic Mirror Model

The performance of Inductive Power Transfer Systems (IPTS) is strictly influenced by the behavior of the magnetic flux in the core and the magnetic field distribution of the coils, therefore an analytic method for studying the behavior of magnetic fields is needed in order to design efficient coils. Usually, the analysis of the coils in inductive power transfer systems is carried out using simulation tools based on the FEM, and no equation-based models for the study of magnetic field distribution and magnetic flux density of a coil are used. Consequently, no information about which parameters can be manipulated to optimize the IPT system is explicitly available to the researchers.

For this type of analysis, an Improved Magnetic Mirror Model (IM^3) was introduced in (Lee et al., 2013), where the complexity of the system is reduced with the assumption of infinite core permeability. This model is similar to the image charge model used for the calculation of the electric field of a perfect conductor (Ulaby & Ravaioli, 2015), but in this case, the magnetic flux density over a core plate and its points of maximum are the objectives of the study. All the results obtainable by IM^3 analysis can be verified using FEM simulations to prove the validity of this method. In the following paragraphs, both mono and dual coils are analysed with this equation-based tool, with the aim to obtain useful expressions for IPT system design. The results of FEM simulations based on ANSYS® Maxwell® carried out in (Sohn et al., 2015) will be compared to the analytic results in order to verify the quality of the proposed model.

Mono coil systems with a pick-up coil are mainly used in OLEVs, because they require a long track inductor (of which the length l_0 is assumed infinite), which has a width ω_0 typically lower than 1 m. As it can be seen in (Lee et al., 2013), the width of the pick-up coil is slightly larger than the one of the mono coil in order to provide better lateral tolerance with respect to horizontal displacements.

Due to the complexity of the study, the system will be analysed in various steps, starting from the analysis of the mono coil alone. In (Rim & Mi, 2017) shows a cross-section of the mono coil with its open core plate. The magnetic flux is distorted because of the permeability of the core, and it is difficult to calculate its expression analytically.

Due to the high permeability of the core, the magnetic field is perpendicular to the core surface, and, consequently, the problem can be solved with a magnetic mirror model in which the mirror current generates a magnetic field that compensates the one produced by the source current on the x-axis. In (Lee et al., 2013) shows the relation between the core permeability and the magnetic flux density vector angle.

As it can be seen, the flux becomes almost perpendicular for a permeability that exceeds 1000 H m^{-1}. Considering that the materials used in practice usually have a permeability ranging between 2000 and 5000 H m^{-1}, this assumption is applicable for most systems. The magnitude of the magnetic flux density can be expressed as:

$$B_1 \equiv \left|\boldsymbol{B}_1\right| = \frac{\mu_0 I_s}{2\pi}\sqrt{\left[\frac{y-d_s}{r_s^2}+\gamma_1\frac{y+d_m}{r_m^2}\right]^2+\left[\frac{x}{r_s^2}+\gamma_1\frac{x}{r_m^2}\right]^2} \text{ for } x<\frac{\omega_0}{2}\; 0\le y \tag{41}$$

where

$$r_s = \sqrt{x^2+\left(y-d_s\right)^2} \tag{42}$$

$$r_m = \sqrt{x^2+\left(y-d_m\right)^2} \tag{43}$$

and where γ_1 is the ratio of the mirror current and the source current, d_m is the vertical distance between the mirror current and the core plate, and d_s is the vertical distance between the source current and the core plate. By using FEM simulations, it can be derived that the best fitting values for the previous equations are

$$d_m = d_s \tag{44}$$

$$\gamma_1 \equiv \frac{I_m}{I_s} \cong \left(1-e^{-\frac{\omega_0}{y\alpha_1}}\right) \tag{45}$$

where $\alpha_1 \cong 1.7$ is the curve-fitting coefficient of the mirror current model for the mono coil that can be experimentally obtained from the simulations. The obtained relation agrees with the boundary conditions, because for $y = 0$ the ratio between currents is $\gamma_1 = 1$ for any measurement point along the x -axis. Boundary conditions on the surface can also explain the relationship between the distance of source and mirror currents from the core plate: a consequence of the high permeability of the core plate is that for $y = 0$ the magnetic flux is perpendicular to the core surface, thus $d_m = d_s$ regardless of any other parameter. From the experimental relation, an important feature of γ_1 is highlighted: it does not depend on the distance d, but only on the measurement point y and on the core plate width ω_0. Therefore, the proposed model is different from the ideal magnetic mirror model, where the mirror current does not depend on the observation point. As reported in (Rim and Mi, 2017), this model is accurate for $\left|x\right| < \frac{\omega_0}{4}$ within an error of 5%. However, at the edge of the core, model deviates from the real characteristic with a considerable error. In any case, this is not a practical issue, since the coil is typically designed to be placed at the centre of the plate, where the IM[3] can reasonably approximate the ideal conditions. Another important feature of the proposed IM[3] is the study of the points of maximum of the magnetic flux in the core. This kind of analysis is important in order to keep the magnetic flux below a certain limit and prevent local saturation that can reduce the amount of transferred power. The calculation of these points is not hard, but it can require time, because these local maximum points can be either inside the core or on its surface. The equation of the magnetic flux density on the surface of the core plate can be easily found as follows:

$$B_{1,on} = B_1\Big|_{y=0} = \frac{\mu_0 I_s}{\pi}\frac{x}{d^2 + x^2} \text{ for } \left|x\right| < \frac{w_0}{2} \tag{46}$$

The maximum magnetic flux density can be obtained by setting its derivative with respect to *x* to zero:

$$\frac{\partial B_{1,on}}{\partial x}\Bigg|_{x=x_{1m}} = 0 \rightarrow B_{1m,on} = \frac{\mu_0 I_s}{\pi} \text{ for } x_{1m} = \pm d \tag{47}$$

The local saturation points are in $x = \pm d$, and the saturation increases its value as the distance *d* decreases. The calculation of the magnetic flux density inside the core plate is more complicated, but nonetheless it can be carried out analytically in the following way:

$$B_{1,in} \equiv B_1\Big|_{-d_c<y<0} = \frac{1}{l_0 d_c}\int_x^{\infty} l_0 B_{1,on}\,dx = \frac{\mu_0 I_s}{2\pi d_c}\ln\frac{d^2 + {w_0^2}/{4}}{d^2 + x^2} \text{ for } \left|x\right| < \frac{w_0}{2} \tag{48}$$

In the same way as for the previous case, the local saturations in the core are found as follows:

$$B_{1m,in} \equiv B_1\Big|_{-d_c<y<0} = \frac{\mu_0 I_s}{2\pi d_c}\ln\left(1 + \frac{w_0^2}{4d^2}\right) \tag{49}$$

The local saturation inside the core plate occurs just below the source current, which is in agreement with the FEM simulation results.

As it has been derived, local saturation can occur either inside or on the surface of the core plate, so both the magnetic flux densities obtained previously must satisfy the allowable limit. The main parameter that has to be manipulated in order to obtain the desired value of the magnetic flux density is the thickness of the core plate *d*. Now that the mono coil has been studied individually, the IM[3] analysis can be expanded to the study of a mono coil magnetically coupled to a pick-up coil. The source current between the two core plates is infinitely reflected by each side of the plate with alternating polarity, and the same for its mirror current, which is required in order to obtain a horizontal magnetic field for each plate equal to zero. This behavior is shown in (Lee et al., 2013), where it is important to note that the value of all the mirror currents is the same, because, as it has been said before, the magnitude of the mirror current does not depend on the distance from the plate but only from the observation point.

To determine the magnetic flux density expression between the core plates of the gap *h*, we have to assume a certain number of reflections. In this case, $2n+1$ reflections are assumed, and the resulting magnetic flux density can be obtained as:

$$B_3 \equiv \left|\boldsymbol{B}_3\right| = \left|\frac{\mu_0 I_s}{2\pi}\sum_{k=-n}^{+n}\left[\gamma_{1,2k}\frac{\alpha_x\left(y-d+2hk\right)-\alpha_y x}{x^2+\left(y-d+2hk\right)^2}+\gamma_{1,2k+1}\frac{\alpha_x\left(y+d+2hk\right)-\alpha_y x}{x^2+\left(y+d+2hk\right)^2}\right]\right|$$

$$\text{for } \left|x\right| < \frac{\omega_0}{2},\ 0 \le y \le h \tag{50}$$

The expression of $\gamma_{1,k}$ can be obtained by successive application of IM[3] of an individual mono coil to this two-plate case, in the following way:

$$\gamma_{1,0} = 1$$

$$\gamma_{1,1} = \gamma_1\left(y\right) = \left(1 - e^{-\frac{\omega_0}{\alpha_1 y}}\right),\ \gamma_{1,-1} = \gamma_1\left(h-y\right)$$

...

$$\gamma_{1,k} = \gamma_1^{k-1}\left(h\right)\gamma_1\left(y\right),\ \gamma_{1,-k} = \gamma_1^{k-1}\left(h\right)\gamma_1\left(h-y\right),\ k = 1,2,3,\ldots,2n+1 \tag{51}$$

where γ_1, 0 represents the source current while γ_1, 1 and γ_1, −1 represent respectively the first downward and the first upward mirror currents.

The validity of the magnetic flux density expression found above can be easily checked by comparing it with FEM simulations, as shown in (Rim & Mi, 2017). The value of the magnetic flux density is zero because for $y = h$ the mirror current and the source current are identical, as it can be seen in the previous relation, and they compensate each other. Another important result that has to be highlighted is that the discrepancy between IM[3] and FEM simulation results decreases as *n* increases.

The proposed IM[3] can be applied to a dual coil in the same way it has been applied to the mono coil, so in this Section all the calculations made for the mono coil will be repeated to the dual coil.

As in the previous case, the analysis will be done gradually starting from an individual dual coil system. In this case, there are two source currents in opposite direction, therefore, the IM[3] can be modelled as two mirror currents whose distance from the core plate is the same as for the source currents. Also for the dual coil, the magnetic flux density expression can be obtained by the application of the Ampere's law

$$B_2 \equiv \left|\boldsymbol{B}_2\right| = \frac{\mu_0 I_s}{2\pi}\left[\left(-\frac{y-d}{r_{s1}^2}+\frac{y-d}{r_{s2}^2}\right)\alpha_x + \left(\frac{x+\omega_s}{r_{s1}^2}-\frac{x-\omega_s}{r_{s2}^2}\right)\alpha_y\right] +$$
$$\frac{\mu_0 I_m}{2\pi}\left[\left(-\frac{y-d}{r_{m1}^2}+\frac{y-d}{r_{m2}^2}\right)\alpha_x + \left(\frac{x+\omega_s}{r_{m1}^2}-\frac{x-\omega_s}{r_{m2}^2}\right)\alpha_y\right]$$

$$\text{for } \left|x\right| < \frac{\omega_0}{2},\ 0 \le y \tag{52}$$

where r_{s1}^2, r_{s2}^2, r_{m1}^2, r_{m2}^2 are the distances between the measurement point and the centre of each current source. Just as for the mono coil case, the best values that fit for the IM³ are derived from the FEM simulation results in an experimental way, and the condition obtained is the following:

$$\gamma_2 \equiv \frac{I_m}{I_s} \cong \left(1 - e^{-\frac{w_0+2w_s}{\alpha_2 y}}\right) \text{with } \alpha_2 \cong 2.4 \tag{53}$$

where, similarly to what happened in the mono coil case, α_2 is the curve-fitting coefficient of the mirror current model for the dual coil that can be experimentally obtained from the simulations. An important result obtained from these comparisons is that, just as for the mono coil, the ideal situation $\gamma_2 = 1$ can be achieved for a high value of w_0, and that the IM³ approximates the simulations for the region inside the core plate, that is

$$\left|x\right| < \frac{\omega_0}{2}.$$

However, the curve fitting parameter differs from the one of the mono coil case, so this confirms that the dual coil is a completely different case compared to the mono coil, and it cannot be modelled as the combination of two mono coils. The calculation of the local saturation points proceeds in the same way as for the mono coil: again, it can be found that $\gamma_2 = 1$ for $y = 0$, therefore the magnetic flux density on the surface of the core plate can be derived as:

$$B_{2,on} \equiv B_2\big|_{y=0} = \frac{\mu_0 I_s}{\pi}\left[\frac{w_s + x}{d^2 + \left(w_s + x\right)^2} + \frac{w_s - x}{d^2 + \left(w_s - x\right)^2}\right] \text{for } \left|x\right| < \frac{\omega_0}{2} \tag{54}$$

The points of maximum for the magnetic flux are then calculated by setting its derivative with respect to x to zero:

$$\left.\frac{\partial B_{2,on}}{\partial x}\right|_{x=x_{2m}} = 0 \tag{55}$$

which yields

$$x_{2m} = \pm\left(w_s - d\right) \tag{56}$$

This result is similar to the one obtained for the mono coil, if the distance d is much less than the distance from the two current sources w_s. For the magnetic flux density inside the core plate, an important simplification can be achieved if the symmetry around $x = 0$ is considered:

$$B_{2,in} \equiv B_2\Big|_{-d_c<y<0} = \frac{1}{l_0 d_c}\int_x^{\infty} l_0 B_{2,on}\,dx = \frac{\mu_0 I_s}{2\pi d_c}\ln\left(\frac{d^2+\left(w_s+x\right)^2}{d^2+\left(w_s-x\right)^2}\right) \text{for } 0 \le \left|x\right| \le \frac{w_0}{2} \tag{57}$$

Again, the points of maximum are obtained by setting the derivative to zero, and the result is the following:

$$\left.\frac{\partial B_{2,on}}{\partial x}\right|_{x=x_{3m}} = 0 \rightarrow x_{3m} \cong w_s \text{ for } d \ll w_s \tag{58}$$

Therefore, like in the mono coil case, the point of maximum saturation inside the core plate is just below the source currents (Rim & Mi, 2017). The comparison between the IM[3] and the FEM simulation results, and it can be seen that the discrepancy between the two is negligible. The final study of the dual coil coupled with a pick-up coil can now be done with the same method used for the mono coil (Rim & Mi, 2017).

The magnetic flux density in the gap h can be determined from Ampere's law, as in the previous case:

$$B_4 \equiv \left|\boldsymbol{B}_4\right| = \frac{\mu_0 I_s}{2\pi}\sum_{k=-n}^{+n}\gamma_{2,2k}\left[\left(-\frac{1}{\gamma_{s1,k}^2}+\frac{1}{\gamma_{s2,k}^2}\right)\left(y-d+2hk\right)\alpha_x+\left(\frac{x+w_s}{\gamma_{s1,k}^2}-\frac{x-w_s}{\gamma_{s1,k}^2}\right)\alpha_y\right]+$$
$$\frac{\mu_0 I_m}{2\pi}\sum_{k=-n}^{+n}\gamma_{2,2k+1}\left[\left(-\frac{1}{\gamma_{m1,k}^2}+\frac{1}{\gamma_{m2,k}^2}\right)\left(y-d+2hk\right)\alpha_x+\left(\frac{x+w_s}{r_{m1,k}^2}-\frac{x-w_s}{r_{m2,k}^2}\right)\alpha_y\right]$$

$$\text{for } \left|x\right| < \frac{\omega_0}{2},\ 0 \le y \le h \tag{59}$$

From the comparison between IM[3] and FEM simulation results (Rim & Mi, 2017), it can be seen how this analytic method is in agreement with the simulations in various conditions. In conclusion, the validity of the IM^3 has been proven for both the mono and dual coil coupled with a pick-up coil, and its performance in comparison to FEM simulations confirms its validity as a method for electromagnetic analysis in IPT systems.

COMPUTATIONAL MODELS FOR INDUCTIVE POWER TRANSFER

The models shown so far are very useful tools to perform equation-based analyses of IPT systems, especially when the general behavior of the equivalent circuit parameters is the main research focus. However, it is clear that optimizing an IPT system for practical use is not something that can be done by hand: solving ordinary differential equations is not easy – in fact, in most cases, it is demonstrably impossible to do so – and doing it while considering variations in critical parameters is even more difficult and time-consuming. For this reason, many software applications have been developed in order to exploit computers by making them solve the aforementioned problems in a computational way. This Section contains a brief overview of the main practical applications for numerical methods among current research activities in the context of inductive power transfer.

The choice of separating *iterative numerical methods* from the *finite element method* is purely functional, albeit not completely correct in mathematical terms: in fact, the finite elements method can be carried out with iterative procedures as well, and both these categories are only parts of the greater mathematical area of numerical analysis. In the context of this report, however, it is important to separate the family of iterative methods – typically applied to perform circuit-based simulations and solve ordinary differential equations in order to simulate current or voltage waveforms – from the finite element method, which is instead applied mainly to solve partial differential equations with the aim of simulating the magnetic flux distribution of an object. In terms of practical tools, the first category is usually implemented with software like MATLAB® or the many SPICE-based simulators, whereas the second is related to tools such as ANSYS® Maxwell®, COMSOL Multiphysics® or JMAG®.

Among the various applications for numerical methods in the realm of mathematical physics, the solution of Ordinary Differential Equations (ODEs) is certainly one of the most relevant. Over the last few decades, an incredible wealth of solution methods and commercial software tools have been developed in order to improve the efficiency of computational approaches with respect to the solution of ODEs. This affects electrical engineering as well: in order to study the dynamic characteristics of an electrical system, analytical methods such as the Laplace phasor transform can be applied (Lee et al., 2013). However, these methods can prove to be very complicated and cumbersome, especially as compensation topologies become more and more advanced. In many cases, instead, iterative computational methods are used in order to determine the time-domain variation of electrical quantities starting from a state-space model of the physical configuration. Such models can in turn be directly obtained using analytic tools such as the coupled coils model, typically implemented in the M-model configuration. In (Wang & Wei, 2015), a state-space model for an IPT system which uses an advanced compensation topology is derived starting from the M-model equivalent circuit representation. The obtained state-space representation is then implemented in a computer simulator such as Simulink®, and the time-domain waveforms for the relevant quantities become available with minimal hassle. Iterative methods, in any case, are not limited in application to just the solution of differential equations, but they are widely used in the context of solving linear and nonlinear systems of equations as well. In general, an iterative numerical method is based on the repeated application of approximated solution methods, the results of which are evaluated after every iteration against the desired level of uncertainty, until a satisfying value is reached. This condition is called convergence. The practical implementation of iterative solution methods from a mathematical perspective will not be detailed here. For one thing, numerical analysis has reached such an extreme level of ramification that it would be impossible to give a comprehensive review of all the techniques which are practically exploited in order to perform these kinds of computations. Moreover, these methods are – in general – not specific to electrical engineering studies, but they can be applied in a vast range of areas, therefore it is not really meaningful to explore them in the context of IPT systems.

Across all engineering disciplines, many of the recurring technical challenges are related to the solution of initial value or boundary value problems, that is, problems composed of sets of differential equations and initial or boundary conditions, a good example of which are Maxwell's equations. In practical cases, these problems cannot be solved analytically, which is why numerical solution methods have steadily risen in popularity in the field of mathematical physics since they were introduced, about 80 years ago (Bathe, 2014; Jin, 2014). In this context, the Finite Element Method (FEM) is among the most powerful numerical solution tools that have ever been developed and has been successfully applied to many problems in the realms of structural engineering, fluid dynamics, electromagnetism, and many more. The basic principle of FEM analysis is to partition a generic bounded domain Ω in $\mathfrak{R}^n$ into N non-overlapping (Oden, 2010) – or even overlapping (Bathe, 2014) – sub-domains (the finite elements), where functions can be locally approximated with ease, solve the problem for each individual element, and then recombine the solutions in order to arrive at the solution of the general boundary value problem. As noted in (Oden, 2010), the main feature of FEM analysis is not that it allows to easily represent generic functions on domains in $\mathfrak{R}^n$, but that when applied to boundary value and initial value problems, the summation of all local approximations of the solution over each individual element gives an approximation of the solution over the full domain. This is made possible by the use of weak or variational formulations of given problems, the details of which go beyond the scope of this report (Jin, 2014). Typically, FEM simulations are applied in the context of IPT systems in order to derive an approximation of the magnetic flux distribution, which can then be used to compute physical parameters of the coils, such as self and mutual inductances. It is not uncommon, however, to find it used for determining the current density distribution inside the coils as well, and some formulations (Panchal et al., 2013) of the electromagnetic field allow to simultaneously compute both the magnetic field and current density.

Once the most suitable formulation for Maxwell's equations has been derived, the domain is discretized into a mesh of finite elements. The next step is to select what kind of function will be used to interpolate the initial equation inside each element: while the choice is arbitrary, linear or at most quadratic interpolation functions are the typical choice, since the polynomial order determines the complexity (and therefore the cost in terms of time and computation power) of the remaining part of the analysis (Jin, 2014). Then, an algebraic system of equations is formulated in matrix form using the pre-selected interpolation functions on each of the finite elements, by way of suitable mathematical methods such as Ritz's and Galerkin's, the details of which, once again, are beyond the scope of this report. Finally, this system is solved algebraically, and the results are made available to be visualized or re-used for additional computations.

Several optimization techniques can be implemented in order to improve the accuracy of the results without excessively increasing computation time. A technique that has received great academic attention in the past and is now widely adopted by a range of commercial tools, is the use of adaptive meshes, in which the size of the finite elements is reduced in some areas and increased in others, depending on the relative importance of each region of space. This is, of course, an iterative process, which is usually controlled by measuring the uncertainty connected to the FEM approximation in each individual element: areas that show a tolerable level of uncertainty do not need any additional analysis, whereas areas that need to be approximated with more accuracy are simulated again using smaller elements. This allows to reduce the global uncertainty while using computation resources efficiently. An example of this technique has been applied in (Matsutomo et al., 2012) to the simulation of magnetic flux lines for a one-coil power transmitter. It is clearly visible that the mesh is not subdivided uniformly, but it becomes denser as the flux lines are packed together more tightly.

CASE STUDIES

A selection of case studies related to some of the most relevant design aspects of IPT systems are analyzed with the aim of highlighting practical applications of the various methods discussed above.

The design of an IPT system typically requires different steps. Starting from the definition of the physical characteristics of the coupled inductors, the design process then proceeds to the choice of other circuit elements, depending on the practical application of the system. For the analysis and optimization of the coupled coils there is usually the need to use numerical simulations, in order to study some characteristics of the circuit that otherwise would be difficult or impossible to derive. In (Chang et al., 2014) and (Carmeli et al., 2015), a practical example is given of how an IPT system can be modelled for the design and optimization of the coupled coils through simulations. The first step for the design of the WPT system is the study of coupling coefficient, efficiency and maximum power transferred. For this kind of study, finite element model analysis is used. As explained in (Chang et al., 2014), FEM analysis performs well when the coils have few turns, for an hypothetical coil with an high number of turns a simplified model must be adopted in order to obtain easier calculation for the FEM software but with almost the same level of accuracy. Usually FEM software are used also for the design of the electromagnetic shield due to its accuracy in the approximation of the magnetic field distribution and intensity. Once the coupling factor has been obtained, as well as its relationship with the distance of the two coils and the efficiency of the system, all the other electrical quantities such as currents and voltages must be obtained from simulations. However, for this kind of analysis, FEM software are not considered the best choice due to their high time required for the computation of the quantities. An analysis using a MATLAB® model of the IPTS is therefore preferred for this study in order to spend less time for the calculation of the quantities, considering also that all the calculations have to be done under different conditions.

In (Chang et al., 2014), it can be seen that the coupled coils model is used to implement the WPT system in the simulation software. The same model is used in (Carmeli et al., 2015) for a similar study, and in this case the T-shape configuration of the mutual inductor is selected in order to decouple the inductors and consequently achieve faster computation. This model is, indeed, the most suitable for the computation of the electrical quantities and the analysis of the circuit in simulations, because it takes into consideration all the components without simplification. This means that, for example, a gyrator model, like the one presented in the previous chapter, would not prove useful for a pure analysis of the electrical quantities, but is preferred instead for the design of the compensation circuit and the optimisation of system performance during resonance.

According to the comparison between experimental results and simulations contained in studies such as (Chang et al., 2014) and (Carmeli et al., 2015), a design method based on FEM simulations for optimising the magnetic parameters and the coupling factor, and MATLAB simulations for computing the electrical parameters is proved to be reasonable and valid.

The choice of the compensation circuit is one of the most important steps during the design of an IPT system, since different compensation circuits have different advantages and drawbacks. Usually, the design of the compensation circuit is carried out through the use of an analytical method that involves matrix manipulation and differential equation systems, which can become quite complicated if a lot of conservative components are in the circuit. As explained in "*Gyrator model*", a better method can be used for the sole study of the compensation circuit. A useful case study that can be used to evaluate the merits of the gyrator method can be found in (Zhang et al., 2014). The aim of this paper was to design a voltage-controlled SS compensation circuit for IPT applications. The equivalent circuit proposed for the analysis uses a coupled coils model, whereas the study of the compensation circuit could be made simpler by applying the gyrator model. With the gyrator's properties, the same results can be achieved in a faster way and without the need of solving too many equations. Starting from the coupled inductors depicted in Figure 14a, the intrinsic gyrator can be explicitly represented, as shown in Figure 14b.

An analysis of the circuit can then be done even at frequencies different from the resonant one. The equivalent impedance at the secondary side can be obtained as follows:

$$Z_2 = j\omega L_2 - j\frac{1}{\omega C_2}$$

Figure 14. IPT coils and SS compensation circuit represented with (a) the coupled inductors and (b) the gyrator model

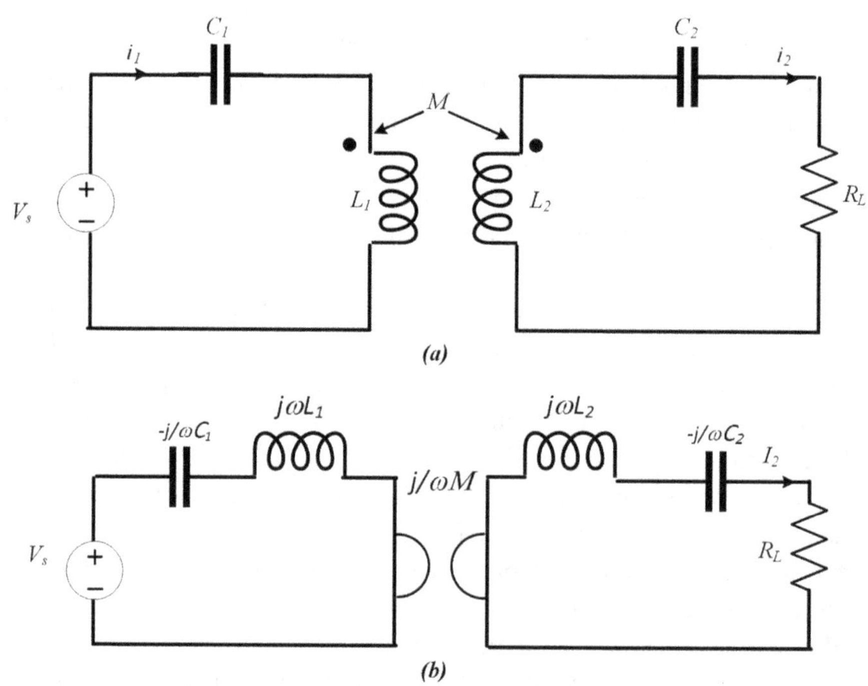

Impedance Z_2 can now be moved to the primary side using the impedance inversion rule (Figure 15a). Thus, the primary current – and consequently the secondary one as well – can be calculated

$$i = \frac{V_s}{Z_1 + \frac{\omega^2 M^2}{Z_2 + R_L}}$$

$$i_2 = \frac{j}{\omega M} V_1 = \frac{j}{\omega M} \cdot i_1 \cdot \frac{(\omega M)^2}{Z_2 + R_L} = \frac{i_1 \cdot j\omega M}{Z_2 + R_L}$$

Now that these quantities have been obtained, the gyrator gain can be computed as follows:

$$G = \frac{i_2}{V_s} = \frac{j\omega M R_L}{Z_1 (Z_2 + R_L) + (\omega M)^2}$$

All the analysis has been done with the assumption that the windings do not have any resistance, however, in case a resistance in the windings is considered, the procedure of the analysis does not change and it is enough to simply add that resistance in the expression obtained above. The main characteristics of this compensation circuit at frequencies different from the resonant one have been found in an easier way compared to the method used in (Zhang et al., 2014) but the major advantages of the gyrator method can be highlighted for the analysis of the compensation circuit at resonant frequency. For the study of the circuit during resonance, the desired resonant frequency must be chosen first. The aim is to have resonance at both the primary and secondary side of the circuit, therefore the frequency needs to have the following value:

$$\omega_f = \frac{1}{\sqrt{L_1 C_1}} = \frac{1}{\sqrt{L_2 C_2}}$$

Figure 15. (a) Equivalent circuit after the application of the impedance inversion rule; (b) In resonant condition, most of the circuit simplifies

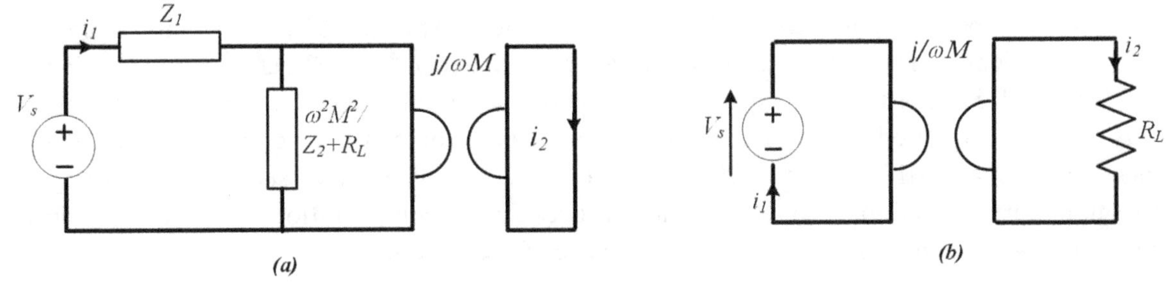

The only parameters that can be modified with the aim to obtain the same resonance frequency at both sides of the compensation circuit are the capacitors. During resonance, the capacitors and the inductors compensate themselves and the resulting equivalent circuit, shown in Figure 32b, does not contain conservative elements, so the analysis becomes much easier. In this case, it can be noted that the secondary current has a different expression and is independent of the load:

$$i_2 = \frac{V_s}{-j\omega_f M}$$

Consequently, the following expression for the gyrator gain can be derived:

$$G \equiv \frac{i_2}{V_s} = \frac{1}{-j\omega_f M}$$

In (Zhang et al., 2014), the voltage gain (G_v) is calculated instead, but for an analysis of the compensation circuit through the gyrator method it is more relevant to highlight the gyrator gain (G). The voltage gain can then be easily derived by multiplying G with the load resistance in the following way:

$$G = G \cdot R_L = \frac{jR_L}{\omega_f M}$$

By using the gyrator model, it is also easy to calculate the power factor of the impedances, in the same way that was presented in "Gyrator model":

$$PF_{Z_1} = PF_{Z_2} = \frac{R_L}{\sqrt{R_L^2 + X_1^2}} = \frac{R_L}{R_L} = 1$$

The results agree with what can be found in literature: the V–SS compensation does not have a reactive power rating and its secondary current is independent of the load in resonance conditions. It is evident how the gyrator method can be useful for this kind of study: in (Zhang et al., 2014), the adaptation of a gyrator method could have led to the same results but in a faster way, without the need of solving complicated systems of differential equations.

With the advancement of scientific research on IPT systems, more and more sophisticated compensation schemes have been proposed and analyzed in the literature. Among these, the *LCC* topology is one of the most efficient, and it has been extensively researched in recent years (Li et al., 2016; Pantic et al., 2011; Li et al., 2015). In order to save space, which is particularly valuable on the vehicle side, integrated designs have been proposed where the main power coil and the compensation coil are stacked on top of one another (Li et al., 2015; Kan et al. 2017; Deng et al., 2017). The main issue with these integrated configurations are the additional couplings that arise between the compensation circuit components and the main coil, which can degrade the total performance of the system.

Therefore, it is interesting to analyses the main modelling tools that are used in the literature to design and optimize these integrated systems, which represent one of the many frontiers of current research efforts in the context of IPT systems. The physical configuration of a typical integrated IPT system is depicted in (Deng et al., 2017). As it can be clearly seen, the compensation circuit is stacked on top of the main coil, which is structured in a DD configuration (Covic et al., 2011), whose name refers to the visual similarity between the coil and the letters "DD". As anticipated above, integrating the main and compensation coils has the unwanted effect of forming couplings between the two circuits. Such a situation can be represented with the equivalent circuit depicted in (Deng et al., 2017). In order to study these effects, the coupled coils model is used in (Kan et al., 2017) and (Deng et al., 2017). While the first paper prefers an M-model representation, the second uses a T-shape equivalent circuit in which the mutual couplings are grouped and expressed in a more compact form.

A first analysis is conducted on the validity of the First Harmonic Approximation (FHA), which is usually assumed true in the context of general circuit design. When higher-order harmonics are included in the analysis, however, it is verified that the integrated design produces stronger high-order harmonics compared to the non-integrated case, which weakens the FHA. Another basic assumption, the Continuous Conduction Mode (CCM) of the output rectifier, is typically assumed to be true, but shows to be weak in the case of integrated LCC design.

In order to minimize the coupling effects, a FEM analysis is then conducted with the aim to evaluate how a displacement of the *DD* compensation coil with respect to the underlying main coil affects the additional coupling coefficients. It is found that a horizontal rotation of 90° of the compensation coil ensures a significant reduction in the unwanted couplings, and effectively allows to neglect their effect.

CONCLUSION

This chapter has explored the current state of the art of modelling techniques used by researchers in the context of inductive power transfer systems. The initial theoretical review of some of these techniques has been validated by considering a selection of relevant case studies. Among the various methodologies, the coupled coils model seems to be the most widely used theoretical tool, often implemented in conjunction with FEM simulations. In fact, while the coupled coils model is very powerful for analysing the electrical parameters of an IPT system, it lacks the ability to perform magnetic flux computations, and therefore to determine crucial electromagnetic parameters such as the self and mutual inductances. In the context of compensation circuit design, a newly proposed model which uses gyrators has been explored. While not as complete as the coupled coils model, the gyrator model has proven to be a versatile tool which allows to greatly simplify the compensation circuit analysis, especially when the performance in resonance conditions is the research objective. The future application of this methodology will mainly depend on how it is received by the scientific community. Another recently proposed method, the improved magnetic mirror model (IM^3), has been analysed in detail. It was found that, compared to numerical simulations, this analytic tool has the advantage of giving detailed information on all the specific parameters that affect the magnetic flux distribution in an IPT system, something that is not as clear in FEM simulations. The precision of this method against computational analysis has demonstrated its validity, and it is likely that this methodology will be implemented in the context of general studies about the physical characteristics of wireless power transfer systems. However, for practical purposes, especially when real-world circuits are designed and tested, it is clear that only numerical methods can provide the accuracy and flexibility needed by researchers. Models such as the IM^3 are very precise, but they are most useful to analyze simplified theoretical cases, where a lot of assumptions are made. The power of numerical methods is that they allow to perform studies on practical configurations as well.

REFERENCES

Barmada, S. (2017). Design and experimental characterization of a combined WPT–PLC system. Wireless Power Transfer, 4(2), 160–170. doi:10.1017/wpt.2017.11

Bathe Klaus-Jürgen. (2014). *Finite Element Procedures* (2nd ed.). Watertown, MA: K. J. Bathe.

Bi, Z., Kan, T., Mi, C. C., Zhang, Y., Zhao, Z., & Keoleian, G. A. (2016). A review of wireless power transfer for electric vehicles: Prospects to enhance sustainable mobility. *Applied Energy*, *179*, 413–425. doi:10.1016/j.apenergy.2016.07.003

Buja, G., Bertoluzzo, M., & Dashora, H. K. (2016). Lumped Track Layout Design for Dynamic Wireless Charging of Electric Vehicles. *IEEE Transactions on Industrial Electronics*, *63*(10), 6631–6640. doi:10.1109/TIE.2016.2538738

Campi, T. (2017). EMC and EMF safety issues in wireless charging system for an electric vehicle (EV). In *2017 International Conference of Electrical and Electronic Technologies for Automotive*. IEEE. 10.23919/EETA.2017.7993214

Carmeli, M. S. (2015). Contactless energy transmission system for electrical vehicles batteries charging. In *2015 International Conference on Clean Electrical Power (ICCEP)*. IEEE. 10.1109/ICCEP.2015.7177536

Catenacci, M., Verdolini, E., Bosetti, V., & Fiorese, G. (2013). Going electric: Expert survey on the future of battery technologies for electric vehicles. *Energy Policy*, *61*, 403–413. doi:10.1016/j.enpol.2013.06.078

Chang, R. (2014). Design of a wireless power transfer system for EV application based on finite element analysis and MATLAB simulation. In *2014 IEEE Conference and Expo Transportation Electrification Asia-Pacific (ITEC Asia-Pacific)*. IEEE. 10.1109/ITEC-AP.2014.6941225

Covic, G. A. (2011). *A bipolar primary pad topology for EV stationary charging and highway power by inductive coupling. In 2011 IEEE Energy Conversion Congress and Exposition* (pp. 1832–1838). IEEE. doi:10.1109/ECCE.2011.6064008

Covic, G. A., & Talbot Boys, J. (2013). Modern Trends in Inductive Power Transfer for Transportation Applications. IEEE Journal of Emerging and Selected Topics in Power Electronics, 1(1), 28–41. doi:10.1109/JESTPE.2013.2264473

Deng, J. (2017). A new integration method with minimized extra coupling effects using inductor and capacitor series-parallel compensation for wireless EV charger. Applied Energy. doi:10.1016/j.apenergy.2017.05.088

International Energy Agency. Global EV Outlook. (2017). *Two million and counting*. IEA.

Jin, J.-M. (2014). *The Finite Element Method in Electromagnetics* (3rd ed.). John Wiley & Sons.

Kalwar, K. A., Aamir, M., & Mekhilef, S. (2015). Inductively coupled power transfer (ICPT) for electric vehicle charging – A review. *Renewable & Sustainable Energy Reviews, 47*, 462–475. doi:10.1016/j.rser.2015.03.040

Kan, T., Nguyen, T.-D., White, J. C., Malhan, R. K., & Mi, C. C. (2017). A New Integration Method for an Electric Vehicle Wireless Charging System Using LCC Compensation Topology: Analysis and Design. *IEEE Transactions on Power Electronics, 32*(2), 1638–1650. doi:10.1109/TPEL.2016.2552060

Kuzey, S., Balci, S., & Altin, N. (2017). Design and analysis of a wireless power transfer system with alignment errors for electrical vehicle applications. *International Journal of Hydrogen Energy, 42*(28), 17928–17939. doi:10.1016/j.ijhydene.2017.03.160

Lee, S., Choi, B., & Rim, C. T. (2013). Dynamics Characterization of the Inductive Power Transfer System for Online Electric Vehicles by Laplace Phasor Transform. *IEEE Transactions on Power Electronics, 28*(12), 5902–5909. doi:10.1109/TPEL.2013.2254500

Lee, W. Y., Huh, J., Choi, S. Y., Thai, X. V., Kim, J. H., Al-Ammar, E. A., ... Rim, C. T. (2013). Finite-Width Magnetic Mirror Models of Mono and Dual Coils for Wireless Electric Vehicles. *IEEE Transactions on Power Electronics, 28*(3), 1413–1428. doi:10.1109/TPEL.2012.2206404

Li, S., Li, W., Deng, J., Nguyen, T. D., & Mi, C. C. (2015). A Double-Sided LCC Compensation Network and Its Tuning Method for Wireless Power Transfer. *IEEE Transactions on Vehicular Technology, 64*(6), 2261–2273. doi:10.1109/TVT.2014.2347006

Li, W., Zhao, H., Deng, J., Li, S., & Mi, C. C. (2016). Comparison Study on SS and Double-Sided LCC Compensation Topologies for EV/PHEV Wireless Chargers. *IEEE Transactions on Vehicular Technology, 65*(6), 4429–4439. doi:10.1109/TVT.2015.2479938

Li, W., Zhao, H., Li, S., Deng, J., Kan, T., & Mi, C. C. (2015). Integrated ${LCC} $ Compensation Topology for Wireless Charger in Electric and Plug-in Electric Vehicles. *IEEE Transactions on Industrial Electronics, 62*(7), 4215–4225. doi:10.1109/TIE.2014.2384003

Maddalena, E. T., & Godoy, R. B. (2017). Model derivation and dynamic analysis of the SPS compensated wireless power transfer system. In *2017 IEEE 8th International Symposium on Power Electronics for Distributed Generation Systems (PEDG)*. IEEE. 10.1109/PEDG.2017.7972495

Matsutomo, S., Noguchi, S., & Yamashita, H. (2012). Adaptive Mesh Generation Method Utilizing Magnetic Flux Lines in Two-Dimensional Finite Element Analysis. *IEEE Transactions on Magnetics, 48*(2), 527–530. doi:10.1109/TMAG.2011.2176471

Mohamed, A. A. S., Berzoy, A., & Mohammed, O. A. (2017). Experimental Validation of Comprehensive Steady-State Analytical Model of Bidirectional WPT System in EVs Applications. *IEEE Transactions on Vehicular Technology, 66*(7), 5584–5594. doi:10.1109/TVT.2016.2634159

Oden, J. (2010). Finite element method. Scholarpedia, 5(5), 9836.

Omori, H. (2012). A novel wireless EV charger using SiC single-ended quasi-resonant inverter for home use. In *2012 15th International Power Electronics and Motion Control Conference (EPE/PEMC)*. IEEE. 10.1109/EPEPEMC.2012.6397524

Panchal, C., Lu, J., & Stegen, S. (2013). Investigation of magnetic flux distribution of EV wireless charging systems. In *2013 Asia-Pacific Symposium on Electromagnetic Compatibility (APEMC)*. IEEE. 10.1109/APEMC.2013.7360652

Pantic, Z., Bai, S., & Lukic, S. M. (2011). ZCS LCC-Compensated Resonant Inverter for Inductive-Power-Transfer Application. *IEEE Transactions on Industrial Electronics*, *58*(8), 3500–3510. doi:10.1109/TIE.2010.2081954

Park, C., Lee, S., Jeong, S. Y., Cho, G.-H., & Rim, C. T. (2015). Uniform Power I-Type Inductive Power Transfer System With DQ-Power Supply Rails for On-Line Electric Vehicles. *IEEE Transactions on Power Electronics*, *30*(11), 6446–6455. doi:10.1109/TPEL.2015.2420372

Paul, C. R. (2006). *Introduction to Electromagnetic Compatibility* (2nd ed.). John Wiley & Sons, Inc.

Raju, S., Wu, R., Chan, M., & Yue, C. P. (2014). Modeling of Mutual Coupling Between Planar Inductors in Wire- less Power Applications. *IEEE Transactions on Power Electronics*, *29*(1), 481–490. doi:10.1109/TPEL.2013.2253334

Rim, C. T. (2018). Wireless Charging of Electric Vehicles. In *Power Electronics Handbook*. Elsevier. doi:10.1016/B978-0-12-811407-0.00038-6

Rim, C. T., & Mi, C. (2017). *Wireless Power Transfer for Electric Vehicles and Mobile Devices*. Chichester, UK: John Wiley & Sons, Ltd. doi:10.1002/9781119329084

Shin, J., Shin, S., Kim, Y., Ahn, S., Lee, S., Jung, G., ... Cho, D.-H. (2014). Design and Implementation of Shaped Magnetic-Resonance-Based Wireless Power Transfer System for Roadway-Powered Moving Electric Vehicles. *IEEE Transactions on Industrial Electronics*, *61*(3), 1179–1192. doi:10.1109/TIE.2013.2258294

Sohn, Y.-H. (2015). Gyrator-Based Analysis of Resonant Circuits in Inductive Power Transfer Systems. IEEE Transactions on Power Electronics, 31(10). doi:10.1109/TPEL.2015.2506737

Swain, A. K., Neath, M. J., Madawala, U. K., & Thrimawithana, D. J. (2012). A Dynamic Multivariable State-Space Model for Bidirectional Inductive Power Transfer Systems. *IEEE Transactions on Power Electronics*, *27*(11), 4772–4780. doi:10.1109/TPEL.2012.2185712

Tellegen, B. D. H. (1948). The gyrator, a new electric network element. Philips Research Reports, 3, 81–101.

Ulaby, F. T., & Ravaioli, U. (2015). *Fundamentals of Applied Electromagnetics* (7th ed.). Pearson.

Vilathgamuwa, D. M., & Sampath, J. P. K. (2015). Wireless Power Transfer (WPT) for Electric Vehicles (EVs) - Present and Future Trends. In Plug In Electric Vehicles in Smart Grids. Ed. by Sumedha Rajakaruna, Farhad Shahnia, and Arindam Ghosh. Power Systems. Singapore: Springer Singapore.

Wang, C. S., Stielau, O. H., & Covic, G. A. (2005). Design Considerations for a Contactless Electric Vehicle Battery Charger. *IEEE Transactions on Industrial Electronics*, *52*(5), 1308–1314. doi:10.1109/TIE.2005.855672

Wang, Z., & Wei, X. (2015). Design Considerations for Wireless Charging Systems with an Analysis of Batteries. *Energies*, *8*(10), 10664–10683. doi:10.3390/en81010664

Zhang, W., Wong, S.-C., Tse, C. K., & Chen, Q. (2014). Analysis and Comparison of Secondary Series- and Parallel-Compensated Inductive Power Transfer Systems Operating for Optimal Efficiency and Load-Independent Voltage-Transfer Ratio. *IEEE Transactions on Power Electronics*, *29*(6), 2979–2990. doi:10.1109/TPEL.2013.2273364

Zhang, Z., & Chau, K. T. (2015). Homogeneous Wireless Power Transfer for Move-and- Charge. *IEEE Transactions on Power Electronics*, *30*(11), 6213–6220. doi:10.1109/TPEL.2015.2414453

Chapter 12
An Inductive Power Transfer System for the Wireless Charging of Electric Vehicles:
Determination of the Magnetic Coupling Factor

Javier Vázquez
University of Castilla-La Mancha, Spain

Pedro Roncero-Sánchez
University of Castilla-La Mancha, Spain

Alfonso Parreño Torres
Castilla-La Mancha Science and Technology Park, Spain

ABSTRACT

This chapter describes two methods whose purpose is to estimate the coupling factor k for the inductive coupling of an experimental prototype designed for the contactless battery charging of electric vehicles. The air gap that separates the two square-shaped coils of the inductive coupling is 150 mm. The first method is analytical and provides an expression with which to calculate the mutual inductance between both coils by solving Neumann's formula, from which k readily follows. The second approach is empirical and combines data from waveforms of currents that were obtained both experimentally and from PSpice simulations, where k is a model parameter, under different loading conditions. The agreement between the analytical and the empirical method is good, as they both yield coupling factor figures that are equal to 0.197 and 0.200, respectively. Both techniques could, therefore, be applied in order to determine the coupling factor of the constructed inductive coupling for other air gaps near to 150 mm, which are also of practical interest in electric vehicle charging.

DOI: 10.4018/978-1-5225-5870-5.ch012

INTRODUCTION

Several countries have recently taken the lead in promoting a transition from gas-powered vehicles to electric vehicles (EVs) in an attempt to reduce CO_2 emissions. One of these countries is Norway which, like various others, has the objective of ensuring that all its new cars will be all-electric by 2025. In fact, a remarkable 28.8% of the new cars sold in Norway in September 2016 were plug-in EVs, and this figure was significantly higher than that of any other country (Lambert, 2016).

The operation of plug-in EVs relies on conductive charging, a charging solution that can currently be regarded as a sound and reliable technology that is supported by many standards. It is, nevertheless, true that unplugged charging solutions are also feasible: Inductive power transfer (IPT) applied to EVs is a wireless power transfer (WPT) scheme that has attained a considerable level of maturity over the past years. IPT technology is particularly appealing to the ordinary vehicle user since, unlike conductive charging, the recharging process of the in-built batteries does not rely on a wired connection between the charging station and the vehicle. It should be added, however, that losses in the magnetic coupling of IPT systems are responsible for a certain reduction in its efficiency, which is relatively low when compared to the corresponding figures that are typically expected in conductive charging. The application of IPT to electric mobility is not limited to street vehicles used for transportation purposes, but also covers a wide range of innovative developments that have been explored in recent years, such as garbage collection (Reatti et al., 2017) and underwater robotics (Allotta, Pugi, Reatti, & Corti, 2017).

This chapter presents work concerning the determination of the magnetic coupling factor of the inductive coupling stage present in an IPT experimental arrangement specifically designed for EV charging. It is structured as follows. The subject is introduced in the following section, in which the key aspect of what is arguably the core of any IPT system is covered, i.e., a magnetic coupling composed of two windings separated by an air gap. A thorough description of the different stages of the IPT prototype is then provided. Two different approaches that lead to the determination of the coupling factor of the magnetic coupling stage are subsequently discussed: on the one hand, an analytical method based on Neumann's formula that yields the mutual inductance between a pair of neighboring closed circuits, and on the other, an alternative empirical technique that relies on both experimental measurements and simulation results obtained with a PSpice circuit model that is a replica of the prototype. Upon comparing the values for the coupling factor obtained using both methods it can be concluded that the PSpice circuit model is a useful tool with which to aid in the determination of the magnetic coupling factor of the constructed coupling stage for the air gap under consideration.

BACKGROUND

On-board chargers that are present in conductive charging schemes have a number of disadvantages that are conveniently addressed by WPT systems, some of which are: the need for a robust wired connection terminated with a plug charger, a galvanic isolation of the on-board electronics, the dimensions of the charger, and the need to ascertain safety considerations that are relevant when the vehicle is operating in wet conditions. A number of WPT manufacturers supply their chargers to members of the automotive industry, such as Audi, BMW, Chrysler, Daimler, Ford, GM, Honda, Mitsubishi and Toyota. Some of these WPT suppliers are Conductix-Wampfler, Evatran, HaloIPT (Qualcomm), Momentum Dynamics and WiTricity. Research initiatives exploring the potential of WPT technology are not limited to manufacturers; universities, government facilities and R&D centers worldwide are, at present, also actively involved in further developing this technology. Relevant examples of such institutions can be found in (Musavi & Eberle, 2014).

While IPT is by no means the only WPT solution available (capacitive power transfer technology is a good example of an alternative that has recently received some attention (Vu et al., 2017)), it can be safely claimed that resonant IPT (RIPT) currently stands out as the most popular WPT technology. RIPT features at least two resonant tanks resonating at a given frequency by means of added capacitors that give rise to several compensation topologies, depending on whether the capacitors are connected in series or in parallel. In order to shed some light on this and other peculiarities of RIPT technology, it would be appropriate to begin by reviewing some of the key elements of the coupling theory on which RIPT is based.

In an IPT system, the self-inductances on the primary and the secondary sides L_p and L_s are magnetically coupled by means of the mutual inductance M. Figure 1 depicts the circuit typically used to determine the equivalent circuit for two mutually coupled coils.

The strength of the magnetic coupling is characterized by the coupling factor k (also called the coupling coefficient), which is a dimensionless figure that is, according to (1), limited between zero and one.

$$k = \frac{M}{\sqrt{L_p L_s}} \tag{1}$$

If k is close to zero, the coils are said to be loosely coupled, and if k is close to one, they are said to be tightly coupled. The coils of ordinary transformers with a core made out of iron are usually tightly coupled, whereas those of air-core transformers are typically loosely coupled. Iron-core transformers can have coupling coefficients approaching unity. The ideal case (that is, $k = 1$) represents a perfect coupling (Bobrow, 1985, p. 799). In this case all the magnetic flux generated by the primary coil flows entirely through the secondary coil and there is, therefore, no leakage flux. In other cases, when $k < 1$,

Figure 1. Electric network model for two mutually coupled coils

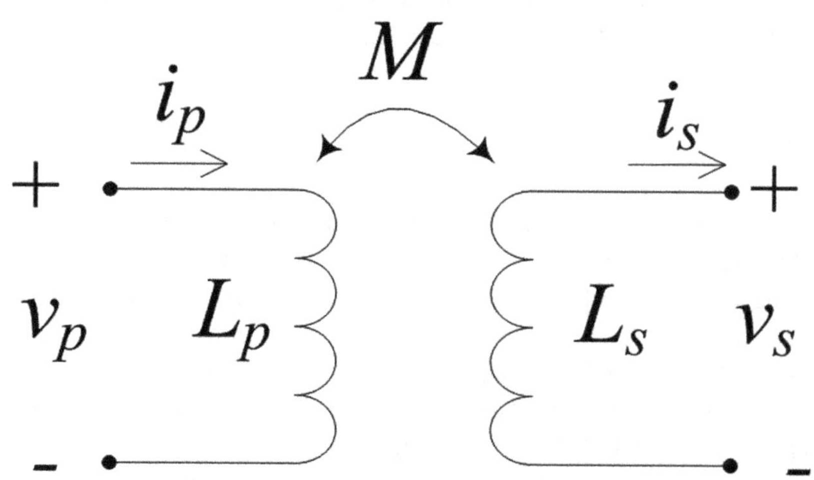

a fraction of the magnetic flux at the primary coil fails to flow through the secondary coil and is thus represented by leakage flux. IPT systems fall into the category of air-core transformers and are, therefore, loosely coupled. Unlike iron-core transformers, the mechanical constraints of IPT designs impose a lack of physical contact between the primary and the secondary coil. In the particular case of inductive charging applied to EVs, the charging station is fitted with the primary winding, which is fixed, whereas the secondary winding is attached to the underside of the vehicle and therefore moves along with it. IPT systems are less efficient than those with tightly-coupled coils: figures in the range from 70% to 80% are usual in such systems (Huh, Lee, Lee, Cho, & Rim, 2011; Pinuela, Yates, Lucyszyn, & Mitcheson, 2013).

Leakage flux can be modeled with leakage inductances L_{p_leak} and L_{s_leak}, which are present on both sides of the magnetic coupling, using the so-called "two-leakage inductance model" (Hesterman, 2007) shown in Figure 2. Apart from the two leakage inductances, the model also includes an ideal transformer with a turns ratio $r = N_p/N_s$ and a magnetizing inductance referred to the primary side, L_M.

Leakage inductances can be expressed in terms of the coupling factor k (Corti et al., 2016; Nahvi & Edminister, 2014):

$$L_{p\_leak} = (1-k)L_p \tag{2}$$

$$L_{s\_leak} = (1-k)L_s \tag{3}$$

Alternatively, they can be written as a function of the mutual inductance M (Córcoles, Pedra, & Salichs, 2004, p. 202; Hurley & Wilcox, 1994):

$$L_{p\_leak} = L_p - rM \tag{4}$$

Figure 2. Two-leakage inductance model of a magnetic coupling

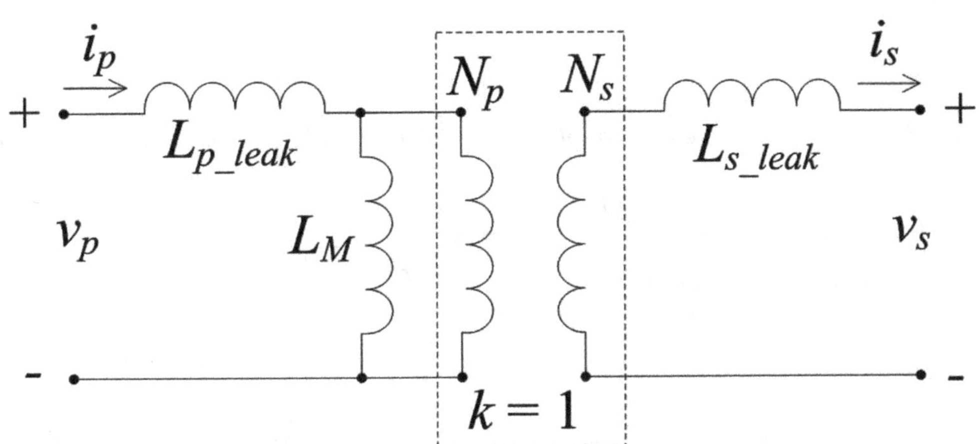

$$L_{s\_leak} = L_s - \frac{M}{r} \quad (5)$$

In either case, it is apparent from expressions (2) to (5) that leakage inductances are, as would be expected, large for loosely coupled coils and small for tightly coupled coils. It should be pointed out that some circuit models of magnetic coupling, other than the two-leakage inductance model, exist and are widely used, and they do not explicitly include leakage inductances. One of these is known as the T-model (Van Schuylenbergh & Puers, 2009, p. 46), which is a practical arrangement for circuit analysis since none of its three inductors is coupled. Yet another model, which is intended to be analyzed specifically in the frequency domain, features current-controlled voltage generators on the primary and the secondary sides (Liu, Ng, Lee, & Hui, 2008). Both circuit models are represented in Figure 3.

A circuit analysis of the magnetic coupling carried out with any of the models depicted in Figure 3, when a load resistance R_L is connected to the secondary side, results in an impedance seen from the primary side Z_p that makes two contributions, as stated in equation (6). One is the reactance of the primary coil, X_{Lp}, and the other, Z_r, is termed as reflected impedance.

$$Z_p = \frac{v_p}{i_p} = jX_{L_p} + Z_r \quad (6)$$

Note that Z_r depends on elements that are present on the secondary side:

$$Z_r = \frac{\omega^2 M^2}{R_L^2 + X_{L_s}^2}\left(R_L - jX_{L_s}\right) \quad (7)$$

As it will be demonstrated below, the performance of the IPT system improves considerably if extra capacitors are included on both sides of the magnetic coupling. On the one hand, the active power transferred to the secondary side, P_s, can be derived from the real part of Z_r as follows:

$$P_s = \Re(Z_r)I_{p,rms}^2 = \frac{\omega^2 M^2 R_L}{R_L^2 + X_{L_s}^2} I_{p,rms}^2 \quad (8)$$

Figure 3. Circuit models of two mutually coupled coils: (a) T- model; (b) model with voltage generators

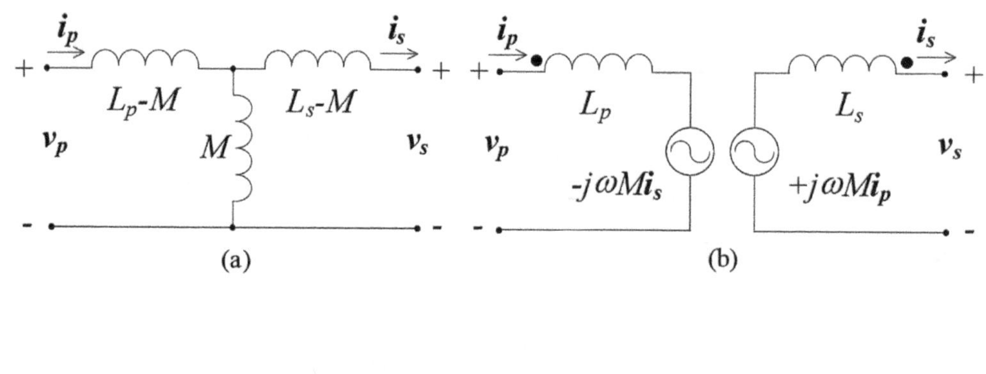

Note that, according to (8), an easy way in which to boost the transferred power consists of reducing or even eliminating the reactance of the secondary coil. This can, in practice, be achieved by adding a compensation capacitor C_s to the secondary side (for example, in series with L_s). The resulting expression for Z_r, which now includes the contribution of C_s, is greatly simplified provided that the system operates at the resonant frequency of the secondary side, which is given by (9).

$$\omega_r = \frac{1}{\sqrt{L_s C_s}} \tag{9}$$

At resonance, both of the reactances that are present on the secondary side (the inductive X_{Ls} and the capacitive X_{Cs}) cancel each other out. Consequently, Z_r is no longer a function of X_{Ls} and neither is P_s, which is larger after the compensation, as desired:

$$P_s = \frac{\omega_r^2 M^2}{R_L} I_{p,rms}^2 \tag{10}$$

On the other hand, the resulting impedance seen from the primary side, when operating at the resonant frequency of the compensated secondary side, can be written as:

$$Z_p(\omega_r) = \frac{\omega_r^2 M^2}{R_L} + jX_{L_p}(\omega_r) \tag{11}$$

Note that the reactive power from the power supply, which occurs because of the presence of an imaginary part in Z_p, is eliminated by compensating the primary self-inductance L_p with a second capacitor C_p. As the IPT system is intended to operate at the resonant frequency of the secondary side, the resonant circuit on the primary side is designed to resonate at exactly the same frequency:

$$\omega_r = \frac{1}{\sqrt{L_s C_s}} = \frac{1}{\sqrt{L_p C_p}} \tag{12}$$

The new expression for Z_p includes an extra term to account for the presence of C_p. Again, at resonance both of the reactances present on the primary side (X_{Lp} and X_{Cp}) cancel each other out and, as a consequence, the resulting expression for Z_p after the compensation has no imaginary part:

$$Z_p(\omega_r) = \frac{\omega_r^2 M^2}{R_L} \tag{13}$$

In summary, if a compensation capacitor is added to the primary side, the reactive power delivered by the power supply is, in practice, reduced, while if another compensation capacitor is connected to the secondary side then the efficiency increases, provided that the system operates at the resonant frequency (Bosshard, Mühlethaler, Kolar, & Stevanovic, 2013). Although the circuit analysis above has been derived under the assumption that both compensation capacitors are connected in series with their respective self-inductances (which results in a topology known as series-series (or S-S)), the conclusions drawn also apply to any of the other three basic compensation topologies: series-parallel (S-P), parallel-series (P-S) and parallel-parallel (P-P). The first and the second capital letters of the abbreviations refer to the compensation type on the primary and the secondary sides, respectively. All four basic topologies are shown in Figure 4. Other compensation schemes that are actually more complex than these four, as they include a variety of passive networks on either side of the magnetic coupling, have also been tested in recent years.

Note that only the S-S and P-P topologies are symmetrical. This is a feature that a vehicle-to-grid (V2G) charging scheme could benefit from, since the power flow needs to be bidirectional in such systems. The S-P and P-S topologies, which lack the necessary symmetry to be useful in V2G applications, must therefore be ruled out. While the P-P topology would still be an interesting choice for battery charging in EVs owing to its current-source behavior, an additional series inductor placed on the primary side would be essential in order to regulate the current supplied by the voltage-source inverter that flows into the parallel resonant tank in ordinary (that is, unidirectional) charging systems (Wang, Stileau, & Covic, 2005). In the case that the power flow is required to be bidirectional, the extra inductor would appear not only on the primary side of the magnetic coupling, but also on its secondary side (Madawala, Neath, & Thrimawithana, 2013). This leads to a bulkier system that is necessarily more expensive, complex and sensitive to electromagnetic compatibility issues. Furthermore, only in the S-S topology is the compensation capacitor on the primary side C_p independent of both the coupling factor and the load, as shown in expression (14). This has implications of practical interest for automotive engineers, as the S-S scheme allows a robust design that is tolerant to misalignments between the primary and the secondary coils and that will necessarily affect the coupling factor. Note that misalignments will probably occur at a contactless charging station.

Figure 4. Four basic compensation topologies: (a) S-S; (b) S-P; (c) P-S; (d) P-P

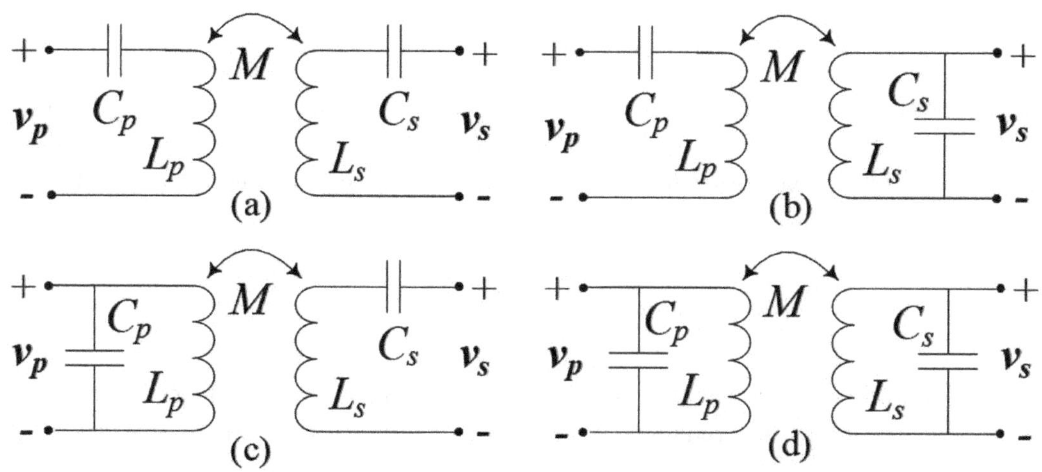

$$C_p = \frac{1}{L_p \omega_r^2} \tag{14}$$

Another advantage of the S-S compensation is its remarkable efficiency, which is approximately 10% greater than that of the S-P topology (Moradewicz & Kazmierkowski, 2009). Recent works have reported the benefits of an S-S configuration for high power IPT applications (Bosshard et al., 2015; Buja, Bertoluzzo, & Mude, 2015; Miskiewicz & Moradewicz, 2011). It should, however, be pointed out that no single topology behaves outstandingly under every operation condition, and the S-S topology is no exception. In fact, it has some limitations that are deserving of attention. For example, the current flowing into the primary side of the magnetic coupling, which is ultimately supplied by an external converter with a rated maximum output current, can undergo a severe increase if the reflected impedance decreases noticeably. This may, according to equations (6) and (7), occur in the following two cases: (a) when the load resistance becomes sufficiently large, or (b) when the mutual inductance decreases below a certain value. The limiting cases of these two scenarios would be an open-circuited secondary side and the lack of a secondary winding, respectively.

EXPERIMENTAL PROTOTYPE

The experimental setup comprises three different blocks, as shown in Figure 5: a supply stage, an S-S compensated RIPT stage and a load stage.

The supply stage is fed from a single-phase grid supply (230 V, 50 Hz). The phase shift control technique was chosen to drive the RIPT stage from an inverter, which obeys the topology of an ordinary H-bridge converter. The inverter outputs a waveform with three voltage levels ($+V_{dc}$, 0, $-V_{dc}$) whose duty cycle and frequency can be set by properly triggering its four switches with additional control electronics.

H-Bridge Converter Operation

The inverter selected for the prototype in order to drive the S-S compensated inductive coupling is a single-phase two-level voltage-source converter (VSC). Figure 6 depicts a circuit diagram of such an

Figure 5. Experimental prototype featuring three stages

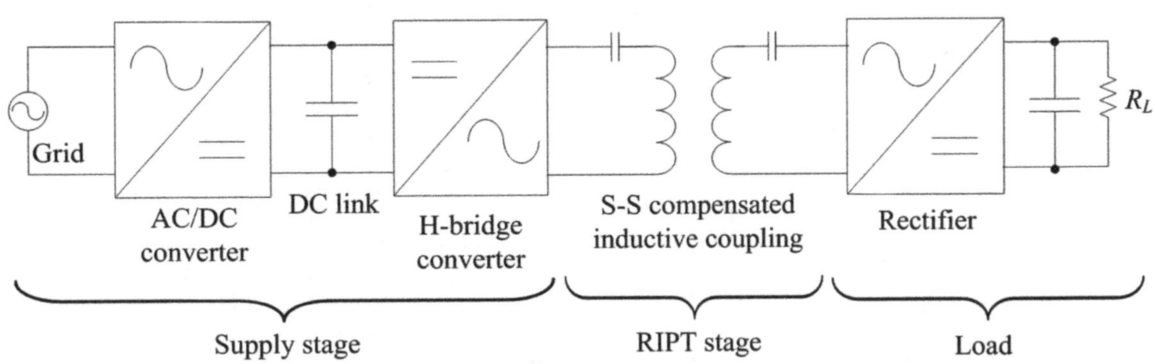

inverter and its four switches, implemented with IGBTs, while Figure 7 shows the commercial device utilized: the three phase VSC SKS 22F B6U + E1CIF + B6CI 13 V12 manufactured by SEMIKRON. Two of its legs are used to form an H-bridge, whereas the remaining leg is left open. Although the inverter delivered by the supplier is intended to be fully operational without the need for extra hardware, a customized PCB was manufactured and attached to its top side. The PCB serves three purposes: protection against overcurrents, the measurement of both the DC link voltage and the two converter leg currents, and the insertion of a fiber optic adapter that is required to interface a real-time dSPACE controller.

The inverter is equipped with a jumper that makes it possible to select a dead time from a few values. A dead time of 500 ns that is present in the gate commands of the inverter transistors was chosen to avoid the risk of a system failure. While the introduction of dead times causes a reduction in the expected duty cycle of the resulting H-bridge voltage waveform, this reduction can be considered negligible in the current work, keeping in mind that the switching frequency of operation is sufficiently high (exactly 18650 Hz, in order to operate the prototype at the resonant frequency of the compensated inductive coupling).

The IC UCC3895 is, meanwhile, a phase-shift PWM controller that outputs the four firing signals required to trigger the transistor gate terminals of the inverter with the objective of producing a controlled converter output voltage by selecting the interval at which the voltage is zero in each period. This interval equals 4α, as Figure 8 illustrates.

The angle α will henceforth be referred to as the zero-voltage angle. By adjusting α to the required value, a phase-shift regulation of the current flowing through the primary side of the coupling was achieved. The aim was to keep the rms primary current approximately equal to 10 A regardless of the load resistance. This is intended to be a safety measure that would help avoid exceeding the rated maximum current of the inverter that drives the inductive coupling.

Figure 6. H-bridge converter: circuit diagram

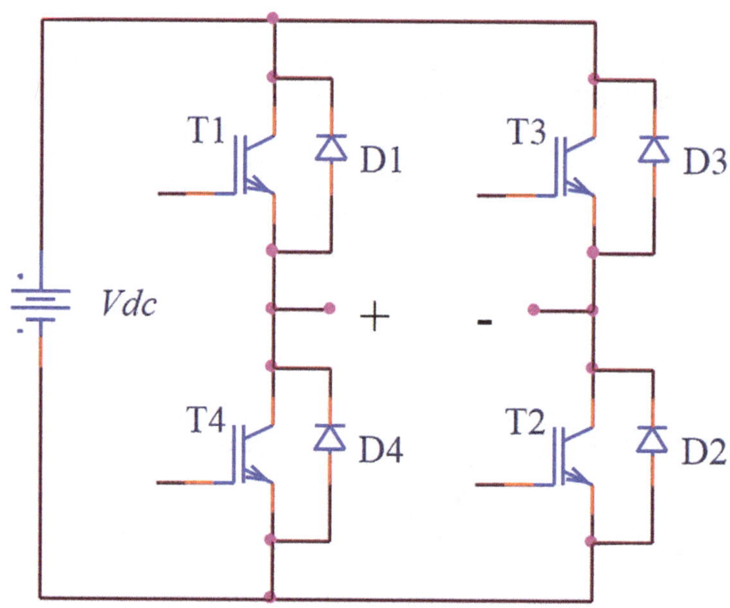

Figure 7. H-bridge converter manufactured by SEMIKRON

Figure 8. H-bridge converter output voltage waveform

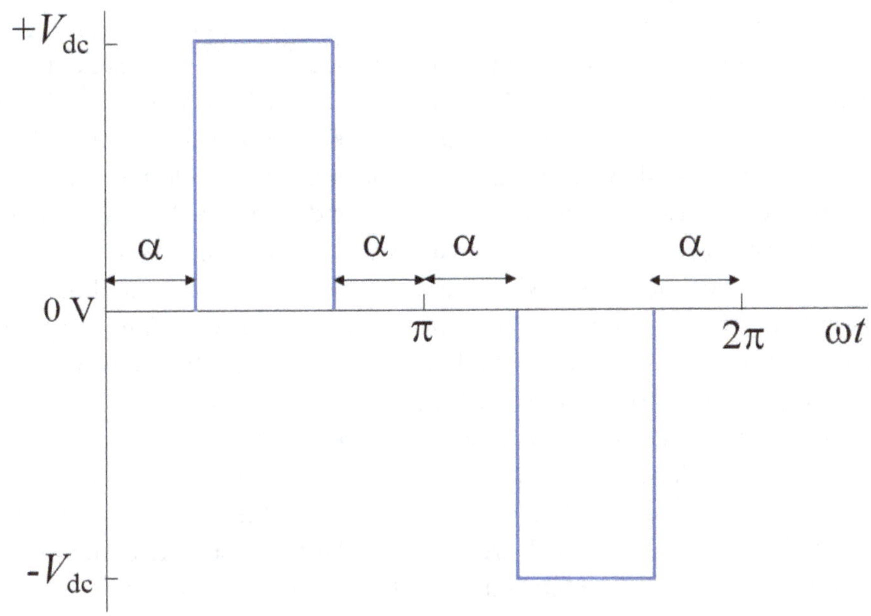

By applying the phase-shift control technique, both the rms voltage V_{rms} and the amplitude of the n-th harmonic V_n of the waveform delivered by the inverter can be expressed as a function of α, as follows from equations (15) and (16) (Hart, 2011, p. 342).

$$V_{rms} = V_{dc}\left(1 - \frac{2\alpha}{\pi}\right)^{1/2} \quad (15)$$

$$V_n = \frac{4V_{dc}}{n\pi}\cos(n\alpha) \quad (16)$$

A dSPACE 1103 control platform was programmed to output an analog value proportional to α. The IC UCC3895 in turn reads that analog value and controls the duty cycle of the inverter output voltage by shifting the switching of one half-bridge with respect to the other half according to α.

Compensated Inductive Coupling

A relevant issue in the design of IPT systems is the shape of the windings. Circular, square and rectangular geometries have been reported to date. For the sake of comparison, let us consider coils of square and circular shapes, and that a circle with a diameter *d* is inscribed in a square with the same side length *d* (a reasonable assumption, taking into account that the coils will ultimately be fitted to the underside of a vehicle). In this case, the area ratio of the square coil over the circular coil is 4/π = 1.27. An analytical model has been used to demonstrate that the mutual inductance *M* between square coils of side length *d* is $(4/\pi)^2$ times larger than between circular coils with a diameter *d* (Raju, Wu, Chan, & Yue, 2014). In a subsequent work, other authors have employed a finite element analysis to show that, in the case under study, i.e., circular and square coils with the same dimensions and a wide range of both coil separations and misalignment distances between coils, the mutual inductance *M* between square-shaped coils is much larger than between circular-shape coils in all cases. In addition to this, the same study proves that there is a steady but moderate drop in *M* as the misalignment increases, and that this tolerance to misalignments is virtually the same for both types of coil geometries (Li, Zhu, Jiang, & Song, 2017). As stated previously, misalignments between windings will arise when the secondary coil, fitted to the vehicle, is driven onto the primary coil, which will be located, presumably buried underground, at the charging station. In the particular case of EV applications, square-shaped coils are, therefore, a convenient choice: not only do they fit better into the rectangular underside of EVs than do circular coils, but they also have a significantly larger coupling factor, provided that the circle is inscribed in the square.

The two square coils manufactured, which are of identical dimensions, are composed of two stacked windings with 11 turns in each coil. The total number of turns in each coil, *N*, therefore amounts to 22, as Figure 9 shows.

Figure 9. Cross-sectional view of the inductive coupling (dimensions are expressed in mm)

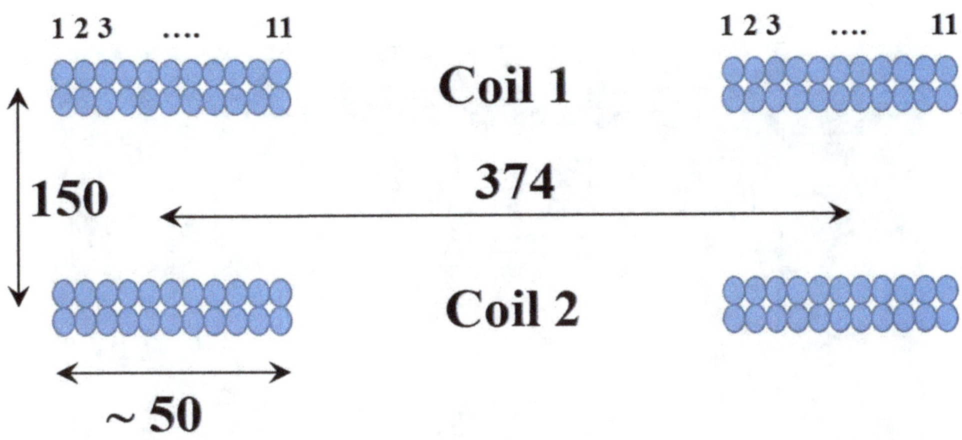

Each coil assembly is housed in a transparent methacrylate enclosure using plates pressing on each side of the coil. The housing was designed with three purposes: to provide mechanical stability, to guarantee a perfect alignment between the coils (that is, coaxiality) and to ensure that both coils rest on parallel planes. Moreover, four screws at the corners of the two pairs of plates allow the separation between coils to be easily modified within a wide range which is appropriate for EV inductive charging (from well below 100 mm to above 200 mm). All of these constructive elements, in addition to the dimensions of the windings, are shown in Figure 10, which is a picture of the actual compensated inductive coupling utilized in the IPT prototype. The 11 turns per row in each coil stretch over a distance of nearly 50 mm. The coil side length, measured at the midpoints of the windings, is 374 mm, and the separation (or air gap) between coils was set to 150 mm throughout the current work. Note also the presence of compensation capacitors, which result from an arrangement combining eight capacitors on each side of the coupling (with four branches including two capacitors each). As the rated capacitance of individual capacitors is 100 nF, the total expected capacitance of the arrangement amounts to 200 nF (although the exact values will deviate from 200 nF owing to tolerances, as will be shown later, in Table 1).

The wire type chosen to be wound was Litz wire with a section of 6.28 mm^2, in an attempt to minimize losses. Note that a key issue in the design process of any IPT system is the number of turns the windings should have. Although the mutual inductance (and consequently the coupling factor) increases with the number of turns, with the consequent improvement in the overall efficiency, some problems arise if the number of turns is too large. On the one hand, the coils grow bulkier and heavier as the number of turns increases; and on the other, the power dissipated in the wire becomes higher and higher. A trade-off should, therefore, be found in order to determine realistic limits for the number of turns. The corresponding figure chosen for the current design is the result of a previous study that sought a target efficiency of 90% for the IPT stage (del Toro, Vázquez & Roncero, 2015). The performance of the charger as regards its efficiency and power is presented below.

Performance Analysis

A circuit model of the charger that is suitable for power calculations is shown in Figure 11. R_p accounts for the resistance of the supply stage added to the stray series resistance of the compensated primary

Figure 10. Inductive coupling with square-shaped windings and compensation capacitors (dimensions are expressed in mm)

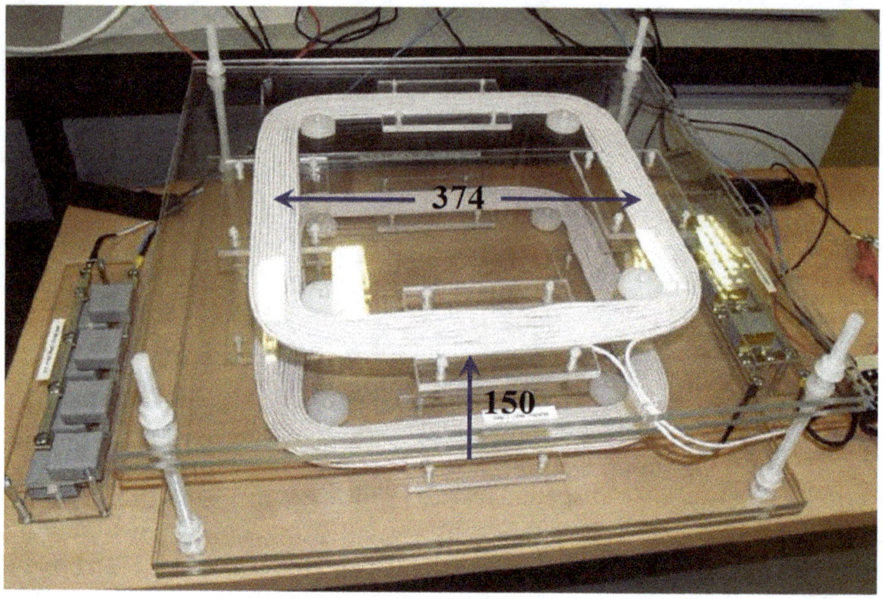

side of the coupling, while R_s accounts for the stray series resistance of the compensated secondary side added to the resistance of the rectifier, and R_L represents the resistance of the battery.

The overall efficiency η of the system is the ratio between the active power P_L absorbed by R_L and the active power P_g supplied by the grid. At resonance, η is:

$$\eta = \frac{P_L}{P_g} = \frac{R_L}{R_L + R_s + (R_L + R_s)^2 \dfrac{R_p}{\omega_r^2 M^2}} \tag{17}$$

Figure 11. Circuit model of the charger

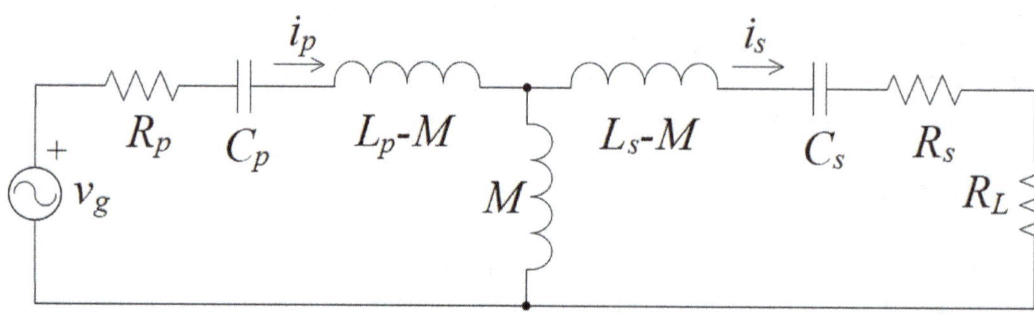

According to (17), it follows that an increase in either ω_r or M leads to an improvement in η. On the contrary, increasing the values of any of the three resistances contributes to its decrease. With regard to the active power transferred to the secondary side at resonance, and considering the resistances added to the model, equation (10) becomes:

$$P_s = \frac{\omega_r^2 M^2}{R_L + R_s} I_{p,rms}^2 = \frac{\omega_r^2 M^2}{R_L + R_s} \frac{V_{p,rms}^2}{Z_p} \tag{18}$$

Upon modifying equations (6) and (7) in order to account for R_p and R_s, and cancelling the reactances when operating at resonance, Z_p is simplified to:

$$Z_p = R_p + \frac{\omega_r^2 M^2}{R_L + R_s} \tag{19}$$

If (19) is substituted in (18), P_s becomes:

$$P_s = \frac{\omega_r^2 M^2 (R_L + R_s)}{\left[(R_L + R_s) R_p + \omega_r^2 M^2\right]^2} V_{p,rms}^2 \tag{20}$$

Equation (20) shows that, for realistic values of the variables involved, a decrease in P_s follows an increase in either ω_r or M for a given voltage $V_{p,rms}$ on the primary side of the coupling. The transferred power will consequently be compromised if the objective of the system design is to maximize the efficiency, as is usually the case.

Measurements carried out with the prototype while keeping $I_{p,rms}$ at around 10 A show that P_L decreases with R_L, from 1548 W for 5 Ω to 455 W for 20 Ω. The overall efficiency reaches 82.5%, although it is above 90% if only the IPT stage is considered. Further details can be found in (del Toro et al., 2015), including the efficiency figures for three different air gaps.

ANALYTICAL DETERMINATION OF THE COUPLING FACTOR

The purpose of this section is to introduce a method that will lead to both the upper and lower limits for the coupling factor k. It concludes by proposing a k value that is halfway between the two limits, which can be regarded as a first estimate of k.

In order to determine k, the self-inductances on the primary and the secondary sides L_p and L_s, along with the mutual inductance M, must be previously known, as follows from (1). To this end, the two self-inductance values were measured using the Agilent 4294A impedance analyzer, which yields not only the required inductances but also their stray resistances. The equivalent circuit provided by the analyzer consists of a self-inductance connected in series with its stray resistance, which is only a fraction of 1 Ω for both L_p and L_s. Although stray elements can be ignored at this stage (they are actually not needed in order to calculate k when applying the proposed analytical method), they will contribute to building an accurate simulation model of the IPT prototype in the following section.

With regard to M, it can be estimated by following an analytical procedure that relies on Neumann's formula. This formula involves a double line integral that provides the mutual inductance between a pair of neighboring closed circuits C_1 and C_2, and obeys the following expression:

$$M_{\text{single\_turn}} = \frac{\mu_0}{4\pi} \oint_{C_1} \oint_{C_2} \frac{d\mathbf{l}_1 \cdot d\mathbf{l}_2}{r} \tag{21}$$

The name $M_{\text{single\_turn}}$ is adopted to emphasize the fact that (21) applies exclusively to two closed circuits (and, as such, they are composed of only one turn each). Observe that $M_{\text{single\_turn}}$ is a purely geometrical quantity since it depends solely on the sizes, shapes and relative positions of the circuits C_1 and C_2 (Griffiths, 1999, p. 311). Upon solving (21) for the particular case of two square-shaped closed circuits that are equal in size, parallel and coaxial, as illustrated in Figure 12, the following result is reported (Grover, 1922).

$$M_{\text{single\_turn}} = 0.008\left(a \ln \frac{a\sqrt{a^2 + h^2} + a^2 + h^2}{ha + h\sqrt{h^2 + 2a^2}} + \sqrt{2a^2 + h^2} - 2\sqrt{a^2 + h^2} + h \right) \tag{22}$$

In (22), a is the side length of the two squares and h represents the distance that separates their parallel planes. $M_{\text{single\_turn}}$ is expressed in microhenries if a and h are in centimeters.

Observe, however, that while the coils of the inductive coupling under consideration are approximately equal, parallel and coaxial, they are not composed of only a single turn but of N. In order to obtain a correct estimate for M, N^2 contributions calculated with (22) must be added. Each contribution results

Figure 12. Equal, parallel and coaxial square-shaped circuits

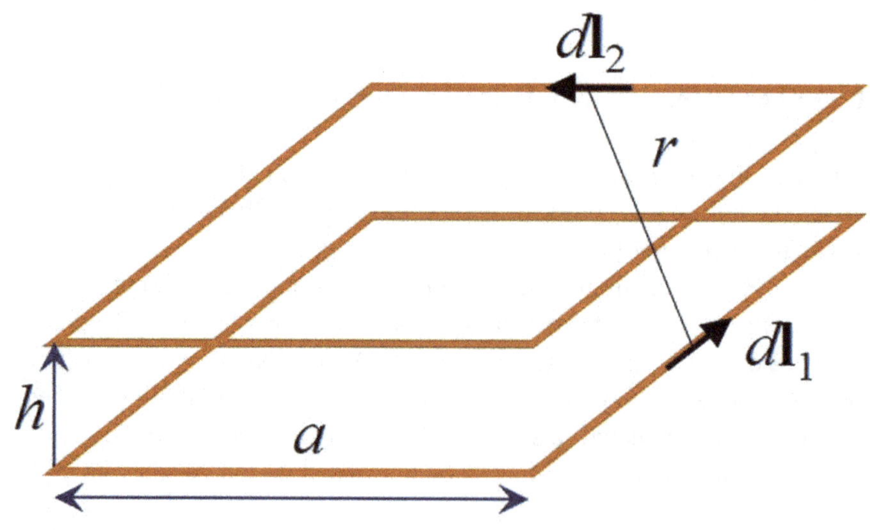

from the interaction between a different pair of turns, where one of the turns belongs to the primary coil and the other to the secondary coil:

$$M = \sum_{i=1}^{i=N} \sum_{j=1}^{j=N} M_{ij} \approx N^2 M_{\text{single\_turn}} \tag{23}$$

Note that, according to (23), the addition of the $N^2 = 484$ contributions from all pairs of turns does not guarantee a very accurate result for M, but rather an approximation. This is a consequence of having applied Neumann's formula in order to calculate the N^2 individual contributions to M, assuming that the calculation is precise in every case. Unfortunately, Neumann's formula does not strictly apply to most of the contributions, since any turn in a coil is perfectly aligned with only two turns in the other coil (owing to the two-layer stack on which the coil construction is based). This means that $M_{\text{single\_turn}}$ is accurate for $2N = 44$ contributions to M, corresponding to those pairs of turns that are equal, parallel and coaxial. With regard to the remaining contributions, which amount to $N^2 - 2N = 440$, some degree of inaccuracy is to be expected because the corresponding pairs of turns are still coaxial and parallel, but unequal. Note, however, that some of those contributions will yield values slightly above the exact result for $M_{\text{single\_turn}}$, while in other cases the opposite will occur. This results in a compensation effect that will prevent the estimate for M obtained with (23) from deviating too much from a truly accurate value.

Taking into account the approximations involved when using (23), both upper and lower limits for M can be calculated by assuming that the side length of every turn is the same. In the case of the upper limit, the side length of all turns in the two coils is supposed to be the longest (which equals the length of the two outer turns in any of the two coils); whereas in the case of the lower limit, the side length will be taken as the shortest (which now corresponds to the length of the two inner turns). Figure 13 illustrates this idea by representing both the outer and the inner turns in a lighter tone than the remaining turns.

The application of this idea to the inductive coupling shown in Figure 9 yields upper and lower limits for M equal to 98.99 μH and 56.70 μH, respectively. If (1) is then solved for k using these two limits, the values obtained are 0.253 and 0.145. Note that the distance between any pair of turns belonging to different coils when using (22) is assumed to be constant and equal to the air gap for all pairs of turns. While this is another small source of error, it can be considered negligible owing to the fact that the coil construction is based on only two layers of turns.

Since the previous procedure provides conservative limits for k, it can be safely claimed that its true value will lie somewhere in between. A first estimate of the actual k can be obtained by once again following the same procedure, but now taking the turn side length to be neither the longest nor the shortest, but the length of the middle turns, as is represented in Figure 14.

In this case, M and k are 76.94 μH and 0.197, respectively. If the corresponding upper and lower limits found for M and k are now averaged, the values obtained are 77.84 μH and 0.199, which are almost equal to those found previously. This similarity should come as no surprise; in fact, it is actually to be expected, as will be observed in the plot of M versus the side length a as follows from (22) and (23), and assuming that the air gap h is equal to 150 mm. The graph shown in Figure 15 reveals an almost linear dependence of M on a, at least within the range of interest defined by the dimensions of the outer turns (421 mm) and the inner turns (326 mm).

Figure 13. Turns selected to determine upper and lower limits for M

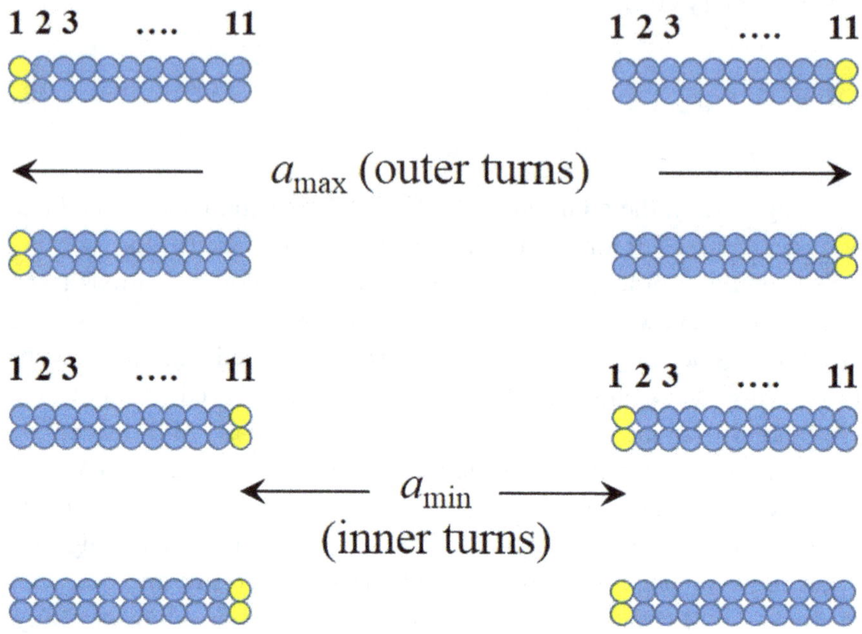

Figure 14. Middle turns selected to determine M

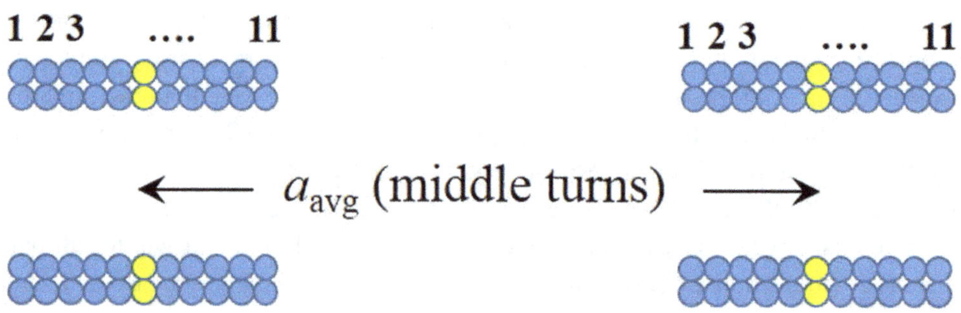

Taking into account the approximations introduced in the analytical method presented in this section, the first estimate obtained for k is probably not very accurate. Nevertheless, it is worth comparing it later with the coupling factor that yields the alternative empirical approach described in the following section. Let us conclude this section by mentioning that the two techniques explained in the current work are by no means the only available methods with which to determine k. For example, the aid of simulation software tools in the construction of a finite element model of the magnetic coupling is yet another way in which a fairly reliable value of k could be provided.

Figure 15. Mutual inductance M versus square side a

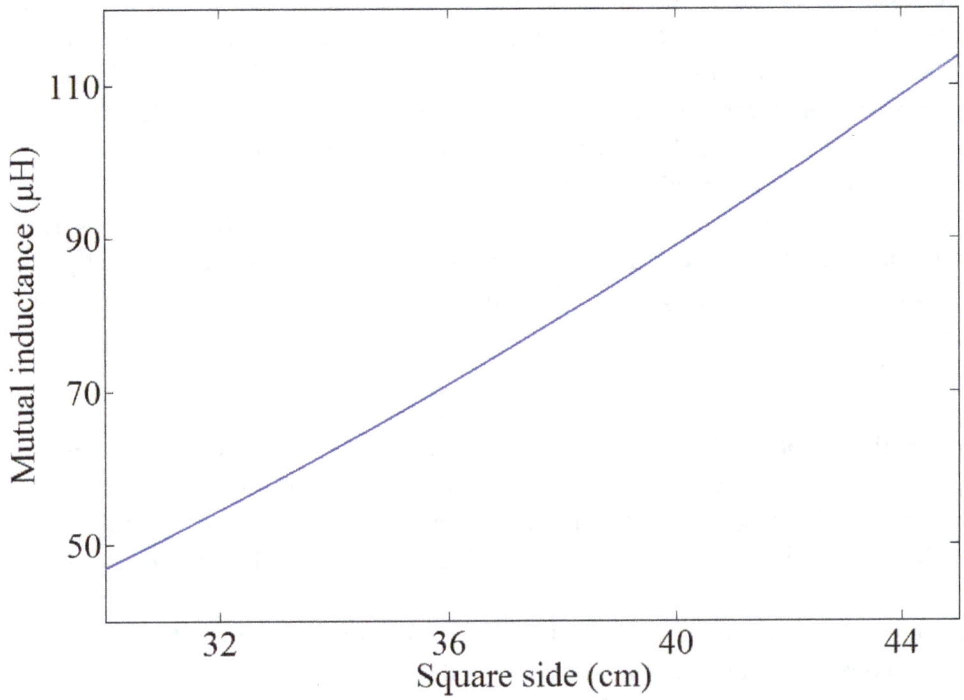

EMPIRICAL DETERMINATION OF THE COUPLING FACTOR

An alternative method with which to figure out k that is not supported by approximated analytical calculations, but rather by experimental measurements, is presented in this section. The proposed empirical determination of k relies on a combination of experimental data obtained from current waveforms on the primary side of the coupling with simulation results provided by a PSpice circuit model that was created as a replica of the prototype.

The underlying idea that the empirical method relies on is finding the best possible match between the experimental current waveforms supplied by the inverter and those generated by the simulation tool. To that end, a number of tentative k values were tried out with PSpice, where k was a key input parameter in the simulation model. The best match was sought by comparing experimental and simulated peak currents for the three load resistances under consideration. It is important to point out that this trial-and-error process had to eventually deliver a single value for k, regardless of the load resistance being tested. This must be so, since k characterizes the magnetic coupling strength between windings for a given air gap and is not, therefore, expected to be influenced by changes in the load.

The experimental measurements will be presented below, prior to describing the simulation model.

Experimental Tests

A number of experimental tests were carried out with the IPT prototype, having previously set the distance between the coils to 150 mm. Different combinations of 10 Ω resistors with a rated maximum power of

800 W each were arranged to obtain load resistances of 5, 10 and 20 Ω. As stated previously, the rms primary current was set manually to approximately 10 A in all the experiments by adjusting α to the necessary value for each load resistance. The current flowing through the primary coil was monitored with the aid of a 50-MHz current probe and simultaneously displayed (along with other variables of interest) using a power analyzer (Voltech PM6000). Figure 16 depicts the waveforms corresponding to both the inverter output voltages and the currents obtained for three load resistances. The corresponding zero-voltage angles are also included below each waveform.

Observe that, as the load resistance R_L increases, the duty cycle of the voltage waveform output produced by the inverter decreases. This decrease in the duty cycle (which necessarily also leads to a drop in the rms voltage) is, according to (15), accompanied by an increase in the zero-voltage angle α.

Simulation Model

The proposed method exploits the fact that the circuit simulator PSpice includes a device that can be used to couple two or more inductors through the coefficient of mutual coupling (a different name used by the simulator that refers to *k*). The software implements a generalized version of the expression used to calculate *k* between a pair of mutually coupled inductors given by (1), in order to take into account the fact that more than two inductors can be simultaneously coupled (since each pair is characterized by a mutual inductance M_{ij} between inductors *i* and *j*). The current work requires the inclusion of only two inductors coupled through a single *M* in the circuit model.

The PSpice model, shown in Figure 17, comprises the three stages of the experimental IPT system: an H-bridge converter, an inductive coupling featuring an S-S compensation and, finally, an ordinary rectifier with a resistive load connected across its terminals.

As can be seen, the H-bridge converter is not implemented with IGBTs, but rather with voltage-controlled switches named SW1 to SW4 and their corresponding diodes (D1u to D4u), which are necessary to make the switches unidirectional. The timing of their four firing signals Control1 to Control4 is conveniently synchronized according to a phase shift control scheme. It is common practice to place antiparallel diodes across switches in those bridges that drive inductive loads, as is the case here, in order to provide an electrical path for the negative currents (diodes D1 to D4). The DC bus voltage in the simulator, denoted by V_{dc}, was carefully adjusted to reflect the existence of a small voltage drop across both the four switches and the diodes in series D1u to D4u (the switches are not ideal and their on-resistance is not, therefore, exactly zero). Upon the introduction of the corrected value for the DC voltage bus, the H-bridge output voltage waveforms generated with PSpice had the same amplitudes as in the experiments, which contributed to improving the accuracy of the circuit model. In addition to this, and with the aim of building a circuit model as accurately as possible, resistive stray elements that are present both in the windings and in the compensating capacitors were added to the PSpice schematic. All of the circuit parameters, including the stray resistances, are summarized in Table 1.

The phase shift action is implemented in the circuit model by generating control signals for the four voltage-controlled switches that are present in the inverter. Figure 18 shows an example of a timing diagram that illustrates the synchronization of the four control signals and the resulting three-level voltage waveform output produced by the inverter for a zero-voltage angle of 60°. Note that the pair of switches one and four, which belong to the left leg of the inverter, are complementarily controlled. The same applies to the pair of switches two and three, located in the right leg. As can be seen, the overlapping

Figure 16. Inverter output voltage and current for three load resistances: experimental waveforms for (a) 5 Ω; (b) 10 Ω; (c) 20 Ω

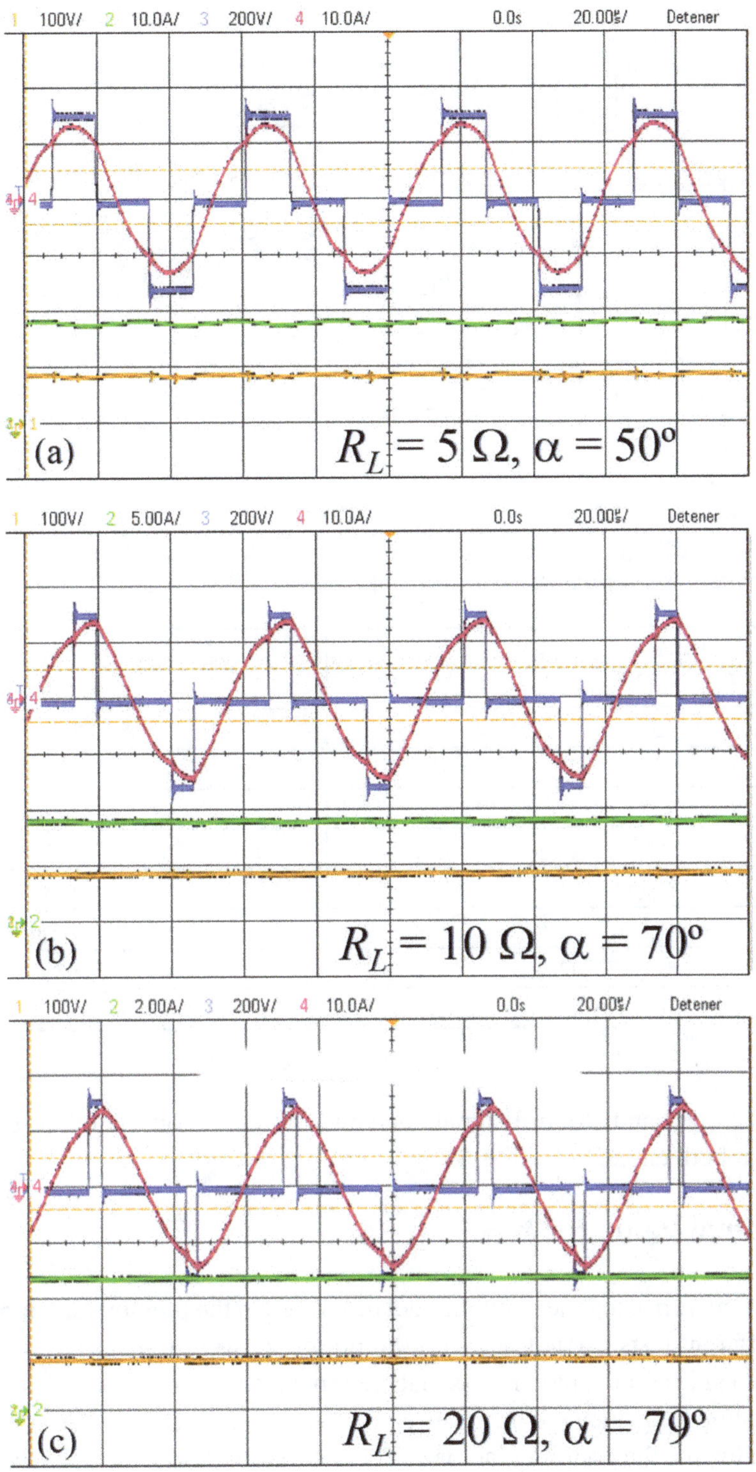

Figure 17. PSpice circuit model of the IPT prototype

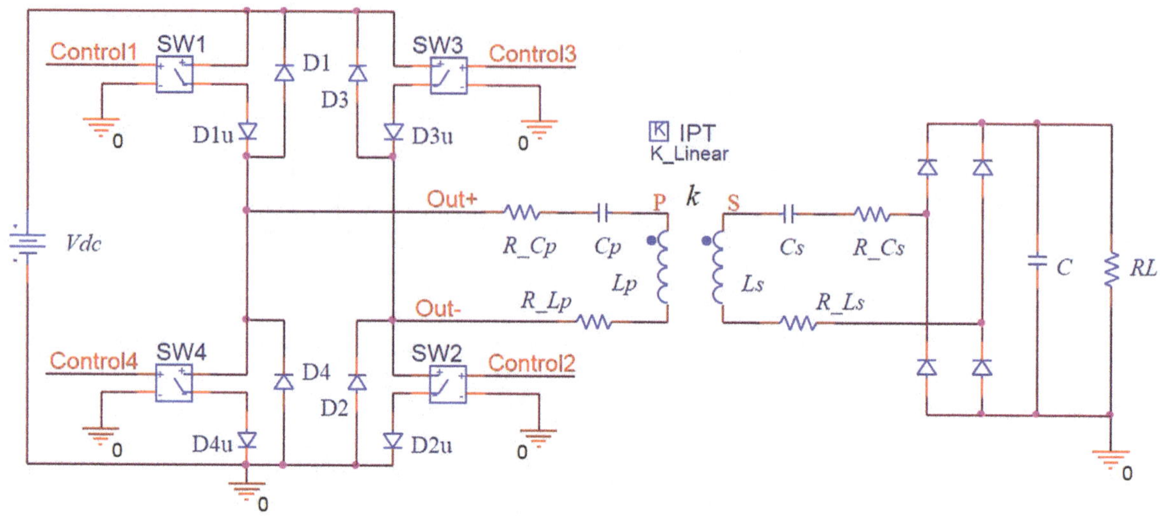

Table 1. Circuit parameters adopted in the PSpice model

Parameter	Value
C_p	186.9 nF
R_C_p	32 mΩ
L_p	393 µH
R_L_p	369 mΩ
C_s	187.0 nF
R_C_s	35 mΩ
L_s	389 µH
R_L_s	367 mΩ
C	18 µF

angle between, on the one hand, control signals one and three and, on the other, between signals two and four, equals 2α in both cases.

Finding the Optimal Value for *k*

Several simulations were run using the circuit model described in the previous section. The experimental conditions of the different tests carried out with the prototype were transferred to each simulation by choosing appropriate values for the parameters that are inputs to the circuit model. These are the DC-bus voltage V_{dc}, the switching frequency f of the inverter, the zero-voltage angle α, the load resistance R_L and, finally, a tentative value for the coupling factor k. The simulation tool processes these parameters, and the resulting voltage and current waveforms at the inverter output are selected and displayed. Table 2 shows, for the three R_L values being tested each time, the peak values obtained from the simulated current

Figure 18. Phase shift action for α = 60º

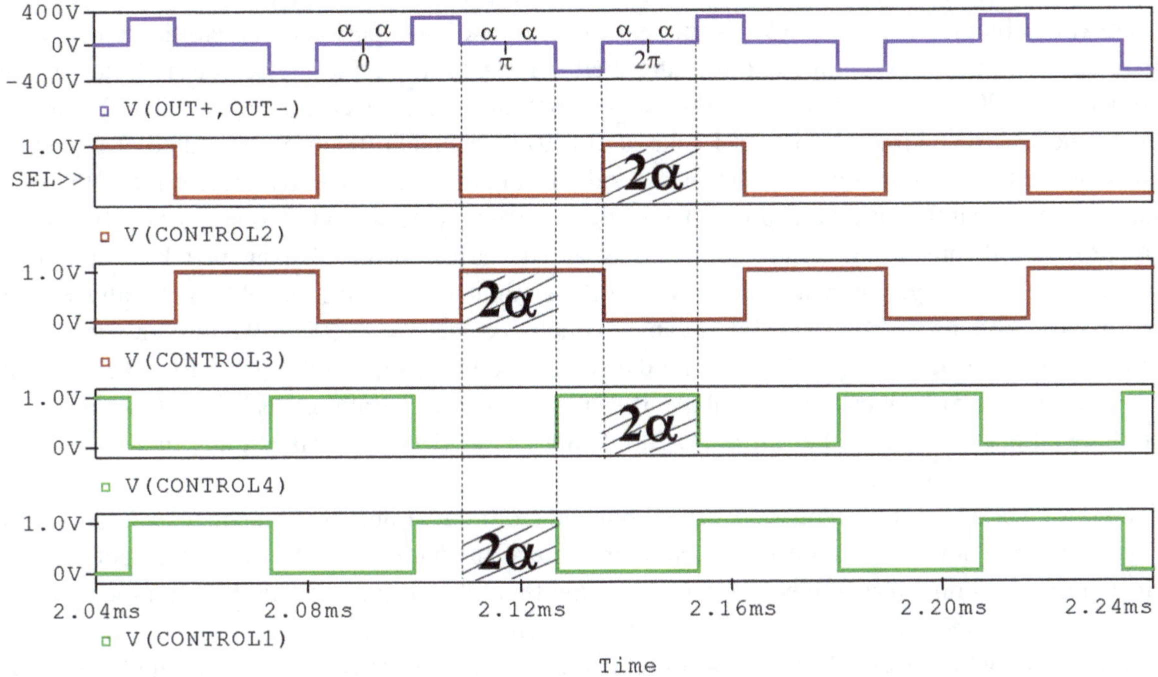

waveforms for a number of tentative *k* values (from 0.206 down to 0.199) that fall in the neighborhood of the optimal *k* value. These simulated peak currents will be compared with the corresponding figures obtained experimentally (which are regarded as target currents). The zero-voltage angle required for every load resistance is also shown in the table.

When the simulated peak currents are below the target values, the corresponding boxes in Table 2 are labeled with the symbol "<", which is intended to visually aid the correct interpretation of the data. Otherwise the data are not accompanied by any particular symbol. The absence of a symbol in any of the boxes from a certain column onwards (reading the columns from left to right) indicates that the best match for *k* will be hit in some of the next columns. Starting with the left-most column ($k = 0.206$), it can be seen that the three simulated peak currents, that are relatively near to the target peak currents are, in all cases, labeled with a "<". This is an indicator that 0.206, despite being close to the optimal *k*

Table 2. Inverter output peak currents for several k values

R_L (Ω)	α (º)	I_{peak} (A) (target)	I_{peak} (A) (simulation)						
5	50	13.30	12.89 <	13.13 <	13.39	13.52	13.66	13.73	13.80
10	70	13.70	12.94 <	13.12 <	13.36 <	13.48 <	13.60 <	13.67 <	13.73
20	79	13.96	13.14 <	13.40 <	13.67 <	13.81 <	13.95 <	14.02	14.09
			0.206	0.204	0.202	0.201	0.200	0.1995	0.199
			coupling factor *k*						

value, is not the best choice. Upon shifting k down to 0.204, all three simulated peak currents get closer to the target values but are also still below them. $k = 0.202$ is the first case in which one of the peak currents (that obtained with $R_L = 5\ \Omega$) is above the corresponding experimental measurement. The two remaining peak currents continue approaching their corresponding target values in the following two attempts ($k = 0.201$ and $k = 0.200$), but do not yet reach them. However, the next k value that follows after another decrementing step of 0.001 (that is, 0.199) reveals that the three simulated peak currents are now above their experimental counterparts (and no labels are consequently present). It is, therefore, possible to state that the optimal coupling factor lies somewhere between 0.200 and 0.199. If k is to be characterized with three significant figures (as is usually the case), and considering that the overall deviations from the three target currents are smaller for the simulated peak currents obtained with $k = 0.200$ than for those obtained with $k = 0.199$, it can be concluded that $k = 0.200$ is the best choice. Adding a fourth significant figure may help to refine the search. To this end, simulations for $k = 0.1995$ show that, in this case, two peak currents are above the target current and only one is below. Once again, the overall deviations are still smaller for $k = 0.200$ than for $k = 0.1995$ (if only slightly this time), which reinforces the choice made.

The three voltage and current waveforms obtained from the simulations, having set k to 0.200 by following the previous trial-and-error process, are shown in Figure 19. They are represented with a dashed and a solid line, respectively. Upon comparing their shapes and amplitudes with those of their experimental counterparts shown in Figure 16, the agreement is notable.

It is interesting to note that the near sinusoidal shape of the current waveforms (particularly in the case of low load resistance values) is in agreement with a low coupling factor since, according to (2), higher-order harmonics are filtered out by a large primary leakage inductance. This prediction is supported by simulations run for a given R_L and two k values that are above the k value that characterizes the actual magnetic coupling. Figure 20 illustrates this point by representing the current waveforms obtained for $k = 0.3$ and $k = 0.5$ with a load resistance of 10 Ω in both cases. It is clear that increasing k for a given R_L results in a decreasing rms current plus greater distortion. The latter can be regarded as a foreseeable consequence of the noticeable presence of higher-order harmonics in the current waveforms, which only occurs when the magnetic coupling is sufficiently tight.

The increasingly growing content of harmonics in the inverter output current waveforms that is apparent for increasing k values raises the question of whether or not there is a limit for k above which the empirical method is no longer reliable as regards determining k with a minimum degree of accuracy. Although further research should be conducted in order to shed some light on this issue, it should be pointed out that the method also worked satisfactorily for an air gap of 125 mm, as can be seen in (Vázquez, Roncero, & Parreño, 2017). In that case, the shape of the waveforms was still quite sinusoidal despite the higher harmonic content, and the k estimate was 0.241 when applying the analytical approach and 0.235 when applying the empirical method. Note that, from a practical perspective, air gaps below 125 mm for EV inductive charging are of little relevance: some authors consider a range of 150-250 mm (Li et al., 2017), while for others, the expected air gap is at least 200 mm (Luca, Reatti, Corti, & Mastromauro, 2017).

Figure 19. Inverter output voltage and current for the three load resistances: simulation waveforms for $k = 0.200$ *and (a) 5 Ω; (b) 10 Ω; (c) 20 Ω*

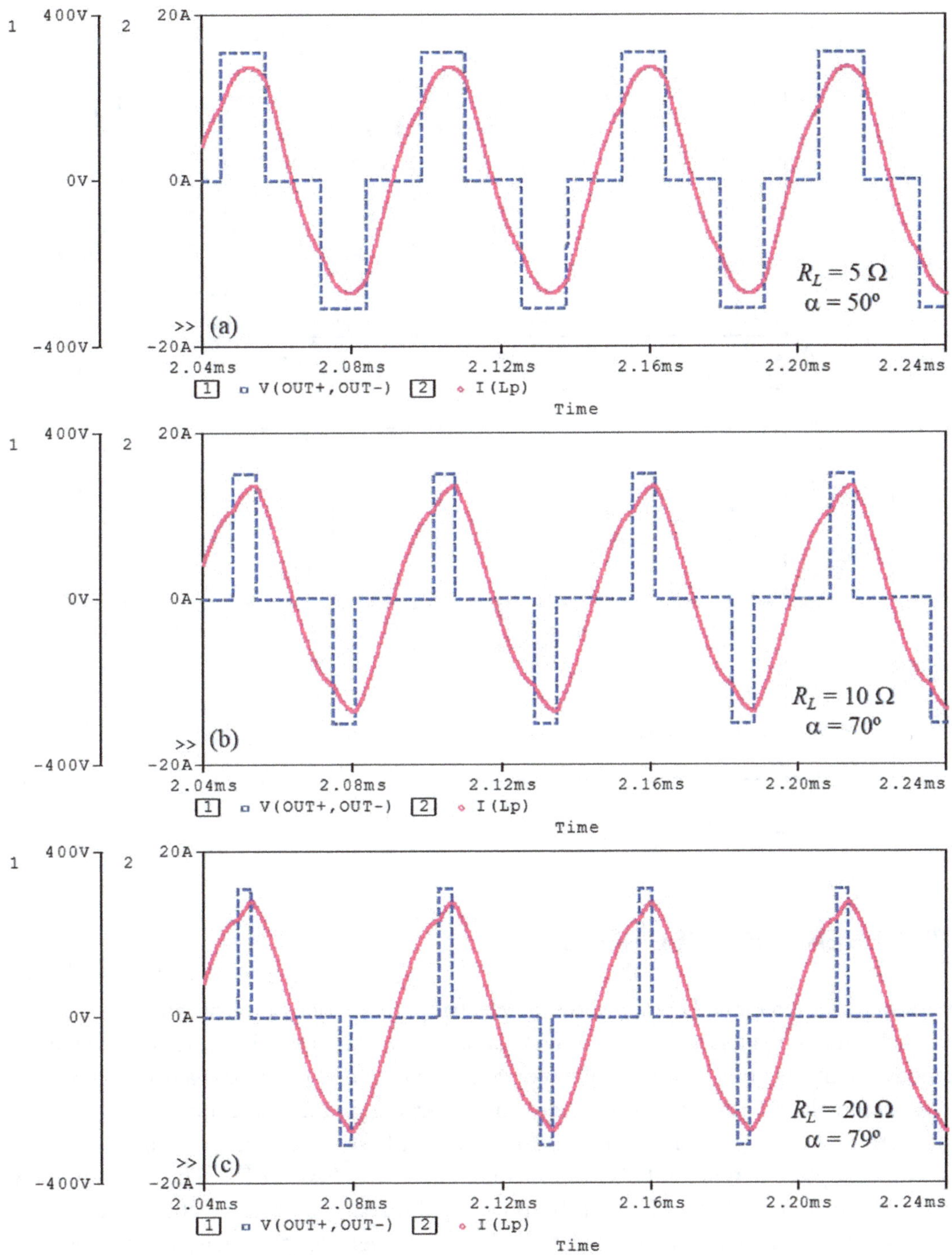

Figure 20. The effect of a tighter coupling on the current waveforms for (a) k = 0.3; (b) k = 0.5

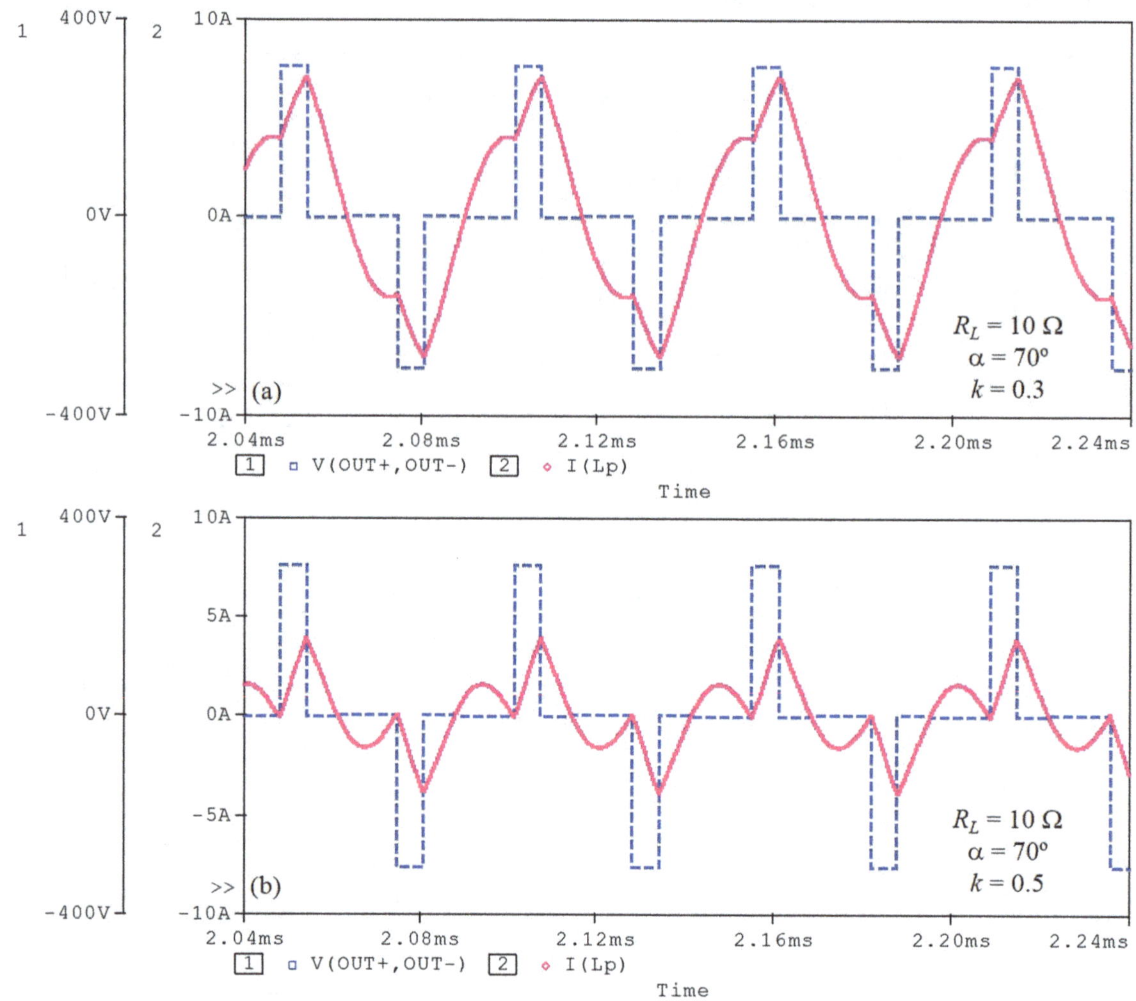

FUTURE RESEARCH DIRECTIONS

Inductive power transfer applied to transportation electrification is one of the promising technologies to have become the focus of attention of numerous organizations and companies worldwide in recent years. The latest developments prove that IPT technology has the potential to become a true competitor in the automotive sector, along with the existing conductive charging, in the near future.

The Society of Automotive Engineers (SAE) released a first issue of a recommended practice for the inductively coupled charging of EVs in 1995, denominated as SAE-J1773. The proposal underwent subsequent revisions in 1999 and 2009, to become a stabilized document in 2014. It establishes the minimum interface compatibility requirements for EV inductively coupled charging in North America, which is intended to transfer power at frequencies much higher than those of power lines. The document includes a large number of specifications that served as a starting point for the early designs of pioneering IPT systems. One of them refers to the fundamental operating frequency for power transfer, which shall range from 130 kHz to 360 kHz (Society of Automotive Engineers, 1999). The proposed range comprises high frequencies with the aim of reducing the size and mass of the on-vehicle part of the magnetic coupling (that is, the secondary side). However, most of the research works to have appeared in recent years report IPT designs that operate well below 100 kHz and do not, therefore, adhere to the range of high switching frequencies suggested by SAE J1773 in the 1999 release, which is a challenging task for design engineers. Two exceptions are the works of (Zhang, Wong, Tse, & Chen, 2014) and (Zimmer, Heinrich, & Parspour, 2014). The former describes the operation of a prototype featuring a converter based on MOSFETs switching at 200 kHz, whereas the latter provides design guidelines for a power unit operating at 140 kHz and suggests the use of SiC-FETs for the physical realization. Both are sound examples of S-S compensated IPT systems.

Several organizations have been developing a wireless charging system standardization in parallel since the release of SAE-J1773. Some of the works have been published in recent years, while others are still in progress. The resulting standards will provide the necessary guidelines and technical requirements for the next generation of IPT systems. One of these organizations is SAE, while others are UL (Underwriters Laboratories), the IEC (International Electrotechnical Commission), and the IEEE (Institute of Electrical and Electronics Engineers) (Hsu, 2013). The SAE International taskforce issued the design standard SAE-J2954 under the title "Wireless Charging of Electric and Plug-in Hybrid Vehicles" in May 2016 (although an initial release of the document was previously published as a guideline as early as 2013). SAE-J2954 establishes a nominal frequency range for light-duty passenger EVs from 81.38 kHz to 90.00 kHz (Li et al., 2017). Note that this is a much narrower range than that proposed by SAE-J1773, and at significantly lower frequencies. The following two recent research works already comply with the new frequency range, and can, therefore, be regarded as good examples of an IPT design methodology that will probably become common practice in the near future. In (Buja et al., 2015) the frequency is set to 85 kHz and the distance between coils varies in the range from 100 to 200 mm, while in (Li et al., 2017), the frequency and the transfer distance are 85.5 kHz and 150 mm, respectively. Both works feature planar ferrite cores in the respective magnetic couplings, whose objective is to enhance the mutual inductance. They also use MOSFETs to implement the switches of the power converters. Unlike IGBTs, whose maximum switching frequency is well below 100 kHz (Mohan, Undeland, & Robbins, 2003, p. 30), MOSFET semiconductors can be switched safely within the frequency range determined by the SAE-J2954 standard. As operation within this range has been a requirement since the SAE standard was released, this will likely unleash an abrupt transition in the converter technology of future IPT designs, moving rapidly from IGBT to MOSFET semiconductors. Another issue that could probably become a trend in the near future is the wide use of SiC in the manufacture of power semiconductors. Power devices made from this compound material, which is characterized by a wide band gap, benefit from a reduced parasitic capacitance, a high frequency switching, low switching losses and a low on-resistance (Millán, Godignon, Perpiñà, Pérez, & Rebollo, 2014). Examples of suppliers that manufacture power modules and components based on SiC are CREE and ROHM.

To continue with the release of new standards, the IEEE Standards Association has initiated some pre-standardization work on dynamic wireless charging, principally motivated by the range of limitations that EVs are currently experiencing and the costs of energy storage for EVs. The aim is to complement SAE-J2954, which deals exclusively with equipment aspects of stationary charging. In turn, UL is working on UL 2750 in order to cover the safety aspects of wireless charging, in parallel with the development of SAE-J2954. Furthermore, the IEC 61980 standard entitled "Electric equipment for the supply of energy to electric road vehicles using an inductive coupling" comprises the following three parts covering different aspects of WPT systems:

- Part 1: General requirements (published in 2015).
- Part 2: Manual connection system using a paddle (published in 2016).
- Part 3: Specific requirements for the magnetic field power transfer systems (published in 2017).

Yet another recent standard that deals with communication issues is SAE-J2847/6, issued in August 2015 and entitled "Communication between wireless charged vehicles and wireless EV chargers". These and other standards that will be released in the near future will have to be taken into account in forthcoming IPT designs.

CONCLUSION

The strength of the magnetic coupling between two mutually coupled coils is characterized by means of the coupling factor k. This factor can be determined in different ways by applying a number of techniques: analytical, experimental, or supported by simulation tools. Two methods with which to estimate the coupling factor of the inductive coupling stage constructed for an IPT experimental prototype devised for the wireless battery charging of EVs are proposed in this chapter. The distance between the two loosely coupled square coils that form the inductive coupling is 150 mm, an air gap that satisfies the specifications of the SAE-J2954 standard.

One of the two methods is principally analytical. It is based on Neumann's formula and provides an expression with which to determine the mutual inductance M between two square-shaped circuits that are coaxial, parallel and of equal size. Keeping in mind that some approximations were introduced in order to obtain M for the real coils used in the prototype, a deviation from the exact M (and, consequently, from the exact k) is to be expected.

The second approach is empirical and relies, on the one hand, on a combination of data obtained from experimental measurements operating the IPT prototype, and on the other, on simulation results delivered by a circuit model built with PSpice. In this case, k is an input parameter that can be properly adjusted in the model to match the experimental peak values of the currents flowing through the primary side of the magnetic coupling for three different load resistances. The shapes of the simulated current waveforms are quite similar to those of their experimental counterparts, regardless of the load resistance, which supports the accuracy of the PSpice model.

The coupling factor obtained with the analytical method is 0.197, whereas the best match resulting from the empirical approach is 0.200. The agreement between the two figures is noticeable and it can, therefore, be claimed that the approximations introduced in the analytical procedure did not affect the result to a large extent, at least in the case of the air gap under study. Both techniques have consequently proven useful to characterize the strength of the magnetic coupling utilized in the prototype, and either could be applied to estimate the coupling factor for the constructed inductive coupling stage in the case of air gaps other than 150 mm that are also of interest in the context of EV battery charging.

ACKNOWLEDGMENT

This research was supported by the Ministry of Economy and Competitiveness of Spain and by the European Regional Development Fund [grant number ENE2015-71417-R (MINECO/FEDER, UE)].

REFERENCES

Allota, B., Pugi, L., Reatti, A., & Corti, F. (2017) Wireless power recharge for underwater robotics. *Proceedings of the 2017 IEEE International Conference on Environment and Electrical Engineering and 2017 IEEE Industrial and Commercial Power Systems Europe*. 10.1109/EEEIC.2017.7977478

Bobrow, L. S. (1985). *Fundamentals of electrical engineering*. Orlando, FL: Saunders College Publishing.

Bosshard, R., Kolar, J. W., Mühlethaler, J., Stevanovic, B., Wunsch, B., & Canales, J. (2015). Modeling and η-α-Pareto optimization of inductive power transfer for electric vehicles. *IEEE Journal of Emerging and Selected Topics in Power Electronics*, *3*, 50–64. doi:10.1109/JESTPE.2014.2311302

Bosshard, R., Mühlethaler, J., Kolar, J. W., & Stevanovic, I. (2013). Optimized magnetic design for inductive power transfer coils. *Proceedings of the 28th Applied Power Electronics Conference and Exposition*, 1812-1818.10.1109/APEC.2013.6520541

Buja, G., Bertoluzzo, M., & Mude, K. N. (2015). Design and experimentation of WPT charger for electric city car. *IEEE Transactions on Industrial Electronics*, *62*(12), 7436–7447. doi:10.1109/TIE.2015.2455524

Córcoles, F., Pedra, J., & Salichs, M. (2004). *Transformadores*. Barcelona, Spain: Edicions UPC.

Corti, F., Grasso, F., Reatti, A., Ayachit, A., Saini, D. K., & Kazimierczuk, M. K. (n.d.). Design of a class-E ZVS inverter with loosely-coupled transformer at fixed coupling coefficient. *Proceedings of the 42nd Annual Conference of the IEEE Industrial Electronics Society*, 5627-5632. doi: 10.1109/IECON.2016.7793285

Del Toro García, X., Vázquez, J., & Roncero-Sánchez, P. (2015). Design, implementation issues and performance of an inductive power transfer system for electric vehicle chargers with series-series compensation. *IET Power Electronics*, *8*(10), 1920–1930. doi:10.1049/iet-pel.2014.0877

Griffiths, D. J. (1999). *Introduction to electrodynamics* (3rd ed.). Upper Saddle River, NJ: Prentice Hall International.

Grover, F. W. (1922). Formulas and tables for the calculation of the inductance of coils of polygonal form. *Scientific Papers of the Bureau of Standards*, *18*(468), 737–762. doi:10.6028/nbsscipaper.141

Hart, D. W. (2011). *Power electronics*. New York, NY: McGraw Hill.

Hesterman, B. (2007, April). *Analysis and modeling of magnetic coupling*. Paper presented at the Denver Chapter of IEEE Power Electronics Society, Boulder, CO.

Hsu, S. (2013). *Alternative fuels & vehicles technology and trend*. Retrieved September 22, 2017 from http://www.csee.net.cn/data/zt_esedc2013/ppt/p4_1.pdf

Huh, J., Lee, S. W., Lee, W. Y., Cho, G. H., & Rim, C. T. (2011). Narrow-width inductive power transfer system for online electrical vehicles. *IEEE Transactions on Power Electronics, 26*(12), 3666–3679. doi:10.1109/TPEL.2011.2160972

Hurley, W. G., & Wilcox, D. J. (1994). Calculation of leakage inductance in transformer windings. *IEEE Transactions on Power Electronics, 9*(1), 121–126. doi:10.1109/63.285502

Lambert, F. (2016). *Norway keeps electric vehicle tax exemption until 2020, positions itself to stay EV leader*. Retrieved September 7, 2017, from https://electrek.co/2016/11/09/norway-keeps-electric-vehicle-tax-exemption-until-2020-positions-itself-to-stay-ev-leader/

Li, Z., Zhu, C., Jiang, J., Song, K., & Wei, G. (2017). A 3-kW wireless power transfer system for sightseeing car supercapacitor charge. *IEEE Transactions on Power Electronics, 32*(5), 3301–3316. doi:10.1109/TPEL.2016.2584701

Liu, X., Ng, W. M., Lee, C. K., & Hui, S. Y. (2008). Optimal operation of contactless transformers with resonance in secondary circuits. *Proceedings of the 23rd Annual Applied Power Electronics Conference and Exposition,* 645-650. doi: 10.1109/APEC.2008.4522790

Luca, P., Reatti, A., Corti, A., & Mastromauro, R. A. (2017) Inductive power transfer: Through a bond-graph analogy, an innovative modal approach. *Proceedings of the 2017 IEEE International Conference on Environment and Electrical Engineering and 2017 IEEE Industrial and Commercial Power Systems Europe*. 10.1109/EEEIC.2017.7977737

Madawala, U. K., Neath, M., & Thrimawithana, D. J. (2013). A power-frequency controller for bidirectional inductive power transfer systems. *IEEE Transactions on Industrial Electronics, 60*(1), 310–317. doi:10.1109/TIE.2011.2174537

Millán, J., Godignon, P., Perpiñà, X., Pérez-Tomás, A., & Rebollo, J. (2014). A survey of wide bandgap power semiconductor devices. *IEEE Transactions on Power Electronics, 29*(5), 2155–2163. doi:10.1109/TPEL.2013.2268900

Miskiewicz, R. M., & Moradewicz, A. J. (2011). Contactless power interface for plug-in electric vehicles in V2G systems. *Bulletin of the Polish Academy of Sciences. Technical Sciences, 59*(4), 561–568. doi:10.2478/v10175-011-0069-z

Mohan, N., Undeland, T. M., & Robbins, W. P. (2003). *Power electronics: Converters, applications and design* (3rd ed.). Hoboken, NJ: Wiley.

Moradewicz, A. J., & Kazmierkowski, M. P. (2009). High efficiency contactless energy transfer system with power electronic resonant converter. *Bulletin of the Polish Academy of Sciences. Technical Sciences, 57*(4), 375–381. doi:10.2478/v10175-010-0141-0

Musavi, F., & Eberle, W. (2014). Overview of wireless power transfer technologies for electric vehicle battery charging. *IET Power Electronics, 7*(1), 60–66. doi:10.1049/iet-pel.2013.0047

Nahvi, M., & Edminister, J. A. (2014). *Electric circuits* (6th ed.). McGraw-Hill Education.

Pinuela, M., Yates, D. C., Lucyszyn, S., & Mitcheson, P. D. (2013). Maximizing DC-to-load efficiency for inductive power transfer. *IEEE Transactions on Power Electronics*, *28*(5), 2437–2447. doi:10.1109/TPEL.2012.2215887

Raju, S., Wu, R., Chan, M., & Yue, C. P. (2014). Modeling of mutual coupling between planar inductors in wireless power applications. *IEEE Transactions on Power Electronics*, *29*(1), 481–490. doi:10.1109/TPEL.2013.2253334

Reatti, A., Corti, F., Pugi, L., Berzi, L., Barbieri, R., Delogu, M., & Pierini, M. Application of induction power recharge to garbage collection service. (2017). *Proceedings of the 2017 IEEE 3rd International Forum on Research and Technologies for Society and Industry*. 10.1109/RTSI.2017.8065961

Society of Automotive Engineers. (1999). *SAE Electric vehicle inductively coupled charging (J1773)*. Warrendale, PA: SAE International.

Van Schuylenbergh K., & Puers, P. (2009). *Inductive powering. Basic theory and application to biomedical systems*. Academic Press. doi: 10.1007/978-90-481-2412-1

Vázquez, J., Roncero-Sánchez, P., & Parreño Torres, A. (2017). Coupling factor of a weak inductive coupling in a 2-kW power transfer system with a 125-mm air gap for electric vehicle chargers. *Proceedings of the 26th International Symposium on Industrial Electronics,* 670-675. 10.1109/ISIE.2017.8001326

Vu, V.-B., Kamal, L. B. M., Tay, J., Pickert, V., Dahidah, M., Logenthiran, T., & Phan, V.-T. (2017). A multi-output capacitive charger for electric vehicles. *Proceedings of the 26th International Symposium on Industrial Electronics,* 565-569. 10.1109/ISIE.2017.8001308

Wang, C.-S., Stielau, O. H., & Covic, G. A. (2005). Design considerations for a contactless electric vehicle battery charger. *IEEE Transactions on Industrial Electronics*, *52*(5), 1308–1314. doi:10.1109/TIE.2005.855672

Zhang, W., Wong, S.-C., Tse, C. K., & Chen, Q. (2014). Design for efficiency optimization and voltage controllability of series-series compensated inductive power transfer systems. *IEEE Transactions on Power Electronics*, *29*(1), 191–200. doi:10.1109/TPEL.2013.2249112

Zimmer, M., Heinrich, J., & Parspour, N. (2014). Design of a 3 kW primary power supply unit for inductive charging systems optimized for the compatibility to receiving units with 20 kW rated power. *Proceedings of the 4th International Electric Drives Production Conference,* 318-322. 10.1109/EDPC.2014.6984415

KEY TERMS AND DEFINITIONS

Compensation Capacitor: A capacitor whose purpose is to be connected either in series or in parallel with a coil in a circuit. The resulting *LC* circuit acts as a resonator at a certain frequency denominated as the resonant frequency of the circuit, in which the reactances of the capacitance and the inductance cancel each other out.

Ideal Transformer: A transformer that is characterized by a perfect magnetic coupling, in which the coupling factor equals unity.

Leakage Inductance: An inductance parameter that is useful as regards modeling loosely-coupled coils, which accounts for the fraction of the magnetic flux generated by one of the coils that does not reach the other coil.

MOSFET: Metal-oxide-semiconductor field-effect transistor. A voltage-controlled device characterized by a high input impedance at its gate and very short switching times (that is, a very high switching speed).

Reflected Impedance: In an inductive coupling, this is the equivalent impedance of the secondary side of the coupling (including the load impedance) as seen from the primary side.

Vehicle-to-Grid (V2G): A system that allows EVs to return electricity to the grid, according to a scheme in which EVs can act not only as loads for the grid but also as generators.

Chapter 13

Focus on OLEV:
On Line Electric Vehicles

Michela Longo
Politecnico di Milano, Italy

Morris Brenna
Politecnico di Milano, Italy

Federica Foiadelli
Politecnico di Milano, Italy

ABSTRACT

Many studies on EVs have been performed in recent years, and various EVs have been developed, like pure battery EVs, hybrid EVs, battery replace EVs, or plug-in hybrid EVs, that use lithium (or polymeric) batteries that can be recharged at home or at a charging station. The biggest challenge to the commercialization of the EV is the battery. The battery problems on electric vehicles can be solved by using roadway-powered electric vehicles (RPEVs). RPEVs do not require heavy and large batteries because they directly get power while moving on a road. These vehicles can take power either in a wired or wireless way. Thus, various wireless power transfer systems (WPTSs) have been developed for RPEVs, and as consequence, new types of RPEV have been developed. WPTS for RPEVs should be able to deliver high power efficiently through a small air gap for avoiding collision between the road and the vehicle.

INTRODUCTION

The environmental pollution caused by fossil fuels is a hot issue around the world in recent years. Since the primary source of energy has been fossil fuels, the concentration of greenhouse gases in the atmosphere, especially CO^2, has increased. These gases lead to the poor air quality, in particular in large cities, and the global warming that can cause ecological calamity such as tropical cyclones, heat waves, drought and extreme tides. International Energy Agency clearly states that the current energy trend is not sustainable environmentally, economically and socially. Therefore, we must devise solutions to achieve the future economic growth without adverse environmental effects.

DOI: 10.4018/978-1-5225-5870-5.ch013

Because of internal combustion engine automobiles are the major source of CO^2, the Electric Vehicle (EV) is the most suitable alternative of petroleum vehicles and it is considered to be an effective solution to reduce pollution and greenhouse gases. Many studies on electric vehicles have been performed in recent years and various EVs have been developed as like pure battery EVs, hybrid EVs, battery replace EVs or plug-in hybrid EVs, which use lithium (or polymeric) batteries that can be recharged at home or at a charging station. The biggest challenge to the commercialization of the EV is the battery. Batteries are heavy, expensive, have short life time and long recharging time: the currently available quick charging time of 20 min is too long for drivers accustomed to rapid fueling. Other problems of batteries are that they need a lot of space, increase too much the weight of the vehicle and they have low energy density, hence, they should be frequently recharged after operating for a short range. As result running efficiency of EV is degraded by these negative effects.

The battery problems on electric vehicles can be solved by using Roadway-Powered Electric Vehicles (RPEVs). RPEVs do not require heavy and large batteries because they directly get power while moving on a road. These vehicles can take power either in a wired or wireless way. A train can be considered a RPEV that take power in a wired way, throughout a pantograph from a contact line. Since the technology has been improved, Wireless Power Transfer System (WPTS) has become more interesting due to the high maintenance of the wired system, as the pantograph. Thus, various Wireless Power Transfer Systems (WPTSs) have been developed for RPEVs and as consequence, new types of RPEV have been developed. WPTS for RPEVs should be able to deliver high power efficiently through a small air gap for avoiding collision between the road and the vehicle. Generally, WPTSs can be divided into three categories: Inductive Power Transfer Systems (IPTSs), Coupled Magnetic Resonance Systems (CMRSs) and Capacitive Power Transfer Systems (CPTSs). Nowadays IPTS is the most appropriated system to transfer power to a Wireless Electric Vehicle (WEV)

FUNDAMENTALS PRINCIPLES OF THE IPTS

In general, IPTs consist on two subsystems as shown in Figure 1: one is in the roadway, it allows to transfer power to the vehicle and consist on a high-frequency inverter, a primary capacitor bank and a power supply rail (or transmitter). The other one in the on-board system to receive power from the roadway system and consist on a pick-up coil (or receiver), a secondary capacitor bank, a rectifier and a regulator for a battery pack (Choi & Jeong, 2015; Choi Rim, 2015).

The electric vehicle can take power from the transmitter in this way:

1. The HF inverter converts electric power at 60Hz from the industrial grid to high frequency. This high frequency current flows into a power supply rail which generate a time-varying magnetic flux by Ampere's law: $\int Hdr = i$ (where H is the magnetic field and i is the supply current);
2. Pick-up coil captures the magnetic flux ϕ of the power supply and in accordance with Faraday's law $v = -\frac{d\varphi}{dt}$, a voltage v is induced across the pick-up coil that is inductively coupled with the supply rail coil;

Figure 1. Configuration of an IPTS for RPEV

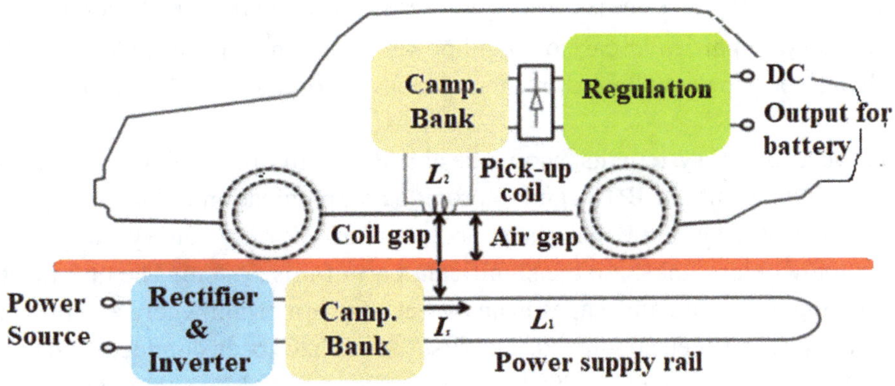

3. The rectifier converts AC to DC power and the regulator control the output voltage which is input voltage of the batteries and the motors. In this way, wireless power transfer is achieved and some of the power is used to feed motors and the other to recharge batteries.
4. Capacitors are used to nullify the leakage reactance due to the leakage magnetic flux that does not link the conductors of the pick-up coil. In this way voltage drop and reactive power are reduced and then the efficiency increases. Depending on the source and load type, capacitors can be connected in series or parallel.

Figure 2 shows the equivalent circuit of IPTS where it is possible to observe V_s is the power supplied by the inverter; L_{l1} and L_{l2} are the leakage inductances of the power supply coil and the pick-up coil respectively, that takes into account the leakage magnetic flux; L_m is the magnetizing inductance; N_1 and N_2 are the number of windings of the two coils that are mutual coupled; C_1 and C_2 are the compensating capacitances connected in series; C_L and R_L represent the load that works at a voltage V_o rectified by a diode bridge.

Figure 2. Example of equivalent circuit for IPTS

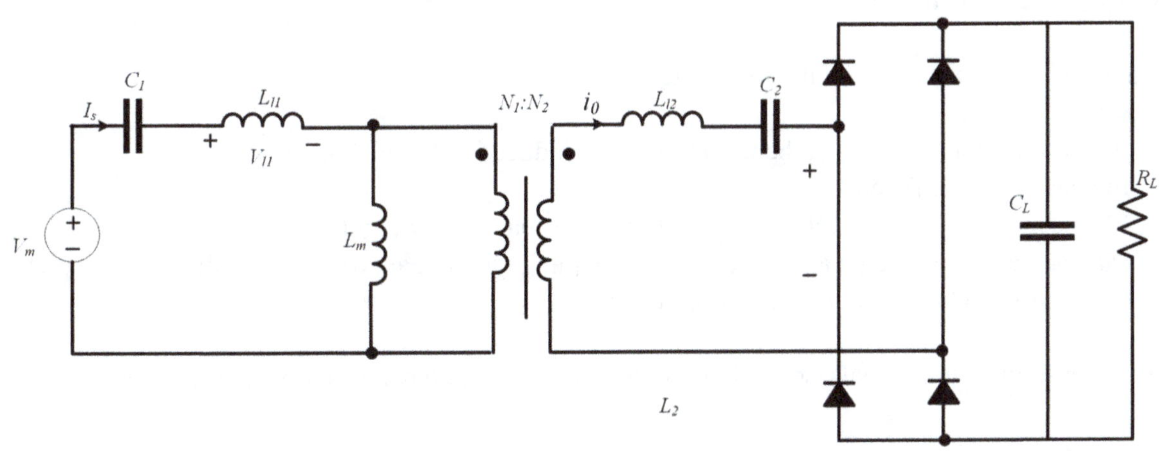

Sometimes resonating LC circuits are used to increase the power transfer efficiency. In this case the switching frequency f_r of the inverter is tuned to the resonant frequency of LC circuits. Actually, resonance is not mandatory, in particular case of small power where the current induced in the leakage inductance is small and so, also the voltage drop among them (Choi & Jeong, 2015; Choi Rim, 2015; Song et al., 2012).

RPEVs also have a control system that should be able to identify the vehicle to switch on and off the inverters, to monitors the health of IPTSs and to provide communication links.

In the past few years, the system RIPT has gained popularity for EV battery charging. The charging pads, which form the wireless links, are the significant part of this system. The optimization of charging pads by employing different core shapes is an aspect very important where extensive studies in the industrial and academic have been conducted. The first studies have been conducted on the conventional designs in particular E-cores (Kim et al., 2001), U-cores (Pedder et al., 1999) and pot cores (Laouamer et al.,1997; Budhia et al., 2011), but for the EV applications, in order to achieve the desired flux path was required to have a greater thickness of the core and for this reason these designs were incompatible. Moreover, these designs are fragile, weighty, expensive and highly sensitive to horizontal misalignments (Budhia et al., 2010). However, it is necessary to consider other problems; in particular, they compromise the ground clearance of the Electric Vehicles or require extensive chassis modification. For this reason, it has been necessary to think for the stationary charging of EVs, different coil structures considering enhanced tolerance to misalignment.

Planar coils are usually classified into two categories (see Figure 3): Polarized Pads (PPs) are multiple coil pads which can couple and create the parallel, perpendicular or both the component of flux (Zaheer et al., 2015); Non-Polarized Pads (NPPs) are single coil pads which can couple and generate only perpendicular components of the flux (including circular pads and square pads).

The initially polarized pad was an H-shaped ferrite bar with double-sided winding (Fujita et al., 2015). In this case, the height of the flux generated was twice the height generated by the PP of the same size giving better tolerance to horizontal misalignments. Moreover, a flux out the back of the pad is generated and for this reason when an aluminium backing plate is added it generates eddy current than that usually expected (Boys & Covic, 2012). Single-sided multiple coil-polarized pads were developed such as: DDP - Double-D Pad, DDQP - Double-D Quadrature Pad and BP - bipolar pad (Boys & Covic, 2014) in order to overcome the limitations of the H-shaped pad.

Design Goals of IPTS

The design issues for IPTSs are the following:

1. Increase magnetic coupling, hence, the voltage induced in the pick-up coil;
2. Increase power efficiency;
3. Makes modules as compact as possible to reduce space and weight;
4. Manage resonance frequency variation and coupling factor change due to misalignment of pick-up coil position, change of air-gap or even temperature;

In order to meet these goals, several important design issues have been improved, in particular:

Figure 3. Different charging pads: (a) CP (b) H-Shaped Pad (c) DD Pad (d) DDQ Pad and (e)Bipolar Pad

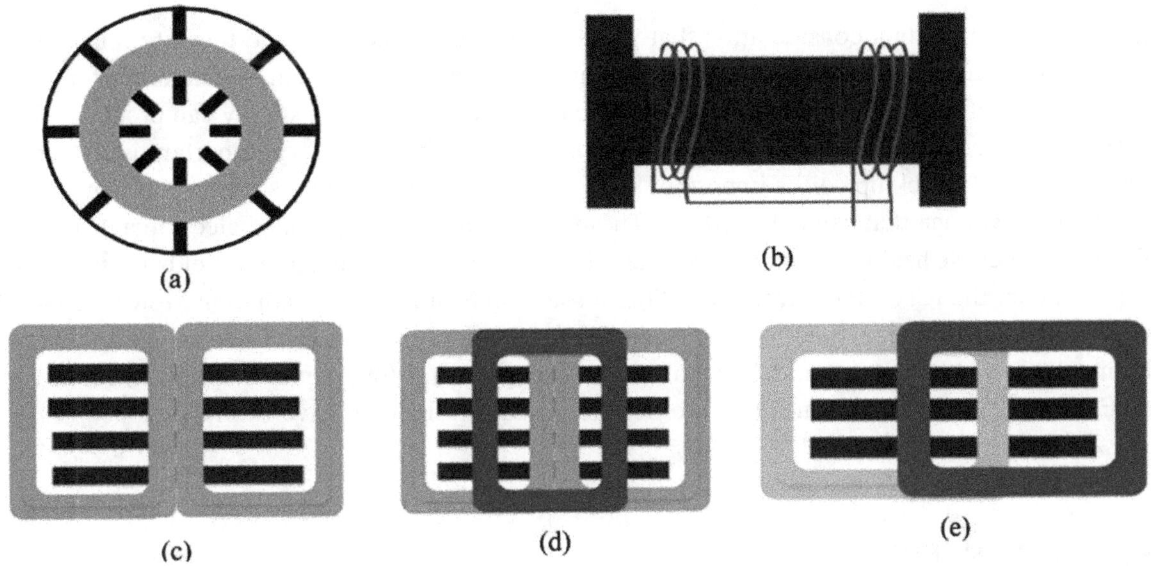

- **Switching Frequency:** A compromise for the inverter and rectifier switching frequency must be done. In fact, higher is f_s higher are the switching losses but the capacitor and coil size can be decreased so much for the same required power. For high values of the switching frequency also the leakage reactance increases then the voltage drops. For this motive f_s cannot be higher than 50 kHz but also lower than 20 kHz because acoustic noise may appear.
- **High Power and Large Current:** Power of hundreds of kilowatts and currents of hundreds of amperes make difficult to handle cables, converters and devices. Due to the high f_s a Litz wire of 30 A rating is needed: Litz wire is a particular wire made of multiple thin wires individually insulated and twisted to reduce the so called "effetto pelle" and then losses.
- **Coil Design:** To increase the magnetic coupling with the pick-up coil and reduce the leakage flux, the design of the coil is crucial for IPTSs. Pick-up coil has to be neither heavy nor thick while the coil of roadway subsystem has a unique design because the cost per km is a critical issue for RPEVs.
- **Segmentation of Power Supply Rail:** IPTS should be many along a road because the power supply rail cannot be infinitely long and needs to be segmented. Each segment has to be switched on and off independently with solid-state switches. Full power is supplied to a segment only when there is an RPEV on that segment. The length of the power supply rail is an important issue because it cannot be too short for economic reasons and also too long due to increased resistance and losses.
- **Resonant Frequency Variations:** Resonant frequency of the road-way and on-board subsystems can vary due to the change of the magnetic coupling cause by lateral misalignments, air-gap variations and longitudinal movements for example. As consequence, coils and inverter need to be able to cope with these frequency variations (Choi & Jeong, 2015).

Safety Consideration

There are others important consideration that we need to take into account and they are related to the safety of pedestrians. Under the road are laid some high voltage cables that can be dangerous for people. In fact, a failure of insulation of these cables or also of the roadway power supply can produce a lethal potential difference for pedestrian. Therefore, the insulation of them is a very important issue and also the insulation of the pick-up coil is important because if high power output is reached, among them is induced a high voltage that can reach 10 kV. These power cables also produce electromagnetic fields (EMF) that can cause health problems to people. For this reason, electromagnetic fields for pedestrians need to be under the ICNIRP (International Commission on Non-Ionizing Radiation Protection) guidelines. ICNIRP is an organization that do researches for the possible effects of non-ionizing radiations on human body and it draw up guidelines to avoid health problems for people. In order to reduce EMF, some EMF cancellation techniques that are not discussed in this topic have been developed (Choi & Jeong, 2015; Suh et al., 2010).

HISTORY OF RPEV'S

The history of RPEV is found in the US patent of the transformer system for electric railways by M.Hutin and M.Leblanc in 1894, and this configuration is quite the same of modern IPTSs. During 1970s, due to the oil crisis, two projects were undertaken in the US to investigate in RPEVs and to minimize the petroleum use of vehicles. These projects gave origin to several prototypes of RPEV but not appropriately operated. In 1992 the university of California, Barkley undertook the Partners for Advanced Transit and Highways (PATH) program, with whom broad investigation, field tests on RPEVs were developed and also the first RPEV buses which achieved an efficiency of 60% with 60 kW power output and a 7.6 cm air-gap. These buses had not been commercialized because they work at low frequency (400 Hz), due to the absences of fast power switches, and because the cost of the high-power rail was around 1M$/Km (Ahm et al, 2010). In addition, RPEV system has low efficiency, acoustic noises, small air-gap and lateral tolerance. Another problem of PATH tests was that the road had some mechanical problems, such as cracks on the asphalt, due to the installation of power transmitters, whose sectional geometry is shown in Figure 4.

Since 2009, OLEV (On-Line Electric Vehicle) project has been conducted by a research team led by Korea Advance Institute of Science and Technology (KAIST) and it has solved many of the problems of PATH project for commercialization. In particular, through OLEV project the operating frequency has been increased at 20 kHz permitting to reach system power efficiency of 83% and the air-gap and the lateral tolerance have been increased at respectively 20 cm and 24 cm (Lee et al., 2010; Yang et al., 2015). Also, the power rail construction cost, which is about the 80% of the total commercialization cost, has been reduced to at least a third of that of the PATH project (Shin et al., 2012).

Figure 4. Section geometry of roadway power transmitter and pick-up coil in path test site

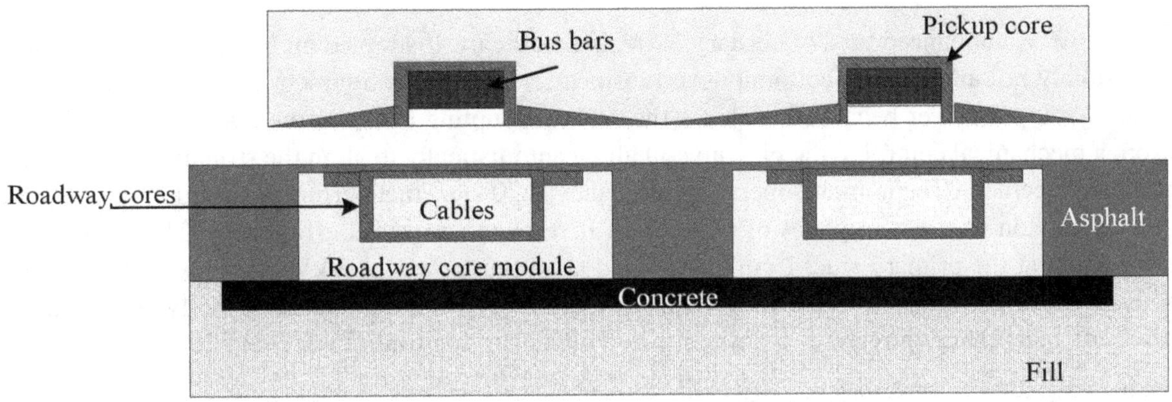

ON-LINE ELECTRIC VEHICLE

Although the name OLEV literally refers to the vehicle, it actually includes the charging units, which are the power transmitters installed along the road (Shin et al., 2012, Lee & Cho, 2015; Ko & Jang, 2013). The critical issues that affect commercialization of OLEV and that have been improved along the years are:

- **Air gap:** It is the distance between transmitter and pick-up coil. Greater is the air-gap, better the vehicle can be driven and lower is the probability to have mechanical impacts with the power rail supply;
- **Lateral misalignment:** Higher is the maximum lateral displacement, better the vehicle can be driven, in fact the driver does not have to stay perfectly centered with the power rail and the power can be transferred efficiently also in case of misalignment;
- **EMF values:** EMF values must respect ICNIRP guidelines to avoid unwanted effects, due to electromagnetic fields, on pedestrians;
- *Overall efficiency;*
- *Initial investment cost;*
- **Deployment time and rail robustness:** Lower is the deployment time and more robust are the power supply modules, lower is the extra cost needed for the infrastructure.
- **Space and weight:** Due to the fact that OLEVs are road vehicles, they need to be small and not heavy, hence, the size and the weight of the components on board is an important issue.

In OLEVs and RPEVs, batteries are used to feed power to electric motors only when the vehicle is running where no power transmitter unit is installed, thus, the size of batteries can be reduced of a lot. The main disadvantage of OLEVs is the initial investment cost, which is significant higher than for example plug-in EVs (Jung et al., 2013). However, a part of this initial cost can be reduced thanks to small batteries and the increased operating time due to the elimination of recharging time. In fact, there is a tradeoff between the capacity of the battery and allocation of power transmitters, as will be explained later in the costs analysis. The distance the vehicle has to travel without getting the power from the power rail supply determines the battery size. KAIST current design goal is to install a battery that can give free driving ranges of 10 km for a bus and 30 km for a car (Yang et al., 2015).

First-Generation OLEV (1G)

The 1G OLEV, announced on 27 February 2009, is a golf cart. It adopted an E-type cores for both the power supply rail and pick-up coil, but as a transformer, lateral misalignment of both of cores reduce drastically output power because it reduces the mutual coupling between the two windings. For that reason, a mechanical control to the pick-up coil allow automatically to align the structures within 3 mm lateral displacement. The nominal operating frequency is 20 kHz, that permitted to significantly reduce the cross section of the core plates of power rail. It reached an overall efficiency of 80% with 100A rated current at the primary side, 1 cm air-gap and an output power of 3 kW per pick-up. The biggest disadvantage of this OLEV was the very small value for air-gap that is not feasible for dynamic charging while stationary charging requires many stoop and high charging time (Park et al., 2010).

Second-Generation OLEV (2G)

The 2G OLEV, announced on 14 July 2009, focused on dramatically improving the air-gap of 1G OLEV. It adopted a U-type mono rail structure and a I-type pick-up coil with a nominal operating frequency of 20 kHz. As shown in Figure 5, the rail core width (W_p) is much smaller than the length of the pick-up coil, so, due to fringe effect increases the effective area of magnetic path. In fact, the effective pick-up width (W_{eff}) increases as the air-gap increases. Therefore, the magnetic flux transferred to the pick-up coil is proportional to the root of the air-gap. This allow to reach an air-gap of 17 cm. A pair of return cable is used to mitigate EMF (Electro Magnetic Field) for the ICNIRP guidelines. The power supply rail is paved with asphalt to provide the same friction force as normal roads and the upward magnetic flux generated from pick-up coil is shielded by an aluminum plate with 5 cm distance. The primary rated current is 200 A and it achieve a maximum efficiency of 72% and a maximum output power of 60 kW with ten pick-up coils (Lee et al., 2010). The big difference between primary core width and the length of pick-up coil desensitize misalignment among the two coils. Indeed, the maximum allowable misalignment is half of the primary core length (W_c), thus, make possible to transfer 50% of total power with 23 cm lateral misalignment and the magnetic flux transferred becomes 70.7% of their maximum.

This technology has been used from Endesa (ENDESA, 2013) for the vehicle initiative consortium for transport operation and road inductive application (VICTORIA) project and adopting the triple charging technologies such as conventional plug-in, stationary, and dynamic charging systems based on an IPTS for RPEVs (Mi et al., 2016). For stationary and dynamic charging, Endesa did not adopt multiple small pads but rather a power supply rail to guarantee several advantages such as its continuous power transfer, low construction cost and time, and simple control characteristics; with a rectangular pick-up, the U-type power supply rail, which used for the 2G OLEVs (Mi et al., 2016).

Third Generation OLEV (3G)

By return cables for limiting EMF in 2G vehicles, construction cost increases and output power is limited. To solve this problem 3G OLEV adopted the W-type power supply rail while pick-up coil remains with flat shape. This structure does not need return cables as shown in Figure 6 because out of the supply core magnetic fields produced by power cables are opposed, hence, the total magnetic field out of the supply core is reduced. The primary core width is still narrow while pick-up coil length is wide, thus, is possible to transfer power with large air-gap. The return path of the magnetic flux in W-type core is

Figure 5. (a) U-type primary coil and I-type pick-up coil, (b) The increased of pick-up area with large air-gap and (c) Maximum allowable lateral misalignment

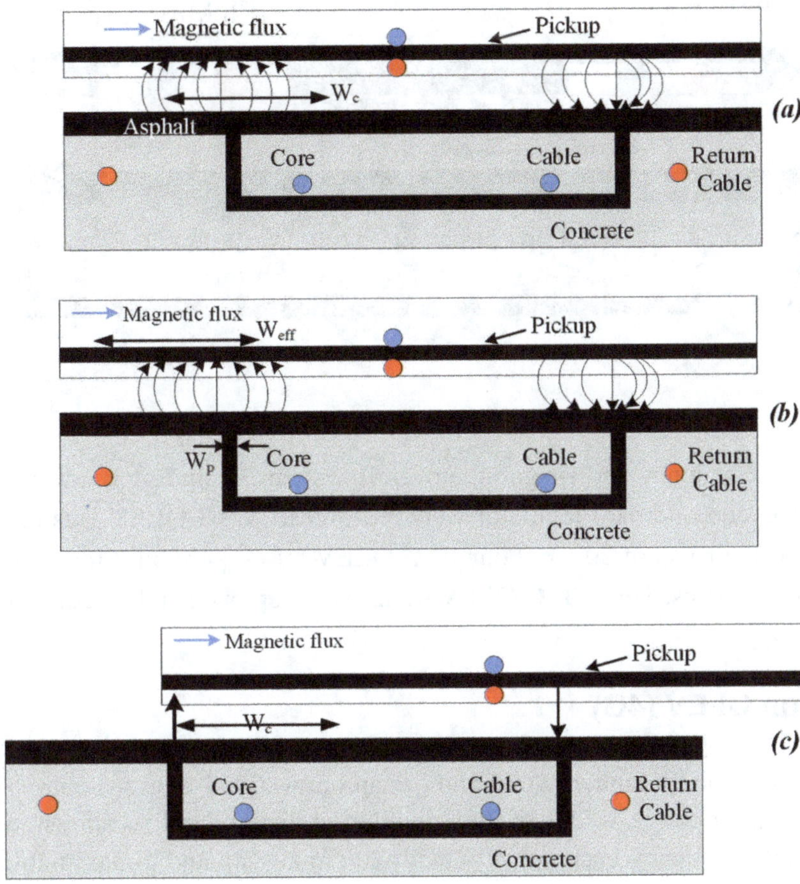

doubled, so the transferred can be increased but the maximum allowable lateral misalignment is a quarter of the length of primary core. The flat pick-up core prohibits the magnetic flux between the power supply rail and pick-up from leaking and so the upward leakage magnetic flux is drastically decreased and the aluminum plate under the vehicle for magnetic field shielding is no longer required. Moreover, the width of power supply rail is decreased to 80 cm which is almost a half of that of the 2G OLEV. The efficiency reached by 3G OLEV SUV is 71% with 17 cm air-gap.

Figure 6 shows the fish bone structure of the primary core; in this way the amount of core is reduced to 1/5 compared to the second generation of OLEV therefore the total magnetic resistance becomes three-quarters smaller. Accordingly, the output power is increased to 15 kW for pick-up coil whereas the decrease of magnetic flux transferred due to bone structure is negligible. For a bone core- gap (X_d) of 7 cm between each W-type core the magnetic flux decreases only by 16%. The power supply rail is reinforced with two iron bars and power cables are protected by fiber reinforced plastic pipes to keep the same endurance of concrete. In addition, during power rail road construction, concretes can percolate down through the bone structure cores. For these reasons the construction of the power rail road can takes a few weeks, thus, is affected by weather and debris and this delay obstruct commercialization of OLEV (Lee et al., 2010).

Figure 6. W-type primary coil and flat pick-up coil

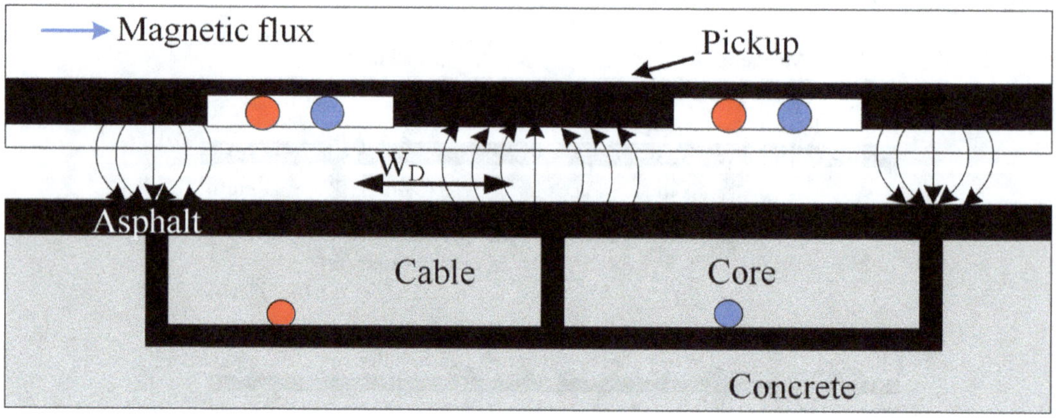

To achieve a higher efficiency and air-gap the overall systems, including roadway inverter, power supply rail, pick-up coil and on-board regulator were redesigned. 3^+G OLEV Bus reach a maximum efficiency of 83% and an air-gap of 20 cm. Four 3^+G OLEV buses were built for tests while six were built for full operation purposes. Three 3^+G OLEV trains were deployed at the Seoul Grand Park (Choi & Jeong, 2015).

Fourth Generation OLEV (4G)

4G OLEV was announced on 31 January 2010 and has an innovative I-type structure of power supply rail with a maximum output power of 25 kW for a double flat pick-up coil, as shown in Figure 7. This lead to reach an Overall efficiency becomes 80% with 20 cm air-gap and a maximum allowable later misalignment of 24 cm. The primary I-type core is divided into many poles that are alternately positive and negative: due to the double pick-up coil the magnetic flux exits only positive magnetic pole of power supply rail and enters only in negative magnetic pole because the magnetic flux enters in the lower pick-up plate and exits form the upper plate. Therefore, this structure allows to reduce leakage magnetic flux because the magnetic flux has a privileged path, hence, increases the transferred one, the induced voltage and the total efficiency. The alternating magnetic poles along the road also leads to drastically reduce EMF for pedestrians.

The I-type power supply rail, whose name stems for its front shape, has only 10 cm width. Another advantage of the 4G OLEV is the reduction of the deployment time within few hours because it adopts a module concept for the I-type power supply rail which include the power supply rail and capacitor banks inside.

Table 1 shows that 4G OLEV do not need FRP (Fiberglass) pipe and concrete to protect power cables because they are placed inside a power module that is robust to high humidity and external mechanical impacts (Choi & Jeong, 2015). Therefore, these modules lead to reduce, besides the deployment time, also the road construction cost. Moreover, the I-type core structure lead to save about 150'000 $ with respect to W-type core, thanks to its small width. Based on what has been said, 4G OLEV have a cost per km of the power rail supply that is about 20% lower than 3G OLEV and then resolve one of the most critical issue for commercialization of OLEV (Choi & Jeong, 2015).

Figure 7. I-type power supply rail and double flat pick-up coil

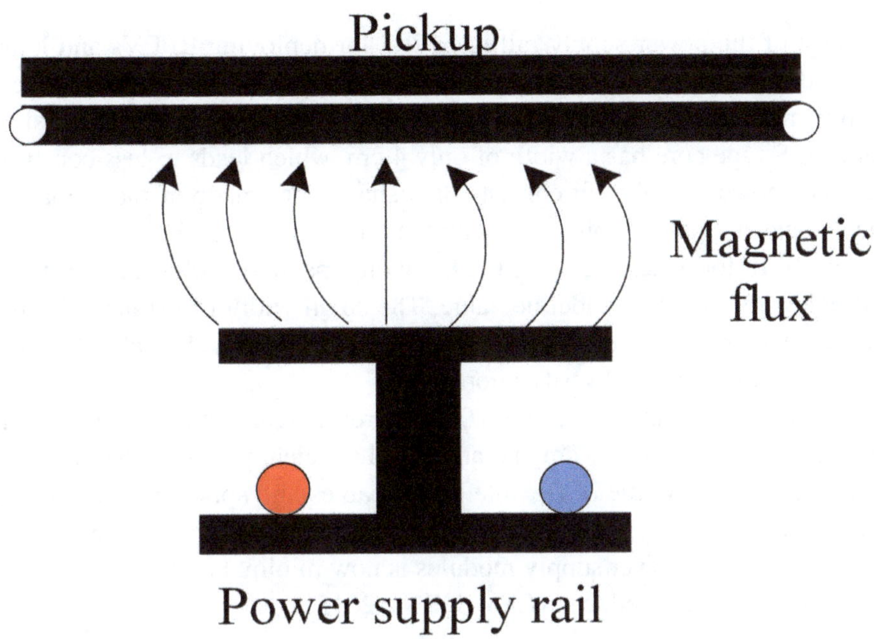

Table 1. Comparison of power supply costs between 3G and 4G OLEV

Unit Price			W-type (3G OLEV)		I-type (4G OLEV)	
Module	*Components*	*Cost [$/km]*	*Num./km*	*Sum [$/km]*	*Num./km*	*Sum [$/km]*
Inverter [100 kW]	-	55,000	4	220,000	4	220,000
Cable	Supply Cable	54,000	2	108,000	4.4	237,600
	Common Cable	54,000	2	108,000	2	108,000
Cable Protector	FRP Pipe	15,000	4	60,000	0	0
	FRP Package	26,000	0	0	1	25,920
Core	Rib type	250,000	1	250,000	0	0
	Square type	96,000	0	0	1	96,000
Capacitor	Cap.Box	5,000	40	200,000	0	0
	Module type	300	0	0	500	150,000
Road Construction	Concrete	60,000	1	60,000	0	0
	Asphal Uncover	2,000	1	2,000	0.5	1,000
	Asphalt cover	11,000	1	11,000	0	0
Misc.	-	50,000	1	50,000	0.2	10,000
Total				**1,069,000**		**854,520**

Fifth Generation OLEV (5G)

The construction cost of the power supply rail is critical for deploying RPEVs and long construction time results in more traffic blocks and extra costs. To further reduce these costs and increase the robustness of power supply rail, ultra slim S-type primary core was proposed, where the S-shape came from its front view. In fact, S-type core has a width of only 4 cm, which leads to less construction cost and deployment time. The power supply rail consists of S-shape core made of many magnetic poles and power cables. Each magnetic pole is connected with the others with a ferrite core plates. This structure significantly reduces EMF for pedestrians due to alternating magnetic poles like fourth-generation and so EMF are well below the ICRNP guideline value. The small width of primary S-type core and the length of double flat pick-up coil allow to an increase of the maximum allowable lateral misalignment that become 30 cm from experimental verifications.

The 5G OLEV become operating in October 2012 and reach a maximum output power of 22 kW for each pick-up coil while an air-gap of 20 cm and an overall efficiency of 71%. Moreover, S-type model makes it easier to fold itself by virtue of flexible power cables and so, connecting power cables is no longer necessary after being deployed because modules can be connected at the factory. Therefore, the impact of deploying the S-type power supply modules is now minimal and the change of existing road operation condition is also greatly reduced (Choi & Rim, 2015).

The S-type power supply modules includes S-type power supply rail, a transparent module cover and an aluminum box for better transfer heat of capacitor banks to the ground. The aluminum box, which need to be isolated from capacitors, also plays a role in cancelling leakage magnetic flux generated by power supply rails. Moreover, each module, that has two magnetic poles, is serially connected to capacitor bank that is installed in an adjacent module to reduce high voltage stress due to the great self-inductance of power supply rail. In general, thinner power cables are used in order to increase ampere-turns of power supply rail but also conduction losses increase.

For commercialization, power supply modules should be robust to high humidity and external mechanical impacts for at least 10 years. As a remedy, modules can be filled with epoxy to reinforce them and protect them from mechanical stresses. As a result, the advantages of S-type power supply rail are principally three:

1. Substantial reduction in construction cost and time of commercialization and deployment thanks to the S-type power rail and the compact power modules;
2. Large lateral tolerance;
3. Further reduction of EMF generated by power supply rail for pedestrians due to the opposite magnetic polarity of adjacent poles.

Sixth Generation OLEV (6G)

Although RPEVs are free from battery problem and ready for commercialization, they are widely used so far due to the high initial investment cost. To cope with this problem, we need a strong motivation for building the infrastructure for RPEVs: make Stationary Charging Electric Vehicles (SCEVs) compatible with IPTS for RPEVs.

In fact, in spite of SCEVs require batteries, they will be widely deployed all over the world because they are a great solution of engine combustion vehicles and in the near future they will replace wired electric vehicles chargers. This is the reason why the 6G OLEV was developed. The first two buses become operative on 6 August 2013 and they have 85% efficiency, 17 cm air- gap and a total power of 100 kW. 6G OLEV adopted a power supply rail that is the same as U- and W-type power supply rail of 3G and 3^{+} G OLEV, but in this case, there is no core plate.

The coreless power supply rail can produce a uniform magnetic field along a road and a rectangular pick-up coil, the same used for SCEVs and compatible with this power supply rail, get uniform output power when moving on the road. Therefore, 6G OLEV can both take power while moving on a road with an IPTS system and take power, for recharging batteries, from a recharging station for SCEVs in a stationary mode.

The self-inductance of a coreless power supply rail, that is the sum of the leakage inductance and the magnetizing inductance $L_1 = L_{l1} + L_{m1}$, can become half of the self-inductance of a power supply rail with core plates as shown in Fig. 19(a). In fact, the leakage inductance increases due to the increase of leakage magnetic flux, but the magnetizing inductance decreases because the magnetic permeability of iron is higher than air, hence, the magnetizing reluctance of air is higher than the magnetizing reluctance of iron. Since the voltage drop across the leakage inductance is $V_{l1} = jwL_{l1}I_s$, this voltage stress is reduced and then the operating frequency can be increased from 20 kHz to 85 kHz. In this way, it is possible to reach SAE J2954 standard for SCEVs and also to increase the lateral tolerance with respect to the conventional power supply rails. However, also the mutual inductance of power rail decreases with coreless power supply rail along an air-gap (Choi & Rim, 2015).

WPTSS APPLIED TO RAILWAYS

The term OLEV refers only for Roadway Powered EVs developed by KAIST but Wireless Power Transfer System technology can be applied also to railways and not only to road vehicles (Rim, 2018). In fact, there are many advantages for using WPTSs for railways:

- The absence of the catenary system and pantograph lead to: a reduction of construction costs of 15%, a reduction of tunnel cross-section area of 20% and a reduction of air resistance, that consume a lot of energy, allowing higher speed;
- Maintenance is reduced due to the absence of mechanical contact and also the maintenance work is easier because it is at ground level: these factors lead a reduction of 52% of maintenance costs;
- The WPTS is a more environment-friendly system because of the lack of dust produced by the wearing of contacts;
- The collection capacities of WPTS does not depend on speed and weather conditions, allowing the system to be less susceptible to failure;
- WPTS is more safe and reliable, because there are no high voltage cables exposed to the general public;
- WPTS is helpful in achieving aesthetically improved cities because no overhead power cable modules are required.

The Korea Railroad Research Institute (KRRI) and KAIST developed in 2013 a large-capacity WPTS for the wireless low-floor tram that takes power both when stopped and when moved. The supply cables feed power to the tram at 800 V and 60 kHz reaching 200 kW of feeding capacity. Because trams travel on fixed rails, the air-gap can be lower and so wireless low-floor tram reach 86% efficiency, that is higher than OLEV buses, with 30 mm air-gap (Su & Cho, 2017; Rim, 2018).

In 2014 KRRI and KAIST developed a WPTS for high-speed train that reach 1 MW power transferred and 82.7% efficiency with 60 kHz frequency and 5 cm air-gap. They used the train HEMU-430X (High-speed Electric Multiple Unit-430 Km/h experiment) as target train and they installed four pick-up coils, that work at 2810 V. The problem found is that when the load resistance decreases also the efficiency dramatically decrease.

There are other research teams that have developed trams that take power with a WPTS while moving: Bombardier research team through PRIOMOV project has developed some IPTSs for the dynamic charging of trams and buses since 2010 (Park et al., 2010). In particular, they manufacture a tram that was deployed in Augsburg, Germany. The tram uses an operating frequency of 20 kHz and reach a power of 250 kW with 6 cm of air gap and few centimeters of lateral displacement. The peculiarity of PRIMOV trams is that it uses a three-phase power system supply because of its high power, unlike OLEVs that reach smaller powers. This high-power delivery capacity may offset the initial investment cost of HF inverter and power supply coils but it is not clear that multiphase power systems for RPEVs are better than single-phase systems because of they are more complex, and they need more power cables.

Bombardier had also deployed PRIMOV buses, that reach 200 kW of maximum power, in Brunswick and Mannheim, Germany and in Bruges, Belgium since 2013 (Song et al., 2013). While Germany is at the stage of pre-commercialization for light train, Korea achieved the laboratory testing stage for tram and high-speed trains but there are some barriers facing the development of WPTSs for railways:

- Low power transfer efficiency;
- High energy consumption;
- A feeder system needs to be installed along the whole route;
- Excessive facilities are required because of the high level of reactive power.

Cross-Segmented Power Rails

As previously sad, power supply rail is segmented into few sub rails that are activated only when the vehicle is on the roadway in order to not waste energy and reduce EMF for pedestrians. Each sub rail can be activated by providing it with HF current through a switch box from an inverter.

There are three types of segmented power rail:

1. The first type of segmented power rail is the centralized switching type, composed of a few sub rails, a bundle of power cables and a centralized switch box. Through the switch box, one of several pairs of supply cables is connected to the inverter at a time and it is able to supply power to the OLEV. The problem of this solution is that it need too many bundle of power cables.
2. To reduce the number of power cables, the second type of segmented power rail has been developed. This solution has only one pair of power supply cables, but multiple switch boxes located between two sub rails. All the switch boxes are connected to the inverter and they are activated through a control signal cable.

3. Recently has been proposed a cross-segmented power supply rail called X-rail. X-rail consist of segmented sub rails, autocompensation switch boxes, control signal lines and roadway harnesses. In this solution, sub rails are not separated to each other's but they are connected through the autocompensation switch boxes. Thus, the inverter can activate more than one sub rail at a time and then can drive multiple OLEVs.

Moreover, the total length of power cables in X-rail is reduced compared with the other two segmented power rails. In (Choi & Jeong, 2015), it is possible to observe that the length for X-rail's power cables is half or more with respect to the other technologies. Hence, also the power rail construction cost and conductive power losses decrease proportionally as consequence.

ECONOMIC ANALYSIS

In 2010 KAIST team done a comparison of the infrastructure and the energy consumption costs of 3G OLEV bus versus other EVs and ICE (Internal Combustion Engine) vehicles. As it said previously OLEVs have high infrastructure costs but at the same time, these costs are lower with respect to other EVs, while ICE and then HEV, that are hybrid vehicles with both electric and internal combustion motors, need no money for infrastructure because they can move on asphalt roads. For evaluate energy consumption costs, KAIST team considered a driving distance of 13'286 km per year. OLEVs lead to a reduction of energy consumption cost of 863 $ in a year with respect to ICE, in fact the cost of electric energy in a year is only 391 $ considering a consumption of 4'429 kWh. PEVs lead to the best energy cost reduction in a year but for these vehicles the initial infrastructure cost is higher than OLEVs, about 0.45 billion of $ more (Suh et al., 2010). Another issue that is important to take into account is that this analysis was done with the 3G OLEV but with the following generation OLEV, KAIST has reduced anymore the initial cost of the infrastructure.

Initial Investment Cost Analysis

Another type of economic analysis that is presented is an initial investment cost analysis of public transportation systems operating with wireless charging electric vehicles (EVs) (Rim, 2018; Mouhrim et al., 2016).

There are three types of wireless charging systems:

1. Stationary Wireless Charging (SWC).
2. Quasi-dynamic Wireless Charging (QWC).
3. Dynamic Wireless Charging (DWC).

In SWC charging happens only when the vehicle is parked or idle, in QWC power is transferred when the vehicle is moving slowly or in stop-and-go mode, and in DWC power is supplied even when the vehicle is in motion. The stationary solution was first introduced in order to eliminate the cable connection of plug-in EVs. The DWC system is more advanced and it is the one used for OLEV, but this does not necessarily mean that DWC will replace the SWC and QWC solutions. In fact, the costs and benefits of each system depend on the fleet size, service range and other cost factors, such as battery prices or charging installation costs. Thus, depending on the operational conditions and costs, the SWC solution may be more competitive than DWC, or vice versa. The goal is to compare the initial investment costs for installing the three types of wireless charging (SWC, QWC and DWC) in a new public transportation system. The focus is on public transit -applications, because most of current wireless charging EV applications have been developed and tested for public transit and no for personal vehicles. A significant part of the initial investment analysis is the comparison about costs of different wireless charging systems. This comparison is important, not only because for each type of charging system the costs of the chargers and components are different, but also because the solution used determines the cost of the batteries and the charging infrastructure.

Example: consider a route which is served by a fleet of 10 vehicles. If a stationary charging is used for this service, every vehicle of the fleet need a battery large enough to complete the required circulation and, as a charger needs to be installed only at the base station for an SWC system, the costs of the batteries will be larger than cost of the charger. Suppose now that the QWC type is used, where chargers are installed at the various bus stations, resulting in a cost reduction of batteries. Supposing to adopt a DWC system, batteries will be smaller, but the costs of the charging infrastructure will increase.

This shows that the costs of the batteries and the charging infrastructure can radically change depending on which type of wireless charging solution is used. Thus, the decision about which type of wireless charging use is not so easy and immediate because is difficult consider the combined costs of the batteries and the charging infrastructure. Analysing the costs of QWC and DWC systems is complicated, because in planning for a QWC or DWC-based transit system, the issue of charging infrastructure allocation should be considered together with the size of the batteries. The position of the charging stations determines how frequently the vehicles are charged and where be allocated, in order to supply enough electric power to the EV. Another fact to be considered in the allocation of charging infrastructure is due to the battery size as the battery capacity determines the maximum amount of energy that can be stored. In fact, if the battery capacity is low, the per-charging distance will be short. Consequently, for small batteries if frequent charges are required, the number of charging stations along the route will increase.

A specific optimization model is used to deal with this fact, evaluating the minimum initial investments and so the optimal allocation of the charging stations in relation to battery capacity for the minimum total investment required for dynamic and quasi-dynamic solutions (Park et al., 2010).

For this analysis, two different routing scenarios are considered, and the actual data collected from the OLEV systems currently operating on the KAIST campus and in Gumi City are applied.

Is known that the costs of ownership, lifetime cost analyses and any other return-on-investment analyses are important, but the initial investment cost needs to be evaluated before any other investment and cost analysis. Therefore, this analysis can be considered as the first step toward further investment and analysis (Jang et al., 2016).

Operations and Scenarios

The scenarios used for the initial investment cost analysis are different. It is considered that each of the three solutions operate under the following conditions:

- Passengers are served by identical EVs.
- EVs circulate on a predefined service route, and one complete circulation of the route is called "service".

- EVs start and finish each service at the base station.
- Once a service is completed, the vehicle remains inactive at the base station for a certain amount of time, which is called "resting time".
- EVs follow a regulated speed.

Actually, these conditions are not so much different from the operational rules of the existing public transit systems for any urban environment. These conditions fit with the rules used in the operations of the KAIST OLEVs at KAIST campus and at Gumi City, on which the analysis is based, so a realistic cost figures and actual data can be utilized in the analysis. The KAIST campus shuttle and the Gumi City urban transit are characterized as "short" and "medium" routing services, respectively, as indicated in Table 2. For the Gumi City case, the 34.448-km service route connects the east and west ends of the city through the city centre, with 23 stops and a base station. The KAIST campus route circulates a 3.374-km service with eight bus stops. This KAIST shuttle case may represent a transit service for short-travel distances with frequent stops within limited areas, such as theme parks, school bus routes or in-town transit services.

Characteristics of EV Types

Charging methods, places and cost structures are compared, for each of the three types of charging solution (Jang et al., 2016). In SWC the vehicles are charged at the base station, as shown in Figure 8(a). For QWC, Figure 8(b), the vehicle is charged after a service or while a stop-and-go at bus stations. It is assumed that the identical chargers are installed only at the bus stops and the base station. However, the major investment cost of a QWC-based system consists of the cost of the batteries and the chargers present in the bus stations (Kim et al., 2013).

For DWC, as shown in Figure 8(c), charging is done while the vehicle is in motion. The charging infrastructure involves power supply rails that are installed anywhere on the route and the main cost is for the batteries and the power supply rails.

Economic System Design Method

The goal of this analysis is to compare the initial costs of different EV types for each scenario, and not to evaluate the absolute investment amount. Assuming that each type of EV uses the same type of vehicle units, the vehicle unit cost can be excluded. The fleet of vehicles (which comprises the costs of batteries, vehicle units and other charging components) and the charging infrastructure are characterized as the "energy logistics cost", which accounts for more than 80% of the total cost of an EV based transit

Table 2. Bus category according to service distance

	KAIST	Gumi City
Service category	Short-distance	Medium-distance
Total service distance	3.374 km	34.448 km
Number of stations	8	23
Total energy requirement	5.60 kWh	39.96 kWh

Figure 8. Charging allocation properties for each type of EV: (a) SWC-EV (b) QWC-EV (c) DWC-EV

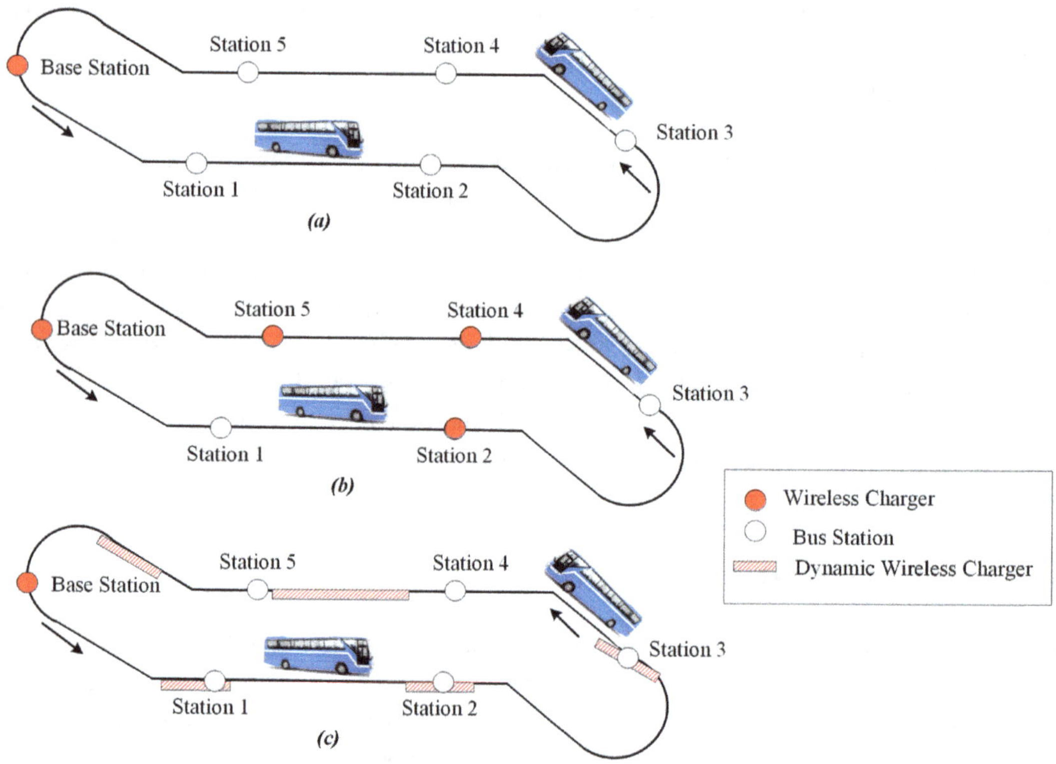

system. Therefore, understanding the cost structure of the energy logistics is critical for deciding on investments in EV-based transit systems.

The optimization models (which estimates the minimum initial investment cost for each type of EV for a given service route) developed to deal with the complex trade-off between the costs of battery size and charging infrastructure and the relative calculations is not reported because results are the most important thing, which are shown in (Jang et al., 2016): the three-dimensional bar graphs of INV_{min} for different values of C_b and C_{in} for the Gumi City case at different k values, which is the fleet vehicles number.

For each graph, the *x* and *y*-axes represent the battery cost coefficient (C_b) and the charging infrastructure cost efficiency (C_{in}), respectively. The *z*-axis represents the minimum investment cost (INV_{min}), which shows the most economical EV type for the corresponding battery and infrastructure cost coefficients. In these graphs, if the most economical solution is the SWC type, QWC type or DWC type the bars are respectively white, gray or black. It can be seen that dynamic solution is the better solution when the cost of the battery is high and the infrastructure cost is low, while the stationary vehicles are preferred when the infrastructure cost is high and the battery cost is low. Given that the infrastructure cost remains the same, and if the battery cost increase, the DWC solution becomes more economical because this type of vehicles requires smaller batteries than the stationary ones.

However, when the infrastructure cost increase but the battery cost remains the same, the SWC solution becomes cheaper, because it has the least expensive charging infrastructure between the three types. As easily predictable, the QWC solution is competitive as a compromise between stationary and dynamic solutions. In (Jang et al., 2016; Choi et al., 2014), another comparison at different *k* values that means that changes the fleet size has been reported. A larger fleet means, in dynamic solution, that the charging infrastructure is shared with a greater number of vehicles, which have smaller batteries than SWC, so the all infrastructure is optimized. The same analysis for the KAIST campus case has been done. For most of scenarios, the economical choice is the SWC (in fact, the service route is short, so a small battery can hold enough power to complete a service without recharging). Interestingly, the DWC type becomes competitive when the number of EVs is more than 19. In practice, it may not make sense to run 19 or more EVs for such a short route (in fact, the KAIST OLEV on the KAIST campus was installed mainly to validate the reliability of the system). DWC is the more economical solution for medium-range service, as in the Gumi City case. QWC type is never shown as the most competitive solution.

Sensitivity Analysis for Coefficient *k*

Figure 9 represent the total investment cost (y-axes) in function of the number of vehicles (x-axes) for the Gumi City case.

The cost of the stationary solution increases linearly, as the total cost of the system is proportional to the total size of the battery. The charger is installed only at the base station, so this cost is fixed even if the number of EVs increases, resulting in the energy logistics cost linearly proportional to k.

For the dynamic solution, the slope of the cost-line is relatively smaller than the SWC case, as a growing number of EVs serves to increase optimization, so that smaller batteries are more economical. Using smaller batteries, the rate of change in cost against k decreases. When $k = 9$ and $k = 17$, the cost lines for SWC-QWC and for QWC-DWC cross, respectively. This means that if k is less than nine the better solution is the stationary one; if k is between nine and 17, is the quasi-dynamic solution to be more competitive, while dynamic one is competitive when k is more than 17.

CONCLUSION AND FUTURE PERSPECTIVES

It can be seen that a DWC system is beneficial when the battery costs are high, and the costs of charging infrastructure are low. In the opposite case, SWC is more beneficial.

DWC is not economical for a short distance service as the wireless charging infrastructure is more expensive than the conventional plug-in chargers, so only a short-range service that requires a limited wireless charging infrastructure might be believed economical. It is important to remember that there is a minimum fleet size needed for DWC systems at current cost figures and the current cost figures for DWC are subject to change, as DWC is still in the early stages of commercialization. For example, most of the wireless charging components present in the OLEV vehicle are custom made because there is no standardization of parts and components. Thus, if these components will be made with a mass production process, the dynamic solution would become more competitive.

A qualitative chart between charging investment cost and battery cost is shown in Figure 10. Considering the past and current trends in technology development, it is suggested a movement in cost, represented by the vector.

The rate of the cost decline in batteries has been very slow for the last several decades. However, if the parts will be mass-produced and if standardized construction methods will be introduced, there is a significant opportunity for reducing the cost of the charging infrastructure, which decreasing will be faster than the costs of batteries one.

Figure 9. Economic fleet – scale analysis for the Gumi city case

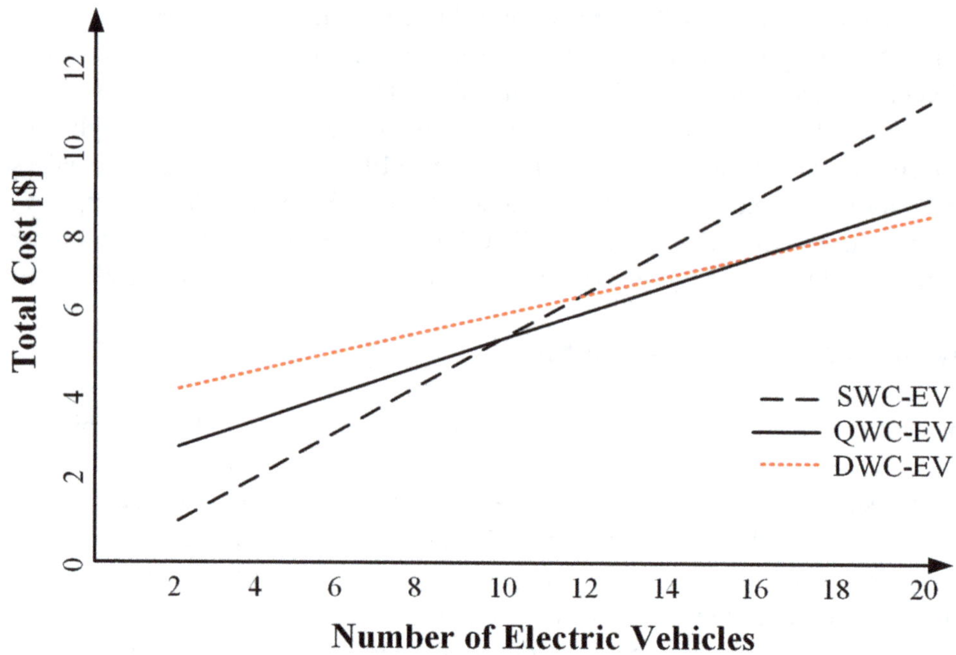

Figure 10. Possible future cost trend

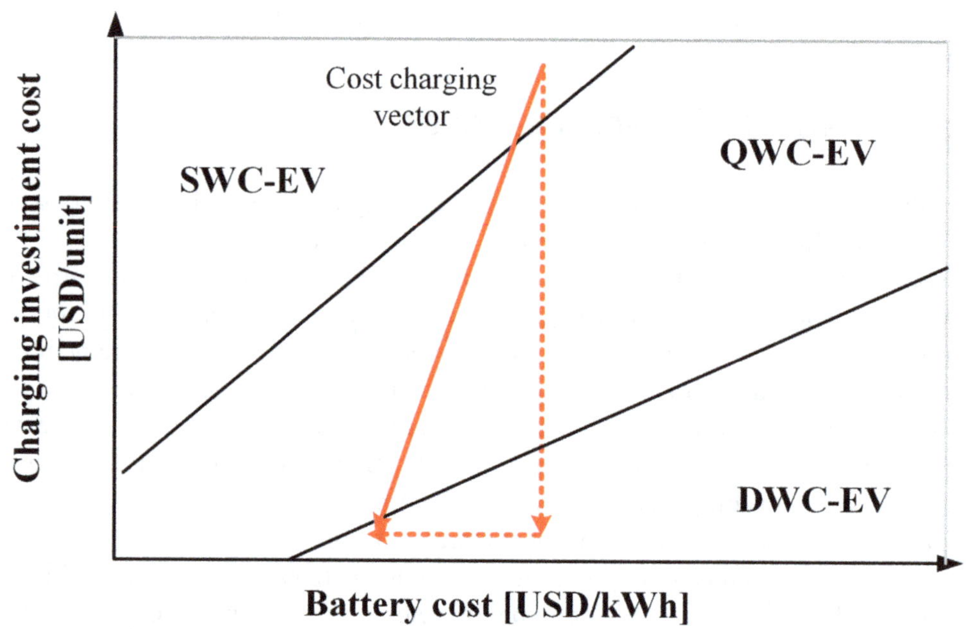

The On-Line Electric Vehicle can be a great solution to reduce pollution and the global warming. As shown previously, the biggest problem is the initial cost, in particular for the infrastructure, due to the characteristics of components that are custom made. For these reasons, OLEV will be in the future a solution to internal combustion engine vehicles especially for public transportation and even if the costs of OLEV's components will decrease, it is difficult that it will substitute private cars. In fact, there are other EV, such as PHEV, that are more suitable for substituting ICE private cars in the future because they travel on the existing roads and the costs of batteries is decreasing rapidly. Nowadays OLEV buses can be implemented almost only in closed environments, such as a university campus, due to the feasibility of installing the infrastructure even if it is not economical expedient because of the short distance of travel. However, in the future, thanks to a reduction of costs, that can be reached for example with a mass production of the components, it can be implemented for buses that travel between two cities and then it can be very appropriate for public transport with long distances service route. WPTS technology is also suitable for railways like trams and light trains especially, but also for high speed train. For railways, the deployment of power transmitters is easier and this technology lead to many advantages as is described previously. In the future, WPTSs for trams and trains will be a good solution of catenary systems reducing costs, maintenance and increasing safety and the environmental and aesthetic impact, in particular in urban cities.

REFERENCES

Ahn, S., Pak, J., Song, T., Lee, H., Byun, J.-G., Kang, D., ... Kim, J. (2010, July). Low frequency electromagnetic field reduction techniques for the On-Line Electric Vehicle (OLEV). *2010 IEEE International Symposium on Electromagnetic Compatibility*, 625-630.

Bi, Z., Kan, T., Mi, C. C., Chang, Y., Zao, Z., & Keoleian, G. A. (2016). A review of wireless power transfer for electric vehicles: Prospects to enhance sustainable mobility. *Applied Energy*, *179*, 413–425. doi:10.1016/j.apenergy.2016.07.003

Boys, J. T., & Covic, G. A. (2012). *Inductive power transfer systems (IPT) fact sheet: no. 1 – basic concepts.* Available to: https://www.qualcomm.com/documents/inductive-power-transfer-systems-ipt-fact-sheet-no-1-basic-concepts

Boys, J. T., & Covic, G. A. (2014). *Inductive power transfer systems (IPT) fact sheet: no. 2 – Magnetic circuits for powering electric vehicles.* Available to: https://www.qualcomm.com/media/documents/files/ipt-fact-sheet.pdf

Budhia, M., Covic, G., & Boys, J. (2010). A new IPT magnetic coupler for electric vehicle charging systems. *Proc. Annual Conference on IEEE Industrial Electronics Society*, 2487-2492.

Budhia, M., Covic, G. A., & Boys, J. T. (2011). Design and optimization of circular magnetic structures for lumped inductive power transfer systems. *IEEE Transactions on Power Electronics*, *26*(11), 3096–3108. doi:10.1109/TPEL.2011.2143730

Chen, F., Taylor, N., & Kringros, N. (2015). Electrification of roads: Opportunities and challengs. *Applied Energy, 150*, 109–119. doi:10.1016/j.apenergy.2015.03.067

Choi, S. Y., Beom, W., Seog, G., Jeong, T., & Rim, C. T. (2014, May). Trends of Wireless Power Transfer Systems for Roadway Powered Electric Vehicles. *2014 IEEE 79th Vehicular Technology Conference (VTC Spring)*, 1-5.

Choi, S.Y., & Jeong, S.Y. (2015). Advanced in Wireless Power Transfer Systems for Roadway-Powered Electric Vehicles. *IEEE Journal of Emerging and Selected Topics in Power Electronics, 3*(1), 18-36.

Choi, S. Y., & Rim, C. T. (2015). Recent Progress in Developments of On-Line Electric Vehicle. *6th International Conference on Power Electronics Systems and Applications (PESA)*, 1-8. 10.1109/PESA.2015.7398964

ENDESA. (2013). *General Concepts of VICTORIA Project in Endesa Website*. Available to: http://www.endesa.com/en/saladeprensa/noticias/wireless-en-route-charging-electric-buses

Fujita, T., Yasuda, T., & Akagi, H. (2015). A moving wireless power transfer system applicable to a stationary system. *Proc. IEEE Energy Conversion Congress and Exposition*, 4943–4950. 10.1109/ECCE.2015.7310357

Jang, Y. J., Jeong, S., & Lee, M. S. (2016). Initial Energy Logistics Cost Analysis for Stationary, Quasi-Dynamic, and Dynamic Wireless Charging Public Transportation Systems. *Energies, 9*(7), 483. doi:10.3390/en9070483

Jeong, S., Jang, Y. J., & Kum, D. (2014, October). Design optimization of the OLEV system considering battery lifetime. *17th International IEEE Conference on Intelligent Transportation System (ITSC)*, 2492-2498.

Jung, S., Lee, J.H., Jang, G., & Lee, H. (2013, July). Peak power measurement and substation capacity calculation for DC On-Line Electric Vehicle system. *IEEE Power & Energy Society General Meeting*, 1-5.

Kim, C. G., Seo, D. H., You, J. S., Park, J. H., & Cho, B. H. (2001). Design of a contactless battery charger for cellular phone. *IEEE Transactions on Industrial Electronics, 48*(6), 1238–1247. doi:10.1109/41.969404

Kim, Y., Shin, S., Jung, G., Lee, C., Lee, S., Shin, J., ... Chung, S. (2013, May). Phase analysis of currents and voltages in the On-Line Electric Bus (OLEB) system. *12th International Conference on Environment and Electrical Engineering*, 445-449.

Ko, Y. D., & Jang, Y. J. (2013). The Optimal System Design of the Online Electric Vehicle Utilizing Wireless Power Transmission Technology. *IEEE Transactions on Intelligent Transportation Systems, 14*(3), 1255–1265. doi:10.1109/TITS.2013.2259159

Laouamer, R., Brunello, M., Ferrieux, J. P., Normand, O., & Buchheit, N. (1997). A multi-resonant converter for non-contact charging with electromagnetic coupling. *Proc. IEEE International Conference on Industrial Electronics, Control, and Instrumentation* (Cat. No.97CH36066), 792–797.

Lee, H.-J., & Cho, D.-H. (2015, September). Economic analysis based on the Interrrelationships of the OLEV System Components. *2015 IEEE 18th International Conference on Intelligent Transportation System*, 2677-2682.

Lee, S., Huh, J., Park, C., Choi, N.-S., Cho, G.-H., & Rim, C.-T. (2010). On-Line Electric Vehicle using inductive power transfer system. 2010 IEE Energy Conversion Congress and Exposition, Atlanta, GA.

Mi, C. C., Buja, G., Choi, S. Y., & Rim, C. T. (2016). Modern Advances in Wireless Power Transfer Systems for Roadway Powered Electric Vehicles. *IEEE Transactions on Industrial Electronics*, *63*(10), 6533–6545. doi:10.1109/TIE.2016.2574993

Mouhrim, N., Alaoui, A., El, H., & Boukachour, J. (2016, May). Optimal allocation of wireless power transfer system for electric vehicles in a multipath environment. *3rd International Conference on Logistics Operations Management (GOL)*, 1-7. 10.1109/GOL.2016.7731684

Park, M.-H., Shin, E.-G., Lee, H.-R., & Suh, I.-S. (2010, October). Dynamic model and control algorithm of HVAC system for OLEV® application. *ICCAS*, *2010*, 1312–1317.

Pedder, D. A. G., Brown, A. D., & Skinner, J. A. (1999). A contactless electrical energy transmission system. *IEEE Transactions on Industrial Electronics*, *46*(1), 23–30. doi:10.1109/41.744372

Rim, C. T. (2018). *Wireless Charging of Electric Vehicles* (4th ed.). Power Electronics Handbook.

Shin, J., Song, B., Lee, S., Shin, S., Kim, Y., Jung, G., & Jeon, S. (2012, March). Contactless Power Transfer Systems for On-Line Electric Vehicle (OLEV). *2012 IEEE International Electric Vehicle Conference*, 1-4. 10.1109/IEVC.2012.6183255

Song, B., Shin, J., Chung, S., Shin, S., Lee, S., Kim, Y., Jung, G., & Jeon, S. (2013, May). Design of a pickup with compensation winding for on-line electric vehicle (OLEV). *2013 IEEE Wireless Power Transfer (WPT)*, 60-62.

Song, B., Shin, J., Lee, S., Shin, S., Kim, Y., Jeon, S., & Jung, G. (2012, March). Design of a High Power Transfer Pickup for On-line Vehicle Electric Vehicle (OLEV). *IEEE International Electric Vehicle Conference*, 1-4. 10.1109/IEVC.2012.6183285

Suh, N.P., & Cho, D.H. (2017). *The On-line Electric Vehicle_Wireless Electric Ground Transportation Systems*. Springer International Publishing ©Elsevier Ltd.

Suh, N.P., Cho, D.H., & Rim, C.T. (2010, April). *Design of On-Line Electric Vehicle*. Plenary lecture at the 2010 CIRP Design Conference, Nantes, France.

Wolterink, S., & Bauer, P. (2014, June). High-range On-line Electric Vehicles Powered by Inductive Power Transfer. *IEEE Transportation Electrification Conference and Expo (ITEC)*, Dearborn, MI. 10.1109/ITEC.2014.6861781

Yang, Y. Y., Jeon, S., & Ko, Y. D. (2015). System optimization of the On-Line Electric Vehicle operating in a closed environment. *Computers & Industrial Engineering*, *80*, 222–235. doi:10.1016/j.cie.2014.12.004

Zaheer, A., Hao, H., Covic, G. A., & Kacprzak, D. (2015). Investigation of multiple decoupled coil primary pad topologies in lumped IPT systems for interoperable electric vehicle charging. *IEEE Transactions on Power Electronics*, *30*(4), 1937–1955. doi:10.1109/TPEL.2014.2329693

Compilation of References

Abiri, P., Abiri, A., Packard, R. R. S., Ding, Y., Yousefi, A., Ma, J., ... Hsiai, T. K. (2017). Inductively powered wireless pacing via a miniature pacemaker and remote stimulation control system. *Scientific Reports*, *7*(1), 6180. doi:10.103841598-017-06493-5 PMID:28733677

Aditya, K., & Williamson, S. S. (2014). Design Considerations for Loosely Coupled Inductive Power Transfer (IPT) System for Electric Vehicle Battery Charging – A Comprehensive Review. *IEEE Transportation Electrification Conference*. 10.1109/ITEC.2014.6861764

Agarwal, K., Jegadeesan, R., Guo, Y. X., & Thakor, N. V. (2017). Wireless power transfer strategies for implantable bioelectronics: Methodological review. *IEEE Reviews in Biomedical Engineering*.

Ahlbom, A., Bergqvist, U., & Bernhardt, J. H. (1998). Guidelines for limiting exposure to time-varying electric, magnetic, and electromagnetic fields (up to 300 GHz). *Health Physics*, *74*(4), 494–522. PMID:9525427

Ahn, D., & Ghovanloo, M. (2016). Optimal design of wireless power transmission links for millimeter-sized biomedical implants. *IEEE Transactions on Biomedical Circuits and Systems*, *10*(1), 125–137. doi:10.1109/TBCAS.2014.2370794 PMID:25616074

Ahn, D., & Hong, S. (2015). Wireless Power Transfer Resonance Coupling Amplification by Load Modulation Switching Controller. *IEEE Transactions on Industrial Electronics*, *62*(2), 898–908. doi:10.1109/TIE.2014.2336627

Ahn, S., Pak, J., Song, T., Lee, H., Byun, J.-G., Kang, D., ... Kim, J. (2010, July). Low frequency electromagnetic field reduction techniques for the On-Line Electric Vehicle (OLEV). *2010 IEEE International Symposium on Electromagnetic Compatibility*, 625-630.

Ahn, S., Pak, J., Song, T., Lee, H., Byun, J.-G., Kang, D., & Kim, J. (2011). Low frequency electromagnetic field reduction techniques for the on-line electric vehicle (OLEV). In *Proceedings of the IEEE International Symposium on Electromagnetic Compatibility*. Fort Lauderdale, FL: IEEE.

Aldhaher, S., Luk, P. C. K., & Whidborne, J. F. (2014). Electronic Tuning of Misaligned Coils in Wireless Power Transfer Systems. *IEEE Transactions on Power Electronics*, *29*(11), 5975–5982. doi:10.1109/TPEL.2014.2297993

Allota, B., Pugi, L., Reatti, A., & Corti, F. (2017) Wireless power recharge for underwater robotics. *Proceedings of the 2017 IEEE International Conference on Environment and Electrical Engineering and 2017 IEEE Industrial and Commercial Power Systems Europe*. 10.1109/EEEIC.2017.7977478

Almada, M., Blanca-Pimentel, L., Block, J., Gonzalez, J., Valenta, C., & Durgin, G. (2012). Power Conversion Gain as a design metric for RFID Systems. *2012 IEEE Antennas and Propagation Society International Symposium*, 1-2. 10.1109/APS.2012.6349367

Amghar, B., Darcherif, A.-M., Barbot, J.-P., & Boukhetala, D. (2014). Modelling and control of series resonant converter for high voltage applications. *Proceedings of IEEE Energycon 2014*, 216-221. DOI: 10.1109/ENERGYCON.2014.6850431

Aoki, Otsuka, & Idogaki Shibata. (1997). Microwave energy transmission system for microrobot. *IEICE Transactions on Electronics*, *80c*(2), 303–308.

Arai, H. (2011). Wireless Power Transfer System. *Journal of the Korean Institute of Electromagnetic Engineering and Science*, *11*(3).

Arrawatia, Maryam, & Kumar. (2011). RF Energy Harvesting System from Cell Towers in 900MHz Band. *National Conference on Communications*, 1-5.

Artan, N. S., Li, X., Patel, R., Ning, C., Ludvig, N., & Chao, H. J. (2011). Multi-layer coils for efficient transcutaneous power transfer. *Proceedings of Annual International Conference of the IEEE Engineering in Medicine and Biology Society*, 3031-3034. 10.1109/IEMBS.2011.6090829

Artan, N. S., Patel, R. C., Ning, C., & Chao, H. J. (2012). High-efficiency wireless power delivery for medical implants using hybrid coils. *Proceedings of Annual International Conference of the IEEE Engineering in Medicine and Biology Society*, 1683-1686. 10.1109/EMBC.2012.6346271

Austrin, L., & Engdahl, G. (2004). Modeling of a three-phase application of a magnetic amplifier. *Proc. of the 24th International Congress of the Aeronautical Sciences*, 1-8.

Ayachit, Reatti, Saini, & Kazimierczuk. (2016). Design of choke inductor in class-E ZVS power amplifier. *Proc. of 42nd Annual Conference of IEEE Industrial Electronics Society*, 5621-5626.

Azambuja, R., Brusamarello, V. J., Haffner, S., & Porto, R. W. (2014). Analysis and Optimization of an Inductive Power Transfer with a Randomized Method. *IEEE Transactions on Instrumentation and Measurement*, *63*(5), 1145–1152. doi:10.1109/TIM.2013.2296397

Baikova, E., Romba, L., Melicio, R., & Valtchev, S. (2018b). Simulation and experiment on electric field emissions generated by wireless energy transfer. In L. M. Camarinha-Matos, K.O. Adu-Kankam, & M. Julashokri (Eds.), Technological Innovation for Resilient Systems (pp. 243-251). Cham, Switzerland: Springer. doi:10.1007/978-3-319-78574-5_23

Baikova, E. N., Romba, L., Valtchev, S. S., Melicio, R., Fernão Pires, V., Krusteva, A., & Gigov, G. (2018a). Electromagnetic field generated by a wireless energy transfer system: Comparison of simulation to measurement. *Journal of Electromagnetic Waves and Applications*, *32*(5), 554–571. doi:10.1080/09205071.2017.1399832

Baikova, E. N., Valtchev, S. S., Melicio, R., Krusteva, A., & Pires, V. F. (2016). Study of the electromagnetic interference generated by wireless power transfer systems. *International Review of Electrical Engineering*, *11*(5), 526–534.

Baikova, E. N., Valtchev, S. S., Melicio, R., & Pires, V. M. (2016a). Electromagnetic interference from a wireless power transfer system: experimental results. In *Proceedings of the International Conference on Renewable Energies and Power Quality (ICREPQ'2016)*. Madrid, Spain: ICREPQ. 10.24084/repqj14.560

Baikova, E., Valtchev, S., Melicio, R., Krusteva, A., & Gigov, G. (2016c). Study on electromagnetic emissions from wireless energy transfer. In *Proceedings of the IEEE 17th International Conference on Power Electronics and Motion Control (PEMC)*, Varna, Bulgaria: IEEE. 10.1109/EPEPEMC.2016.7752045

Baikova, E., Valtchev, S., Melicio, R., Krusteva, A., & Pires, V. F. (2016b). Study of the electromagnetic interference generated by wireless power transfer systems. *International Review of Electrical Engineering*, *11*(5), 526–534.

Bao, D., Zou, Z., Nejad, M. B., Qin, Y., & Zheng, L.-R. (2018). *A Wirelessly Powered UWB RFID Sensor Tag with Time-Domain Analog-to-Information Interface. IEEE Journal of Solid-State Circuits.*

Bareiss, M., Krenz, P. M., Szakmany, G. P., Tiwari, B. N., Kalblein, D., Orlov, A. O., ... Lugli, P. (2013). Rectennas revisited. *IEEE Transactions on Nanotechnology*, *12*(6), 1144–1150. doi:10.1109/TNANO.2013.2281373

Barili, A., Brambilla, A., Cottafava, G., & Dallago, E. (1988). A simulation model for the saturable reactor. *IEEE Transactions on Industrial Electronics*, *35*(2), 301–306. doi:10.1109/41.192663

Barmada, S. (2017). Design and experimental characterization of a combined WPT–PLC system. Wireless Power Transfer, 4(2), 160–170. doi:10.1017/wpt.2017.11

Barman, S., Rez, A., & Kumara, N. (2015). Wireless powering by magnetic resonant coupling: Recent trends in wireless power transfer system and its applications. *Renewable & Sustainable Energy Reviews*, *51*, 1525–1552. doi:10.1016/j.rser.2015.07.031

Bathe Klaus-Jürgen. (2014). *Finite Element Procedures* (2nd ed.). Watertown, MA: K. J. Bathe.

Baua, S. K., Amineh, R., & Artan, N. S. (2018). Wireless power transfer with multi-layer planar spiral coils as secondary coils. *Proceedings of IEEE Wireless Power Transfer Conference (WPTC).*

Beh, T. C., Imura, T., Kato, M., & Hori, Y. (2010). Basic study of improving efficiency of wireless power transfer via magnetic resonance coupling based on impedance matching. *Proceedings of IEEE ISIE*, 2011-2016.

Beh, T. C., Kato, M., Imura, T., Oh, S., & Hori, Y. (2013). Automated impedance matching system for robust wireless power transfer via magnetic resonance coupling. *IEEE Transactions on Industrial Electronics*, *60*(9), 3689–3698. doi:10.1109/TIE.2012.2206337

Biederman, W., Yeager, D., Narevsky, N., Koralek, A., Carmena, J., Alon, E., & Rabaey, J. (2013). A fully-integrated, miniaturized (0.125 mm^2) 10.5 μW wireless neural sensor. *IEEE Journal of Solid-State Circuits*, *48*(4), 960–970. doi:10.1109/JSSC.2013.2238994

Bienkowski, P., & Trzaska, H. (2012). *Electromagnetic measurements in the near field* (A. Duffy, Ed.). SciTech Publishing, Inc. doi:10.1049/SBEW042E

Bin Alam, S., Ullah, M. S., & Moury, S. (2013). Design of a Low Power 2.45 GHz RF Energy Harvesting Circuit for Rectenna. *International Conference on Electronics & Vision*, 1-4.

Bi, Z., Kan, T., Mi, C. C., Zhang, Y., Zhao, Z., & Keoleian, G. A. (2016). A review of wireless power transfer for electric vehicles: Prospects to enhance sustainable mobility. *Applied Energy*, *179*, 413–425. doi:10.1016/j.apenergy.2016.07.003

Bjorninen, T., Muller, R., Ledochowitsch, P., Sydanheimo, L., Ukkonen, L., Maharbiz, M., & Rabaey, J. M. (2012). Design of wireless links to implanted brain-machine interface systems. *IEEE Antennas and Wireless Propagation Letters*, *11*, 1663–1666. doi:10.1109/LAWP.2013.2239252

Boaventura, A., & Carvalha, N. (2011). Maximizing DC power in energy harvesting circuits using multisine excitation. *IEEE MTT-S International Microwave Symposium Digest*, 1 - 4. 10.1109/MWSYM.2011.5972612

Bobrow, L. S. (1985). *Fundamentals of electrical engineering*. Orlando, FL: Saunders College Publishing.

Bosshard, R., Kolar, J. W., Mühlethaler, J., Stevanovic, B., Wunsch, B., & Canales, J. (2015). Modeling and η-α-Pareto optimization of inductive power transfer for electric vehicles. *IEEE Journal of Emerging and Selected Topics in Power Electronics*, *3*, 50–64. doi:10.1109/JESTPE.2014.2311302

Bosshard, R., Kolar, J. W., & Wunsch, B. (2014). Accurate finite-element modelling and experimental verification of inductive power transfer coil design. In *Proceedings of the 29th IEEE Annual Applied Power Electronics Conference and Exposition (APEC)*. Fort Worth, TX: IEEE.

Bosshard, R., Muhlethaler, J., Kolar, J. W., & Stevanovic, I. (2013). Optimized magnetic design for inductive power transfer coils. In *Proceedings of the 28th Applied Power Electronics Conference and Exposition (APEC 2013)*. Long Beach, CA: IEEE. 10.1109/APEC.2013.6520541

Boys, J. T., & Covic, G. A. (2012). *Inductive power transfer systems (IPT) fact sheet: no. 1 – basic concepts*. Available to: https://www.qualcomm.com/documents/inductive-power-transfer-systems-ipt-fact-sheet-no-1-basic-concepts

Boys, J. T., & Covic, G. A. (2014). *Inductive power transfer systems (IPT) fact sheet: no. 2 – Magnetic circuits for powering electric vehicles*. Available to: https://www.qualcomm.com/media/documents/files/ipt-fact-sheet.pdf

Boys, J. T., Covic, G. A., & Green, A. W. (2000). Stability and Control of Inductively Coupled Power Transfer Systems. *IEE Proceedings. Electric Power Applications*, *147*(1), 37–43. doi:10.1049/ip-epa:20000017

Boys, J. T., Hu, A. P., & Covic, G. A. (2000). Critical Q Analysis of a current-fed resonant converter for ICPT applications. *Electronics Letters*, *36*(17), 1440–1442. doi:10.1049/el:20001030

Brown, W. C., & Triner, J. F. (1982). Experimental thin-film, etched-circuit rectenna. IEEE MTT-S Int, Microwave Symp.Dig., 185-187.

Brown. (1984). The history of power transmission by radio waves. *IEEE Trans. MTT*, *32*(9), 1230-1242.

Brown. (1986). A microwave powered, long duration, high altitude platform. *MTT- S International Microwave Symposium Digest*, *86*(1), 507-510.

Brown, W. C. (1957). Description and Operating Characteristics of the Platinotron – A New Microwave Tube Device. *Proceedings of the IEEE*, *45*(9), 1209–1222.

Brown, W. C. (1974). The Technology and Application of Free-Space Power Transmission by Microwave Beam. *Proceedings of the IEEE*, *62*(1), 11–25. doi:10.1109/PROC.1974.9380

Brown, W. C. (1984). The History of Power Transmission by Radio Waves. *IEEE Transactions on Microwave Theory and Techniques*, *MTT-32*(9), 1230–1242. doi:10.1109/TMTT.1984.1132833

Brusamarello, V. J., Blauth, Y. B., Azambuja, R., Müller, I., & Sousa, F. R. (2013). Power Transfer with an Inductive Link and Wireless Tuning. *IEEE Transactions on Instrumentation and Measurement*, *62*(5), 924–931. doi:10.1109/TIM.2013.2245041

Budhia, M., Covic, G., & Boys, J. (2010). A new IPT magnetic coupler for electric vehicle charging systems. *Proc. Annual Conference on IEEE Industrial Electronics Society*, 2487-2492.

Budhia, M., Covic, G. A., & Boys, J. T. (2011). Design and optimization of circular magnetic structures for lumped inductive power transfer systems. *IEEE Transactions on Power Electronics*, *26*(11), 3096–3108. doi:10.1109/TPEL.2011.2143730

Buettner, M., Prasad, R., Philipose, M., & Wetherall, D. (2009). Recognizing Daily Activities with RFID-Based Sensors. *11th International Conference on Ubiquitous Computing*.

Buja, G., Bertoluzzo, M., & Dashora, H. K. (2016). Lumped Track Layout Design for Dynamic Wireless Charging of Electric Vehicles. *IEEE Transactions on Industrial Electronics*, *63*(10), 6631–6640. doi:10.1109/TIE.2016.2538738

Buja, G., Bertoluzzo, M., & Mude, K. N. (2015). Design and experimentation of WPT charger for electric city car. *IEEE Transactions on Industrial Electronics*, *62*(12), 7436–7447. doi:10.1109/TIE.2015.2455524

Caccami & Marrocco. (2018). Electromagnetic Modeling of Self-tuning RFID Sensor Antennas in Linear and Nonlinear Regimes. *IEEE Transactions on Antennas and Propagation*, 1-1.

Campi, T. (2017). EMC and EMF safety issues in wireless charging system for an electric vehicle (EV). In *2017 International Conference of Electrical and Electronic Technologies for Automotive*. IEEE. 10.23919/EETA.2017.7993214

Cannon, B. L., Hoburg, J. F., Stancil, D. D., & Goldstein, S. C. (2009). Magnetic resonant coupling as a potential means for wireless power transfer to multiple small receivers. *IEEE Transactions on Power Electronics*, *24*(7), 1819–1825. doi:10.1109/TPEL.2009.2017195

Carmeli, M. S. (2015). Contactless energy transmission system for electrical vehicles batteries charging. In *2015 International Conference on Clean Electrical Power (ICCEP)*. IEEE. 10.1109/ICCEP.2015.7177536

Carrara, S., Cavallini, A., Ghoreishizadeh, S., Olivo, J., & De Micheli, G. (2012). Developing highly-integrated subcutaneous biochips for remote monitoring of human metabolism. *IEEE Sensors*, 1-4.

Casanova, J. J., Low, Z. N., & Lin, J. (2009). A loosely coupled planar wireless power system for multiple receivers. *IEEE Transactions on Industrial Electronics*, *56*(8), 3060–3068. doi:10.1109/TIE.2009.2023633

Catenacci, M., Verdolini, E., Bosetti, V., & Fiorese, G. (2013). Going electric: Expert survey on the future of battery technologies for electric vehicles. *Energy Policy*, *61*, 403–413. doi:10.1016/j.enpol.2013.06.078

Catrysse, M., Hermans, B., & Puers, R. (2003). An inductive power system with integrated bidirectional data transmission. *Proceedings of XVII Eurosensors*, 843-846.

Cavallini, A., Jost, T. R., Ghoreishizadeh, S. S., Olivo, J., de Beeck, M. O., Gorissen, B., ... Carrara, S. (2015). A Subcutaneous Biochip for Remote Monitoring of Human Metabolism: Packaging and biocompatibility assessment. *IEEE Sensors Journal*, *15*(1), 417–424. doi:10.1109/JSEN.2014.2339638

Cellular Operators Association of India. (n.d.). *Annual Report, 2013-2014*. Retrieved from http://www.dot.gov.in/sites/default/files/AR%202013-14%20English%20%282%29_1.pdf

Chae, H., Yeager, D., Smith, J., & Fu, K. (2007). Maximalist Cryptography and Computation on the WISP UHF RFID Tag. *Proceedings of the Conference on RFID Security*.

Chakaravarthi, Prasath, Philip, Jayaganthan, Velmurugan, & Arunachalam. (2018). Reusable Passive Wireless RFID Sensor for Strain Measurement on Metals. *IEEE Sensors Journal*, 1-1.

Chang, R., Quan, L., Zhu, X., Zong, Z., & Zhou, H. (2014). Design of a wireless power transfer system for EV application based on finite element analysis and MATLAB simulation. In *Proceedings of the IEEE Conference and Expo Transportation Electrification Asia-Pacific (ITEC Asia-Pacific)*. Beijing, China: IEEE. 10.1109/ITEC-AP.2014.6941225

Chaudhri, R., Lester, J., & Borriello, G. (2008). An RFID Based System for Monitoring Free Weight Exercises. *The 6th ACM Conference on Embedded Networked Sensor Systems*. 10.1145/1460412.1460488

Chen, C. J., Chu, T. H., Lin, C. L., & Jou, Z. C. (2010). A study of loosely coupled coils for wireless power transfer. *IEEE Transactions on Circuits and Wystems. II, Express Briefs*, *57*(7), 536–540. doi:10.1109/TCSII.2010.2048403

Chen, F., Taylor, N., & Kringros, N. (2015). Electrification of roads: Opportunities and challengs. *Applied Energy*, *150*, 109–119. doi:10.1016/j.apenergy.2015.03.067

Cheng, K. W. E., & Lu, Y. (2003). Development of a contactless power converter. *IEEE ICIT*, *2*, 786–791.

Chen, K., & Zhao, Z. (2013). Analysis of the double-layer printed spiral coil for wireless power transfer. *IEEE Journal of Emerging and Selected Topics in Power Electronics*, *1*(2), 114–121. doi:10.1109/JESTPE.2013.2272696

Cheon, S., Kim, Y.-H., Kang, S.-Y., Lee, M. L., Lee, J.-M., & Zyung, T. (2011). Circuit-Model-Based Analysis of a Wireless Energy-Transfer System via Coupled Magnetic Resonances. *IEEE Transactions on Industrial Electronics*, *58*(7), 2906–2914. doi:10.1109/TIE.2010.2072893

Chinthavali & Wang. (2016). Sensitivity Analysis of a Wireless Power Transfer (WPT) System for Electric Vehicle Application. *IEEE Energy Conversion Congress and Exposition (ECCE).*

Choi, S. Y., Beom, W., Seog, G., Jeong, T., & Rim, C. T. (2014, May). Trends of Wireless Power Transfer Systems for Roadway Powered Electric Vehicles. *2014 IEEE 79th Vehicular Technology Conference (VTC Spring)*, 1-5.

Choi, S.Y., & Jeong, S.Y. (2015). Advanced in Wireless Power Transfer Systems for Roadway-Powered Electric Vehicles. *IEEE Journal of Emerging and Selected Topics in Power Electronics, 3*(1), 18-36.

Choi, B., Nho, J., Cha, H., Ahn, T., & Choi, S. (2004). Design and Implementation of Low-Profile Contacless Battery Charger Using Planar Printed Circuit Board Windings as Energy Transfer Device. *IEEE Transactions on Industrial Electronics*, *51*(1), 140–147. doi:10.1109/TIE.2003.822039

Choi, S. Y., Gu, B. W., Jeong, S. Y., & Rim, C. T. (2015). Advances in wireless power transfer systems for roadway-powered electric vehicles. *IEEE Journal of Emerging and Selected Topics in Power Electronics*, *3*(1), 18–36. doi:10.1109/JESTPE.2014.2343674

Choi, S. Y., & Rim, C. T. (2015). Recent Progress in Developments of On-Line Electric Vehicle. *6th International Conference on Power Electronics Systems and Applications (PESA)*, 1-8. 10.1109/PESA.2015.7398964

Choi, W. P., Ho, W. C., Liu, X., & Hui, S. Y. R. (2010). Comparative Study on Power Conversion Methods for Wireless Battery Charging Platform. In *Proceedings of 14th International Power Electronics and Motion Control Conference (EPE-PEMC 2010)*. IEEE. 10.1109/EPEPEMC.2010.5606675

Chopra, S., & Bauer, P. (2011). Analysis and design considerations for a contactless power transfer system. *Telecommunications Energy Conference*. 10.1109/INTLEC.2011.6099774

Chou, C.-K., D'Andrea, J., & Petersen, R. (2005). *IEEE Standard for Safety Levels with Respect to Human Exposure to Radio Frequency Electromagnetic Fields, 3 kHz to 300 GHz*. New York, NY: Institute of Electrical and Electronics Engineers.

Christ, A., Douglas, M. G., Roman, J. M., Cooper, E. B., Sample, A. P., Waters, B. H., & Kuster, N. (2013). Evaluation of wireless resonant power transfer systems with human electromagnetic exposure limits. *IEEE Transactions on Electromagnetic Compatibility*, *55*(2), 265–274.

Christ, A., Douglas, M., Nadakuduti, J., & Kuster, N. (2013). Assessing human exposure to electromagnetic field from wireless power transmission systems. *Proceedings of the IEEE*, *101*(6), 1482–1493. doi:10.1109/JPROC.2013.2245851

Colella & Catarinucci. (2018). Electromagnetic Design of UHF RFID Tags Enabling a Novel Method to Retrieve Sensor Data. *IEEE Journal of Radio Frequency Identification*, 1-1.

Córcoles, F., Pedra, J., & Salichs, M. (2004). *Transformadores*. Barcelona, Spain: Edicions UPC.

Corti, F., Grasso, F., Reatti, A., Ayachit, A., Saini, D. K., & Kazimierczuk, M. K. (n.d.). Design of a class-E ZVS inverter with loosely-coupled transformer at fixed coupling coefficient. *Proceedings of the 42nd Annual Conference of the IEEE Industrial Electronics Society*, 5627-5632. doi: 10.1109/IECON.2016.7793285

Covic, G. A., & Talbot Boys, J. (2013). Modern Trends in Inductive Power Transfer for Transportation Applications. IEEE Journal of Emerging and Selected Topics in Power Electronics, 1(1), 28–41. doi:10.1109/JESTPE.2013.2264473

Covic, G. A. (2011). *A bipolar primary pad topology for EV stationary charging and highway power by inductive coupling. In 2011 IEEE Energy Conversion Congress and Exposition* (pp. 1832–1838). IEEE. doi:10.1109/ECCE.2011.6064008

Curty, J., Joehl, N., Krummenacher, F., Dehollain, C., & Declercq, M. (2005). A model for μ-power rectifier analysis and design. *Circuits and Systems, IEEE Transactions on, 52*(12), 2771–2779. doi:10.1109/TCSI.2005.854294

Dai, J., & Ludois, D. C. (2015). A Survey of Wireless Power Transfer and a Critical Comparison of Inductive and Capacitive Coupling for Small Gap Applications. *IEEE Transactions on Power Electronics, 30*(11), 6017–6029. doi:10.1109/TPEL.2015.2415253

Degenford, J. E., Sirkis, M. D., & Steier, W. H. (1964). The Reflecting Beam Waveguide. *IEEE Transactions on Microwave Theory and Techniques, 12*(4), 445–453. doi:10.1109/TMTT.1964.1125845

Del Toro García, X., Vázquez, J., & Roncero-Sánchez, P. (2015). Design, implementation issues and performance of an inductive power transfer system for electric vehicle chargers with series-series compensation. *IET Power Electronics, 8*(10), 1920–1930. doi:10.1049/iet-pel.2014.0877

Deng, J. (2017). A new integration method with minimized extra coupling effects using inductor and capacitor series-parallel compensation for wireless EV charger. Applied Energy. doi:10.1016/j.apenergy.2017.05.088

Department of Telecommunications. (n.d.). *Ministry of Communications and Information Technology, Annual Report, 2012-2013*. Retrieved from http://meity.gov.in/writereaddata/files/Annual%20Report%202012-13.pdf

Dias, J.-V., Kim, H., & Jang, D. (2011). Computer model for railway inductive power supply using Valtchev model. *Proceedings of International Conference on Electrical Machines and Systems (ICEMS)*. 10.1109/ICEMS.2011.6073510

Dieckerhoff, S., Ryan, M. J., & Doncker, R. W. D. (1999). Design of an IGBT-Based LCL-Resonant Inverter for High-Frequency Induction Heating. In *Conference Record of the 1999 IEEE Industry Applications Conference. Thirty-Forth IAS Annual Meeting (Cat. No.99CH36370)*. IEEE. 10.1109/IAS.1999.806017

Din, N. M., Chakrabarty, C. K., Bin Ismail, A., Devi, K. K. A., & Chen, W. Y. (2012). Design of RF Energy Harvesting System for Energizing Low Power Devices. *Progress in Electromagnetics Research, 132*, 49–69. doi:10.2528/PIER12072002

Dobkin, D. M. (2007). *The RF in RFID: Passive UHF RFID in Practice*. New York: Elsevier.

Dobkin, D. M. (2008). *The RF in RFID. Passive UHF RFID in Practice*. Elsevier.

Dukju, A., & Songcheol, H. (2013). A study on magnetic field repeater in wireless power transfer. *IEEE Transactions on Industrial Electronics, 60*(1), 360–371. doi:10.1109/TIE.2012.2188254

Durgin. (2008). *The Hidden Benefits of Backscatter Radio and RFID at 5.8 GHz*. URSI 2008.

Durgin, G., Valenta, C., Akbar, M., Morys, M., Marshall, B., & Lu, Y. (2013). Modulation and Sensitivity Limits for Backscatter Receivers. *IEEE International Conference on RFID*, 124 - 130. 10.1109/RFID.2013.6548145

Ekman, C. K. (2011). On the synergy between large electric vehicle fleet and high wind penetration – an analysis of the Danish case. *Renewable Energy, 36*(2), 546–553. doi:10.1016/j.renene.2010.08.001

Electronic and mechanical improvement of the receiving terminal of a free-space microwave power transmission system. (1977). Raytheon Company, Wayland, MA, Tech.Rep.PT-4964, NASA REP. CR135194.

Elliot, G. A. J., Covic, G. A., Kacprzak, D., & Boys, J. T. (2006). A New Concept: Asymmetrical pick-ups for inductively coupled power transfer monorail systems. *IEEE Transactions on Magnetics, 42*(10), 3389–3391. doi:10.1109/TMAG.2006.879619

ENDESA. (2013). *General Concepts of VICTORIA Project in Endesa Website*. Available to: http://www.endesa.com/en/saladeprensa/noticias/wireless-en-route-charging-electric-buses

EPCglobal. (2008). *EPC RF Identity Protocols Class-1 Generation-2 UHF RFID Protocol for Communications Version 1.2.1*. Author.

Eteng, A. A., Rahim, S. K. A., Leow, C. Y., Jayaprakasam, S., & Chew, B. W. (2017). Low-power near-field magnetic wireless energy transfer links: A review of architectures and design approaches. *Renewable & Sustainable Energy Reviews*, *77*, 486–505. doi:10.1016/j.rser.2017.04.051

Etinger, A., Pilossof, M., Litvak, B., Hardon, D., Einat, M., Kapilevich, B., & Pinhasi, Y. (2017). Characterization of a Schottky Diode Rectenna for Millimeter Wave Power Beaming Using High Power Radiation Sources. *Acta Physica Polonica A*, *131*(5), 1280–1284. doi:10.12693/APhysPolA.131.1280

European Commission. (2015). Opinion on Potential health effects of exposure to electromagnetic fields (EMF). Luxembourg: Scientific Committee on Emerging and Newly Identified Health Risks (SCENIHR).

Fabbri, G., Tarquini, G., Pasquali, L., Anniballi, L., Odoardi, S., Teodori, S., & Santini, E. (2014). Impact of V2G/G2V technologies on distributed generation systems. *Proc. of the 23th IEEE International Symposium on Industrial Electronics*, 1-4. 10.1109/ISIE.2014.6864867

Fahim, ElBatt, Mohamed, & Al-Ali. (2018). Towards Extended Bit Tracking for Scalable and Robust RFID Tag Identification Systems. *IEEE Access*, 1-1.

Far Eastern Electronic Toll Collection Co. Ltd. (n.d.). Retrieved from https://www.fetc.net.tw/en/

Fernandes, R., Boaventura, A. S., Carvalho, N., & Matos, J. (2011). Increasing the range of wireless passive sensor nodes using multisines. *IEEE International Conference on RFID-Technologies and Applications*, 549 - 553. 10.1109/RFID-TA.2011.6068599

Fernández, L. P., San Román, T. G., Cossent, R., Domingo, C. M., & Frías, P. (2011). Assessment of the impact of plug-in electric vehicles on distribution networks. *IEEE Transactions on Power Systems*, *26*(1), 206–213. doi:10.1109/TPWRS.2010.2049133

Finkenzeller, K. (2003). *RFID Handbook: Fundamentals and Applications in Contactless Smart Cards and Identification*. John Wiley and Sons. doi:10.1002/0470868023

Franco, S. (2001). *Design with Operational Amplifiers and Analog Integrated Circuits*. New York: McGraw-Hill.

Freedman, N. (1973). Raytag, an electronic remote data readout system. *Proceedings Carnahan Conference Electronic Crime Countermeasures*, 104 - 107.

Fujino & Ogimura. (2004). A rectangular parabola rectenna with elliptical beam for sps test satellite experiment. *Proc. of the Institute of Electronics, Information and Communication Engineers*, 1-10.

Fujita, Kaya, Kunimi, Ishii, Ogihata, & Kusaka Fujino. (1996). A dual polarization microwave power transmission system for microwave propelled airship experiment. *Proc. de ISAP*, *96*(2), 393–396.

Fujita, T., Yasuda, T., & Akagi, H. (2015). A moving wireless power transfer system applicable to a stationary system. *Proc. IEEE Energy Conversion Congress and Exposition*, 4943–4950. 10.1109/ECCE.2015.7310357

Fu, W., Zhang, B., & Qiu, D. (2009). Study on frequency-tracking wireless power transfer system by resonant coupling. *Proceedings of IEEE 6th International Conference on Power Electronics and Motion Control*, 2658-2663.

Gendron, M., Hazel, B., Boudreault, E., Champliaud, H., & Pham, X.-T. (2017). Modelling circular inductors coupled to a semi-infinite magnetic medium considering the proximity effect. *Journal of Electromagnetic Waves and Applications*, *31*(13), 1232–1254. doi:10.1080/09205071.2017.1336491

Gigov, G., Krusteva, A., & Valtchev, S. (2016). Experimental study of wireless inductive system for electric vehicles batteries charging. *Proceedings of 17th IEEE-PEMC2016 IEEE International Power Electronics and Motion Control Conference*, 286-290. 10.1109/EPEPEMC.2016.7752012

Glaser, P. E. (1968). Power from the sun: Its future. *Science*, *162*(3856), 857–861. doi:10.1126cience.162.3856.857 PMID:17769070

Gong, F., Cong, Y., Wei, Z., Yin, B., Chi, H., & Luan, X. (2016). Analysis of the multi-layer printed spiral coil for wireless power transfer system used in medical implants. *7th Asia Pacific International Symposium on Electromagnetic Compatibility*, 460-462.

Gope, Lee, & Quek. (2018). Lightweight and Practical Anonymous Authentication Protocol for RFIDSystems Using Physically Unclonable Functions. *IEEE Transactions on Information Forensics and Security*, 1-1.

Goßler, C., Bierbrauer, C., Moser, R., Kunzer, M., Holc, K., Pletschen, W., ... Schwarz, U. T. (2014). GaN-based micro-LED arrays on flexible substrates for optical cochlear implants. *Journal of Physics. D, Applied Physics*, *47*(20), 205401. doi:10.1088/0022-3727/47/20/205401

Goubau, G., & Schwering, F. (1961). On the Guided Propagation of Electromagnetic Wave Beams. *IRE Transactions on Antennas and Propagation*, *9*(3), 248–256. doi:10.1109/TAP.1961.1144999

Green, A. W., & Boys, J. T. (1994). *10 kHz Inductively Coupled Power Transfer Concept and Control*. London: Institution of Electrical Engineers.

Griffiths, D. J. (1999). *Introduction to electrodynamics* (3rd ed.). Upper Saddle River, NJ: Prentice Hall International.

Grifin, J. D. (2009). *High-frequency modulated-backscatter communication using multiple antennas* (Ph.D. dissertation). Georgia Institute of Technology.

Grifin, J. D., & Durgin, G. D. (2009). Multipath Fading Measurements for Multi-Antenna Backscatter RFID at 5.8 GHz. *IEEE International Conference on RFID*, 322 - 329. 10.1109/RFID.2009.4911197

Grifin, J. D., & Durgin, G. D. (2009, April). Complete link budgets for backscatter radio and RFID systems. *IEEE Antennas & Propagation Magazine*, *51*(2), 11–25. doi:10.1109/MAP.2009.5162013

Grover, F. W. (1922). Formulas and tables for the calculation of the inductance of coils of polygonal form. *Scientific Papers of the Bureau of Standards*, *18*(468), 737–762. doi:10.6028/nbsscipaper.141

Guidi, F., Decarli, N., Dardari, D., Mani, F., & D'Errico, R. (2018). *Millimeter-Wave Beamsteering for Passive RFID Tag Localization. IEEE Journal of Radio Frequency Identification.*

Gujar, M., Datta, A., & Mohanty, P. (2013). Smart Mini Grid: An Innovative Distributed Generation based Energy System. *IEEE ISGT*, 1-5.

Gutmann, R. J., & Gworek, R. B. (1979). Yagi-uda receiving elements in microwave power transmission system rectennas. *The Journal of Microwave Power*, *14*(4), 313–320. doi:10.1080/16070658.1979.11689166

Ha, S. (2016). *Silicon integrated high-density electrocortical and retinal neural interfaces* (PhD Dissertation). University of California, San Diego, CA.

Hagerty, J. A., Lopez, N. D., Popovic, B., & Popovic, Z. (2000). *Broadband rectenna arrays for randomly polarized incident waves*. IEEE. doi:10.1109/EUMA.2000.338757

Hao, J., Suxia, L., Junmin, Z., Shysheng, L., & Shahnasser, H. (2011) Design and optimization of printed spiral coils in wireless power transfer for biomedical implants. *Proceedings of 4th International Conference on Biomedical Engineering and Informatics*, 1064-1067.

Harouni, Z., Cirio, L., Osman, L., Gharsallah, A., & Picon, O. (2011). A Dual Circularly Polarized 2.45GHz Rectenna for Wireless Power Transmission. *IEEE Antennas and Wireless Propagation Letters*, 10.

Harris, D. (1952). *Radio transmission systems with modulatable passive responder*. U.S. Patent 2 927 321.

Hart, D. W. (2011). *Power electronics*. New York, NY: McGraw Hill.

Hasan, A., Peterson, A. F., & Durgin, G. D. (2011b). Reected electro-material signatures for self-sensing passive RFID sensors. *IEEE International Conference on RFID*, 62 - 69.

Hasan, A., Peterson, A., & Durgin, G. (2011a). Feasibility of Passive Wireless Sensors Based on Reected Electro-Material Signatures. *ACES Journal of Applied Electromagnetics*, *25*(6), 552–560.

Haus, H., & Huang, W. (1991). Coupled-mode theory. *Proceedings of the IEEE*, *79*(10), 1505–1518. doi:10.1109/5.104225

Hernandez, V. H., Gehrt, A., Reuter, K., Jing, Z., Jeschke, M., Schulz, A. M., & … . (2014). Optogenetic stimulation of the auditory pathway. *The Journal of Clinical Investigation*, *124*(3), 1114–1129. doi:10.1172/JCI69050 PMID:24509078

Hesterman, B. (2007, April). *Analysis and modeling of magnetic coupling*. Paper presented at the Denver Chapter of IEEE Power Electronics Society, Boulder, CO.

Hewlett-Packard. (1982). Alimentations modulaires 50 W pour les constructeurs OEM (revue du Hewlett-Packard). *Toute l'électronique*, *471*, 43-48.

Hilshey, A. D., Hines, P. D. H., Rezaei, P., & Dowds, J. R. (2013). Estimating the impact of electric vehicle smart charging on distribution transformer aging. *IEEE Transactions on Smart Grid*, *4*(2), 905–913. doi:10.1109/TSG.2012.2217385

Hochmair, E. S. (1984). System Optimization for Improved Accuracy in Transcutaneous Signal and Power Transmission. *IEEE Transactions on Biomedical Engineering*, *BME-31*(2), 177–186. doi:10.1109/TBME.1984.325404 PMID:6546736

Ho, J., Yeh, A., Neofytou, E., Kim, S., Tanabe, Y., Patlolla, B., ... Poon, A. (2014). Wireless power transfer to deep-tissue microimplants. *Proceedings of the National Academy of Sciences of the United States of America*, *111*(22), 7974–7979. doi:10.1073/pnas.1403002111 PMID:24843161

Holcomb, D., Burleson, W., & Fu, K. (2007). Initial SRAM State as a Fingerprint and Source of True Random Numbers for RFID Tags. *Proceedings of the Conference on RFID Security*.

Hsu, S. (2013). *Alternative fuels & vehicles technology and trend*. Retrieved September 22, 2017 from http://www.csee.net.cn/data/zt_esedc2013/ppt/p4_1.pdf

Hsu, J.-U. W., Hu, A. P., & Swain, A. (2012). Fuzzy logic-based directional full-range tuning control of wireless power pickups. *IET Power Electronics*, *5*(6), 773–781. doi:10.1049/iet-pel.2011.0364

Huang, F.-J., Yo, T.-C., Lee, C.-M., & Luo, C.-H. (2012). Design of Circular Polarization Antenna with Harmonic Suppression for Rectenna Application. *IEEE Antennas and Wireless Propagation Letters*, 11.

Huang, L., & Hu, A. P. (2015). Defining the mutual coupling of capacitive power transfer for wireless power transfer. *Electronics Letters*, *51*(22), 1806–1807. doi:10.1049/el.2015.2709

Huang, R., & Zhang, B. (2015). Frequency, Impedance Characteristics and HF Converters of Two-Coil and Four-Coil Wireless Power Transfer. *IEEE Journal of Emerging and Selected Topics in Power Electronics*, *3*(1), 177–183. doi:10.1109/JESTPE.2014.2315997

Huang, X. L., Wang, W., Tan, L. L., Zhao, L. L., & Zhou, Y. L. (2012). Study of transmission performance on strong coupling wireless power transfer system in free position. *Proc. of the 31st Progress in Electromagnetics Research Symposium*, 669-676.

Huh, J., Lee, S. W., Lee, W. Y., Cho, G. H., & Rim, C. T. (2011). Narrow-width inductive power transfer system for online electrical vehicles. *IEEE Transactions on Power Electronics*, *26*(12), 3666–3679. doi:10.1109/TPEL.2011.2160972

Hui, S. Y. R., & Ho, W. W. C. (2005). A New Generation of Universal Contacless Battery Charging Platform for Portable Consumer Electronic Equipment. *IEEE Transactions on Power Electronics*, *20*(3), 620–627. doi:10.1109/TPEL.2005.846550

Hui, S. Y. R., Zhong, W., & Lee, C. K. (2014). A Critical Review of Recent Progress in Mid-Range Wireless Power Transfer. *IEEE Transactions on Power Electronics*, *29*(9), 4500–4511. doi:10.1109/TPEL.2013.2249670

Hung-Yu, S., Jia-You, L., & Tsung-Wen, C. (2011). Study of contactless inductive charging platform with core array structure for portable products. *Proceedings on International Conference on Consumer Electronics, Communications, and Networks,* 756-759.

Hurley, W. G., & Kassakian, J. G. (1979). Induction Heating of Circular Ferromagnetic Plates. *IEEE Transactions on Magnetics*, *MAG-15*(3), 1174–1181. doi:10.1109/TMAG.1979.1145874

Hurley, W. G., & Wilcox, D. J. (1994). Calculation of leakage inductance in transformer windings. *IEEE Transactions on Power Electronics*, *9*(1), 121–126. doi:10.1109/63.285502

Hyeon-Chang, S., JinWook, K., Do-Hyeon, K., Kwan-Ho, K., & Young-Jin, P. (2011). Self-resonant coil with coaxial-like capacitor for wireless power transfer. *Proceedings of Asia-Pacific Microwave Conference*, 90-93.

Ibrahim, A., & Kiani, M. (2016). A figure-of-merit for design and optimization of inductive power transmission links for millimeter-sized biomedical implants. *IEEE Transactions on Biomedical Circuits and Systems*, *10*(6), 1100–1111. doi:10.1109/TBCAS.2016.2515541 PMID:28055825

IEEE Standard C95.1. (2005). *IEEE Standard for Safety Levels with Respect to Human Exposure to Radio Frequency Electromagnetic Fields, 3 kHz to 300 GHz*. USA: IEEE.

Imura, T., & Hori, Y. (2011). Maximizing air gap and efficiency of magnetic resonant coupling for wireless power transfer using equivalent circuit and neumann formula. *IEEE Transactions on Industrial Electronics*, *58*(10), 4746–4752. doi:10.1109/TIE.2011.2112317

Imura, T., Okabe, H., & Hori, Y. (2009). Basic experimental study on helical antennas of wireless power transfer for electric vehicles by using magnetic resonant couplings. In *Proceedings of the IEEE Vehicle Power and Propulsion Conference (VPPC '09)*. Dearborn, MI: IEEE. 10.1109/VPPC.2009.5289747

Instruments, T. (2008). *Mixed Signal Microcontroller*. Retrieved from http://www.ti.com/lit/ds/slas632/slas632.pdf

Instruments, T. (2014). *ULP Meets Energy Harvesting: A Game-Changing Combination for Design Engineers*. Available: http://www.ti.com/lsds/ti/apps/alternative_energy/harvesting

International Agency for Research on Cancer (IARC). (2013). *Non-ionizing radiation, part 2: radiofrequency electromagnetic fields*. Lyon, France: International Agency for Research on Cancer.

International Commission on Non-Ionizing Radiation Protection (ICNIRP). (2010). Guidelines for limiting exposure to time-varying electric, magnetic and electromagnetic fields (1 Hz – 100 kHz). *Health Physics*, *99*, 818–836. PMID:21068601

International Energy Agency. Global EV Outlook. (2017). *Two million and counting*. IEA.

Isaev, Y. N., Vasileva, O. V., Budko, A. A., & Lefebvre, S. (2016). *Parameters assessment of the inductively-coupled circuit for wireless power transfer*. Paper presented at the International Conference on Mechanical Engineering, Automation and Control Systems, Moscow, Russia.

Ito, T., Fujino, Y., & Fujita, M. (1993). Fundamental experiment of a rectenna array for microwave power reception. *IEICE Trans. Commun. E (Norwalk, Conn.)*, *76-B*(12), 1508–1513.

Jang, Y. J., Jeong, S., & Lee, M. S. (2016). Initial Energy Logistics Cost Analysis for Stationary, Quasi-Dynamic, and Dynamic Wireless Charging Public Transportation Systems. *Energies*, *9*(7), 483. doi:10.3390/en9070483

Jang, Y., & Jovanovic, M. M. (2003). A Contactless Electrical Energy Transmission System for Portable Telephone Battery Chargers. *IEEE Transactions on Industrial Electronics*, *50*(3), 520–527. doi:10.1109/TIE.2003.812472

Jang, Y., Jovanović, M., & Ruiz, J., & Liu, G. (2015). Series-resonant converter with reduced frequency-range control. *Proceedings of IEEE Applied Power Electronics Conference and Exposition (APEC)*, 1453-1460. DOI: 10.1109/APEC.2015.7104539

Jeong, S., Jang, Y. J., & Kum, D. (2014, October). Design optimization of the OLEV system considering battery lifetime. *17th International IEEE Conference on Intelligent Transportation System (ITSC)*, 2492-2498.

Jian, M. S., Chen, Y. L., & Li, S. J. (2015). Wireless Power Charger with Dynamic Pivoting Antenna Module for Various Wireless Sensor. *13th EHAC'15*, 110-113.

Jian, M. S., Chen, Y. L., Tong R. W., Lin, Y. H., Cheng C. (2017). Various Green Energy Sources in Smart Grid Power Network based on Cloud. *ICACT*, 660-663.

Jian, M. S., Fang, Y. C., Tong R. W., Lin, Y. H., (2016) Wireless Green Energy Power Transmission System Based on Assembly Method and Pivoting Antenna Module. *ICASI*, 1-4.

Jian, M. S., Chou, T. Y., & Hsu, S. H. (2009). RFID Encryption/Decryption Technology Aided Multimedia and Data Intellectual Property Protection. *WSEAS Trans. on Communications*, *8*(7), 734–743.

Jian, M. S., Chou, T. Y., & Hsu, S. H. (2012). Mobility corresponding location-aware information services based on embedded RFID platform. *International Journal of Ad Hoc and Ubiquitous Computing*, *9*(2), 122–131. doi:10.1504/IJAHUC.2012.045553

Jian, M. S., & Hsu, S. H. (2009). Aware Public/Personal Diversity of Information Services based on embedded RFID Platform. *Proc. of ICACT*, 1145-1150.

Jian, M. S., & Wu, J. S. (2013). *RFID Applications and Challenges (Chapter) of Radio Frequency Identification from System to Applications*. In-Tech.

Jin, J.-M. (2010). *Theory and computation of electromagnetic fields*. John Wiley and Sons. doi:10.1002/9780470874257

Jin, J.-M. (2014). *The Finite Element Method in Electromagnetics* (3rd ed.). John Wiley & Sons.

Jolani, F., Yu, Y., & Chen, Z. (2015). Enhanced planar wireless power transfer using strongly coupled magnetic resonance. *Electronics Letters*, *51*(2), 173–175. doi:10.1049/el.2014.4104

Joseph, A. H. (2003). *Nonlinear Circuits and Antennas for Microwave Energy Conversion* (Ph.D. dissertation). Department of Elect. and Comp. Eng., University of Colorado.

Joshi, S., & Moddel, G. (2016, May). Simple Figure of Merit for Diodes in Optical Rectennas. *IEEE Journal of Photovoltaics, 6*(3), 668–672. doi:10.1109/JPHOTOV.2016.2541460

Jovanovic, M., Liu, K.-H., Oruganti, R., & Lee, F. (1987). State-plane analysis of qasi-resonant converters. *IEEE Transactions on Power Electronics, 2*(1), 36–44. doi:10.1109/TPEL.1987.4766330

Jow, U. M., & Ghovanloo, M. (2007). Design and Optimization of Printed Spiral Coils for Efficient Transcutaneous Inductive Power Transmission. *IEEE Transactions on Biomedical Circuits and Systems, 1*(3), 193–202. doi:10.1109/TBCAS.2007.913130 PMID:23852413

Jow, U., & Ghovanloo, M. (2009). Modeling and optimization of printed spiral coils in air, saline, and muscle tissue environments. *IEEE Transactions on Biomedical Circuits and Systems, 3*(5), 339–347. doi:10.1109/TBCAS.2009.2025366 PMID:20948991

Jr. Cleveland, R. F., & Ulcek, J. L. (1999). *Questions and Answers about Biological Effects and Potential Hazards of Radiofrequency Electromagnetic Fields*. Washington, DC: Federal Communications Commission.

Jui-Hung, C., Lin, D.-B., Weng, K.-L., & Li, H.-J. (2012). Novel T-shape Slot Couple Feed Dual Circular Polarized Rectenna. *Proceedings of ISAP*.

Jung, S., Lee, J.H., Jang, G., & Lee, H. (2013, July). Peak power measurement and substation capacity calculation for DC On-Line Electric Vehicle system. *IEEE Power & Energy Society General Meeting*, 1-5.

Kalwar, K. A., Aamir, M., & Mekhilef, S. (2015). Inductively coupled power transfer (ICPT) for electric vehicle charging – A review. *Renewable & Sustainable Energy Reviews, 47*, 462–475. doi:10.1016/j.rser.2015.03.040

Kan, T., Nguyen, T.-D., White, J. C., Malhan, R. K., & Mi, C. C. (2017). A New Integration Method for an Electric Vehicle Wireless Charging System Using LCC Compensation Topology: Analysis and Design. *IEEE Transactions on Power Electronics, 32*(2), 1638–1650. doi:10.1109/TPEL.2016.2552060

Karalis, A., Joannopoulos, J., & Soljacic, M. (2007). Efficient wireless nonradiative mid-range energy transfer. *Annals of Physics, 323*(1), 34–48. doi:10.1016/j.aop.2007.04.017

Kassiri, H., Bagheri, A., Soltani, N., Abdelhalim, K., Jafari, H. M., Salam, M. T., ... Genov, R. (2014, September). Inductively-powered direct-coupled 64-channel chopper-stabilized epilepsy-responsive neurostimulator with digital offset cancellation and tri-band radio. *40th EEE European Solid State Circuits Conference (ESSCIRC)*, 95-98. 10.1109/ESSCIRC.2014.6942030

Kazimierczuk, M. K. (2014). *High-Frequency Magnetic Components* (2nd ed.). Chichester, UK: John Wiley.

Kiani, M., Jow, U., & Ghovanloo, M. (2011). Design and optimization of a 3-coil inductive link for efficient wireless power transmission. *IEEE Transactions on Biomedical Circuits and Systems, 5*(6), 579–591. doi:10.1109/TBCAS.2011.2158431 PMID:21922034

Kim, C. G., Seo, D. H., You, J. S., Park, J. H., & Cho, B. H. (2001). Design of a Contacless Battery Charger for Cellular Phone. *IEEE Transactions on Industrial Electronics, 48*(6), 1238–1247. doi:10.1109/41.969404

Kim, M., Lee, D. S., & Youn, M. J. (1991). A new state feedback control of resonant converters. *IEEE Transactions on Industrial Electronics, 38*(3), 173–179. doi:10.1109/41.87584

Kim, S., Ho, J. S., Chen, L. Y., & Poon, A. S. Y. (2012). Wireless power transfer to a cardiac implant. *Applied Physics Letters*, *101*(7), 073701. doi:10.1063/1.4745600

Kim, S., Park, H.-H., Kim, J., Kim, K., & Ahn, S. (2014). Design and analysis of a resonant reactive shield for a wireless power electric vehicle. *IEEE Transactions on Microwave Theory and Techniques*, *62*(4), 1057–1066. doi:10.1109/TMTT.2014.2305404

Kim, T., Artan, N. S., Viventi, J., & Chao, H. J. (2012). Spatiotemporal compression for efficient storage and transmission of high-resolution electrocorticography data. *Proceedings* of *2012 Annual International Conference of the IEEE*, 1012-1015.

Kim, Y. G., & Vinck, H. (2018). *Anti-Collision Algorithms for FM0 Code and Miller Subcarrier Sequence in RFID Applications. IEEE Transactions on Vehicular Technology.*

Kim, Y., Shin, S., Jung, G., Lee, C., Lee, S., Shin, J., ... Chung, S. (2013, May). Phase analysis of currents and voltages in the On-Line Electric Bus (OLEB) system. *12th International Conference on Environment and Electrical Engineering*, 445-449.

Kissin, M. L. G., Boys, J. T., & Covic, G. A. (2009). Interphase Mutual Inductance in Polyphase Inductive Power Transfer Systems. *IEEE Transactions on Industrial Electronics*, *56*(7), 2393–2400. doi:10.1109/TIE.2009.2020076

Klesser, H., & Klaassens, J. B. (1986). Transformer-induced low-frequency oscillations in the series-resonant converter. *Proceedings of IEEE Power Electronics Specialists Conference PESC*, 236-245. 10.1109/PESC.1986.7415567

Kline, M., Izyumin, I., Boser, B., & Sanders, S. (2011). Capacitive power transfer for contactless charging. In *Proceedings of the 26th IEEE Annual Applied Power Electronics Conference and Exposition.* Fort Worth, TX: IEEE.

Ko, Roy, Smith, Lee, & Cho. (2007). An Enhanced RFID Multiple Access Protocol for Fast Inventory. *IEEE Globecom 2007*, 4575 - 4579.

Koelle, A., Depp, S., & Freyman, R. (1975, August). Short-range radio-telemetry for electronic identification using modulated RF backscatter. *Proceedings of the IEEE*, *63*(8), 1260–1261. doi:10.1109/PROC.1975.9928

Koo, G. (2010). *Signal constellations of a retrodirective array phase modulator (RAPM)* (Master's thesis). Georgia Institute of Technology.

Ko, W. H., Liang, S. P., & Fung, C. D. F. (1977). Design of Radio-Frequency Powered Coils for Implant Instruments. *Medical & Biological Engineering & Computing*, *15*(6), 634–640. doi:10.1007/BF02457921 PMID:203785

Ko, Y. D., & Jang, Y. J. (2013). The Optimal System Design of the Online Electric Vehicle Utilizing Wireless Power Transmission Technology. *IEEE Transactions on Intelligent Transportation Systems*, *14*(3), 1255–1265. doi:10.1109/TITS.2013.2259159

Kumar, A., Mirabbasi, S., & Chiao, M. (2009). Resonance-based wireless power delivery for implantable devices. *Proceedings of IEEE Biomedical Circuits and Systems Conference*, 25-28. 10.1109/BIOCAS.2009.5372092

Kurs, A., Karalis, A., Moffatt, R., Joannopoulos, J. D., Fisher, P., & Soljačić, M. (2007). Wireless power transfer via strongly coupled magnetic resonances. *Science*, *317*(5834), 83–86. doi:10.1126cience.1143254 PMID:17556549

Kuzey, S., Balci, S., & Altin, N. (2017). Design and analysis of a wireless power transfer system with alignment errors for electrical vehicle applications. *International Journal of Hydrogen Energy*, *42*(28), 17928–17939. doi:10.1016/j.ijhydene.2017.03.160

Laakso, I., & Hirata, A. (2013). Evaluation of the induced electric field and compliance procedure for a wireless power transfer system in an electrical vehicle. *Physics in Medicine and Biology*, *58*(21), 7583–7593. doi:10.1088/0031-9155/58/21/7583 PMID:24107570

Lambert, F. (2016). *Norway keeps electric vehicle tax exemption until 2020, positions itself to stay EV leader*. Retrieved September 7, 2017, from https://electrek.co/2016/11/09/norway-keeps-electric-vehicle-tax-exemption-until-2020-positions-itself-to-stay-ev-leader/

Landt, J. (2001). *Shrouds of Time - The History of RFID*. Retrieved from http://www.transcore.com/pdf/AIM%20shrouds_of_time.pdf

Landt, J. (2005, October/November). The History of RFID. *IEEE Potentials*, *24*(4), 8–11. doi:10.1109/MP.2005.1549751

Laouamer, R., Brunello, M., Ferrieux, J. P., Normand, O., & Buchheit, N. (1997). A multi-resonant converter for non-contact charging with electromagnetic coupling. *Proc. IEEE International Conference on Industrial Electronics, Control, and Instrumentation* (Cat. No.97CH36066), 792–797.

Laskovski, A. N., Yuce, M. R., & Dissanayake, T. (2011). Stacked spirals for biosensor telemetry. *IEEE Sensors Journal*, *11*(6), 1484–1490. doi:10.1109/JSEN.2010.2091123

Lazzi, G. (2005). Thermal effects bioimplants. *IEEE Engineering in Medicine and Biology Magazine*, *24*(5), 75–81. doi:10.1109/MEMB.2005.1511503 PMID:16248120

Lee, H.-J., & Cho, D.-H. (2015, September). Economic analysis based on the Interrrelationships of the OLEV System Components. *2015 IEEE 18th International Conference on Intelligent Transportation System*, 2677-2682.

Lee, S., Huh, J., Park, C., Choi, N.-S., Cho, G.-H., & Rim, C.-T. (2010). On-Line Electric Vehicle using inductive power transfer system. 2010 IEE Energy Conversion Congress and Exposition, Atlanta, GA.

Lee, C. K., Zhong, W. X., & Hui, S. Y. R. (2012). Effects of magnetic coupling of nonadjacent resonators on wireless power domino-resonator systems. *IEEE Transactions on Power Electronics*, *27*(4), 1905–1916. doi:10.1109/TPEL.2011.2169460

Lee, D. S., Liu, Y. H., & Lin, C. R. (2012). A Wireless Sensor Enabled by Wireless Power. *Sensors (Basel)*, *12*(12), 16116–16143. doi:10.3390121216116 PMID:23443370

Lee, H., Naishadham, K., Tentzeris, M., & Shaker, G. (2010). A Novel Highly Sensitive Antenna-Based Smart Skin Gas Sensor Utilizing Carbon Nanotubes and Inkjet Printing. *IEEE Antennas and Propagation Symposium*, 1593-1596.

Lee, S., Choi, B., & Rim, C. T. (2013). Dynamics Characterization of the Inductive Power Transfer System for Online Electric Vehicles by Laplace Phasor Transform. *IEEE Transactions on Power Electronics*, *28*(12), 5902–5909. doi:10.1109/TPEL.2013.2254500

Lee, W. S., Son, W. I., Oh, K. S., & Yu, J. W. (2011). Contactless energy transfer systems using antiparallel resonant loops. *IEEE Transactions on Industrial Electronics*, *60*(1), 350–359. doi:10.1109/TIE.2011.2177611

Lee, W. Y., Huh, J., Choi, S. Y., Thai, X. V., Kim, J. H., Al-Ammar, E. A., ... Rim, C. T. (2013). Finite-Width Magnetic Mirror Models of Mono and Dual Coils for Wireless Electric Vehicles. *IEEE Transactions on Power Electronics*, *28*(3), 1413–1428. doi:10.1109/TPEL.2012.2206404

Leung, H. Y., McCornick, D., Budgett, D., & Hu, A. P. (2013). Design methodology for inductive power transfer systems targeting high power implantable devices. In *Proceedings of the IEEE International Symposium on Circuits and Systems (ISCAS 2013)*. Beijing: IEEE. 10.1109/ISCAS.2013.6572457

Lewis, P. M., Ayton, L. N., Guymer, R. H., Lowery, A. J., Blamey, P. J., Allen, P. J., ... Rosenfeld, J. V. (2016). Advances in implantable bionic devices for blindness: A review. *ANZ Journal of Surgery*, *86*(9), 654–659. doi:10.1111/ans.13616 PMID:27301783

Lim, Y., Tang, H., Lim, S., & Park, J. (2014). An Adaptive Impedance Matching Network Based on a Novel Capacitor Matrix for Wireless Power Transfer. *IEEE Transactions on Power Electronics*, *29*(8), 4403–4413. doi:10.1109/TPEL.2013.2292596

Lin, J. C. (2013). Wireless power transfer for mobile applications, and health effects. *IEEE Antennas & Propagation Magazine*, *55*(2), 250–253. doi:10.1109/MAP.2013.6529362

Li, S., Li, W., Deng, J., Nguyen, T. D., & Mi, C. C. (2015). A Double-Sided LCC Compensation Network and Its Tuning Method for Wireless Power Transfer. *IEEE Transactions on Vehicular Technology*, *64*(6), 2261–2273. doi:10.1109/TVT.2014.2347006

Li, S., & Mi, C. C. (2015). Wireless power transfer for electric vehicle applications. *IEEE Journal of Emerging and Selected Topics in Power Electronics*, *3*(1), 4–17. doi:10.1109/JESTPE.2014.2319453

Liu, Hu, Nair, & Covic. (2010). 2-D Alignment Analysis of Capacitive Coupled Contactless Power Transfer Systems. *IEEE ECCE Conference*.

Liu, D., Hu, H., & Georgakopoulos, S. V. (2017). Misalignment Sensitivity of Strongly Coupled Wireless Power Transfer Systems. *IEEE Transactions on Power Electronics*, *32*(7), 5509–5519. doi:10.1109/TPEL.2016.2605698

Liu, D., Li, F., Zou, X., Liu, Y., Hui, X., & Tao, X. (2011). New analysis and design of a RF rectifier for RFID and implantable devices. *Sensors (Basel)*, *11*(June), 6494–6508. doi:10.3390110706494 PMID:22163968

Liu, X., & Hui, S. Y. R. (2007a). Equivalent Circuit Modeling of a Multilayer Planar Winding Array Structure for use in a Universal Contacless Battery Charging Platform. *IEEE Transactions on Power Electronics*, *22*(1), 21–29. doi:10.1109/TPEL.2006.886655

Liu, X., & Hui, S. Y. R. (2007b). Simulation Study and Experimental Verification of a Universal Contactless Battery Charging Platform with Localized Charging Features. *IEEE Transactions on Power Electronics*, *22*(6), 2202–2210. doi:10.1109/TPEL.2007.909301

Liu, X., Ng, W. M., Lee, C. K., & Hui, S. Y. (2008). Optimal operation of contactless transformers with resonance in secondary circuits. *Proceedings of the 23rd Annual Applied Power Electronics Conference and Exposition,* 645-650. doi: 10.1109/APEC.2008.4522790

Li, W., Zhao, H., Deng, J., Li, S., & Mi, C. C. (2016). Comparison Study on SS and Double-Sided LCC Compensation Topologies for EV/PHEV Wireless Chargers. *IEEE Transactions on Vehicular Technology*, *65*(6), 4429–4439. doi:10.1109/TVT.2015.2479938

Li, W., Zhao, H., Li, S., Deng, J., Kan, T., & Mi, C. C. (2015). Integrated ${LCC} $ Compensation Topology for Wireless Charger in Electric and Plug-in Electric Vehicles. *IEEE Transactions on Industrial Electronics*, *62*(7), 4215–4225. doi:10.1109/TIE.2014.2384003

Li, Z., Zhu, C., Jiang, J., Song, K., & Wei, G. (2017). A 3-kW wireless power transfer system for sightseeing car supercapacitor charge. *IEEE Transactions on Power Electronics*, *32*(5), 3301–3316. doi:10.1109/TPEL.2016.2584701

Lok Sabha Secretariat. (2014). *Norms for the Setting Up of Telecom Towers, its Harmful Effects and Setting Up of Security Standards in Expansion of Telecom Facilities*. New Delhi: Ministry of Communications and Information Technology.

Low, Z. N., Chinga, R. A., Tseng, R., & Lin, J. (2009). Design and test of a high power high-efficiency loosely coupled planar wireless power transfer system. *IEEE Transactions on Industrial Electronics*, *56*(5), 1801–1812. doi:10.1109/TIE.2008.2010110

Luca, P., Reatti, A., Corti, A., & Mastromauro, R. A. (2017) Inductive power transfer: Through a bondgraph analogy, an innovative modal approach. *Proceedings of the 2017 IEEE International Conference on Environment and Electrical Engineering and 2017 IEEE Industrial and Commercial Power Systems Europe*. 10.1109/EEEIC.2017.7977737

Ludwig, R., & Bretchko, P. (2000). *RF Circuit Design – Theory and Applications*. Prentice Hall.

Luo, Y., He, J., & Liu, H. (2014). Application of the Distributed Generation, Micro and Smart Power Grid in the Urban Planning. *IEEE-Cyber*, 634-637.

Lu, X., Wang, P., Niyato, D., Kim, D. I., & Han, Z. (2016). D. In Kim, Z. Han, Wireless charging technologies: Fundamentals, standards, and network applications. *IEEE Communications Surveys and Tutorials*, *18*(2), 1413–1452. doi:10.1109/COMST.2015.2499783

Lyshevski, S. E. (2000). Resonant converters: Nonlinear analysis and control. *IEEE Transactions on Industrial Electronics*, *47*(4), 751–758. doi:10.1109/41.857955

Madawala, U. K., Neath, M., & Thrimawithana, D. J. (2013). A power-frequency controller for bidirectional inductive power transfer systems. *IEEE Transactions on Industrial Electronics*, *60*(1), 310–317. doi:10.1109/TIE.2011.2174537

Maddalena, E. T., & Godoy, R. B. (2017). Model derivation and dynamic analysis of the SPS compensated wireless power transfer system. In *2017 IEEE 8th International Symposium on Power Electronics for Distributed Generation Systems (PEDG)*. IEEE. 10.1109/PEDG.2017.7972495

Maghami, M. H., Sodagar, A. M., Lashay, A., Esfahani, H. R., & Riazi-Esfahani, M. (2014). Visual prostheses: The enabling technology to give sight to the blind. *Journal of Ophthalmic & Vision Research*, *9*(4), 494. doi:10.4103/2008-322X.150830 PMID:25709777

Malaysian Communications and Multimedia Commission. (2013). *Requirements for Devices using Ultra-Wideband (UWB) Technology Operating in the frequency bands of 30MHz to 960MHz, 2.17GHz to 10.6GHz, 21.65GHz to 29.5GHz and 77GHz to 81GHz", SKMM SRSP-549*. UWB.

Marian, V., Menudier, C., Thévenot, M., Vollaire, C., & Verdier, J. (2011). Efficient design of rectifying antennas for low power detection. *International Microwave Symposium*. doi:10.1109/MWSYM.2011.5973270

Mateu, L., & Moll, F. (2005). Review of energy harvesting techniques and applications for microelectronics. *Proceedings of the SPIE Microtechnologies for the New Millennium*, 359-73. 2005.

Matsutomo, S., Noguchi, S., & Yamashita, H. (2012). Adaptive Mesh Generation Method Utilizing Magnetic Flux Lines in Two-Dimensional Finite Element Analysis. *IEEE Transactions on Magnetics*, *48*(2), 527–530. doi:10.1109/TMAG.2011.2176471

Mbacke, Mitton, & Rivano. (2018). A survey of RFID readers anticollision protocols. *IEEE Journal of Radio Frequency Identification*, 1-1.

Mbombolo, S., & Park, C. (2011). An improved detector topology for a rectenna. *Proceedings of the IMWS-IWPT*, 23-26. 10.1109/IMWS.2011.5877083

McSpadden, J. O., & Chang, K. (1994). A dual polarized circular patch rectifying antenna at 2.45 GHz for microwave power conversion and detection. *IEEE MTT-S International Microwave Symposium Digest*, 1749-52. 10.1109/MWSYM.1994.335103

McSpadden, J. O., Fan, L., & Chang, K. (1998, December). Design and experiments of a high-conversion efficiency 5.8-GHz rectenna. *IEEE Transactions on Microwave Theory and Techniques, VOL., 46*(12), 2053–2060. doi:10.1109/22.739282

Medeiros, R., Valtchev, S., & Valtchev, S. (2012). The efficient and stable charging of electric vehicle batteries: simplified instantaneous regulation. Springer Verlag Series: Technological Innovation for Value Creation. *IFIP Advances in Information and Communication Technology, 372*, 486-493.

Mendel, J. M. (1995). Fuzzy logic systems for engineering: A tutorial. *Proceedings of the IEEE, 83*(3), 345–377. doi:10.1109/5.364485

Meng, M., Ibrahim, A., & Kiani, M. (2015). Design considerations for ultrasonic power transmission to millimeter-sized implantable microelectronic devices. *Proceedings of IEEE Biomedical Circuits and Systems Conference*. 10.1109/BioCAS.2015.7348295

Mi, C. C., Buja, G., Choi, S. Y., & Rim, C. T. (2016). Modern Advances in Wireless Power Transfer Systems for Roadway Powered Electric Vehicles. *IEEE Transactions on Industrial Electronics, 63*(10), 6533–6545. doi:10.1109/TIE.2016.2574993

Michaelides, E. E. (2012). *Alternative Energy Sources. In Green Energy and Technology*. Springer.

Millán, J., Godignon, P., Perpiñà, X., Pérez-Tomás, A., & Rebollo, J. (2014). A survey of wide bandgap power semiconductor devices. *IEEE Transactions on Power Electronics, 29*(5), 2155–2163. doi:10.1109/TPEL.2013.2268900

Miskiewicz, R. M., & Moradewicz, A. J. (2011). Contactless power interface for plug-in electric vehicles in V2G systems. *Bulletin of the Polish Academy of Sciences. Technical Sciences, 59*(4), 561–568. doi:10.2478/v10175-011-0069-z

Mohamed, A. A. S., Berzoy, A., & Mohammed, O. A. (2017). Experimental Validation of Comprehensive Steady-State Analytical Model of Bidirectional WPT System in EVs Applications. *IEEE Transactions on Vehicular Technology, 66*(7), 5584–5594. doi:10.1109/TVT.2016.2634159

Mohamed, A., Berzoy, A., De Almeida, F., & Mohammed, O. (2017). Modeling and Assessment Analysis of Various Compensation Topologies in Bidirectional IWPT System for EV application. *IEEE Transactions on Industry Applications, 53*(5), 4973–4984. doi:10.1109/TIA.2017.2700281

Mohan, N., Undeland, T. M., & Robbins, W. P. (2003). *Power electronics: Converters, applications and design* (3rd ed.). Hoboken, NJ: Wiley.

Mohan, N., Undeland, T., & Robbins, W. (2003). *Power electronics (converters, applications and design)* (3rd ed.). New York: John Wiley & Sons.

Monti, G., Terricone, L., & Spartano, M. (2011). X-band planar rectenna. *IEEE Antennas and Wireless Propagation Letters, 10*, 1116–1118. doi:10.1109/LAWP.2011.2171029

Moorey, C., Holderbaum, W., & Potter, B. (2015). Investigation of High-Efficiency Wireless Power Transfer Criteria of Resonantly-Coupled Loops and Dipoles through Analysis of the Figure of Merit. *Energies, 8*(10), 11342–11362. doi:10.3390/en81011342

Moradewicz, A. J., & Kazmierkowski, M. P. (2009). High efficiency contactless energy transfer system with power electronic resonant converter. *Bulletin of the Polish Academy of Sciences. Technical Sciences, 57*(4), 375–381. doi:10.2478/v10175-010-0141-0

Mouhrim, N., Alaoui, A., El, H., & Boukachour, J. (2016, May). Optimal allocation of wireless power transfer system for electric vehicles in a multipath environment. *3rd International Conference on Logistics Operations Management (GOL)*, 1-7. 10.1109/GOL.2016.7731684

Murliky, L., Porto, R. W., & Brusamarello, V. J. (2017). Multivariable Optimization Method for Inductive Power Transfer in Wireless Sensors Nodes. In *Proceedings of the IEEE International Instrumentation and Measurement Technology Conference (I2MTC 2017)*. Torino: IEEE. 10.1109/I2MTC.2017.7969690

Musavi, F., & Eberle, W. (2014). Overview of wireless power transfer technologies for electric vehicle battery charging. *IET Power Electronics*, *7*(1), 60–66. doi:10.1049/iet-pel.2013.0047

Nahvi, M., & Edminister, J. A. (2014). *Electric circuits* (6th ed.). McGraw-Hill Education.

National Collaborating Centre for Environmental Health. (2013). *Radiofrequency toolkit for environmental health practitioners*. Vancouver, Canada: British Columbia Centre for Disease Control.

Nguyen, V., & Dhyanchand, J. (1987). An implementation of current-mode control for a series-resonant DC-DC converter. *Proceedings of IEEE Applied Power Electronics Conference APEC*, 266-273. 10.1109/APEC.1987.7067161

Nguyen, V., Dhyanchang, J., & Thollot, P. (1986). Steady-state and small signal analysis of series-resonant converter: a novel graphical approach. *Proceedings of IEEE Industrial Application Society Conference IAS*, 692-701.

Nikitin, P., & Rao, K. V. S. (2008). Antennas and Propagation in UHF RFID Systems. *IEEE International Conference on RFID*, 277 - 288. 10.1109/RFID.2008.4519368

Niu, W.-Q., Chu, J.-X., Gu, W., & Shen, A.-D. (2013). Exact Analysis of Frequency Splitting Phenomena of Contactless Power Transfer Systems. *IEEE Transactions on Circuits and Systems. I, Regular Papers*, *60*(6), 1670–1677. doi:10.1109/TCSI.2012.2221172

Oden, J. (2010). Finite element method. Scholarpedia, 5(5), 9836.

Omori, H. (2012). A novel wireless EV charger using SiC single-ended quasi-resonant inverter for home use. In *2012 15th International Power Electronics and Motion Control Conference (EPE/PEMC)*. IEEE. 10.1109/EPEPEMC.2012.6397524

Ong, C. S., Karim, M. F., & Ong, L. C. (2010). A Compact 2x2 Circularly Polarized Antenna Array for Energy Harvesting. *Proceedings of Asia-Pacific Microwave Conference*.

Ostaffe, H. (2012). Retrieved August 16, 2012, from http://www.powercastco.com/uhf-rfid-sensing-passive-rfid-wireless-sensor-tags/

Paing, Morroni, Dolgov, Shin, Brannan, Zane, & Popovic. (2007). Wirelessly-powered wireless sensor platform. *Proceedings of the 37th European Microwave Conference*, 999-1002.

Panchal, C., Lu, J., & Stegen, S. (2013). Investigation of magnetic flux distribution of EV wireless charging systems. In *2013 Asia-Pacific Symposium on Electromagnetic Compatibility (APEMC)*. IEEE. 10.1109/APEMC.2013.7360652

Pantic, Z., Bai, S., & Lukic, S. M. (2011). ZCS LCC-Compensated Resonant Inverter for Inductive-Power-Transfer Application. *IEEE Transactions on Industrial Electronics*, *58*(8), 3500–3510. doi:10.1109/TIE.2010.2081954

Pantic, Z., Lee, K., & Lukic, S. M. (2014). Multifrequency Inductive Power Transfer. *IEEE Transactions on Power Electronics*, *29*(11), 5995–6005. doi:10.1109/TPEL.2014.2298213

Pantic, Z., & Lukic, S. M. (2012). Framework and topology for active tuning of parallel compensated receivers in power transfer systems. *IEEE Transactions on Power Electronics*, *27*(11), 4503–4513. doi:10.1109/TPEL.2012.2196055

Park, C., Lee, S., Jeong, S. Y., Cho, G.-H., & Rim, C. T. (2015). Uniform Power I-Type Inductive Power Transfer System With DQ-Power Supply Rails for On-Line Electric Vehicles. *IEEE Transactions on Power Electronics*, *30*(11), 6446–6455. doi:10.1109/TPEL.2015.2420372

Park, M.-H., Shin, E.-G., Lee, H.-R., & Suh, I.-S. (2010, October). Dynamic model and control algorithm of HVAC system for OLEV® application. *ICCAS, 2010*, 1312–1317.

Patil, D., & Agarwal, V. (2015). Compact Onboard Single-Phase EV Battery Charger With Novel Low-Frequency Ripple Compensator and Optimum Filter Design. *IEEE Transactions on Vehicular Technology*, *65*(4), 1948–1956. doi:10.1109/TVT.2015.2424927

Paul, C. R. (2006). *Introduction to Electromagnetic Compatibility* (2nd ed.). John Wiley & Sons, Inc.

Pavone, D., Buonanno, A., D'Urso, M., & Corte, F. D. (2012). Design Considerations for Radio Frequency Energy Harvesting Devices. *Progress in Electromagnetics Research B*, *45*, 19–35. doi:10.2528/PIERB12062901

Pedder, D. A. G., Brown, A. D., & Skinner, J. A. (1999). A contactless electrical energy transmission system. *IEEE Transactions on Industrial Electronics*, *46*(1), 23–30. doi:10.1109/41.744372

Pereirinha & Trovão. (2012). Multiple energy sources hybridization: the future of electric vehicles? In *New Generation of Electric Vehicles*. Intech.

Perlo, P., & Meyer, G. (2012). ICT for the fully electric vehicle, research needs and challenges ahead (2nd ed.). European Union.

Pinto, R., Lopresto, V., & Genovese, A. (2017). *Human exposure to wireless power transfer systems: a numerical dosimetric study*. Paper presented at the 11th European Conference on Antennas and Propagation (EUCAP), Paris, France. 10.23919/EuCAP.2017.7928254

Pinuela, M., Yates, D. C., Lucyszyn, S., & Mitcheson, P. D. (2013). Maximizing DC-to-load efficiency for inductive power transfer. *IEEE Transactions on Power Electronics*, *28*(5), 2437–2447. doi:10.1109/TPEL.2012.2215887

Poon, A., O'Driscoll, S., & Meng, T. (2010). Optimal frequency for wireless power transmission into dispersive tissue. *IEEE Transactions on Antennas and Propagation*, *58*(5), 1739–1750. doi:10.1109/TAP.2010.2044310

Porto, R. W., Brusamarello, V. J., Müller, I., Riano, F. L. C., & Sousa, F. R. (2017a). Wireless power transfer for contactless instrumentation and measurement. *IEEE Instrumentation & Measurement Magazine*, *20*(4), 49–54. doi:10.1109/MIM.2017.8006394

Porto, R. W., Brusamarello, V. J., Müller, I., & Sousa, F. R. (2015). Design and characterization of a power transfer inductive link for wireless sensor network nodes. In *Proceedings of the IEEE International Instrumentation and Measurement Technology Conference (I2MTC 2015)*. Pisa: IEEE. 10.1109/I2MTC.2015.7151454

Porto, R. W., Brusamarello, V. J., Pereira, L. A., & Sousa, F. R. (2017b). Fine tuning of an inductive link through a voltage-controlled capacitance. *IEEE Transactions on Power Electronics*, *32*(5), 4115–4124. doi:10.1109/TPEL.2016.2598284

Porto, R. W., Murliky, L., Oliveira, G., & Brusamarello, V. J. (2017c). Multivariable Automatic Compensation Method for Inductive Power Transfer System. In *Proceedings of the 2017 IEEE International Conference on Environment and Electrical Engineering and 2017 IEEE Industrial and Commercial Power Systems Europe (EEEIC / I&CPSEurope)*. Milan: IEEE. 10.1109/EEEIC.2017.7977659

Publitek European Editors. (2013). *Tune In, Charge Up: RF Energy Harvesting Shows its Potential*. Available: http://www.digikey.com

Rahim, R. A., Hassan, S. I. S., Malek, F., & Junita, M. N. (2012a). A 2.45 GHz Circular Patch Antenna with Harmonic Suppression for Wireless Power Transmission. *IEEE Colloquium on Humanities, Science &Engineering Research (CHUSER)*.

Rahim, R. A., Hassan, S. I. S., Malek, F., Junita, M. N., Jamlos, M. F., & Azizan, M. N. (2012b). A 2.45GHz Harmonic Suppression Rectangular Patch Antenna. *International Symposium on Computer Applications and Industrial Electronics (ISCAIE)*. 10.1109/ISCAIE.2012.6482105

Rahim, R. A., Syed, I. S. H., Malek, F., Nordin, J. M., & Hassan, H. F. (2013). Harmonics Suppression Single-fed Dual-Circularly Polarized Microstrip Patch Antenna for Future Wireless Power Transmission. *IACSIT International Journal of Engineering and Technology, 5*(5).

Rajagopal, S., & Khan, F. (2011). Multiple receiver support for resonance coupling based wireless charging. *Proc. IEEE International Conference Communications Workshops*, 1-5.

Raju, S., Wu, R., Chan, M., & Yue, C. (2014). Modeling of mutual coupling between planar inductors in wireless power applications. *IEEE Transactions on Power Electronics, 1*, 114–121.

Raju, S., Wu, R., Chan, M., & Yue, C. P. (2014). Modeling of Mutual Coupling Between Planar Inductors in Wire- less Power Applications. *IEEE Transactions on Power Electronics, 29*(1), 481–490. doi:10.1109/TPEL.2013.2253334

Ramrakhyani, A. K., Mirabbasi, S., & Chiao, M. (2011). Design and optimization of resonance-based efficient wireless power delivery systems for biomedical implants. *IEEE Transactions on Biomedical Circuits and Systems, 5*(1), 48–63. doi:10.1109/TBCAS.2010.2072782 PMID:23850978

Ransford, B., Clark, S., Salajegheh, M., & Fu, K. (2008). Getting things done on computational rfids with energy-aware checkpointing and voltage-aware scheduling. *USENIX Workshop on Power Aware Computing and Systems.*

Rappaport, T. S. (2001). *Wireless Communications: Principle and Practice* (2nd ed.). Prentice Hall.

Ravikiran & Keshri. (2017). Comparative Evaluation of S-S and P-S Topologies for Wireless Charging of Electrical Vehicles. *IEEE Industrial and Electronics society Conference.*

Reatti, A., Corti, F., Pugi, L., Berzi, L., Barbieri, R., Delogu, M., & Pierini, M. Application of induction power recharge to garbage collection service. (2017). *Proceedings of the 2017 IEEE 3rd International Forum on Research and Technologies for Society and Industry*. 10.1109/RTSI.2017.8065961

Redlarski, G., Lewczuk, B., Żak, A., Koncicki, A., Krawczuk, M., Piechocki, J., & Gradolewski, D. (2015). The influence of electromagnetic pollution on living organisms: Historical trends and forecasting changes. *BioMed Research International, 2015*, 1–18. doi:10.1155/2015/234098 PMID:25811025

Reese, J., Brown, K., & Zhen, J. (2012). *RF Energy Harvesting* (Graduate Thesis). University of Cincinnati.

Richard, C. (1973, December). Johnson, H. Allen Ecrer, J. Searcy Hollis, Determination of far-field antenna patterns from near-field-field measurements. *Proceedings of the IEEE, 61*(12), 1668–1694. doi:10.1109/PROC.1973.9358

Rifai, A. B., & Hakami, M. A. (2014). Health hazards of electromagnetic radiation. *The Journal of Bioscience and Medicine, 2*(08), 1–12. doi:10.4236/jbm.2014.28001

Rim, C. T. (2018). *Wireless Charging of Electric Vehicles* (4th ed.). Power Electronics Handbook.

Rim, C. T. (2018). Wireless Charging of Electric Vehicles. In *Power Electronics Handbook*. Elsevier. doi:10.1016/B978-0-12-811407-0.00038-6

Rim, C. T., & Mi, C. (2017). *Wireless Power Transfer for Electric Vehicles and Mobile Devices*. Chichester, UK: John Wiley & Sons, Ltd. doi:10.1002/9781119329084

Rindorf, L., Lading, L., & Breinbjerg, O. (2008). Resonantly coupled antennas for passive sensors. *Proceedings of IEEE Sensors*, 1611–1614.

Riviere, S., Alicalapa, F., Douyere, A., & Lan Sun Luk, J. D. (2010). A Compact Rectenna Device at Low Power Level. *Progress in Electromagnetics Research C*, *16*, 137–146. doi:10.2528/PIERC10071604

Romba, Baikova, Borges, Valtchev, & Melicio. (2018). Wireless battery charger for ev with circular or planar coils: comparison. In *Technological Innovation for Resilient Systems*. Springer.

Romba, Valtchev, & Melicio. (2017). Single-phase wireless power transfer system controlled by magnetic core reactors at transmitter and receiver. *Technological Innovation for Smart Systems*, 419-428.

Romba, L. F., Baikova, E. N., Valtchev, S. S., & Melicio, R. (2016). Electric vehicle battery charger: wireless power transfer system controlled by magnetic core reactor. *Proc. of the Conference on Electronics, Telecommunications and Computers*, 1-2.

Romba, L. F., Valtchev, S. S., Melicio, R., Mudrov, M. V., & Ziuzev, A. M. (2017). Electric vehicle battery charger controlled by magnetic core reactor to wireless power transfer system. *Proc. of the IEEE 17th International Conference on Environment and Electrical Engineering*, 1-6. 10.1109/EEEIC.2017.7977782

Rossetto, L. (1992). A simple control technique for series resonant converters. *Proceedings of IEEE Power Electronics Specialists Conference PESC'92*, *2*, 787-792. 10.1109/PESC.1992.254803

Roundy, S., Wright, P. K., & Rabaey, J. M. (2003, July). A study of low level vibrations as a power source for wireless sensor nodes. *Computer Communications*, *26*(11), 1131–1144. doi:10.1016/S0140-3664(02)00248-7

Roy, S., Jandhyala, V., Smith, J., Wetherall, D., Otis, B., Chakraborty, R., ... Sample, A. (2010, September). RFID: From Supply Chains to Sensor Nets. *Proceedings of the IEEE*, *98*(9), 1583–1592. doi:10.1109/JPROC.2010.2053690

Ruehli, A. E. (1974). Equivalent circuit models for three-dimensional multiconductor systems. *IEEE Transactions on Microwave Theory and Techniques*, *22*(3), 216–221. doi:10.1109/TMTT.1974.1128204

Rueter, D. (2016). Induction coil as a non-contacting ultrasound transmitter and detector: Modelling of magnetic fields for improving the performance. *Ultrasonics*, *65*, 200–210. doi:10.1016/j.ultras.2015.10.003 PMID:26522956

SAE Standard J2954 "Wireless Power Transfer for Light-Duty Plug-In/ Electric Vehicles and Alignment Methodology" 2016.

Sample, A. P., Meyer, D. A., & Smith, J. R. (2011). Analysis, Experimental Results, and Range Adaptation of Magnetically Coupled Resonators for Wireless Power Transfer. *IEEE Transactions on Industrial Electronics*, *58*(2), 544–554. doi:10.1109/TIE.2010.2046002

Sample, A., Braun, J., Parks, A., & Smith, J. (2011). Photovoltaic enhanced UHF RFID tag Antennas for dual purpose energy harvesting. *IEEE International Conference on RFID*, 146 - 153. 10.1109/RFID.2011.5764615

Sample, A., & Smith, J. (2009). Experimental results with two wireless power transfer systems. *IEEE Radio and Wireless Symposium*, 16 - 18. 10.1109/RWS.2009.4957273

Scorcioni, Larcher, Bertacchini, & Vincetti. (2013). An Integrated RF Energy Harvester for UHF Wireless Powering Applications. *IEEE Conf. on Wireless Power Transfer*, 92-95.

Senjuti, S. (2013). *Design and optimization of efficient wireless power transfer links for implantable biotelemetry systems*. University of Western Ontario.

Sennouni, M. A., Zbitou, J., Tribak, A., Benaissa, A., & Latrach, M. (2013). *Circular Polarized Square Patch Antenna Array for Wireless Power Transmission*. IEEE. doi:10.1109/IRSEC.2013.6529698

Seo, D., Carmena, J. M., Rabaey, J. M., Alon, E., & Maharbiz, M. M. (2013). *Neural dust: an ultrasonic, low power solution for chronic brain-machine interfaces*. arXiv preprint arXiv:1307.2196

Seung-Hwan, L., & Lorenz, R. D. (2011). Development and validation of model for 95%-efficiency 220-W wireless power transfer over a 30-cm air gap. *IEEE Transactions on Industry Applications*, *47*(6), 2495–2504. doi:10.1109/TIA.2011.2168555

Shashank & Inman. (2009). *Energy Harvesting Technologies* (Vol. 21). New York: Springer.

Shi, Guo, Yan, Yu, Luo, & Li. (2018). Optimizing Directional Reader Antennas Deployment in UHF RFID Localization System by Using a MPCSO Algorithm. *IEEE Sensors Journal*, 1-1.

Shin, J., Shin, S., Kim, Y., Ahn, S., Lee, S., Jung, G., ... Cho, D.-H. (2014). Design and Implementation of Shaped Magnetic-Resonance-Based Wireless Power Transfer System for Roadway-Powered Moving Electric Vehicles. *IEEE Transactions on Industrial Electronics*, *61*(3), 1179–1192. doi:10.1109/TIE.2013.2258294

Shin, J., Song, B., Lee, S., Shin, S., Kim, Y., Jung, G., & Jeon, S. (2012, March). Contactless Power Transfer Systems for On-Line Electric Vehicle (OLEV). *2012 IEEE International Electric Vehicle Conference*, 1-4. 10.1109/IEVC.2012.6183255

Shinohara, Kunimi, Miura, Matsumoto, & Fujiwara. (1998). Open experiment of microwave power experiment with automatically target chasing system (Japanese). *IEICE Trans. B-II, J81-B-II,* (6), 657-661.

Shinohara, N. (2006). *Wireless power transmission for solar power satellite*. Retrieved from http://www.sspi.gatech.edu/wptshinohara.pdf

Shreshta, Noh, & Choi. (2013). Comparative Study of Antenna Designs for RF Energy Harvesting. *International Journal of Antennas and Propagat.*

Sim, C.-Y.-D., Chang, M.-H., & Chen, B.-Y. (2014). Microstrip-Fed Ring8 Slot Antenna Design with Wideband Harmonic Suppression. *IEEE Transactions on Antennas and Propagation*, *62*(9), 4828–4832. doi:10.1109/TAP.2014.2330595

Sirbu, I., & Mandache, L. (2017). Comparative Analysis of Different Topologies for Wireless Power Transfer System. *EEEIC Conference*. 10.1109/EEEIC.2017.7977863

Smith, J. R. (2011). *WISP (Wireless Identification and Sensing Platform)*. Retrieved 2011, from https://sensor.cs.washington.edu/WISP.html

Smith, J., Sample, A., Powledge, P., Mamishev, A., & Roy, S. (2006). A wirelessly powered platform for sensing and computation. *8th International Conference on Ubiquitous Computing (Ubicomp)*, 495 - 506. 10.1007/11853565_29

Society of Automotive Engineers. (1999). *SAE Electric vehicle inductively coupled charging (J1773)*. Warrendale, PA: SAE International.

Sodagar, A. M., Wise, K. D., & Najafi, K. (2009). A wireless implantable microsystem for multi- channel neural recording. *IEEE Transactions on Microwave Theory and Techniques*, *57*(10), 2565–2573. doi:10.1109/TMTT.2009.2029957

Sohn, Y.-H. (2015). Gyrator-Based Analysis of Resonant Circuits in Inductive Power Transfer Systems. IEEE Transactions on Power Electronics, 31(10). doi:10.1109/TPEL.2015.2506737

Son, S. H., & Jones, G. G. (2014). *Electronic device stand*. US Patent 8737064.

Song., B., Shin, J., Chung, S., Shin, S., Lee, S., Kim, Y., Jung, G., & Jeon, S. (2013, May). Design of a pickup with compensation winding for on-line electric vehicle (OLEV). *2013 IEEE Wireless Power Transfer (WPT)*, 60-62.

Song, B., Shin, J., Lee, S., Shin, S., Kim, Y., Jeon, S., & Jung, G. (2012, March). Design of a High Power Transfer Pickup for On-line Vehicle Electric Vehicle (OLEV). *IEEE International Electric Vehicle Conference*, 1-4. 10.1109/IEVC.2012.6183285

Souesme, B., Cheron, Y., & Metz, M. (1989). Study of a control method to gain the best dynamic performances of the series resonant converter. *Proceedings of European Power Electronics Conference EPE'89*, 1041-1047.

Spanik, P., Frivaldsky, M., Dragona, P., & Jaros, V. (2016). Analysis of Proper Configuration of Wireless Power Transfer System for Electric Vehicle Charging. *ELEKTRO Conference*. 10.1109/ELEKTRO.2016.7512071

Staebler, P. (2017). *Human exposure to electromagnetic fields: from extremely low frequency (ELF) to radiofrequency*. John Wiley and Sons. doi:10.1002/9781119384533

Stockman, H. (1948, October). Communication by means of reected power. *Proceedings of the IRE*, *36*(10), 1196–1204. doi:10.1109/JRPROC.1948.226245

Strauch, L., Pavlin, M., & Bregar, V. B. (2015). Optimization, design, and modelling of ferrite core geometry for inductive wireless power transfer. *International Journal of Applied Electromagnetics and Mechanics*, *49*(1), 145–155. doi:10.3233/JAE-150029

Stronks, H. C., & Dagnelie, G. (2014). The functional performance of the ARGUS II retinal prosthesis. *Expert Review of Medical Devices*, *11*(1), 23–30. doi:10.1586/17434440.2014.862494 PMID:24308734

Suh, N.P., & Cho, D.H. (2017). *The On-line Electric Vehicle_Wireless Electric Ground Transportation Systems*. Springer International Publishing ©Elsevier Ltd.

Suh, N.P., Cho, D.H., & Rim, C.T. (2010, April). *Design of On-Line Electric Vehicle*. Plenary lecture at the 2010 CIRP Design Conference, Nantes, France.

Summerer, L., & Purcell, O. (2009). *Concepts for wireless energy transmission via laser. Europeans Space Agency (ESA)*. Advanced Concepts Team.

Sutardja, C., & Rabaey, J. M. (2018). Isolator-Less Near-Field RFID Reader for Sub-Cranial Powering/Data Link of Millimeter-Sized Implants. *IEEE Journal of Solid-State Circuits*, 1-11.

Swain, A. K., Neath, M. J., Madawala, U. K., & Thrimawithana, D. J. (2012). A Dynamic Multivariable State-Space Model for Bidirectional Inductive Power Transfer Systems. *IEEE Transactions on Power Electronics*, *27*(11), 4772–4780. doi:10.1109/TPEL.2012.2185712

Tak, Y., Park, J., & Nam, S. (2009). Mode-based analysis of resonant characteristics for near-field coupled small antennas. *IEEE Antennas and Wireless Propagation Letters*, *8*, 1238–1241. doi:10.1109/LAWP.2009.2036133

Talapko, D., & Tesnjak, S. (2014). Influence of Distributed Power Generation Sources on Improvement of Power Supply Availability in Telecom Infrastructure. *IEEE DPSP*, 12-21.

TDK Tech Notes. (2017). *Wireless Power Transfer Products*. Retrieved from: https://product.tdk.com/info/en/products/wirelesscharge/ technote/tpo/index.html

Tellegen, B. D. H. (1948). The gyrator, a new electric network element. Philips Research Reports, 3, 81–101.

Tesla, N. (1904). The transmission of electric energy without wires. *The Thirteenth Anniversary Number of the Electrical World and Engineer*.

Tesla, N. (1914). *Apparatus for Transmitting Electrical Energy*. US patent n. 1119732.

Tesla, N. (1904). *Experiments with Alternate Current of High Potential and High Frequency*. McGraw.

Thomas, S., & Reynolds, M. (2010). QAM backscatter for passive UHF RFID tags. *IEEE International Conference on RFID*, 210 - 214.

Tomar & Gupta. (2012). Wireless power Transmission: Applications and Components. *International Journal of Engineering, 1*(5).

Treffers, M. (2015). History, Current Status and Future of the Wireless Power Consortium and the Qi Interface Specification. *IEEE Circuits and Systems Magazine*, *15*(2), 28–31. doi:10.1109/MCAS.2015.2418973

Trinkaus, G. (2000). *Magnetic amplifiers another lost technology*. High Voltage Press.

Trotter, M. S. (2011). *Range finding in passive wireless sensor networks using power optimized waveforms* (Ph.D. dissertation). Georgia Institute of Technology.

Trotter, M. S., & Durgin, G. D. (2010). Survey of range improvement of commercial rfid tags with power optimized waveforms. *IEEE International Conference on RFID*, 195 - 202. 10.1109/RFID.2010.5467265

Trotter, M. S., Grifin, J. D., & Durgin, G. D. (2009). Power-optimized waveforms for improving the range and reliability of RFID systems. *IEEE International Conference on RFID*, 80 - 87. 10.1109/RFID.2009.4911196

Trotter, M., Valenta, C., Koo, G., Marshall, B., & Durgin, G. (2012). Multi-Antenna Technologies for Enabling Passive RFID Tags and Sensors at Microwave Frequencies. *IEEE International Conference on RFID*, 1 - 7.

Trumpis, M., Insanally, M., Zou, J., Elsharif, A., Ghomashchi, A., Artan, N. S., ... Viventi, J. (2017). A low-cost, scalable, current-sensing digital headstage for high channel count μECOG. *Journal of Neural Engineering*, *14*(2), 026009. doi:10.1088/1741-2552/aa5a82 PMID:28102827

Turton, H., & Moura, F. (2008). Vehicle-to-grid systems for sustainable development: An integrated energy analyses. *Technological Forecasting and Social Change*, *75*(8), 1091–1108. doi:10.1016/j.techfore.2007.11.013

Ulaby, F. T., & Ravaioli, U. (2015). *Fundamentals of Applied Electromagnetics* (7th ed.). Pearson.

Valenta, C., & Durgin, G. (2013). Link Budgets for Backscatter Radio and RFID Systems Using Power-optimized Waveforms. *IEEE International Symposium on Antennas & Propagation*, 237-244. 10.1109/APS.2013.6711817

Valenta, C., Graf, P., Trotter, M., Koo, G., Durgin, G., Daly, W., & Schafer, B. (2010b). Transient Backscatter Channel Measurements at 5.8 GHz Across High-Voltage Insulation Gaps. *AMTA 2010 Symposium.*

Valenta, C., Graf, P., Trotter, M., Koo, G., Durgin, G., & Schafer, B. (2010a). Backscatter Channel Measurements at 5.8 GHz Across High-Voltage Corona. *IEEE Sensors Conference*, 2400. 10.1109/ICSENS.2010.5690964

Valtchev, S. (1988). *High-frequency DC-DC 2.5 kW converter (a super-resonant mode voltage booster for the telephone exchanges battery rooms, with 15 output stages in parallel). Industrial Contract of TU Sofia and The Plant for Power Supplies.* Report, Pernik.

Valtchev, S., Miletiev, R., Arnaoudov, R., & Valtchev, S. (2009). Control strategy for efficient operation of super-resonant SLSR (contactless) converters. *Proceedings of ICEST'09, Veliko Turnovo*, 527-530. http://pessoa.fct.unl.pt/ssv/ICEST_2009_Valtchev.pdf

Valtchev, S., Neves-Medeiros, R., Krusteva, A., Gigov, G., & Avramov, P. (2014). A wireless energy transceiver based on induction heating equipment. *Proceedings of 16th International PEMC Conference and Exposition, Antalya, Turkey*, 1017-1023. 10.1109/EPEPEMC.2014.6980642

Valtchev, S. (1980). *Transistor generators for high-frequency (440 kHz) induction heating. Investigation Contract Report, VMEI, Dept. Power Electronics*. TU Sofia.

Valtchev, S. (1990). Some regulation characteristics of pulse-width modulated series resonant power conversion. *Proceedings of Power Electronics and Motor Control Conference PEMC'90*, 83-87.

Valtchev, S. (2001). *High-frequency super-resonant series resonant DC/DC converters (converters of 3 kW, 5 kW, 10 kW, 17 kW, some for contactless energy transfer)*. Industry Consultancy Contract.

Valtchev, S., Brandisky, K., Borges, B., & Klaassens, J. B. (2009). Resonant Contactless Energy Transfer with Improved Efficiency. *IEEE Transactions on Power Electronics*, *24*(3), 685–699. doi:10.1109/TPEL.2008.2003188

Valtchev, S., Jorge, L. R., Craciunescu, A., & Brandisky, K. (2012). High frequency wireless energy for improving the EV autonomy. *Proceedings of International Symposium on Electric Vehicle and 4-th Annual Conference of Polish Society for Environment Friendly Vehicles*.

Valtchev, S., & Klaassens, J. B. (1990). Efficient resonant power conversion. *IEEE Transactions on Industrial Electronics*, *IE–37*(6), 490–495. doi:10.1109/41.103453

Valtchev, S., Klaassens, J. B., & Wesenbeeck, M. P. N. (1995). Super-resonant converter with switched resonant inductor with PFM-PWM control. *IEEE Transactions on Power Electronics*, *10*(6), 760–765. doi:10.1109/63.471296

Valtchev, S., Medeiros, R., Valtchev, S., & Klaassens, J. B. (2011). An instantaneous regulation for the wired and wireless super-resonant converters. *Proceedings of International Telecommunication Energy Conference INTELEC*. 10.1109/INTLEC.2011.6099875

Valtchev, S., & Valtchev, S. (2012). Experimental study of the energy portions control in series resonant converter. *Proceedings of 13th International Conference OPTIM*. 10.1109/OPTIM.2012.6231994

Van Schuylenbergh K., & Puers, P. (2009). *Inductive powering. Basic theory and application to biomedical systems*. Academic Press. doi: 10.1007/978-90-481-2412-1

Vázquez, J., Roncero-Sánchez, P., & Parreño Torres, A. (2017). Coupling factor of a weak inductive coupling in a 2-kW power transfer system with a 125-mm air gap for electric vehicle chargers. *Proceedings of the 26th International Symposium on Industrial Electronics,* 670-675. 10.1109/ISIE.2017.8001326

Vilathgamuwa, D. M., & Sampath, J. P. K. (2015). Wireless Power Transfer (WPT) for Electric Vehicles (EVs) - Present and Future Trends. In Plug In Electric Vehicles in Smart Grids. Ed. by Sumedha Rajakaruna, Farhad Shahnia, and Arindam Ghosh. Power Systems. Singapore: Springer Singapore.

Viventi, J., Kim, D.-H., Vigeland, L., Frechette, E. S., Blanco, J. A., Kim, Y.-S., ... Litt, B. (2011, December). Flexible, foldable, actively multiplexed, high-density electrode array for mapping brain activity in vivo. *Nature Neuroscience*, *14*(12), 1599–1605. doi:10.1038/nn.2973 PMID:22081157

Vu, V.-B., Kamal, L. B. M., Tay, J., Pickert, V., Dahidah, M., Logenthiran, T., & Phan, V.-T. (2017). A multi-output capacitive charger for electric vehicles. *Proceedings of the 26th International Symposium on Industrial Electronics,* 565-569. 10.1109/ISIE.2017.8001308

Wageningen, D., & Staring, T. (2010). The Qi Wireless Power Standard. In *Proceedings of the International Power Electronics and Motion Control Conference (EPE-PEMC 2010)*. IEEE. 10.1109/EPEPEMC.2010.5606673

Wahjudi, A. (1992). *Modelvorming en simulatie van een super-resonante omzetter met een hoogspanningstransformator* (M.Sc. thesis). Faculteit der Elektrotechniek, TU Delft.

Wang, G., Liu, W., Sivaprakasam, M., Zhou, M., Weiland, J. D., & Humayun, M. S. (2006). A dual band wireless power and data telemetry for retinal prosthesis. *Proceedings of IEEE EMBS Confeferce*, 28-38.

Wang, C., Stielau, O. H., & Covic, G. A. (2005). Design Considerations for a Contacless Electric Vehicle Battery Charger. *IEEE Transactions on Industrial Electronics*, *52*(5), 1308–1314. doi:10.1109/TIE.2005.855672

Wang, W. C., Tai, C. C., Wu, S. J., & Liu, Z. Y. (2015). A hybrid genetic algorithm with fuzzy logic controller for wireless power transmission system of electric vehicles. In *Proceedings of 2015 International Conference on Industrial Technology*. IEEE. 10.2991/itms-15.2015.255

Wang, Z., & Wei, X. (2015). Design Considerations for Wireless Charging Systems with an Analysis of Batteries. *Energies*, *8*(10), 10664–10683. doi:10.3390/en81010664

Wei, W. B., Liu, Q. Z., Yin, Y. Z., & Zhou, H. J. (2007). *Reconfigurable Microstrip Patch Antenna with Switchable Polarization* (Vol. 75). Progress in Electromagnetics Research.

Wei, W., & Qiang, F. (2011). Design and simulation of printed spiral coil used in wireless power transmission systems for implant medical devices. *Proceedings of Annual International Conference of the IEEE Engineering in Medicine and Biology Society*, 4018-4021. 10.1109/IEMBS.2011.6090998

Wen, F., & Huang, X. (2017). Human exposure to electromagnetic fields from parallel wireless power transfer systems. *International Journal of Environmental Research and Public Health*, *14*(2), 1–15. doi:10.3390/ijerph14020157 PMID:28208709

Wilson, B. S. (2017). The modern cochlear implant: A triumph of biomedical engineering and the first substantial restoration of human sense using a medical intervention. *IEEE Pulse*, *8*(2), 29–32. doi:10.1109/MPUL.2017.2649039 PMID:28328495

Wireless Power Consortium (WPC). (2016a). *The Qi Wireless Power Transfer System Power Class 0 Specification Parts 1 and 2: interface definitions*. Version 1.2.2. Wireless Power Consortium. Retrieved from: https://www.wirelesspowerconsortium.com/

Wireless Power Consortium (WPC). (2016b). *The Qi Wireless Power Transfer System Power Class 0 Specification Part 4: reference designs*. Version 1.2.2. Wireless Power Consortium. Retrieved from: https://www.wirelesspowerconsortium.com/

Wolterink, S., & Bauer, P. (2014, June). High-range On-line Electric Vehicles Powered by Inductive Power Transfer. *IEEE Transportation Electrification Conference and Expo (ITEC)*, Dearborn, MI. 10.1109/ITEC.2014.6861781

Won, D. H., Kim, H. S., & Jang, B. J. (2010). 13.56 MHz wireless power transfer system using loop antennas with tunable impedance matching circuit. *Journal of Korean Electromagnetic Engineering Society*, *21*(5), 519–527. doi:10.5515/KJKIEES.2010.21.5.519

Wu, C. C., Jian, M. S., & Chou, T. Y. (2011). Environmental Affection Based RFID Potential Bio-Disease Tracking/Tracing System, WSEAS Transactions on Systems. *WSEAS Transactions on Systems*, *10*(2), 38–48.

Wu, Y., Liu, P., Qiu, J., & Dai, G. (2015). Enable sustainable sensor networks with non-contact charging: Efficient deployment of energy hubs. *Int. J. of Sensor Networks*, *18*(3/4), 172–181. doi:10.1504/IJSNET.2015.070406

Xiao, Aflaki, Lang, Zhang, Tao, Py … Chang. (2018). Printed UHF RFID Reader Antennas for Potential Retail Applications. *IEEE Journal of Radio Frequency Identification*, 1-1.

Xu, D. Q., Joós, G., Lévesque, M., & Maier, M. (2013). Integrated V2G, G2V, and renewable energy sources coordination over a converged fiber-wireless broadband access network. *IEEE Transactions on Smart Grid*, *4*(3), 1381–1390. doi:10.1109/TSG.2013.2253337

Xue Chin, Q. (2005). Design of a 5.8-Ghz. rectenna incorporating a new patch antenna. *IEEE Antennas and Wireless Propagation Letters*, *4*, 175–178. doi:10.1109/LAWP.2005.846434

Yang, Crisp, Penty, & White. (2018a). RFID Enabled Health Monitoring System for Aircraft Landing Gear. *IEEE Journal of Radio Frequency Identification*, 1-1.

Yan, G., Ye, D., Zan, P., Wang, K., & Ma, G. (2007). Micro-robot for endoscope based on wireless power transfer. *Proceedings of International Conference in Mechatronics Automation*, 3577-3581. 10.1109/ICMA.2007.4304140

Yang, L., Rida, A., Vyas, R., & Tentzeris, M. (2007). RFID Tag and RF Structures on a Paper Substrate Using Inkjet-Printing Technology. *IEEE Transactions on Microwave Theory and Techniques*, *55*(12), 2894–2901. doi:10.1109/TMTT.2007.909886

Yang, T., Xiang, J., Wang, Y., Tan, X., Wang, J., Yan, N., ... Min, H. (2018b). *An Active Tag Using Carrier Recovery Circuit for EPC Gen2 Passive UHF RFID Systems. IEEE Transactions on Industrial Electronics.*

Yang, Y. Y., Jeon, S., & Ko, Y. D. (2015). System optimization of the On-Line Electric Vehicle operating in a closed environment. *Computers & Industrial Engineering*, *80*, 222–235. doi:10.1016/j.cie.2014.12.004

Yee, Y. F., & Chakrabarty, C. K. (2007). Phase detection using ad8302 evaluation board in the superheterodyne microwave interferometer for line average plasma electron density measurements. *Measurement*, *40*(9), 849–853. doi:10.1016/j.measurement.2007.02.005

Yeh, A., Ho, J., Tanabe, Y., Neofytou, E., Beygui, R., & Poon, A. (2013). Wirelessly powering miniature implants for optogenetic stimulation. *Applied Physics Letters*, 103.

Yeoh, W (2010). *Wireless power transmission (WPT) application at 2.4 GHz in common network*. Academic Press.

Yeo, J., & Kim, D. (2009). Harmonic Suppression Characteristic of a CPW-Fed Circular Slot Antenna Using Single Slot on a Ground Conductor. *Progress in Electromagnetics Research Letters*, *11*, 11–19. doi:10.2528/PIERL09072108

Yin, M., Borton, D., Aceros, J., Patterson, W., & Nurmikko, A. (2013). A100-channel hermetically sealed implantable device for chronic wireless neurosensing applications. *IEEE Transactions on Biomedical Circuits and Systems*, *7*(2), 115–128. doi:10.1109/TBCAS.2013.2255874 PMID:23853294

Yoo & Chang. (1992). Theoretical and experimental development of 10 and 35 GHz rectennas. *IEEE Transactions on Microwave and Techniques, 40*(6).

Young-Ho, S., & Chang, K. (2002). A High Efficiency Dual-Frequency Rectenna for 2.45GHz and 5.8GHz Wireless Power Transmission. *IEEE Transactions on Microwave Theory and Techniques*, *50*(7).

Zaheer, A., Hao, H., Covic, G. A., & Kacprzak, D. (2015). Investigation of multiple decoupled coil primary pad topologies in lumped IPT systems for interoperable electric vehicle charging. *IEEE Transactions on Power Electronics*, *30*(4), 1937–1955. doi:10.1109/TPEL.2014.2329693

Zeng, F. G. (2017). Challenges in improving cochlear implant performance and accessibility. *IEEE Transactions on Biomedical Engineering*, *64*(8), 1662–1664. doi:10.1109/TBME.2017.2718939 PMID:28650803

Zeng, F. G., Rebscher, S., Harrison, W., Sun, X., & Feng, H. (2008). Cochlear implants: System design, integration, and evaluation. *IEEE Reviews in Biomedical Engineering*, *1*, 115–142. doi:10.1109/RBME.2008.2008250 PMID:19946565

Zhang, Liu, Meng, Wu, Lee, Xu, … Kim. (2014). Design of a Compact Planar Rectenna for Wireless Power Transfer in the ISM Band. *International Journal of Antennas and Propagation.*

Zhang, W., & Chris Mi, C. (2015). Compensation Topologies of High Power Wireless Power Transfer System. *IEEE Transactions on Vehicular Technology*, *65*(6), 4768–4778. doi:10.1109/TVT.2015.2454292

Zhang, W.-H., Cheong, P., Lu, W.-J., & Tam, K.-W. (2018). *Planar Endfire Circularly Polarized Antenna for Low Profile Handheld RFID Reader. IEEE Journal of Radio Frequency Identification.*

Zhang, W., Wong, S.-C., Tse, C. K., & Chen, Q. (2014). Analysis and Comparison of Secondary Series- and Parallel-Compensated Inductive Power Transfer Systems Operating for Optimal Efficiency and Load-Independent Voltage-Transfer Ratio. *IEEE Transactions on Power Electronics*, *29*(6), 2979–2990. doi:10.1109/TPEL.2013.2273364

Zhang, W., Wong, S.-C., Tse, C. K., & Chen, Q. (2014). Design for efficiency optimization and voltage controllability of series-series compensated inductive power transfer systems. *IEEE Transactions on Power Electronics*, *29*(1), 191–200. doi:10.1109/TPEL.2013.2249112

Zhang, X., Ho, S. L., & Fu, W. N. (2011). Quantitative analysis of a wireless power transfer cell with planar spiral structures. *IEEE Transactions on Magnetics*, *47*(10), 3200–3203. doi:10.1109/TMAG.2011.2147768

Zhang, Y., Zhao, Z., & Chen, K. (2014). Frequency-Splitting Analysis of Four-Coil Resonant Wireless Power Transfer. *IEEE Transactions on Industry Applications*, *50*(4), 2436–2445. doi:10.1109/TIA.2013.2295007

Zhang, Z., & Chau, K. T. (2015). Homogeneous Wireless Power Transfer for Move-and- Charge. *IEEE Transactions on Power Electronics*, *30*(11), 6213–6220. doi:10.1109/TPEL.2015.2414453

Zhou, D. D., Dorn, J. D., & Greenberg, R. J. (2013). *The ARGUS II retinal prosthesis system: An overview.* Paper presented at the 2013 IEEE International Conference on Multimedia and Expo Workshops (ICMEW), San Jose, CA. 10.1109/ICMEW.2013.6618428

Zhou, D., & Greenbaum, E. (2009). *Implantable Neural Prostheses 1*. New York: Springer.

Zhu, C., Liu, K., Yu, C., Ma, R., & Cheng, H. (2008). Simulation and experimental analysis on wireless energy transfer based on magnetic resonances. *Proc. IEEE Conference Vehicle Power and Propulsion*, 1-4.

Zimmer, M., Heinrich, J., & Parspour, N. (2014). Design of a 3 kW primary power supply unit for inductive charging systems optimized for the compatibility to receiving units with 20 kW rated power. *Proceedings of the 4th International Electric Drives Production Conference,* 318-322. 10.1109/EDPC.2014.6984415

About the Contributors

Alicia Triviño-Cabrera is a Telecommunication Engineer (2002) and a Computer Science Engineer (2008) by the University of Málaga (Spain). Her Thesis, which was defended in 2007, focused on wireless technologies. She has authored 50+ papers in international conferences, prestigious journals and books concerning wireless networks, wireless power transfer and electric vehicles. Part of her research work has been conducted in prestigious centers such as Samsung Advanced Institute of Technology in South Korea (2004), Laboratory of Communications at the University of Coimbra (2007), Orange Labs R&D in France (2009) and Aston University in the United Kingdom (2017). In 2010, she started teaching at the University of Malaga (Spain) where she is currently an Associate Professor.

José Antonio Aguado received his Industrial Engineering (major power systems) degree and PhD from the University of Malaga in 1997 and 2001, respectively. He is Professor and Department Head at the Electrical Engineering Department, University of Málaga in Spain. He has led 38 publicly-funded research and consultancy projects in the area of operation and planning of smartgrids, renewable energies and wireless power transfer. He has authored 70+ papers in journals, book chapters, and conferences. He actively participates in scientific associations such as IEEE where he is Vice-President of the IEEE Power and Energy Spanish Section.

***

Reza Amineh is currently an Assistant Professor at the Department of Electrical and Computer Engineering, NYIT. Prior to NYIT, he was a Principal Scientist in the Department of Sensor Physics at Halliburton Co. He received the Ph.D. degree in electrical engineering from McMaster University, Canada, in 2010. He was a Post-Doctoral Fellow with University of Toronto and McMaster University, from 2012 to 2013 and from 2010 to 2012, respectively. He was a Ph.D. Intern with the Advanced Technology Group, BlackBerry, in 2009. He has authored/co-authored over 55 journal and conference papers, and a book chapter. He contributed in over 40 patent disclosures in applied electromagnetics while working in Halliburton Co. His research interests include Applied Electromagnetics. Dr. Amineh was a recipient of the Banting Post-Doctoral Fellowship from the Government of Canada, in 2012-2013 and the Ontario Ministry of Research and Innovation (OMRI) Post-Doctoral Fellowship in 2010-2011. During his Ph.D. program, he was awarded the McMaster Internal Prestige Scholarship Clifton W. Sherman for two consecutive years. He has co-authored an Honorable Mention Paper presented at the IEEE Symposium on Antennas and Propagation, and the International Union of Radio Science, in 2008. He has also co-authored a paper selected among the journal of Inverse Problems' "Highlights Collection of 2010". Dr. Amineh is a senior member of IEEE.

N. Sertac Artan is an Assistant Professor of Electrical and Computer Engineering at New York Institute of Technology (NYIT). He got his Ph.D. degree in Electrical Engineering from New York University (formerly Polytechnic University). Prior to joining NYIT, Dr. Artan was on the faculty of the New York University School of Engineering. He also worked as an ASIC (Application Specific Integrated Circuit) Design Engineer and designed integrated circuits for commercial, academic and military applications. Dr. Artan served in the organizing committees of the ACM Conference on Security and Privacy in Wireless and Mobile Networks (WiSec), Workshop on RFID Security (RFIDSec), IEEE Sarnoff Symposium, and ACM/IEEE Symposium on Architectures for Networking and Communications Systems (ANCS). Dr. Artan's current research interests include low-power embedded systems and VLSI circuits for implantable medical devices and high-speed circuits for network security.

Agasthya Ayachit received his B.S degree in electrical and electronics Engineering from a VTU-affiliated institution in Bangalore, India in 2009. He obtained his M.S. degree in electrical engineering from Wright State University, Dayton, OH, USA in 2010 after which he served as a lecturer at Pennsylvania State University. He has interned at GE Global Research, Mediatek, Inc., and Google, Inc., where he worked on the design, development, and testing of high frequency power systems. He is pursuing his Ph.D. degree in electrical engineering at Wright State University. He has also served as a graduate teaching assistant for power electronic courses at the Department of Electrical Engineering. His research areas include modeling of dc-dc PWM and resonant converters, wireless power charging, magnetic component design, and RF power amplifiers. He is currently working as a systems engineer at Mercedes-Benz Research and Development of North America and deeply involved in electric vehicle charging strategies.

Elena Baikova has a MSc Degree in Electrical Power Systems and Grids completed in 1981 in the Moscow Power Engineering Institute. Since 1980 until 1992 is an Professor in the Technical School of Light Industry in Moscow. Granted the equivalence degree in Electrical Engineering and Computer 1999 by the Instituto Superior Técnico, Technical University of Lisboa. Recognition of the Master's degree in Electrical and Computer Engineering from Universidade Nova de Lisboa (UNL) in 2009. Actually, she is a PhD student in Electrical and Computer Engineering at the Faculty of Science and Technology, Universidade Nova de Lisboa. She is teaching at the Superior School of Technology of Setubal Polytechnic Institute. She is a collaborator at the research center CTS-UNINOVA/UNL with interest in the wireless power transfer for electric vehicles, electromagnetic resonant coupling and electromagnetic emissions.

Morris Brenna is an Associate Professor in the Electric Power System sector at the Department of Energy of Politecnico di Milano. He has made special studies on power quality, converter controls and power supply of electric systems for transportation.

Valner Brusamarello received the B.E. degree in electrical engineering and the M.Sc. degree in material science from the Universidade Federal do Rio Grande do Sul (UFRGS), Porto Alegre, Brazil, in 1992 and 1996, respectively, and the Dr. degree in material science from the Universidade Federal de Santa Catarina, Florianópolis, Brazil, in 2000. He is currently a Professor in the Department of Electrical Systems of Automation and Energy, UFRGS. His main research interests include the area of applied electronic instrumentation, energy harvesting systems and wireless power transmission.

Fabio Corti received his B.S. degree in Electronics and Telecommunications Engineering from the University of Florence, School of Engineering Florence, Italy in 2014. He obtained his M.S. degree in Electrical engineering from University of Florence, School of Engineering, in 2016. He is currently a Ph.D Student under a program totally funded By Magneti Marelli S.p.A. His research areas include modeling of DC-DC PWM and resonant converters, wireless power charging, magnetic component design, and RF power amplifiers.

Daniel da Costa (S'04-M'08-SM'14) was born in Fortaleza, Ceará, Brazil, in 1981. He received the B.Sc. degree in Telecommunications from the Military Institute of Engineering (IME), Rio de Janeiro, Brazil, in 2003, and the M.Sc. and Ph.D. degrees in Electrical Engineering, Area: Telecommunications, from the University of Campinas, SP, Brazil, in 2006 and 2008, respectively. His Ph.D thesis was awarded the Best Ph.D. Thesis in Electrical Engineering by the Brazilian Ministry of Education (CAPES) at the 2009 CAPES Thesis Contest. From 2008 to 2009, he was a Postdoctoral Research Fellow with INRS-EMT, University of Quebec, Montreal, QC, Canada. Since 2010, he has been with the Federal University of Cear\'{a}, where he is currently an Assistant Professor. Prof. da Costa is currently Editor of the IEEE Communications Surveys and Tutorials, IEEE Access, IEEE Transactions on Vehicular Technology, EURASIP Journal on Wireless Communications and Networking, and KSII Transactions on Internet and Information Systems. He has also served as Associate Technical Editor of the IEEE Communications Magazine. From 2012 to 2017, he was Editor of the IEEE Communications Letters. He has served as Guest Editor of several Journal Special Issues. He has been involved on the organization of several conferences. He is currently the Latin American Chapters Coordinator of the IEEE Vehicular Technology Society. Also, he acts as a Scientific Consultant of the National Council of Scientific and Technological Development (CNPq), Brazil and he is a Productivity Research Fellow of CNPq. From 2012 to 2017, he was Member of the Advisory Board of the Ceará Council of Scientific and Technological Development (FUNCAP), Area: Telecommunications. Prof. da Costa is the recipient of three conference paper awards. He received the Exemplary Reviewer Certificate of the IEEE Wireless Communications Letters in 2013, the Exemplary Reviewer Certificate of the IEEE Communications Letters in 2016, the Certificate of Appreciation of Top Associate Editor for outstanding contributions to IEEE Transactions on Vehicular Technology in 2013, 2015 and 2016, and the Exemplary Editor Award o IEEE Communications Letters in 2016. He is Distinguished Lecturer of the IEEE Vehicular Technology Society. He is a Senior Member of IEEE, Member of IEEE Communications Society and IEEE Vehicular Technology Society.

Fernando de Sousa received the B.Eng. and M.Sc. degrees in electrical engineering from the Federal University of Campina Grande, Campina Grande, Brazil, in 1996 and 2000, respectively, and the Ph.D. degree from Ecole Nationale Supérieure des Télécommunications, Paris, France, in 2004. From 1996 through 2000, he worked at the industry, in consumer electronics companies such as Sharp and Whirpool. In 2005, he worked at Federal University of Santa Catarina as a Visiting Professor. From 2006 to 2009, he was with Federal University of Rio Grande do Norte, where was the Head of the Microelectronics and Embedded Systems Laboratory. Since 2009, he has been with Federal University of Santa Catarina, heading a research laboratory focused on RF application and wireless instrumentation. He is the coordinator of the IC design activities at the Brazilian National Institute on Nano and Microelectronics Systems. His main research interests include RF electronics, radio-frequency integrated circuit design, wireless power transfer, and wireless instrumentation electronics.

Haiyang Ding (M'13) was born in Hebei, China, in 1980. He received the B.Sc. degree in communications engineering from the Xi'an Communications Institute, Xi'an, China, in 2003, the M.Sc. degree (Hons.) in electrical engineering from the Beijing University of Technology, Beijing, China, in 2006, and the Ph.D. degree (Hons.) in telecommunications engineering from Xidian University, Xi'an, in 2013. Since 2013, he has been with the State Key Laboratory of Integrated Service Networks, Xidian University, where he is currently an Associate Professor. His research interests include channel modeling and characterization, cooperative communications, and cognitive radio systems. He serves as the Editor of the KSII Transactions on Internet and Information Systems. He was a recipient of the Research in Motion Wireless Research Scholarship from Xidian University in 2012, the IEEE Communications Letters Exemplary Reviewer Certificate in 2013, the IEEE TRANSACTIONS ON COMMUNICATIONS Exemplary Reviewer Certificate in 2014, and the Excellent Ph.D. Dissertation Certificate of Shaanxi Province in 2015.

Federica Foiadelli is an Associate Professor in the Electric Power System sector at the Department of Energy of Politecnico di Milano. She has made special studies on electric safety, renewable energy applications and automation in electric systems for transportation.

Jianhua Ge was born in Jiangsu, China, in 1961. He received the B.Sc. and Ph.D. degrees in communications engineering from Xidian University, Xi'an, China, in 1982 and 1990, respectively. He is currently a Professor and the Deputy Director with the State Key Laboratory of Integrated Services Networks, Xidian University, Xi'an. He has worked on digital television (DTV) standardization as a DTV Technical Expert. His current research interests include digital video broadcasting system and multiple-input-multiple-output, and mobile communication techniques.

Ming-Shen Jian is an assistant professor of Dept. Computer Science and Information Engineering at National Formosa University. Ming-Shen Jian's current research interests are in the area related to IOT application, Big Data, optimal solution, Intelligent System, and cloud computing. He received B.S. degree at Electrical and Control Engineering in National Chiao Tung University and Ph.D. at the Department of Computer Science and Engineering of the National Sun Yet-Sen University in Taiwan, 2007, investigating resource management in 3G mobile communication systems.

Marian Kazimierczuk received the M.S., Ph.D., and D.Sci. degrees in electronics engineering from the Department of Electronics, Warsaw University of Technology, Warsaw, Poland, in 1971, 1978, and 1984, respectively. He was a Teaching and Research Assistant with the department of Electronics, Institute of Radio Electronics, Warsaw University of Technology, from 1972-1978, where he was an Assistant Professor from 1978-1984. In 1984, he was a Project Engineer with Design Automation Inc., Lexington, MA, USA. From 1984-1985, he was a Visiting Professor with the Department of Electrical and Computer Engineering, Virginia Polytechnic Institute and State University, Blacksburg, VA, USA. Since 1985, he has been with the Department of Electrical Engineering, Wright State University, Dayton, OH, USA, where he is currently a University Distinguished Professor. He is the author/co-author of 7 books, 200 journal papers, and 250 conference papers, and holds 8 patents. He has served as an Associate Editor for IEEE TRANSACTIONS ON CIRCUITS AND SYSTEMS and Journal of Circuits, Systems and Computers, and serves as an Associate Editor for IEEE TRANSACTIONS ON INDUSTRIAL ELECTRONICS, International Journal of Circuits Theory and Applications, and IET Circuits, Devices and Systems. His research interests include high-frequency high-efficiency switching-mode tuned power amplifiers, resonant and pulse width modulated dc/dc power converters, dc/ac inverters, high-frequency rectifiers, electronic ballasts, modeling and control of converters, high-frequency magnetics, wireless power transfer, renewable energy sources, and wide-band gap power semiconductor devices.

Mukesh Khandelwal received his BTech degree in Electronics and Communication Engineering and his MTech degree in Digital Communication from GGSIP University, India, and his PhD degree in Electronics Engineering from the Indian Institute of Technology (IIT-ISM), Dhanbad. He is currently working as an Assistant Professor in the Department of Electronics and Communication Engineering at Bhagwan Parshuram Institute of Technology, India, which is affiliated to Guru Gobind Singh Indraprastha University. Mukesh is also an editor at Enago Crimson Interactive Pvt. Ltd, USA and Editor in Cambridge Scholars Publishing, UK. He has published 33 research papers in refereed international journals and serves as a reviewer for various renowned journals including IEEE MTTs, IEEE Access, IET Microwave Antenna and Propagation, IET Electronics Letters Wireless Personal Communication, the International Journal of Electronics and Communications, and the Journal of Electromagnetic Waves and Applications. He has contributed to various edited research books. He was awarded with a Research Fellowship by the Indian Department of Science and Technology (Govt of India) in 2013. His current research interests focus on microstrip antennas, metamaterials, 5-G antennas, UWB antennas, rectennas, reconfigurable antennas, multiband antennas, circularly polarized antennas, antenna array, defected ground structure, microwave components, electromagnetics, wireless communication, telecommunication, RADAR and satellite communication.

Michela Longo is an Assistant Professor in the Electric Power System sector at the Department of Energy of Politecnico di Milano. She has made special studies on electric vehicles, renewable energy applications and automation in electric systems for transportation.

Rui Melício was born in Lisboa, Portugal, and received the MSc in 2004, from Instituto Superior Técnico, Lisboa, Portugal, received the PhD in 2010 from Universidade da Beira Interior, Portugal, received the Post-PhD in 2012 from Fundação para a Ciência e a Tecnologia, Lisboa, Portugal and Habilitation in 2014 from the Universidade de Évora, Portugal. He is member of IDMEC research laboratory at Instituto Superior Técnico, Universidade de Lisboa and an Assistant Professor with Habilitation at Universidade de Évora. His research interests include: power systems management and operation; power electronics and its applications; renewable energies; smart grids; mechatronics. He is a Senior Member of the IEEE Industrial Electronics Society.

Lucas Murliky received Bachelor in Computer Engineering by Universidade Estadual do Rio Grande do Sul - UERGS (2015) and Master in Electrical Engineering by Federal University of Rio Grande do Sul - UFRGS (2017), where he is currently working toward the Ph.D. degree. His main research interests are wireless power transfer, electronic instrumentation, measurement systems and minimal operate systems.

Alfonso Parreño Torres received an M.Sc. degree in Electronic Engineering from the University of Valencia, Valencia, Spain, in 2006, and a Ph.D. from the University of Castilla-La Mancha, Ciudad Real, Spain, in 2016. He is with the Castilla-La Mancha Science and Technology Park, Albacete, Spain, and with the University of Castilla-La Mancha as a part-time assistant professor. His research interests include control, power electronics, power quality, and signal processing algorithms for electrical power systems.

Rodrigo Porto received the B.Eng. and M.Sc. degrees in electrical engineering from the Universidade Federal do Rio Grande do Sul, Porto Alegre, Brazil, in 2002 and 2009, respectively, where he is currently working toward the Ph.D. degree in electronic instrumentation. His main research interests include electronic instrumentation, measurement systems, and wireless power transfer.

Alberto Reatti received the degree in electronics engineering, from the University of Florence, Florence, Italy, in 1988 and the Ph.D. degree 1993, from the University of Bologna. In 1992, he was an Associate Researcher in the Department of Electrical Engineering, Wright State University, Dayton, OH, USA. In 1994, he was a Visiting Professor in the Department of Electronics, University of Siena, Siena, Italy. For several years, he has also been a Visiting Professor in the Department of Electronics, University of Florence, where he is currently an Associate Professor at the Department of Information Engineering. His research interests include high-frequency, resonant and pulse-width modulated dc-dc power converters, dc-ac inverters, high-frequency rectifiers, electronic ballasts, modeling and control of converters, high-frequency magnetic, and power semiconductor devices, and renewable power sources. He has published more than 100 technical papers, including papers appeared in IEEE Transactions and Journals. He also holds two patents. One of this patents allowed him to be selected as a member of a University of Florence spin off company. Dr. Reatti served as a Chairman of the Power Electronic and Power Systems Committee of IEEE CAS Society and as an Associate Editor of the IEEE TRANSACTIONS ON CIRCUITS AND SYSTEMS I: REGULAR PAPERS. He is also responsible for agreements of cultural and scientific cooperation among University if Florence and other Country Research Institutions.

Luis Romba received the BSc in Electronics and Telecommunications and the degree in Electrical Power Distribution and Power Control Systems from Instituto Superior de Engenharia, Lisbon, Portugal. He received the MSc by Faculdade de Ciencias e Tecnologia (FCT), Universidade Nova de Lisboa (UNL), Portugal in Renewable Energy and Sustainable Energy Systems. He is currently pursuing the Ph.D. degree at FCT/UNL. His professional carrier was developed to a great extend in radar, navigation, communications, and support equipment engineering. He is now a researcher at the research center CTS-UNINOVA of FCT/UNL with a special interest in the wireless power transfer for electric vehicles, applying electromagnetic resonant coupling and magnetic core reactor.

Pedro Roncero-Sánchez Elipe received his Ph.D. in Industrial Engineering from the University of Castilla-La Mancha, Spain, in 2004. From 2006 to the end of 2007, he carried out postdoctoral research at the University of Glasgow, Scotland, in the field of the control of power electronic converters for the amelioration of power quality. From 2012, he has been an Associate Professor at the School of Industrial Engineering at the University of Castilla-La Mancha, where he conducts several R&D projects. His research interests include the control of power electronic converters, power quality, renewable energy systems, energy storage devices and wireless power transfer systems. Dr. Roncero-Sánchez is a Senior Member of the IEEE.

Stanimir Valtchev IEEE Member'90-Senior Member'08, Professor Auxiliar of Power Electronics, Dept. of Electr. Engg of the Faculty of Science and Technology, Universid. Nova de Lisboa, Full Professor of Burgas Free University, Bulgaria, Best graduate (MSc) of 1974, Technical University Sofia (TUS), in semiconductor and electronics technology, military service in radars, researcher in medical equipment and power converters and laser supplies, Auxiliary Director of the Centre of Robotics of TUS. In 1987 worked in TU Delft as Assistant Professor, since 1988 Assistant Professor in TUS (Power Supplies and Converters), Deputy Dean of TUS. Working on HF resonant power converters, IEEE Meritorious Paper Award, 1997, Best Paper Award from PEMC2014. Since 1994, teaching and consulting in Portugal and in the Netherlands. Based on a versatile experience, the research includes energy conversion, energy harvesting, wireless energy transfer, electric vehicles, energy management and storage, smart grids, tunnel FET devices and biosensors. He was the General Chair of the IEEE-PEMC2016 conference and since 2012 he is Invited Full Professor of Burgas Free University, Bulgaria, Chair of Subcommittee in IES of IEEE.

Svilen Valtchev is currently Assistant Professor of Mathematics at Polytechnic Institute of Leiria, Portugal and, since 2003, researcher at Center for Computational and Stochastic Mathematics (CEMAT-IST) at University of Lisbon, Portugal. He earned his Bachelor (pre-Bologna, 5 years) in Applied Mathematics and Computer Science in 2002 and his PhD in Mathematics in 2008, both from Technical University of Lisbon, Portugal. His main research interest is related to the development and application of Numerical Methods for Partial Differential Equations. In particular, he is currently working on the analysis of Meshless (Mesh-free) numerical methods for acoustic and elastic wave propagation problems.

Javier Vázquez del Real received an M.Sc. degree in Physics from the University of Valencia, Valencia, Spain, in 1992; a Master's degree in Telecommunications from the Polytechnic University of Madrid, Madrid, Spain, in 2001; and a Ph.D. from the University of Castilla-La Mancha, Ciudad Real, Spain, in 2006. After obtaining pre-doctoral grants to work with the Research Centre for Optoelectronic Devices at Alcatel in Stuttgart, Germany and with the Valencian Institute for Agricultural Research in Valencia, he held engineering positions in multinational corporations, such as Bosch and Valeo. In 2001, he joined the University of Castilla-La Mancha, where he is currently a Lecturer in Electronics. He has participated in post-doctoral visits to Newcastle University, Newcastle upon Tyne, U.K., and to the IK4-Research Alliance in the Basque Country, Spain. His current research interests cover power electronics and sensors.

Chen Yen-Lung is a master degree student of Dept. Computer Science and Information Engineering at National Formosa University. His current research interests are in the area related to Internet Of Things (IOT) and intelligent system. He received B.S. degree at Computer Science and Information Engineering at National Formosa University, 2014.

Index

P

R

S

V

W

www.ingramcontent.com/pod-product-compliance
Lightning Source LLC
LaVergne TN
LVHW080534300926
849162LV00007B/1608